PHAGOCYTE FUNCTION

CYTOMETRIC CELLULAR ANALYSIS

Series Editors

J. Paul Robinson
Purdue University Cytometry Laboratories
Purdue University
West Lafayette, Indiana

George F. Babcock
Department of Surgery
University of Cincinnati College of Medicine
Cincinnati, Ohio

New Volume in Series

Phagocyte Function: A Guide for Research and Clinical Evaluation
J. Paul Robinson and George F. Babcock, *Volume Editors*

Forthcoming Volumes in Series

Immunophenotyping
Carlton C. Stewart and Janet K.A. Nicholson, *Volume Editors*

Emerging Tools for Single Cell Analysis: Advances in Optical Measurement Technologies
Gary Durack and J. Paul Robinson, *Volume Editors*

PHAGOCYTE FUNCTION

A Guide for Research and Clinical Evaluation

Edited by

J. PAUL ROBINSON
Purdue University Cytometry Laboratories
Purdue University
West Lafayette, Indiana

GEORGE F. BABCOCK
Department of Surgery
University of Cincinnati College of Medicine
Cincinnati, Ohio

Wiley-Liss

A JOHN WILEY & SONS, INC. PUBLICATION

New York • Chichester • Weinheim • Brisbane • Singapore • Toronto

This book is printed on acid-free paper. ♾

Published simultaneously in Canada.

While the authors, editors, and publisher believe that drug selection and dosage and the specification and usage of equipment and devices as set forth in this book are in accord with current recommendations and practice at the time of publication, they accept no legal responsibility for any errors or omissions, and make no warranty, expressed or implied, with respect to material contained herein. In view of ongoing research, equipment modifications, changes in governmental regulations, and the constant flow of information relating to drug therapy, drug reactions, and the use of equipment and devices, the reader is urged to review and evaluate the information provided in the package insert or instructions for each drug, piece of equipment, or device for, among other things, any changes in instructions or indication of dosage or usage and for added warnings and precautions.

The cover image is a transmission electron micrograph of lipid-laden macrophage and neutrophil obtained by bronchopulmonary lavage of a rat. Micrograph courtesy of John J. Turek, Ph.D., Purdue University, School of Veterinary Medicine, Department of Basic Medical Sciences, W. Lafayette, IN

Library of Congress Cataloging-in-Publication Data:

Phagocyte function : a guide for research and clinical evaluation /
edited by J. Paul Robinson, George F. Babcock.
p. cm. — (Cytometric cellular analysis)
"A Wiley-Liss publication."
ISBN 0-471-12364-1 (cloth : alk. paper)
1. Phagocytes—Research—Methodology. 2. Phagocytosis—Research—Methodology. 3. Cell receptors—Research—Methodology. 4. Flow cytometry. 5. Confocal microscopy. I. Robinson, J. Paul. II. Babcock, George F., 1948– . III. Series.
[DNLM: 1. Phagocytes—physiology. 2. Receptors, Immunologic—physiology. 3. Flow Cytometry. QW 690 P5316 1998]
QR185.8.P45P47 1998
616.07′99—dc21
DNLM/DLC
for Library of Congress 98-12095

Printed in the United States of America.

10 9 8 7 6 5 4 3 2 1

Contents

Preface

During the late nineteenth and early twentieth centuries an intense debate developed between the supporters of the "cellular" and "humoral" theories of immunity. One major proponent of the cellular theory was Elie Metchnikoff, who proposed that the body's phagocytes were a primary defense system in immunity. The studies of Metchnikoff and Paul Ehrlich in both vertebrate and non-vertebrate animals provided a wealth of information about defense against invading organisms, for which they were jointly awarded the Nobel prize in Medicine in 1908. Although the importance of phagocytes in the host's defense against invading organisms slowly gained acceptance in the scientific/medical community during the early twentieth century, the intricate links between neutrophils, monocytes/macrophages, and virtually all other components of the immune system were essentially completely unknown. In the past 10 to 15 years one of the most exciting aspects of phagocytic research has been the new functional and cytochemical characterizations made possible by cytometry-based technologies. These observations, along with quirks of nature such as chronic granulomatous disease, have given investigators a better understanding of the interactions between various components of the immune system. The ability to measure cellular receptors, cytochemical systems, enzyme pathways, and ionic changes in the intracellular environment and to document organelle functions has often been due to cytometric advances.

The crude measurements of light bursts made by Baldridge and Gerard in the 1930s or Hirsch's fascinating cinematographic observations of phagocytosis in the 1960s are simplistic compared to the opportunities offered by today's sophisticated tools. Flow cytometry, confocal microscopy, and image analysis have opened the way to a tremendous increase in our understanding of the mechanisms that allow phagocytic cells to play so many roles in normal and abnormal physiology. This volume highlights some of the knowledge gained as a result of cytometric capabilities and where possible, links this knowledge to the clinical implications.

This volume covers the nature of receptor–cell interactions, detailing several mechanisms for which clinical importance has been established. FcγR receptors are now known to play a crucial role in functional communication between the humoral and cellular arms of the immune system. The knowledge now being extracted from the role of these receptors will no doubt be translated into valuable therapeutics. Several chapters are devoted to the role of cellular receptors and signal transduction. There are also chapters detailing the cellular mechanisms following activation of oxygen and nitrogen metabolic pathways and the relationships between neutrophils and other cells, particularly endothelial cells. These respiratory burst mechanisms have been studied for many years in the neutrophil, yet new pathways of activation and response are continually discovered. Several chapters of this volume are devoted to the functional role of monocytes/macrophages in health and disease, discussing in particular the role of these cells in a number of clinical syndromes as well as the role of macrophages in tissue rejection, trauma, and wound healing. To some extent we have attempted to bring together parallel research on both neutrophils and macrophages. Although the systems are well defined and quite different they share a number of common functional roles and metabolic pathways.

J. Paul Robinson, Purdue University
George F. Babcock, University of Cincinnati

Contributors

J. Wesley Alexander
Division of Transplantation
University of Cincinnati College of Medicine
Cincinnati, OH

George F. Babcock
Department of Surgery
University of Cincinnati College of Medicine and Shriners Burns Institute
Cincinnati, OH

Robert C. Bass
Department of Surgery
University of Cincinnati College of Medicine and Shriners Burns Institute
Cincinnati, OH

Robert Bjerknes
Department of Pediatrics
University of Bergen
Haukeland Hospital
Bergen, Norway

Robert J. Burnett
Department of Surgery
University of Cincinnati College of Medicine and Shriners Burns Institute
Cincinnati, OH

Wayne O. Carter
Pfizer Medical Technology Group
Groton, CT

J. David Chambers
The Salk Institute
San Diego, CA

Bruce H. Davis
Department of Clinical Pathology
William Beaumont Hospital
Royal Oak, MI

John J. Finlay-Jones
Department of Microbiology and Infectious Diseases
School of Medicine
Flinders University of South Australia
Adelaide, Australia

Julie A. Gegner
University of San Diego
San Diego, CA

David G. Greenhalgh
Shriners Hospital for Children, Northern California
Sacramento, California

Paul M. Guyre
Department of Physiology
Dartmouth Medical School
Lebanon, NH

PRUE H. HART
Department of Microbiology and Infectious Diseases
School of Medicine
Flinders University of South Australia
Adelaide, Australia

WOLFGANG KELLERMANN
Institut für Anästhesiologie
Klinikum Grosshadern der Universität
München, Germany

PADMA KUMAR NARAYANAN
SmithKline Beecham
King of Prussia, PA

ALICE N. NEELY
Departments of Surgery and Physiology and Biophysics
University of Cincinnati College of Medicine and Shriners Burns Institute
Cincinnati, OH

JOHN P. NOLAN
National Flow Cytometry Resource
Los Alamos National Laboratory
Los Alamos, NM

CORA K. OGLE
Division of Transplantation
University of Cincinnati College of Medicine and Shriners Burns Institute
Cincinnati, OH

GENEVA M. OMANN
Departments of Surgery and Biological Chemistry
University of Michigan Medical School
and Veterans Administration Medical Center
Ann Arbor, MI

J. PAUL ROBINSON
Purdue University Cytometry Laboratories
Purdue University
West Lafayette, IN

GORDON D. ROSS
Department of Pathology
University of Louisville
Louisville, KY

GABRIEL ROTH
Institut für Anästhesiologie
Klinikum Grosshadern der Universität
München, Germany

JULIE ROWE
Department of Microbiology and Infectious Diseases
School of Medicine
Flinders University of South Australia
Adelaide, Australia

ELIZABETH R. SIMONS
Department of Biochemistry
Boston University School of Medicine
Boston, MA

LARRY A. SKLAR
National Flow Cytometry Resource
Los Alamos National Laboratory
Los Alamos, NM
and Department of Cytometry
University of New Mexico School of Medicine
Albuquerque, NM

JOSEPH S. SOLOMKIN
Department of Surgery
University of Cincinnati College of Medicine and Shriners Burns Institute
Cincinnati, OH

BRIAN P. THORNTON
Department of Pathology
University of Louisville
Louisville, KY

PETER S. TOBIAS
Department of Immunology
The Scripps Research Institute
La Jolla, CA

JOHN F. VALENTE
Division of Transplantation
University of Cincinnati College of Medicine
Cincinnati, OH

GÜNTER K. VALET
Department of Cell Biochemistry
Max-Planck-Institut für Biochemie
Martinsried, Germany

VÁCLAV VĚTVIČKA
Department of Pathology
University of Louisville
Louisville, KY

PAUL K. WALLACE
Department of Microbiology
Dartmouth Medical School
Lebanon, NH

PHAGOCYTE FUNCTION

1

Analysis of the Phagocyte Membrane Lectin CR3 Using Fluorescence-Labeled Polysaccharides and Flow Cytometry

Gordon D. Ross
Department of Pathology, University of Louisville, Louisville, Kentucky

Václav Větvička
Department of Pathology, University of Louisville, Louisville, Kentucky

Brian P. Thornton
Department of Pathology, University of Louisville, Louisville, Kentucky

STRUCTURE AND FUNCTION OF CR3

This chapter focuses on methods developed for flow cytometry analysis of the lectin site of the leukocyte β_2-integrin known as CR3, Mac-1, or CD11b/CD18. CR3 is the most important phagocyte receptor used for recognition of bacteria. It is capable of responding to three types of molecules that may be present on microorganisms: (1) the opsonic C3 fragments iC3b, C3dg, and C3d; (2) certain polysaccharides that contain glucose (β-glucans), mannose, or *N*-acetyl-D-glucosamine (NADG); and (3) opsonic IgG through formation of a transmembrane signaling complex with phosphatidylinositol-linked FcγRIIIB (CD16). The lectin site of CR3 is key to these

Phagocyte Function: A Guide for Research and Clinical Evaluation, Edited by J. Paul Robinson and George F. Babcock. ISBN 0-471-12364-1

responses, and novel flow cytometry techniques incorporating fluorescence-labeled polysaccharides have been a major development that allowed both molecular mapping of the lectin site and definition of its sugar specificity.

Although the structure of CR3 is known, the molecular basis for many of its functions is unknown. The functions of CR3 can be divided into the two major categories of cellular adhesion and cytotoxicity. This chapter concentrates on cytotoxic functions that involve specific recognition of pathogenic surfaces, and summarizes briefly the function of CR3 in mediating leukocyte adhesion to host cell surfaces. Within each category of functions, multiple types of counter-receptors or ligands are recognized via two binding site domains. As is typical for an integrin, CR3 is capable of both outside-in and inside-out signaling, and has several functional states of activation in which binding sites are either hidden or exposed, as well as states in which its cytoplasmic tail is either linked or unlinked to the filamentous actin cytoskeleton.

Structure of CR3 and the β_2-Integrins

The three β_2-integrins each consist of two noncovalently linked polypeptides known as α and β subunits (Table 1). A single type of β subunit, designated β_2-integrin or CD18, is encoded by a gene on chromosome 21 and shared by all three membrane complexes. The 95-kDa CD18 polypeptide is linked to one of three distinct α subunits encoded by three linked genes on chromosome 16 (Corbi et al., 1988). The first β_2-integrin, the "lymphocyte function-associated antigen" or LFA-1 molecule, has an α subunit of 175 kDa known either as α_L or CD11a. LFA-1 is expressed on all leukocyte types and has a primary function in mediating intercellular adhesion between leukocytes and other host cells via a family of counter-receptors, the "intercellular adhesion molecules," abbreviated ICAM-1, ICAM-2, and ICAM-3.

The second member of the β_2-integrins, CR3, has an α subunit of 165 kDa known as α_M or CD11b. CR3 was originally described as both an iC3b-receptor, "complement receptor type three," and Mac-1, a membrane marker of monocytes and macrophages. CR3 is expressed on nearly all neutrophils, monocytes, eosinophils, and natural killer cells, but on only a minor subset of B cells ($CD5^+$ cells) (De la Hera et al., 1988) and fewer T cells (Muto et al., 1993). Tissue macrophages express less CR3 than monocytes, but CR3 is a specific marker of immune-activated $CD8^+$ T cells (McFarland et al., 1992; Nielsen et al., 1994).

The third member of the β_2-integrin/CD18 family is CR4 or p150,95. CR4 is expressed on the same leukocyte types that express CR3, but usually at a lower level than CR3. Only tissue macrophages express more CR4 than CR3 per cell (Hogg et al., 1986). The 150-kDa α-subunit of CR4 is CD11c or α_X. Much less is known about the ligands and functions of CR4 as compared to the other two β_2-integrins, largely because most leukocytes express little CR4 and there is some uncertainty in distinguishing its functions from those of LFA-1 and CR3.

A vast array of monoclonal antibodies (mAbs) that recognize the distinct α subunits of the β_2-integrins have been reported; many of these are listed in Table 1. When using these mAbs it is important to note that they recognize distinct epitopes, only some of which have been mapped to specific polypeptide regions (Diamond et al.,

TABLE 1. The CD18 Family of Leukocyte β_2-Integrins

Molecule Names	Ligand Specifity	Structure	Available mAbs to α-Chains	Cellular Distribution
LFA-1, CD11a/CD18, or $\alpha_L\beta_2$-integrin	ICAM-1, ICAM-2, ICAM-3	175-kDa α-chain (CD11a), 95-kDa β-chain (CD18)	TS 1/22.1.1.13, G-25.2, MHM24, MEM-25, -30, -83, -94–95	All leukocyte types
CR3, Mac-1, CD11b/CD18, Mo1, or $\alpha_M\beta_2$-integrin	iC3b, C3dg, C3d, LPS, ICAM-1, fibrinogen, factor X, β-glucans, certain mannose polysaccharides, NADG-glycosaminoglycans, heparan sulfate, FcRIIIB, urokinase-plasminogen activator receptor	165-kDa α-chain (CD11b), 95-kDa β-chain (CD18)	Anti-Mac-1, -Mo1, -Leu15, OKM1, OKM9, OKM10, MN-41, 44, 903, 904, 60.1, Vim-12, M522, LPM19c, 14B6E.2, 5A4.C5, TMG65, LM2/1, CBRM1/1, /2, /4, /5, /9, /10, /13, /16, /17, /18, /20, /21, /22, /23, /24, /25, /26, /28, /29, /30, /31, /32, /33, /34, CBRN3/11	Monocytes/macrophages, neutrophils, eosinophils, NK cells, $CD5^+$B cells, activated $CD8^+$T cells
CR4, p150,95, CD11c/CD18, or $\alpha_X\beta_2$-integrin	iC3b, fibrogen	150-kDa α-chain (CD11c), 95-kDa β-chain (CD18)	Anti-LeuM5 (SHCL3), L-29, Ki-M1, KB23, Bu-15, 3.9, BLY6	Monocytes/macrophages, neutrophils, eosinophils, NK cells, activated B cells

1993). Some mAbs may block one or more receptor functions, others may activate receptor function, and some others may have no effect on receptor function.

Host Defense Functions of Phagocytic Cells Dependent upon CR3

CR3 and LFA-1 make up an important adhesion molecule system used by phagocytic cells for migrating out of blood vessels to sites of infection, the process known as diapedesis (Carlos and Harlan, 1990; Springer, 1990; Hogg, 1992; Smith, 1993). Patients with an inherited deficiency of the three CD18 family members suffer from a disease known as leukocyte adhesion deficiency (LAD) and have repeated life-threatening bacterial infections caused largely by the inability of neutrophils to migrate into sites of infection (Anderson and Springer, 1987; Arnaout et al., 1987; Todd and Freyer, 1988; Walport and Ross, 1989; Smith, 1993). Inflammatory signals such as C5a or lipopolysaccharide activate both CR3 and LFA-1 through inside-out signaling to reveal a hidden high-affinity binding site for ICAM-1 expressed on endothelial cells. This form of activated CR3 mediates firm attachment to host cells bearing ICAM-1 without mediating cytotoxicity of the host cells. By contrast, CR3 can also be activated to mediate both firm attachment and a potent cytotoxic reaction in response to bacteria bearing both opsonic iC3b and cell wall polysaccharides that bind to the lectin site of CR3. Such cytotoxic reactions include phagocytosis, respiratory burst (Newman et al., 1984; Ross et al., 1985; Cain et al., 1987), release of granule enzymes such as myeloperoxidase (Pryzwansky et al., 1991), and cytokines such as TNF-α (Abel and Czop, 1992) and the leukotriene LTB_4 (Czop and Austen, 1985; Czop et al., 1988; Lew et al., 1991; Graham et al., 1993; Au et al., 1994; Petersen et al., 1994). Both these forms of "activated CR3" differ in their functional capacity and possibly also exposed ligand binding sites, but the molecular mechanism for each type of receptor activation is unknown. On neutrophils, CR3 also forms a transmembrane complex with FcγRIIIB that serves as the major neutrophil receptor for particle- or immune complex-associated IgG (Graham et al., 1989; Zhou et al., 1992; Petty and Todd, 1993; Zhou et al., 1993; Stöckl et al., 1995). In the absence of CR3, the phosphatidylinositol-linked FcγRIIIB is largely incapable of transmembrane signaling for phagocytosis (Krauss et al., 1994). Only by forming an association with transmembrane CR3 [or alternatively, transmembrane FcγRII, CD32, (Salmon et al., 1991; Naziruddin et al., 1992; Edberg and Kimberly, 1994; Zhou and Brown, 1994)] is FcγRIIIB capable of initiating a cytotoxic response to IgG-opsonized target cells.

Ligands Recognized by CR3 Before and After Activation

Key to the functions of CR3 is its ability to bind an array of ligands, such as host cell ICAM-1 or yeast cell wall β-glucan (Table 2). Protein ligands such as iC3b, fibrinogen, ICAM-1, and clotting factor X have binding sites located within a loop of protein formed near the *N*-terminus and known as the I-domain (Fig. 1). Although the I-domain was at first claimed to have an β_1-integrin-like RGD specificity (Wright et al., 1987; Russell and Wright, 1988), this was later disputed when proteins such as iC3b were shown to bind via non-RGD containing sites (Taniguchi-Sidle and Isenman, 1992). Furthermore, the observation that iC3b can bind to the I-domain of rest-

TABLE 2. Ligand Binding Sites and Functions of CR3

Activation state	Ligand	Specificity	CD11b Binding Site	Function
Resting	EC3bi/C3dg	iC3b/C3dg/C3d	I-domain	Adherence
	Zymosan	β-glucan or SZP mannose polysaccharide	Lectin site	Phagocytosis, respiratory burst, degranulation, release of TNF-α and LTB_4
	FcγRIIIB	?	Lectin site	Phagocytosis, respiratory burst, degranulation
	E. coli	LPS	?	Phagocytosis
Activated	Endothelial cell	ICAM-1	I-domain	Adherence
	Fibrinogen	?	I-domain	Adherence
	Clotting factor X	?	I-domain	Adherence
	Neutrophil	?	I-domain	Aggregation
	K562 cell	?	?	Cytotoxicity

ing CR3, whereas ICAM-1 or fibrinogen only binds to the I-domain of activated CR3, indicates that there is probably more than one protein ligand-binding site contained in the I-domain. The synthesis of an I-domain peptide representing the iC3b-binding site (Ueda et al., 1994), as well as the crystallization of the entire I-domain (Lee et al., 1995), should allow identification of the other proteins that bind to this site.

Located *C*-terminal to the I-domain is a polysaccharide-binding lectin site (Fig. 1). Although at first thought to be specific for β(1,3)-glucans (Ross et al., 1987), the CR3 lectin was recently shown to have a broader sugar specificity using fluorescein isothiocyanate-labeled polysaccharides and flow cytometry (Thornton et al., 1996). The lectin site not only binds bacteria and yeast cell wall polysaccharides triggering phagocytosis or extracellular cytotoxicity, but also is responsible for the attachment and transmembrane signaling function of phosphatidylinositol-linked neutrophil receptors such as FcγRIIIB and the urokinase-plasminogen activator receptor (Petty and Todd, 1993; Zhou et al., 1993; Xue et al., 1994; Stöckl et al., 1995). Despite its rather broad reactivity with polysaccharides, CR3 appears to have only one lectin site, because all polysaccharides that bind to CR3 were shown to cross-inhibit binding of the others, again using flow cytometry techniques (Thornton et al., 1996).

Upregulation of CR3 Membrane Density Versus Activation and Priming

In addition to possessing different binding sites that are either exposed or hidden depending on cellular activation state, the membrane density of neutrophil CR3 molecules can also be increased rapidly from a cytoplasmic pool of preformed receptors present in specific granules stored just below the membrane surface (Singer et al., 1989; Pryzwansky et al., 1991; Sengelov et al., 1993). Cellular activating events such

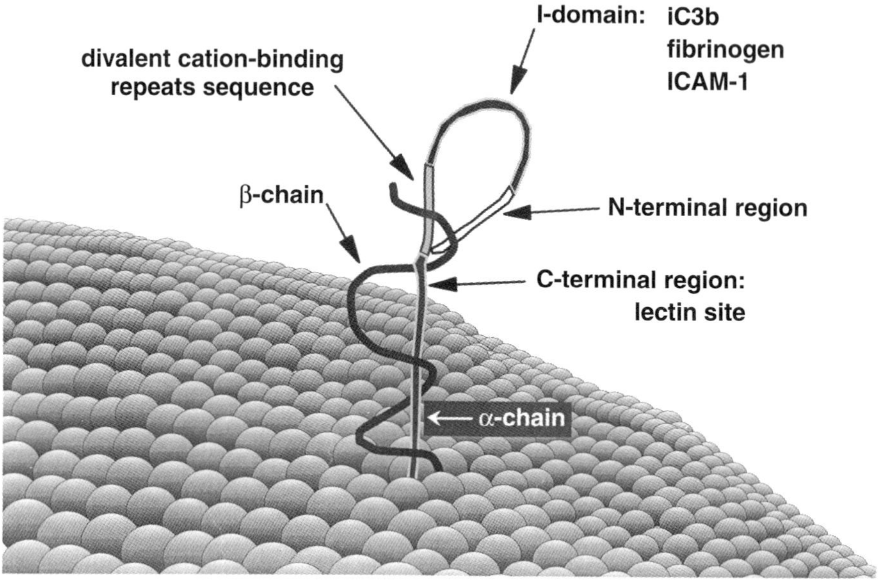

Fig. 1. Possible three dimensional structure of CR3 showing the major defined regions of its α-subunit, CD11b. Both the α and β subunits are transmembrane and have short cytoplasmic tails. CR3 is believed to have a typical integrin structure, including an α-subunit loop formed between the *N*-terminal region and the divalent cation-binding repeats sequence. This α-chain loop should allow CR3 to form contacts with particles bearing the protein ligands that bind to the I-domain: iC3b, fibrinogen, and ICAM-1. The lectin site of CR3 is believed to be located in the *C*-terminal region adjacent to the divalent cation-binding repeats sequence and near the point where the *N*-terminus forms the looped epitope recognized by the mAbs CBRM1/32 and M522.

as triggering the receptors for C5a or lipopolysaccharide, or cross-linking CR3 itself with a particulate polysaccharide, cause a rapid upregulation in the number of CR3 expressed per cell. With neutrophils this has been shown to result from the fusion of specific granules containing CR3 with the plasma membrane (Pryzwansky et al., 1991). Natural killer cells are able to upregulate their CR3 rapidly in a manner similar to neutrophils, except that C5a and lipopolysaccharide are not stimulating agents (Werfel et al., 1991; Muto et al., 1993). Such newly expressed CR3 greatly increases particle binding capacity and adherence to certain cell surfaces, and can appear to change the specificity or affinity of CR3 just because the large increase in receptor number makes a weak adhesion reaction more readily demonstrable. On the other hand, such rapidly upregulated receptors derived from granule pools do not become attached to the actin cytoskeleton and therefore are incapable of mediating functions requiring cytoskeletal attachment such as particle phagocytosis or homotypic aggregation (Buyon et al., 1988; Pryzwansky et al., 1991). Nevertheless, bacterial phagocytosis requires a cooperativity of both forms of CR3. Prior to ingestion, bacteria must be captured by the phagocyte via its high density of mobile CR3 that are unlinked to the cytoskeleton and therefore able to aggregate rapidly at the site of particle contact. CR4 functions in a similar manner on macrophages (Ross et al., 1992). Once bacteria have been captured on the cell surface by the large and mobile CR3 subset, then attachment occurs secondarily to the less numerous and immobile subset of CR3 that are connected to filamentous actin and able to mediate phagocytosis.

One of the most important functions of CR3 is the role that it plays as a C3 receptor in the clearance and destruction of microorganisms or immune complexes opsonized with C3. It has only recently become appreciated that microbial cell wall polysaccharides play an essential role in this process by priming CR3 for subsequent cytotoxic reactions against C3-opsonized targets. Target cells such as erythrocytes or tumor cells that are opsonized with iC3b/C3dg, but lack CR3-priming polysaccharides, are resistant to CR3-dependent cytotoxicity, despite mediating avid CR3-dependent adherence. This may represent a control mechanism that prevents an attack on host cells that have become opsonized with C3 as bystanders to an immune reaction. However, one consequence is that the CR3 of macrophages and natural killer cells does not mediate cytotoxicity of tumor cells, even though most cancer cell types activate complement and are heavily coated with opsonizing iC3b and C3dg fragments. Similarly, priming with small-molecular-weight β-glucans did not stimulate cytokine secretion, whereas natural killer cell secretion of IFN-α, IFN-γ, and TNF-α was stimulated when primed CR3 was later activated by iC3b-target cells. Both CR3 priming by β-glucans and subsequent cytokine release triggered by iC3b-targets were blocked by anti-CD11b antibodies.

The attachment of phagocyte CR3 to C3-opsonized bacteria or yeast triggers a rapid series of cytotoxic events including a respiratory burst, degranulation, and phagocytosis. The essential role played by cell wall polysaccharides on iC3b-opsonized bacteria can be appreciated when it is compared to the near absent response stimulated by erythrocytes opsonized with iC3b. The response to bacteria, although very rapid, is actually a multi-step reaction that involves (1) avid adhesion of the bacteria via the opsonizing iC3b to the mobile subset of membrane CR3; (2) cell wall polysaccharide attachment to the lectin site of CR3, resulting in priming of the immobile subset of CR3; and (3) activation of the primed receptor to mediate a cytotoxic response. CR3 priming can be stimulated without receptor activation by incubating phagocytes with a small soluble polysaccharide such as a yeast β-glucan that binds to the lectin site of individual CR3 molecules without cross-linking the receptor. Priming of CR3 is tyrosine kinase-dependent and results in exposure of the activation epitope recognized by the mAb CBRM1/5, but is not associated with degranulation, a respiratory burst, or any other morphological marker of phagocyte activation. Nevertheless, phagocytes or natural killer cells bearing polysaccharide-primed CR3 are able to mediate their full range of cytotoxic functions in response to iC3b-opsonized target cells such as erythrocytes and tumor cells that otherwise do not stimulate CR3-dependent responses (Větuička et al., 1996).

Activation of Tyrosine Kinases and Phosphorylation of Paxillin

Triggering of CR3 for cytotoxic responses has been shown to activate the tyrosine kinases p125^{fak} (Fuortes et al., 1994) and p58^{fgr} (Berton et al., 1994), both of which may play a role in the phosphorylation of the CR3-associated cytoplasmic protein paxillin (Fuortes et al., 1994; Graham et al., 1994). In neutrophils from LAD patients who lack all of the β_2-integrins, no type of neutrophil stimulation is able to stimulate paxillin phosphorylation, suggesting that paxillin phosphorylation is absolutely tied to CR3 function (Graham et al., 1994). As part of the activation process for phagocytosis, the cytoplasmic domain of the β subunit may become linked to actin in a manner resembling β_1- and β_3-integrins using the interlinking actin-binding proteins α-actinin, vinculin, and talin (Pavalko and LaRoche, 1993; Rabb et al., 1993). However, demonstration that such a focal adhesion type linkage is functional in CR3-dependent cytotoxic function has not yet been accomplished. It is unknown whether the same or different tyrosine phosphorylation events are involved in CR3 priming versus CR3 activation.

ANALYSIS OF CR3 EXPRESSION AND FUNCTION BY FLOW CYTOMETRY

Flow cytometry has been a major tool for identification of novel mAbs to leukocyte membrane antigens, and after the most recent International Leukocyte Antigen Workshop there are now more than 150 recognized CD antigens. Monoclonal antibodies that distinguish activated forms of membrane antigens offer an additional perspective for linking membrane antigen expression to cellular function, but so far only two, mAb 24 (Picker et al., 1993) and CBRM1/5 (Diamond and Springer, 1993), have been identified for the β_2-integrins. Monoclonal antibodies to specific domains of CD11b have also proved useful for pinpointing binding sites for the various functional ligands of CR3 (Diamond et al., 1993; Thornton et al., 1996).

Flow cytometry analysis of membrane receptors using soluble fluorescence-labeled ligands has not been a widely used technique. Nevertheless, if soluble pure ligands are available that bind with high affinity, these could potentially be labeled with fluorescein isothiocyanate (FITC) for analysis of binding characteristics by flow cytometry. Specific fluorescence of a receptor by a FITC ligand is defined as staining that is saturable and reversible, and thus inhibitable by excess unlabeled ligand. Negative background staining is thus defined as the staining obtained with the FITC ligand in the presence of a 50-fold molar excess of homologous unlabeled ligand. Major advances in defining the lectin site of CR3 were made by examining FITC-labeled polysaccharides and their staining characteristics, as well as unlabeled polysaccharides and their ability to block staining by mAbs to specific epitopes of CR3 (Thornton et al., 1996). It appears likely that the binding characteristics of several other CR3 ligands (e.g., fibrinogen, factor X) could also be analyzed for their binding characteristic by flow cytometry.

Method for Polysaccharide Staining Analysis of CR3 by Flow Cytometry

Using flow cytometry and FITC-labeled polysaccharides, specific surface staining for CR3 is readily demonstrated with all leukocyte types (Fig. 2). Staining is shown with

SZP–FITC, a complex mannose polysaccharide isolated from zymosan, the cell wall fraction from baker's yeast (*Saccharomyces cerevisiae*). Using ^{125}I-polysaccharide generated with Iodogen following introduction of a tyramine residue on the reducing end (Cheong and Hahn, 1991), ^{125}I-SZP was shown to have a binding affinity for CR3 of 6.7×10^{-8} *M* (Thornton et al., 1996). Unlabeled SZP was also shown to be the most effective inhibitor of CR3 staining by any of the polysaccharide–FITC conjugates (Table 3). Accordingly, SZP–FITC was selected as the fluorescence probe for flow cytometry analysis of the lectin site of CR3. However, similar flow cytometry staining results were obtained with two commercially available β-glucan preparations. Table 4 lists the polysaccharides that have been tested for CR3-specific staining of neutrophils by flow cytometry. Laminarin, a seaweed-derived β-glucan of ~12 kDa, is available from Sigma Chemical Co., St. Louis, Missouri. Another β-glucan, from baker's yeast, was previously obtained from Molecular Probes, Eugene, Oregon, in either unlabeled or FITC-coupled form, but unfortunately this product was later withdrawn from the market. With unlabeled polysaccharides, FITC coupling is

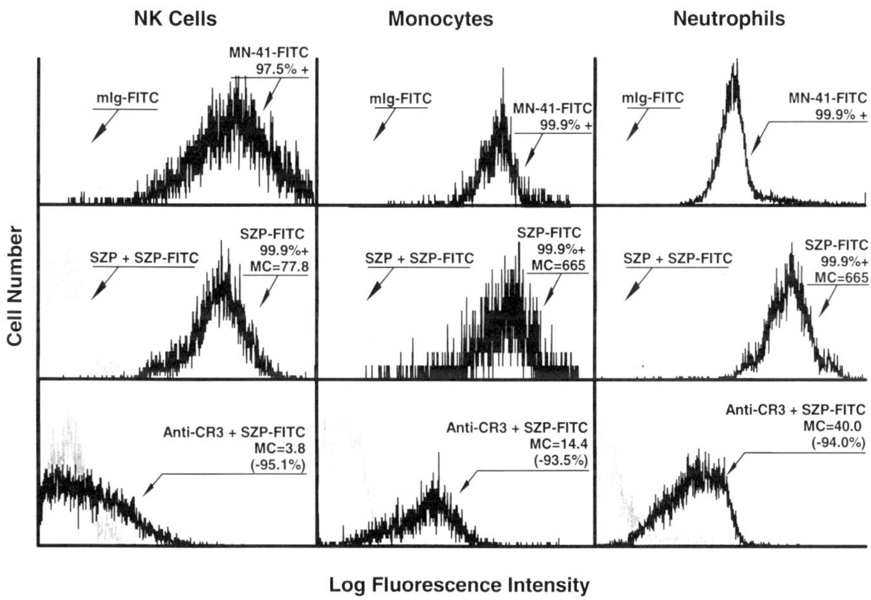

Fig. 2. CR3 specificity of leukocyte staining by polysaccharides. Isolated human peripheral blood natural killer (NK) cells, monocytes, and neutrophils were stained similarly by either an anti-CR3 mAb to CD11b (MN-41–FITC) or isolated soluble zymosan polysaccharide (SZP–FITC). Nonspecific mouse IgG (mIg–FITC) served as the nonspecific staining control for MN-41–FITC and is shown by the gray line in the top histograms. For SZP–FITC staining, the nonspecific staining control was defined as SZP–FITC staining in the presence of a 50-fold excess of unlabeled SZP (SZP + SZP–FITC), and is shown as the gray line in the middle and lower histograms. Both the mAb to CR3 and SZP–FITC stained virtually all these leukocyte types, and each produced similar staining profiles. Treatment of leukocytes with anti-CR3 prior to staining with SZP–FITC reduced the total fluorescence staining by ~95% (lower histograms). Similar CR3-dependent staining was obtained with seaweed β-glucan–FITC (laminarin) and the β-glucan–FITC of proprietary origin sold by Molecular Probes, Inc., Eugene, Oregon.

TABLE 3. Sugar Inhibition of SZP–FITC or β-Glucan–FITC Staining of CR3

	Hexose Concentration Causing 50% Inhibition of Staining by	
Unlabeled Sugar	2 μg/ml SZP–FITC	10 μg/ml Laminarin–FITC
SZP	0.4 μg/ml	0.2 μg/ml
Molecular Probes β-glucan	5 μg/ml	3 μg/ml
Laminarin (seaweed β-glucan)	5 μg/ml	2 μg/ml
Barley β-glucan	31 μg/ml	3 μg/ml
Lentinan (mushroom β-glucan)	60 μg/ml	3 μg/ml
Yeast carboxymethyl β-glucan	25 μg/ml	6 μg/ml
Yeast α-mannan	>100 μg/ml[a]	>100 μg/ml[a]
Dextran	>100 μg/ml[a]	>100 μg/ml[a]
N-Acetyl-D-Glucosamine (NADG)	50 m*M*	150 m*M*
α-Methylmannoside	100 m*M*	250 m*M*

[a]No inhibition detected at 100 μg/ml.

carried out by the identical methods used for coupling antibodies, except that FITC to polysaccharide conjugation ratios cannot be easily calculated (Winchester and Ross, 1986). As with antibodies, the optimal conjugation ratio is determined empirically by pilot experiments with individual types of polysaccharides. Specific staining is always defined as the staining by FITC–polysaccharide that is inhibitable by a 50- to 100-fold molar excess of unlabeled polysaccharide. With excess coupling to FITC, there may be an increase in nonspecific staining that lowers the net specific staining. With each cell type analyzed, a control tube for negative background staining is always included in which the cells are incubated in a mixture of polysaccharide–FITC plus excess unlabeled homologous polysaccharide. For example, with laminarin–FITC, maximum specific staining of neutrophils was achieved with 10 μg/ml laminarin–FITC, and thus the negative background control consisted of neutrophils stained with a mixture of 10 μg/ml laminarin–FITC plus 500 μg/ml of unlabeled laminarin. Staining was carried out in HBSS containing 1% bovine serum albumin (H-BSA), but similar results were obtained when staining was carried out in H-BSA containing 10 m*M* EDTA or PBS/BSA, because the CR3 lectin site does not require divalent cations (Thornton et al., 1996).

Staining Intensity and Upregulation of CR3

Staining intensity depends on the surface density of CR3, and thus on the degree of cellular activation and upregulation of surface CR3. Neutrophil CR3 upregulation occurs rapidly when any of the isolation buffers or gradient media contains even trace contamination with endotoxin (Berger et al., 1984). With neutrophils isolated with lipopolysaccharide-free buffers and media, the number of CR3 per cell is very low ($\sim 3\text{–}5 \times 10^3$ per cell), and staining intensity with FITC-labeled mAbs to CR3 or polysaccharides is also low. However, when neutrophils are isolated with the usual

TABLE 4. Staining of Neutrophil CR3 with Different Polysaccharides Coupled to FITC

Type of FITC-labeled Polysaccharide	Polysaccharide–FITC Concentration Required for Maximum Staining(μg/ml)	Percent Inhibition of Specific Fluorescence by	
		SZP (50 μg/ml)	MN-41 anti-CR3 (50 μg/ml)
Dextran–FITC	No specific staining	n.d.[a]	n.d.
α-Mannan–FITC	No specific staining	n.d.	n.d.
Barley β-glucan–FITC	10	92.3	94.5
Yeast carboxymethyl β-glucan–FITC	5	87.8	78.9
Seaweed β-glucan (laminarin)–FITC	10	91.5	89.9
Mushroom β-glucan (lentinan)–FITC	5	96.3	81.0
Molecular Probes β-glucan–FITC	2	100	96.1
SZP–FITC	2	99.8	77.4

[a]n.d.: not done.

buffers and media that are not specifically prepared to be lipopolysaccharide-free, the surface density of CR3 is usually 10-fold higher and much more readily detectable by fluorescence flow cytometry. Small polysaccharides such as SZP or laminarin that do not cross-link CR3 do not trigger neutrophil CR3 upregulation (Větvička et al., 1996), whereas CR3 upregulation is stimulated by large soluble β-glucans (e.g., lentinan) or particulate β-glucan, even with natural killer cells that are unresponsive to lipopolysaccharide (Muto et al., 1993).

Analysis of Lectin Site Sugar Specificity by Flow Cytometry

The sugar specificity of CR3 was demonstrated both with individual types of polysaccharides labeled with FITC (Table 4) and with serial dilutions of unlabeled polysaccharides (Table 3) that were examined for reduction in linear fluorescence intensity of staining by SZP–FITC or β-glucan–FITC (Thornton et al., 1996). A specificity for β-glucans was readily shown by staining with FITC-labeled pure β-glucans from various fungi or seaweed (Table 4). An additional specificity for polysaccharides containing mannose was suggested by staining with SZP, some preparations of which were ~95% mannose and <5% glucose. However, no specific staining with yeast α-mannan was detectable, and moreover, unlabeled α-mannan did not inhibit staining by SZP–FITC or β-glucan–FITC (Table 3). Analysis of monosaccharides for blockade of staining found no inhibition by mannose or glucose (Thornton et al., 1996), but significant inhibition of both β-glucan–FITC and SZP–FITC by α-methylmannoside, a pure mannose-containing sugar (Table 3). This finding, in addition to the observation that SZP had the highest affinity for CR3 and yet consisted primarily of mannose, indicated that a single lectin site in CR3 was probably responsible for binding certain polysaccharides containing either mannose or glucose. Finally, early work on the CR3 lectin site and phagocytosis of zymosan had shown inhibition by NADG (Ross et al., 1985), and this was confirmed by flow cytometry analysis that showed inhibition of both SZP–FITC and β-glucan–FITC staining by NADG (Table 3).

Molecular Mapping of the Lectin Site by Flow Cytometry

Mapping of the lectin site was attempted by three methods using flow cytometry. First, SZP–FITC staining was evaluated with CHO cells expressing recombinant CR3 in which portions of the CD11b α subunit were replaced with homologous portions of CD11c, the α subunit of CR4. Second, mAbs to defined regions of CD11b were tested for their ability to inhibit CR3 staining by SZP–FITC. Third, unlabeled SZP was tested for its ability to inhibit staining by mAbs to defined regions of CD11b.

Tests for SZP–FITC staining of CHO cells expressing molecular chimeras of CR3/CR4 indicated that the lectin site was not contained within the I-domain region of CD11b (Thornton et al., 1996) that had previously been shown to contain the binding sites for the known protein ligands of CR3 using the same CHO cell clones (Diamond et al., 1993). However, SZP–FITC staining occurred with CR4 molecules containing either of two distinct regions of CR3 located *C*-terminal to the I-domain: (1) the divalent cation-binding region and (2) the region of CD11b located *C*-terminal to the cation-binding region (Fig. 1). The possible existence of two distinct lectin sites was not supported by either the sugar inhibition studies or the Scatchard plot analysis of ^{125}I-SZP binding, each of which indicated a single type of lectin site. Although CHO cells expressing native CR4 did not exhibit lectin activity, one hypothesis that would explain these results is the existence of a hidden lectin site within CR4 that mapped to a site in the *C*-terminal domain of CD11c that was homologous to the lectin site in CD11b. Similar iC3b-binding sites in the I-domains of CR3 and CR4 have previously been demonstrated, and the possibility of other shared structures appears likely because CR4 has >80% sequence homology with CR3. The putative hidden lectin site in CR4 might be exposed in the molecular chimera in which the cation-binding region of CD11c was substituted for the homologous region of CD11b. This might occur if the folding of CD11b and CD11c were controlled by the divalent cation-binding region. A putative loop in CD11b is thought to be formed when the *N*-terminus binds to a site in the divalent cation-binding region (Fig. 1). Thus substitution of the cation-binding region of CD11c with the homologous region of CD11b may change the folding of CD11c so as to expose a cryptic lectin site located *C*-terminal to the cation binding region. Confirmation of this hypothesis will require demonstration of similar lectin sites in *C*-terminal domain peptides derived from CD11b and CD11c.

Few or no useful data for mapping the lectin site were obtained in experiments that examined the inhibition of SZP–FITC staining by mAbs to defined epitopes of CD11b. Monoclonal antibodies to each of the defined regions of CD11b produced comparable inhibition of SZP–FITC staining. This was unexpected, because the same panel of mAbs had been used as one strategy to map the protein ligand binding sites to the I-domain (Diamond et al., 1993).

The most useful lectin site mapping data came from experiments that tested the ability of unlabeled SZP to inhibit staining by mAbs to defined regions of CD11b.

Possibly because of its high affinity for CR3, SZP blocked selectively the staining of neutrophil CR3 by 11 mAbs to domains of CD11b located *C*-terminal to the divalent cation-binding region. By contrast, unlabeled SZP had no effect on CR3 staining by six mAbs to I-domain epitopes or the single available mAb to the divalent-cation binding region. Two mAbs thought to react with an epitope formed where the *N*-terminal folds back onto the cation-binding region (CBRM1/32 and M522) were also partially inhibited (62 and 38%) by the prior saturation of CR3 with SZP. However, 7 out of 11 mAbs to domains located *C*-terminal to the cation-binding region were inhibited >70%, and one of these mAbs, CBRM1/16, was 85% inhibited from staining CR3 by SZP. Similar inhibition profiles were obtained when several of the mAbs were examined for inhibition of CR3 staining by unlabeled β-glucan or NADG. Thus the mAb blockade experiments supported the hypothesis that a single lectin site was located *C*-terminal to the divalent cation-binding region of CD11b (Thornton et al., 1996).

Functional Significance of the Lectin Site of CR3

The lectin site of CR3 is responsible for priming the receptor for cytotoxic responses. Bacterial or yeast cell wall polysaccharides that bind to the lectin site, causing aggregation of membrane CR3, stimulate neutrophil or monocyte/macrophage phagocytosis, respiratory bursts, and release of cytotoxic granules. C3 fragments deposited onto microorganisms work cooperatively with polysaccharides in stimulating cytotoxic responses. Although iC3b ligation to the I-domain does not prime or activate CR3, it does produce avid adhesion of the microorganism to the phagocyte, thereby facilitating secondary recognition of polysaccharides by the lectin site. This appears to be particularly important with yeast or bacteria on which the critical polysaccharides are either poorly exposed or of low affinity for the CR3 lectin site. Bactericidal activity of neutrophils is inefficient with most bacteria without opsonizing iC3b. Microorganisms, tumor cells, or virus-infected host cells that lack cell surface polysaccharides that prime CR3 are bound by CR3-bearing phagocytic cells or natural killer cells but are not killed. The importance of polysaccharide priming for responses to microorganisms may explain the relatively broad sugar specificity of the CR3 lectin site. In particular, β-glucans are not common surface polysaccharides on bacteria or yeast, and thus the ability of the CR3 lectin to bind polysaccharides containing mannose and NADG, as well as glucose, is probably significant. The cytotoxic activity of neutrophils toward the majority of iC3b-opsonized bacteria suggests that additional sugars that prime CR3 are yet to be identified.

The lectin site of CR3 is apparently also utilized for neutrophil cytotoxic responses involving IgG-opsonized bacteria and the phosphatidylinositol-linked FcγRIIIB. Recent reports suggest that FcγRIIIB can bind to the lectin site of CR3 and prime the receptor in the same way as a bacterial polysaccharide (Graham et al., 1989; Zhou et al., 1992; Petty and Todd, 1993; Zhou et al., 1993; Stöckl et al., 1995). This may be the explanation for the phagocytic and cytotoxic response to IgG-opsonized erythrocytes or tumor cells that lack membrane polysaccharides capable of directly priming CR3. Responses to soluble immune complexes are also inhibited by sugars such as

NADG that disrupt the membrane complexes of CR3 and FcγRIIIB (Sehgal et al., 1993).

Use of Polysaccharides for Immunotherapy of Cancer

Although tumor cells lack polysaccharides capable of priming CR3, they are frequently coated with opsonizing Ig and iC3b/C3dg by virtue of an immune response to tumor-specific antigens. Soluble polysaccharides are thus able to prime the CR3 of neutrophils, macrophages, and natural killer cells to mediate the cytotoxicity of tumor cells naturally targeted with iC3b/C3dg that would otherwise be resistant to such CR3-mediated attack. This may explain the action of soluble β-glucans from yeast, mushrooms, and bacteria that have been used successfully for more than 30 years for therapy of cancer in experimental animals and later patients (Diller et al., 1963; Nakahara et al., 1964; Chihara et al., 1969; Di Luzio et al., 1979; Rose et al., 1984; Morikawa et al., 1985; Kosaka et al., 1990; Chen and Hasumi, 1993; Suzuki et al., 1994).

REFERENCES

Abel G, Czop JK (1992): Stimulation of human monocyte β-glucan receptors by glucan particles induces production of TNF-α and IL-1β. Int J Immunopharmacol 14:1363–1373.

Anderson DC, Springer TA (1987): Leukocyte adhesion deficiency: an inherited defect in the Mac-1, LFA-1, and p150,95 glycoproteins. Annu Rev Med 38:175–194.

Arnaout MA, Dana N, Pitt J, Todd RF III (1987): Deficiency of two human leukocyte surface membrane glycoproteins (Mo1 and LFA-1). Fed Proc 44:2664–2670.

Au B, Williams TJ, Collins PD (1994): Zymosan-induced IL-8 release from human neutrophils involves activation via the CD11b/CD18 receptor and endogenous platelet-activating factor as an autocrine modulator. J Immunol 152:5411–5419.

Berger M, O'Shea J, Cross AS, Folks TM, Chused TM, Brown EJ, Frank MM (1984): Human neutrophils increase expression of C3bi as well as C3b receptors upon activation. J Clin Invest 74:1566–1571.

Berton G, Fumagalli L, Laudanna C, Sorio C (1994): β2 integrin-dependent protein tyrosine phosphorylation and activation of the *FGR* protein tyrosine kinase in human neutrophils. J Cell Biol 126:1111–1121.

Buyon JP, Abramson SB, Philips MR, Slade SG, Ross GD, Weissmann G, Winchester RJ (1988): Dissociation between increased surface expression of gp165/95 and homotypic neutrophil aggregation. J Immunol 140:3156–3160.

Cain JA, Newman SL, Ross GD (1987): Role of complement receptor type three and serum opsonins in the neutrophil response to yeast. Complement Inflamm 4:75–86.

Carlos TM, Harlan JM (1990): Membrane proteins involved in phagocyte adherence to endothelium. Immunol Rev 114:1–24.

Chen J-T, Hasumi K (1993): Activation of peritoneal macrophages in patients with gynecological malignancies by sizofiran and recombinant interferon-gamma. Biotherapy 6:189–194.

Cheong J-J, Hahn MG (1991): A specific, high-affinity binding site for the hepta-β-glucoside elicitor exists in soybean membranes. Plant Cell 3:137–147.

Chihara G, Maeda Y, Hamuro J, Sasaki T, Fukuoka F (1969): Inhibition of mouse sarcoma 180 by polysaccharides from Lentinus edodes (Berk.) sing. Nature 222:687–688.

Corbi AL, Larson RS, Kishimoto TK, Springer TA, Morton CC (1988): Chromosomal location of the genes encoding the leukocyte adhesion receptors LFA-1, Mac-1, and p150,95. J Exp Med 167:1597–1607.

Czop JK, Austen KF (1985): Generation of leukotrienes by human monocytes upon stimulation of their β-glucan receptor during phagocytosis. Proc Natl Acad Sci USA 82:2751–2755.

Czop JK, Puglisi AV, Miorandi DZ, Austen KF (1988): Perturbation of β-glucan receptors on human neutrophils initiates phagocytosis and leukotriene B_4 production. J Immunol 141:3170–3176.

De la Hera A, Alvarez-Mon M, Sánchez-Madrid F, Martinez-A C, Durantez A (1988): Co-expression of Mac-1 and p150,95 on $CD5^+$ B cells. Structural and functional characterization in a human chronic lymphocytic leukemia. Eur J Immunol 18:1131–1134.

Di Luzio NR, Williams DL, McNamee RB, Edwards BF, Kitahama A (1979): Comparative tumor-inhibitory and anti-bacterial activity of soluble and particulate glucan. Int J Cancer 24:773–779.

Diamond MS, Garcia-Aguilar J, Bickford JK, Corbi AL, Springer TA (1993): The I domain is a major recognition site on the leukocyte integrin Mac-1 (CD11b/CD18) for four distinct adhesion ligands. J Cell Biol 120:1031–1043.

Diamond MS, Springer TA (1993): A subpopulation of Mac-1 (CD11b/CD18) molecules mediates neutrophil adhesion to ICAM-1 and fibrinogen. J Cell Biol 120:545–556.

Diller IC, Mankowski ZT, Fisher ME (1963): The effect of yeast polysaccharides on mouse tumors. Cancer Res 23:201–208.

Edberg JC, Kimberly RP (1994): Modulation of Fcγ and complement receptor function by the glycosylphosphatidylinositol-anchored form of FcγRIII. J Immunol 152:5826–5835.

Fuortes M, Jin W, Nathan C (1994): β2 integrin-dependent tyrosine phosphorylation of paxillin in human neutrophils treated with tumor necrosis factor. J Cell Biol 127:1477–1483.

Graham IL, Anderson DC, Holers VM, Brown EJ (1994): Complement receptor 3 (CR3, Mac-1, integrin $\alpha_M\beta_2$, CD11b/CD18) is required for tyrosine phosphorylation of paxillin in adherent and nonadherent neutrophils. J Cell Biol 127:1139–1147.

Graham IL, Gresham HD, Brown EJ (1989): An immobile subset of plasma membrane CD11b/CD18 (Mac-1) is involved in phagocytosis of targets recognized by multiple receptors. J Immunol 142:2352–2358.

Graham IL, Lefkowith JB, Anderson DC, Brown EJ (1993): Immune complex-stimulated neutrophil LTB_4 production is dependent on β_2 integrins. J Cell Biol 120:1509–1517.

Hogg N (1992): Roll, roll, roll your leucocyte gently down the vein. . . . Immunol Today 13:113–115.

Hogg N, Takacs L, Palmer DG, Selvendran Y, Allen C (1986): The p150,95 molecule is a marker of human mononuclear phagocytes: comparison with expression of Class II molecules. Eur J Immunol 16:240–248.

Kosaka A, Kuzuoka M, Yamashita A, Takino Y, Tanizawa H, Bandou K (1990): A study on the serum concentration of lentinan administered via various routes in tumor bearing mice and in human breast cancer patients: in relation to antitumor effect. In Masihi KN, Lange W (eds): *Immunotherapeutic Prospects of Infectious Diseaes.* Berlin: Springer-Verlag, pp 233–238.

Krauss JC, Poo H, Xue W, Mayo-Bond L, Todd RF III, Petty HR (1994): Reconstitution of antibody-dependent phagocytosis in fibroblasts expressing Fcγ receptor IIIB and the complement receptor type 3. J Immunol 153:1769–1777.

Lee J-O, Rieu P, Arnaout MA, Liddington R (1995): Crystal structure of the A domain from the α subunit of integrin CR3 (CD11b/CD18). Cell 80:631–638.

Lew DB, Leslie CC, Henson PM, Riches DWH (1991): Role of endogenously derived leukotrienes in the regulation of lysosomal enzyme expression in macrophages exposed to β 1,3-glucan. J Leukoc Biol 49:266–276.

McFarland HI, Nahill SR, Maciaszek JW, Welsh RM (1992): CD11b (Mac-1): A marker for $CD8^+$ cytotoxic T cell activation and memory in virus infection. J Immunol 149:1326–1333.

Morikawa K, Takeda R, Yamazaki M, Mizuno D (1985): Induction of tumoricidal activity of polymorphonuclear leukocytes by a linear β-1,3-D-glucan and other immunomodulators in murine cells. Cancer Res 45:1496–1501.

Muto S, Vêtvička V, Ross GD (1993): CR3 (CD11b/CD18) expressed by cytotoxic T cells and NK cells is upregulated in a manner similar to neutrophils following stimulation with various activating agents. J Clin Immunol 13:175–184.

Nakahara W, Fukuoka F, Maeda Y, Aoki K (1964): The host-mediated antitumor effect of some plant polysaccharides. Gann 55:283–288.

Naziruddin B, Duffy BF, Tucker J, Mohanakumar T (1992): Evidence for cross-regulation of FcgammaRIIIB (CD16) receptor-mediated signaling by FcgammaRII (CD32) expressed on polymorphonuclear neutrophils. J Immunol 149:3702–3709.

Newman SL, Devery-Pocius JE, Ross GD, Henson PM (1984): Phagocytosis by human monocyte-derived macrophages. Independent function of receptors for C3b (CR_1) and iC3b (CR_3). Complement Inflamm 1:213–227.

Nielsen HV, Christensen JP, Andersson EC, Marker O, Thomsen AR (1994): Expression of type 3 complement receptor on activated $CD8^+$ T cells facilitates homing to inflammatory sites. J Immunol 153:2021–2028.

Pavalko FM, LaRoche SM (1993): Activation of human neutrophils induces an interaction between the integrin β_2-subunit (CD18) and the actin binding protein α-actinin. J Immunol 151:3795–3807.

Petersen MM, Steadman R, Williams JD (1994): Human neutrophils are selectively activated by independent ligation of the subunits of the CD11b/CD18 integrin. J Leukoc Biol 56:708–713.

Petty HR, Todd RF III (1993): Receptor-receptor interactions of complement receptor type 3 in neutrophil membranes. J Leukoc Biol 54:492–494.

Picker LJ, Treer JR, Nguyen M, Terstappen LWMM, Hogg N, Yednock T (1993): Coordinate expression of β1 and β2 integrin "activation" epitopes during T cell responses in secondary lymphoid tissue. Eur J Immunol 23:2751–2757.

Pryzwansky KB, Wyatt T, Reed W, Ross GD (1991): Phorbol ester induces transient focal concentrations of functional, newly expressed CR3 in neutrophils at sites of specific granule exocytosis. Eur J Cell Biol 54:61–75.

Rabb H, Michishita M, Sharma CP, Brown D, Arnaout MA (1993): Cytoplasmic tails of human complement receptor type 3 (CR3, CD11b/CD18) regulate ligand avidity and the internalization of occupied receptors. J Immunol 151:990–1002.

Rose WC, Reed FC III, Siminoff P, Bradner WT (1984): Immunotherapy of Madison 109 lung carcinoma and other murine tumors using Lentinan. Cancer Res 44:1368–1373.

Ross GD, Cain JA, Lachmann PJ (1985): Membrane complement receptor type three (CR_3) has lectin-like properties analogous to bovine conglutinin and functions as a receptor for zymosan and rabbit erythrocytes as well as a receptor for iC3b. J Immunol 134:3307–3315.

Ross GD, Cain JA, Myones BL, Newman SL, Lachmann PJ (1987): Specificity of membrane complement receptor type three (CR_3) for β-glucans. Complement Inflamm 4:61–74.

Ross GD, Reed W, Dalzell JG, Becker SE, Hogg N (1992): Macrophage cytoskeleton association with CR3 and CR4 regulates receptor mobility and phagocytosis of iC3b-opsonized erythrocytes. J Leukoc Biol 51:109–117.

Russell DG, Wright SD (1988): Complement receptor type 3 (CR3) binds to an Arg-Gly-Asp-containing region of the major surface glycoprotein, gp63, of *Leishmania* promastigotes. J Exp Med 168:279–292.

Salmon JE, Brogle NL, Edberg JC, Kimberly RP (1991): Fcγ receptor III induces actin polymerization in human neutrophils and primes phagocytosis mediated by Fcγ receptor II. J Immunol 146:997–1004.

Sehgal G, Zhang K, Todd RF III, Boxer LA, Petty HR (1993): Lectin-like inhibition of immune complex receptor-mediated stimulation of neutrophils: Effects on cytosolic calcium release and superoxide production. J Immunol 150:4571–4580.

Sengelov H, Kjeldsen L, Diamond MS, Springer TA, Borregaard N (1993): Subcellular localization and dynamics of Mac-1 ($\alpha_m\beta_2$) in human neutrophils. J Clin Invest 92:1467–1476.

Singer II, Scott S, Kawka DW, Kazazis DM (1989): Adhesomes: specific granules containing receptors for laminin, C3bi/fibrinogen, fibronectin, and vitronectin in human polymorphonuclear leukocytes and monocytes. J Cell Biol 109:3169–3182.

Smith CW (1993): Leukocyte-endothelial cell interactions. Semin Hematol 30(suppl 4):45–55.

Springer TA (1990): The sensation and regulation of interactions with the extracellular environment: The cell biology of lymphocyte adhesion receptors. Annu Rev Cell Biol 6:359–402.

Stöckl J, Majdic O, Pickl WF, Rosenkranz A, Prager E, Gschwantler E, Knapp W (1995): Granulocyte activation via a binding site near the C-terminal region of complement receptor type 3 α-chain (CD11b) potentially involved in intramolecular complex formation with glycophosphatidylinositol-anchored FcγRIIIB (CD16) molecules. J Immunol 154:5452–5463.

Suzuki M, Takatsuki F, Maeda YY, Hamuro J, Chihara G (1994): Antitumor and immunological activity of lentinan in comparison with LPS. Int J Immunopharmacol 16:463–468.

Taniguchi-Sidle A, Isenman DE (1992): Mutagenesis of the Arg-Gly-Asp triplet in human complement component C3 does not abolish binding of iC3b to the leukocyte integrin complement receptor type III (CR3, CD11b/CD18). J Biol Chem 267:635–643.

Thornton BP, Větvička V, Pitman M, Goldman RC, and Ross GD (1996): Analysis of the sugar specificity and molecular location of the β-glucan-binding lectin site of complement receptor type III (CD11b/CD18). J Immunol 156:1235–1242.

Todd RF III, Freyer DR (1988): The CD11/CD18 leukocyte glycoprotein deficiency. Hematol Oncol Clin N Am 2:13–31.

Ueda T, Rieu P, Brayer J, Arnaout MA (1994): Identification of the complement iC3b binding site in the β2 integrin CR3 (CD11b/CD18). Proc Natl Acad Sci USA 91:10680–10684.

Větvička V, Thornton BP, and Ross GD (1996): Soluble β-glucan polysaccharide binding to the lectin site of neutrophil or natural killer cell complement receptor type 3 (CD11b/CD18) generates a primed state of the receptor capable of mediating cytotoxicity of iC3b-opsonized target cells. J Clin Invest 98: 50–61.

Walport MJ, Ross GD (1989): Deficiency of the LFA-1 family of molecules. In Asherson GL, Zembala M (eds): *Human Monocytes.* London: Academic Press, Inc. pp 417–427.

Werfel T, Witter W, Götze O (1991): CD11b and CD11c antigens are rapidly increased on human natural killer cells upon activation. J Immunol 147:2423–2427.

Winchester RJ, Ross GD (1986): Methods for enumerating cell populations by surface markers using conventional microscopy. In Rose NR, Friedman H, Fahey JL (eds): *Manual of Clinical Laboratory Immunology.* 3rd ed. Washington, DC: American Society for Microbiology, pp 212–225.

Wright SD, Reddy PA, Jong MTC, Erickson BW (1987): C3bi receptor (complement receptor type 3) recognizes a region of complement protein C3 containing the sequence Arg-Gly-Asp. Proc Natl Acad Sci USA 84:1965–1968.

Xue W, Kindzelskii AL, Todd RF III, Petty HR (1994): Physical association of complement receptor type 3 and urokinase-type plasminogen activator receptor in neutrophil membranes. J Immunol 152:4630–4641.

Zhou M, Brown EJ (1994): CR3 (Mac-1, $\alpha_M\beta_2$, CD11b/CD18) and FcγRIII cooperate in generation of a neutrophil respiratory burst: requirement for FcγRII and tyrosine phosphorylation. J Cell Biol 125:1407–1416.

Zhou M, Todd RF III, van de Winkel JGJ, Petty HR (1993): Cocapping of the leukoadhesin molecules complement receptor type 3 and lymphocyte function-associated antigen-1 with Fc_γ receptor III on human neutrophils: possible role of lectin-like interactions. J Immunol 150:3030–3041.

Zhou M-J, Poo H, Todd RF III, Petty HR (1992): Surface-bound immune complexes trigger transmembrane proximity between complement receptor type 3 and the neutrophil's cortical microfilaments. J Immunol 148:3550–3553.

2

Cytometric Approaches to the Study of Receptors

John P. Nolan
Los Alamos National Laboratory, Los Alamos, New Mexico

J. David Chambers
The Salk Institute, San Diego, California

Larry A. Sklar
Los Alamos National Laboratory, Los Alamos, New Mexico, and University of New Mexico School of Medicine, Albuquerque, New Mexico

Living cells communicate with their environment and with other cells to perform tasks necessary for their survival or the survival of the organism as a whole. It has long been recognized that this communication is mediated by soluble chemical signals and specific cell surface receptors. Over the past 20 years, many of the molecules involved in receiving these signals and transmitting them to the interior of the cell have been identified. However, the nature of the chemical pathways that carry these signals is still the subject of much research and debate.

A major development in the study of receptor–ligand interactions has been the use of radiolabeled ligands. Radioisotopes have provided the sensitivity for detailed analysis of ligand binding to receptors in tissues, cells, and cell membrane preparations. Such measurements have led to the realization that receptors could exist in high-affinity and low-affinity states, that these receptor states were interconverting, and

Phagocyte Function: A Guide for Research and Clinical Evaluation, Edited by J. Paul Robinson and George F. Babcock. ISBN 0-471-12364-1

that receptor agonists and antagonists interacted with these states in different ways. More recently, a number of biochemical and biophysical advances have furthered our understanding of receptor–ligand interactions. The biochemical tools include mutagenesis, which allows the evaluation of the roles of structural domains and specific amino acids in receptor function (Zoller, 1991), and photoaffinity cross-linking, which can identify specific sites of interaction between the ligand and the receptor (Phizicky and Fields, 1995). The biophysical tools include crystallography, to give three-dimensional structural detail (MacArthur et al., 1994), and novel binding methodologies such as surface plasmon resonance spectroscopy, which permits continuous kinetic measurement of ligand binding to purified receptors (Hoyt et al., 1994). Another powerful and complementary tool in the study of receptor-mediated signal transduction is flow cytometry.

Flow cytometric investigation of receptor–ligand interactions combines the power of continuous kinetic analysis, provided by an intrinsic resolution between free and bound ligand, with the study of receptors in their native environment, the intact cell. This allows not only the direct measurement of the rates of ligand association and dissociation (and the receptor affinities which result from those rates), but also the analysis of interconversion of receptor affinity states during the signal transduction process. Changes in ligand affinity and dissociability are associated with such biologically important processes as receptor coupling to signal transducing molecules, receptor desensitization, and receptor internalization. Rapid, multiparameter measurements of individual cells permit the simultaneous evaluation of ligand binding and cell response. In combination with biochemical tools to probe the interactions of receptors with the intracellular signaling machinery, flow cytometry opens an important new window into cell function. In this chapter we review the elements of quantitative flow cytometric analysis of ligand–receptor interactions and examine specific examples of this methodology as applied to phagocyte function.

TRADITIONAL LIGAND BINDING METHODS

The characterization of the binding of a ligand to its receptor has several key elements. We highlight the essential elements of these methods. For a more detailed discussion of cell receptor–ligand binding, the reader is referred to the comprehensive text by Lauffenberger and Linderman (1993). The first requirement is to quantify the amount of bound ligand. Radiolabeled ligands provide quantitative measurements with excellent sensitivity, allowing the use of low levels of receptor (fmoles). However, it is necessary physically to separate free radioligand from bound before quantification. Such heterogeneous assays have several inherent limitations that are discussed below. Next the equilibrium and kinetic features of the system are evaluated. Radioligand equilibrium binding curves have been a major tool for the study of ligand–receptor interactions. Kinetic analysis of ligand–receptor association and dissociation complements the equilibrium measurements of simple systems and provides a means to understand more complicated systems where the assumptions of equilibrium analysis may no longer apply. Computer-aided curve fitting and model evaluation can facilitate the analysis of both equilibrium and kinetic data. Used in combination, these tools can be used to create and evaluate physical models of ligand–receptor interactions.

Equilibrium Binding Measurements Using Radioligands

The starting point for analysis of ligand–receptor interactions is the equilibrium binding isotherm, a determination of specifically bound ligand as a function of ligand concentration. Such experiments typically involve: (1) incubation of receptor and radioligand to allow the system to come to equilibrium; (2) separation of free from bound ligand, usually by centrifugation or filtration; (3) scintillation counting of the sample to quantify the amount of bound radioligand; and (4) estimation of specific binding by subtraction of nonspecific binding (determined in samples containing excess unlabeled ligand) from the total binding.

For the simplest case (Fig. 1A) of a ligand (L) binding specifically to a single class of receptors (R), binding data can be analyzed in terms of a single, reversible equilibrium between free L and R and the ligand–receptor complex (LR):

$$\mathrm{L} + \mathrm{R} \rightleftharpoons \mathrm{LR} \tag{1}$$

This equilibrium can be described by a dissociation constant,

$$K_d = \frac{[\mathrm{L}]\,[\mathrm{R}]}{[\mathrm{LR}]} \tag{2}$$

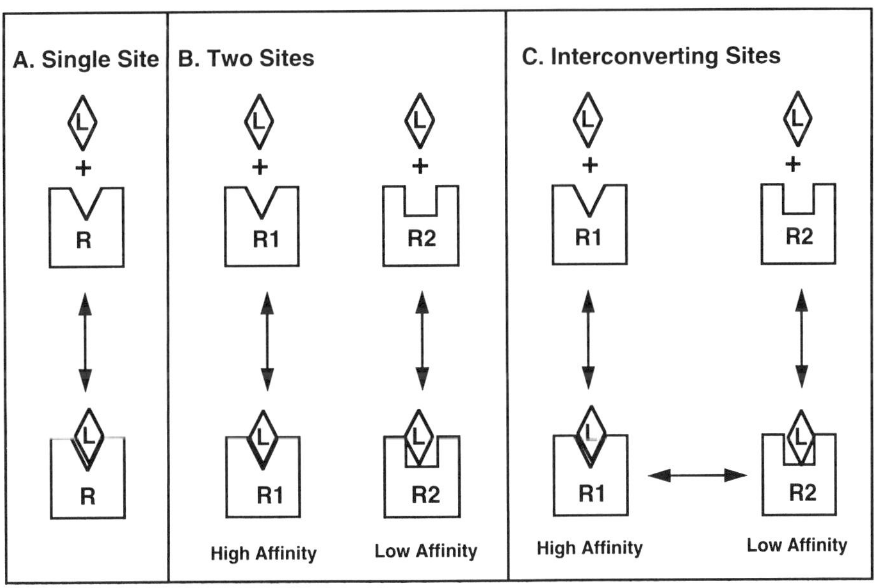

Fig. 1. Models of receptor–ligand interactions. (A) Single receptor class. (B) Two independent receptor classes with different affinities. (C) Two interconverting receptor classes with different affinities.

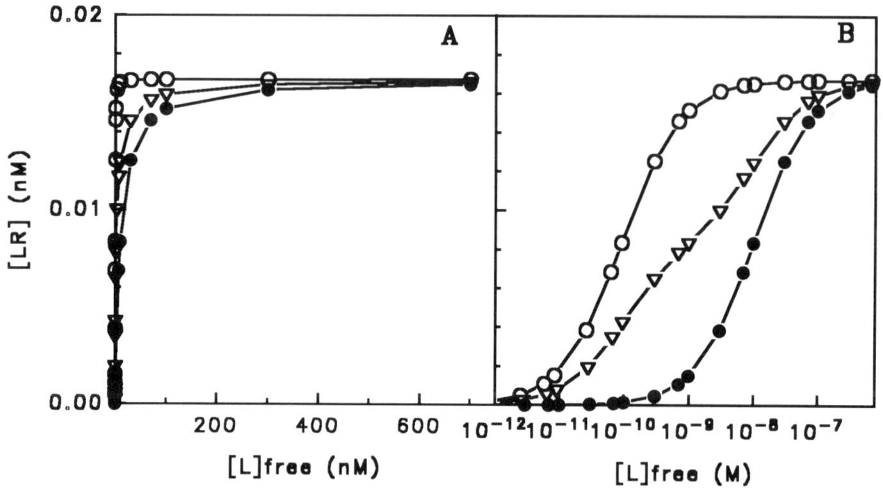

Fig. 2. Equilibrium ligand binding curves. Ligand binding data calculated according to equation 3 for the conditions of receptors per cell at 1×10^5 cells/ml, and a K_d of 0.1 n*M* (○), 10 n*M* (●), or a 50/50 mixture of the two (▽). Presented in (A) linear and (B) semilogarithmic form.

which, under conditions where receptor is present at limiting low dilution, represents the [L] where 50% of R is bound. When [LR] is plotted as a function of [L], the binding curve has the shape of a hyperbola (Fig. 2A) and is described by the equation

$$[\mathrm{LR}] = \frac{[\mathrm{L}]\,[\mathrm{R}]_{\mathrm{tot}}}{K_d + [\mathrm{L}]} \tag{3}$$

where $[\mathrm{R}]_{\mathrm{tot}}$ is the total number of binding sites present. When [L] is plotted on a log axis (Fig. 2B) it is easy to visualize the K_d as the point of 50% receptor occupancy.

For the more complicated but realistic case of two independent classes of receptors with different affinities (Fig. 1B), the number and affinity of each class is not apparent from inspection of the binding curves, though they may be obtained by computer curve fits to the two-site form of the binding curve equation:

$$[\mathrm{LR}] = \frac{[\mathrm{L}][\mathrm{R}_1]_{\mathrm{tot}}}{(K_{d1} + [\mathrm{L}])} + \frac{[\mathrm{L}][\mathrm{R}_2]_{\mathrm{tot}}}{(K_{d2} + [\mathrm{L}])} \tag{4}$$

A commonly used linear transform of the binding data is the Scatchard plot. Scatchard analysis can provide information on the total number of binding sites (*x*-intercept) and the K_d (-1/slope), provided the receptor preparation represents a homogeneous population of independent, noninteracting binding sites. When two classes of receptors are present, the Scatchard plot is biphasic and the K_d and number of receptors in each class can be estimated by analyzed each phase separately. A related and complementary analysis, the Hill plot, requires knowledge of the total number of binding sites and can provide an estimate of the K_d. The Hill coefficient defines whether the requirements of homogeneous, independent binding sites are met. In general, such graphic analyses are useful aids to visualize the data, but have a number of limitations to their quantitative use (Chang et al., 1975; Taylor et al., 1975; Weiland and Molinoff, 1981; Ketelslegers et al., 1984). Estimates of receptor number and K_d are better obtained using computer fitting of the appropriate binding equation (Eq. 3 or 4) to the untransformed data (Motulsky and Ransnas, 1988; Johnson, 1992).

Kinetic Analysis of Receptor–Ligand Interactions

In a cell membrane preparation or a purified, detergent-solubilized system, receptors may indeed represent a homogeneous population of independent binding sites. In the intact cell, membrane-bound receptors almost certainly do not. As will be seen, experimental evidence indicates that the number of receptors and their affinities for ligand change with time during cell activation. Apart from potentially rendering invalid equilibrium binding assumptions (homogeneous, noninteracting, independent sites), these receptor dynamics may also contain information about the biochemical events involved in regulating cell responses. Kinetic approaches to studying ligand–receptor interactions can extract this mechanistic information.

For a single reversible binding event described above, $L + R \rightleftharpoons LR$, the K_d is equal to the ratio of the reverse (dissociation) rate constant (k_{off}) to the forward (association) rate constant (k_{on}):

$$K_d = \frac{k_{off}}{k_{on}} \tag{5}$$

The association rates for small ligands binding to cell surface receptors can approach the diffusion limit (Erickson et al., 1987), in which case the binding affinity is largely determined by the dissociation kinetics. Thus high-affinity ligands are often characterized by slow off rates and low-affinity ligands by high off rates. When the binding rates are not limited by diffusion, it may be argued that conformational changes in either the ligand or the receptor are required for binding to occur. In the case of a ligand–receptor interaction with a K_d of 10^{-10} *M,* and a diffusion-limited bimolecular association rate constant of $1 \times 10^7\ M^{-1}\text{s}^{-1}$, the dissociation rate constant would be 0.001 s^{-1}. For a low-affinity interaction with a K_d of 10^{-8} *M,* and diffusion-limited binding, the k_{off} would be 0.1 s^{-1}. Detailed analyses of the considerations involved in diffusion-limited ligand binding to cell surface receptors have been published (see Model and Omann, 1995, and references therein). In contrast, for larger molecules or antibodies with bivalent interactions, ligand conformational changes involved in receptor binding might affect the overall affinity via both association and dissociation kinetics. These mechanistic distinctions are relevant to understanding the molecular events involved in receptor mediated signaling, and kinetic analysis enables these possibilities to be tested.

The progress of a simple binding reaction can be calculated using the formula for the time-dependent change in ligand bound:

$$\frac{d[\mathrm{LR}]}{dt} = k_{\mathrm{on}}[\mathrm{L}][\mathrm{R}] - k_{\mathrm{off}}[\mathrm{LR}] \tag{6}$$

Because ligand–receptor association is a bimolecular process, the binding rate depends on both [L] and [R], as well as k_{on}. Ligand dissociation depends on k_{off} and [LR]. At the start of a binding experiment, when L is added, $d[\mathrm{LR}]/dt = k_{on}[\mathrm{L}][\mathrm{R}]$ because [LR] is zero. For this reason, plots of initial rates versus [L] are often used to estimate association rate constants. As binding progresses and LR accumulates, the contribution of the reverse reaction becomes significant until, at equilibrium, the reverse rate equals the forward rate, and the net change in binding with time, $d[\mathrm{LR}]/dt$, equals zero. For ligands with diffusion-limited association, where the affinity is dictated by k_{off}, a low-affinity ligand will reach equilibrium faster than a high-affinity ligand (Fig. 3A). If one adds excess unlabeled ligand to the system at equilibrium, there is no available receptor, the forward rate becomes zero, and $d[\mathrm{LR}]/dt = -k_{off}[\mathrm{LR}]$. The dissociation kinetics (Fig. 3B) are that of an exponential decay, with a half-time

$$t_{1/2} = \frac{\ln 2}{k_{\mathrm{off}}} \tag{7}$$

For the high-affinity interaction with a k_{off} of 0.001 s^{-1}, ligand dissociation would have a $t_{1/2}$ of ~ 700 s. For the low-affinity binding with a k_{off} of 0.1 s^{-1}, $t_{1/2}$ would be 7 s.

For two independent receptor classes, R_1 and R_2, with different affinities, the ob-

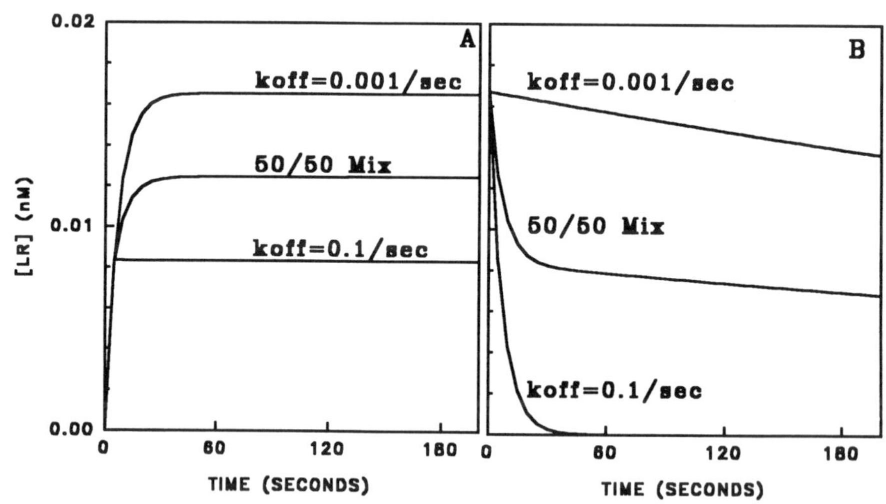

Fig. 3. Ligand binding kinetics. (A) Association kinetics. Calculated for the conditions of Figure 2, using an association rate constant of $1 \times 10^7\ M^{-1}s^{-1}$ and dissociation rate constants of 0.001 s^{-1} (K_d of 10^{-10} *M*), 0.1 s^{-1} (K_d of 10^{-8} *M*), or a 50/50 mixture of the two. (B) Dissociation kinetics, calculated for the conditions in (A).

served rate of binding includes contributions from the association and dissociation rates for each class of receptor:

$$\frac{d[\mathrm{LR}]}{dt} = k_{\mathrm{on1}}[\mathrm{L}][\mathrm{R}_1] + k_{\mathrm{on2}}[\mathrm{L}][\mathrm{R}_2] - k_{\mathrm{off1}}[\mathrm{LR}_1] - k_{\mathrm{off2}}[\mathrm{LR}_2] \quad (8)$$

For ligands with diffusion-limited binding whose affinity is dictated by k_{off}, the fraction of high- and low-affinity receptors will be apparent from the ligand dissociation kinetics (Fig. 3B). As will be described, the use of kinetic models enables the evaluation of interconversion between receptor states in response to ligand binding (Fig. 1C) or other biochemical events, information that is difficult to extract from equilibrium measurements. Thus kinetic approaches have the potential to address issues of macromolecular assembly and the receptor dynamics involved in cellular signal transduction.

Limitations of Radioligand Methods

Although radioligand measurements are currently used by many researchers studying receptor function, heterogeneous assays have inherent limitations. First, methods that require separation of free ligand from bound ligand for every sample are cumbersome for taking multiple time points in kinetic studies compared to a homogeneous assay, which provides continuous data readout. More importantly, separation steps that take even just a few seconds result in the loss of ligand which dissociates rapidly. This limitation is significant when studying low-affinity ligands, which often have fast off rates. By contrast, flow cytometry enables both the homogeneous assay of binding of fluorescent ligands to receptors, and the continuous discrimination of free and bound ligand. These features are ideally suited for kinetic measurements, the binding of ligands with rapid dissociation (low-affinity ligands), and the measurement of rapid association processes (which often occur at saturating ligand concentrations).

FLOW CYTOMETRIC LIGAND BINDING MEASUREMENTS

General Considerations

Real-time analysis of ligand–receptor binding requires continuous detection of bound ligand as a function of time, even at low receptor concentration. For the interaction of a fluorescently labeled ligand with a receptor on an intact cell, flow cytometry is uniquely capable of providing this information. Flow cytometry provides quantitative fluorescence measurements of cell-bound ligand without the need to wash away free ligand. Flow cytometry also permits multiparameter measurement of individual cells and the study of a subset of cells with specific characteristics (for instance, size or expression of surface markers) from the population. Multiparameter analysis allows simultaneous measurement of ligand binding and cell responses (such as ion

TABLE 1. Commonly Used Fluorophores for Cytometry

Fluorophore	Excitation (nm)	Emission (nm)	$\epsilon(cm^{-1}M^{-1})$	Q	Comments
FITC[a]	495	519	7.6×10^4	0.9	pH sensitive
NBD[b]	470	530	2×10^4	0.3	Environment sensitive
BODIPY-FL[c]	503	512	9.0×10^4	<0.9	
TRITC[d]	555	572	4.4×10^5	0.15	
B-Phycoerythrin[e]	546	575	2.4×10^6	0.98	240,000 MW

[a]Fluorescein isothiocyanate (Haugland, 1992).
[b]*N*-7-Nitrobenz-2-oxa-1,3-diazole (Haugland, 1992).
[c]4,4-difluoro-5,7-dimethyl-4-bora-3a,4a-diaza-s-indacene-3-proprionic acid.
[d]Tetramethylrhodamine 5-isothiocyanate (Soper *et al.*, 1993).
[e](Haugland, 1992).

fluxes, membrane potential changes, or shape changes). Advances in the kinetic and spectroscopic capabilities of flow cytometers make possible a direct comparison between cytometric measurements on single cells and a number of complementary bulk phase (cuvette) methods.

Fluorescent Ligands

Classically, receptor types have been identified by observing a cell or tissue response to a physiological stimulus or ligand. The introduction of radioligand technology provided the sensitivity necessary for direct analysis of ligand binding to small quantities of receptor and ultimately the identity of the receptor molecules. As the next logical step in ligand–receptor technology, fluorescently labeled ligands offer the sensitivity for continuous observation of small amounts of receptor. Several considerations go into the selection of the fluorophore: (1) spectral characteristics compatible with flow cytometric detection; (2) ease of conjugation to specific sites on the target molecule; (3) retention of biological activity of the ligand after conjugation; (4) size or molecular mass compatible with the ligand. These issues are discussed in detail elsewhere (Baindur and Triggle, 1994; Haugland, 1992). We highlight a few considerations relevant to flow cytometric analysis of ligand binding.

The fluorescent probe that is still most often used for labeling ligands or other biological molecules is fluorescein isothiocyanate (FITC). Conjugates of FITC are efficiently excited by the 488-nm argon-ion laser line available in most commercial cytometers. Fluorescein isothiocyanate has a high molar absorptivity (Table 1) and a high quantum yield ($Q \sim 0.9$), making it one of the brightest probes available. The Stokes shift, the separation between the absorbance and emission peaks, is small but adequate to detect emitted photons with minimal interference from scattered laser light, when optimized mirror and filter sets are used. The sensitivity of FITC fluorescence to pH and chemical environment can complicate quantitative aspects of the binding analysis, but can also be useful for following receptor processing events such as internalization.

A number of other fluorophore families are in wide use (Table 1). These differ in their spectroscopic properties and may be more or less suitable than FITC for a particular application. In contrast to FITC, BODIPY fluorescence is insensitive to pH. NBD is insensitive to pH but highly sensitive to the polarity of its chemical environment. High extinction coefficients are generally desirable, but when powerful excitation sources (such as lasers) are available, the quantum yield may dictate which probe is brighter. A high extinction coefficient combined with a large spectral overlap between absorbance and emission spectra (probes with small Stokes shifts) can result in efficient density-dependent self-quenching, an undesirable property in most quantitative applications. In principle, a flow cytometer can detect fluorescence from almost any fluorophore, provided the instrument is configured with an appropriate excitation source and emission filters. In practice, experiments on newer commercial cytometers are limited to probes with properties similar to FITC or other fluorophores that are routinely used for immunophenotyping.

Apart from the spectroscopic properties of a probe, it is necessary to consider more obvious features such as its relative size compared to the ligand to be labeled. Phycoerythrin is one of the brightest probes known, but with a molecular weight of 240 kDa, it is also one of the largest. Thus, although phycoerythrin is often used to label large molecules such as antibodies, it would be inappropriate to use phycoerythrin to label a small peptide with a molecular weight of 1 kDa for use in kinetic binding studies. Even when the fluorophore is small, care must be taken to evaluate the effect of conjugation on the binding properties of the ligand. Conjugation of a bulky fluorescent group to a small ligand may alter an important structural feature of the ligand or hinder interaction with the binding pocket.

For example, we have developed a family of fluorescent ligands (peptides ranging from four to seven amino acids) for the formyl peptide receptor. These peptides contain a lysine (serving as a conjugation site) at a single position two to seven amino acids from the formyl group. Whereas the addition of lysine typically reduces the affinity for the receptor by two to three orders of magnitude, FITC conjugation neutralizes the positive charge of the lysine and restores the affinity. Addition of hydrophobic amino acids (phenylalanine) to the carboxy terminus of the peptide further increases the binding affinity and biological activity of the ligand. Although this particular strategy may not be applicable to peptide ligands for other receptors, it may be possible to alter sequence in a manner that provides specific conjugation sites and retains biological activity.

Ligands for cell surface receptors include small peptides and proteins, biopolymers such as polyamines or proteoglycans, and small organic compounds such as steroids or amino acids. The peptides often present the best opportunities for conjugation, and many chemotactic peptides and neuroactive peptides, including some labeled with fluorophore, are available from commercial sources. Protein ligands have been traditionally purified from biological sources before labeling, and problems with the quantity and purity of the starting material can hamper quantitative binding studies. The increasing commercial availability of purified recombinant proteins may alleviate some of these limitations. Still, with large molecules, the possibilities of distinct binding isoforms, glycosylation variants, and heterogeneous fluorophore conjugation must all be considered. The fluorophore to protein ratio (f/p) is an essential characterization because when multiple fluorophores are attached to a single protein molecule, density-dependent fluorescence self-quenching is observed. Other important ligands are amino acids, steroids, or other small organic molecules; strategies for labeling these have been recently reviewed (Banduir and Triggle, 1994). The importance of demonstrating that the affinity and specificity of the fluorescent ligand are similar to the unlabeled ligand cannot be overemphasized. Thus competition with the native ligand for binding as well as similar dose dependencies of biological responses for the labeled and unlabeled ligands are good indicators that the labeled ligand binds to the receptor appropriately.

Discrimination of Free, Bound, and Internalized Ligand

The critical feature that makes flow cytometry a uniquely powerful tool for the study of ligand–receptor interactions is its ability to discriminate cell-associated ligand from free ligand without a wash step. This ability depends on measuring the fluorescence specifically associated with a single cell as it is illuminated by the laser beam. Typically, flow cytometric fluorescence data are acquired only when the detection electronics are triggered by a light scatter signal indicating the presence of a cell. If the sample stream and the laser beam diameters are sufficiently small, the fluorescence from free ligand in the sample volume surrounding the cell is negligible compared to the signal from bound ligand concentrated on the cell surface (Fig. 4).

Murphy (1988) has described the quantitative aspects of these ideas in terms of the ratio of signal from bound ligand to the noise from free ligand. He showed that the theoretical sensitivity depends on the number of receptors per cell, the receptor K_d, and the probe volume defined by the intersection of the laser beam and the sample stream. The sample stream diameter is sensitive to the ratio of the sample flow rate to the sheath flow rate. As sample flow decreases (or as sheath flow increases) the diameter of the sample stream decreases, resulting in a smaller probe volume that contains less free ligand. Cylindrical beam shaping optics can reduce the laser beam height, which also reduces the probe volume. Newer cytometers are likely to have these elements already optimized, and adjustments may not be necessary or even possible. One practical approach to evaluating the discrimination between free and bound ligand uses blank microspheres in the presence of increasing concentrations of free fluorophore. As long as the fluorophore does not bind to the beads, one can determine the contribution of free ligand to the brightness of the beads. Another approach is to define with the cells of interest the relative contribution of nonspecific ligand binding to the cell fluorescence. Our experience with cells with low autofluorescence (leukocytes freshly isolated from blood), and a receptor K_d of 1 n*M*, is that 1000 receptors per cell can be detected with FITC-labeled ligands. There is no real increase in sensitivity as the K_d decreases because the sensitivity is limited by the cellular autofluorescence. However, as the ligand K_d increases, sensitivity is lost because a higher free ligand concentration is required to saturate the receptors. Thus when the K_d is 1 μ*M*, the limit of detection may approach 1,000,000 receptors. In principle, we be-

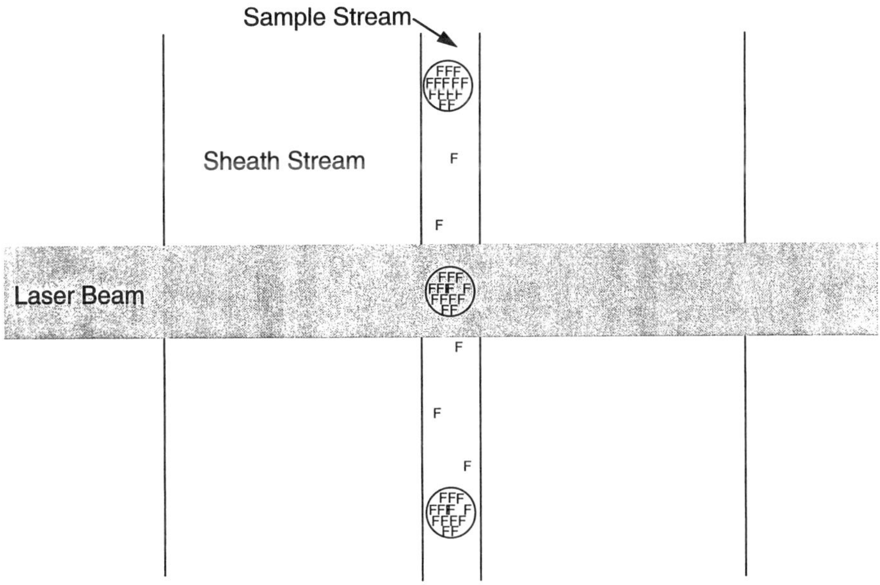

Fig. 4. Flow cytometer point of measurement: the sample stream–laser beam intersection.

lieve that the sensitivity loss at high ligand concentrations can be regained through a rapid dilution of the unbound ligand immediately before measurement. In many other applications, sensitivity is limited by cellular autofluorescence rather than the free ligand. For these situations, Roederer and Murphy (1986) have described a computational approach for autofluorescence correction, which unfortunately has not been widely adopted because of commercial software limitations.

In a living cell, receptor processing may occur rapidly after ligand binding. Receptor internalization results in the removal of ligand–receptor complexes from the cell surface and their delivery to the lysosome. Techniques to follow this processing cytometrically were pioneered by Murphy et al. (1982), who used trypsin to digest FITC–insulin–receptor complexes exposed on the cell surface. Residual ligand fluorescence was presumed to be sequestered inside the cell as the result of endocytosis of the ligand receptor complex. After internalization, the endosomes fuse with lysosomes, where the fluorescence of the FITC ligand is quenched by the lower pH of that organelle. Treating cells with chloroquine increases the lysosomal pH, restoring the fluorescence of lysosomal ligand. Finney and Sklar (1983) took advantage of the pH sensitivity of FITC ligands to follow internalization of FITC–formyl peptides in neutrophils. Rapid addition of dilute acid to the cell suspension results in an extracellular pH decrease that quenches cell surface ligand. Internalized ligand is quenched much more slowly and can therefore be identified. These approaches have been extended to ratiometric determination of the pH of intracellular ligands using combinations of pH-sensitive and -insensitive ligands (Murphy et al., 1984).

Calibration and Quantification

Central to any quantitative analytical method is the use of standards for instrument calibration and the verification of reagent concentrations. For flow cytometry, the standards include solutions containing measured concentrations of fluorophore and solutions of uniform fluorescent particles which have been calibrated against the standard solution (Fay et al., 1991). Commonly, standard solutions are prepared by making a stock solution based on a mass measurement and serial dilution to the desired concentration. However, dyes such as fluorescein are hydrophobic and will stick to tubes and pipette tips at each transfer step. This can result in large errors in the concentration of "known" standards. A preferable method is to measure the absorbance of the solution and calculate the concentration using the extinction coefficient of the dye (Table 2). The absorbance-verified standards can then be used to calibrate the fluorometer for measurements of lower concentrations of reagents, and for calibration of standard particles for instrument calibration. Verification of reagent concentrations is especially critical for flow cytometric ligand binding measurements where a precise knowledge of the concentrations of free and bound ligand is required, just as knowledge of concentration and specific activity of the ligand is required in radioligand binding experiments. This is especially true when the ligand concentration is depleted as it binds to the receptor.

Before stable standard particles were available for instrument calibration, early cytometric studies relied on an isoparametric analysis (Chatelier et al., 1986; Sklar and Finney, 1982). This approach involves measuring binding as a function of *both* [L] and the cell concentration, and gives estimates of receptor number and K_d without calibrating the instrument. This analysis is not readily adapted to kinetic studies involving multiple receptor classes.

Today a number of "standard particles" are available commercially, including some that have been calibrated by the manufacturer. The stability of the standard particles must be verified, and recalibration is an essential requirement for a quantitative interpretation of the flow cytometry measurement. A stable standard bead whose fluorescence in suspension and in the flow cytometer can be compared to the probe of interest in solution will make a valuable addition to the flow cytometry toolbox. When the standardized beads are measured on the cytometer, the fluorescence is expressed as the mean number of fluorophore equivalents per channel (F), and the scale thus

TABLE 2. Calibration for Cytometric Binding Measurements

Solution	Absorbance-verified	Fluorometric	Cytometric
Fluorescein (~100 nM); $\epsilon = 7.25 \times 10^4$ cm^{-1} M^{-1}[a]	X	X	
FITC ligand (~100 nM); $\epsilon = 8.0 \times 10^4$ cm^{-1} M^{-1}[b]	X	X	
FITC ligand (~1 nM)		X	
Standard beads (1×10^6/ml; 1 nM = 6.02×10^5 MESF/bead)		X	X

[a]Fay et al., 1991.

[b]Mercola et al., 1972.

defined is used to calculate the mean equivalent soluble fluorophores (MESF) per cell for an unknown sample (after the appropriate background correction). If the fluorescence properties of the fluorophore used to calibrate the standard particle are identical to those of the receptor-bound labeled ligand, then the mean of number soluble fluorophores per cell is equal to the number of ligand molecules bound. Often this is not the case. For instance, the relative fluorescence yield of FITC conjugated to a small peptide compared to that of free fluorescein (Q_r) is 0.82 (Fay et al., 1991), as calculated from paired absorbance and fluorescence measurements of fluorescein and the FITC conjugate. Additionally, the fluorescence intensity of free ligand (I_f) may be different from the intensity of ligand bound to receptor (I_b). In the case of FITC-labeled formyl peptides, depending on the position of the fluorophore within the ligand, fluorescence has been quenched up to 50% upon binding of labeled ligand to receptor (Fay et al., 1991). Other examples have been observed in which receptor binding enhances NBD-ligand fluorescence (Turcatti et al., 1995). These factors must be included so that the conversion from mean channel number (M) to average number of ligands bound per cell (B) is given by the formula

$$B\left(\frac{\text{molecules}}{\text{cell}}\right) = M\left(\frac{\text{channel number}}{\text{cell}}\right) \times F\left(\frac{\text{fluorescein equivalents}}{\text{channel}}\right) \times \frac{1}{Q_r} \times \frac{I_f}{I_b} \qquad (9)$$

Multiplication of this value by the cell concentration gives the total concentration of bound ligand in the sample tube. Determination of specific binding, the calculation of total receptor number, and estimation of receptor affinity are thereafter analogous to the equilibrium analysis of radioligand–receptor interactions described above.

Kinetic Flow Cytometry

Kinetic measurements by flow cytometry can be made by acquiring discrete histograms at defined time intervals or by continuous data acquisition using time as a parameter. In the first case, the mean fluorescence can be calculated for each histogram and plotted versus the time of histogram acquisition. Typically, histograms can be collected at a rate of up to about six per minute. This is adequate time resolution for measurement of processes having half-times on the order of several minutes.

For continuous data acquisition, time is treated as a parameter and plots of fluorescence versus time can be visualized by a two-parameter histogram. The time parameter can be provided by an internal clock or a voltage ramp of known duration (Martin and Schwartzenruber, 1980). The resolution of the time base depends on the duration of the time ramp and the number of channels assigned to the time parameter. Some commercial cytometer systems have the ability to set the time resolution in the data acquisition software. By contrast, commercial software designers have not facilitated the analysis of cytometric kinetic data. For most cytometric analysis programs, the two-parameter histogram of fluorescence versus time represents the final data output. In practice, the information often desired from such a kinetic experiment is a plot of the mean cell fluorescence versus time. Extracting these data from a two-parameter histogram in FCS data format can be complicated and time consuming. One software package notable for its ability quickly to calculate mean cell fluorescence versus time and to output these data to an ASCII file is IDYLK, a comprehensive cytometry data analysis program developed at the National Flow Cytometry Resource, Los Alamos National Laboratory.

Continuous kinetic measurements are useful for measuring events that occur over the time scale from several minutes to tens of seconds. At shorter time scales, kinetic resolution is limited by the time taken to mix the sample and acquire the first time point. On typical commercial cytometers, this takes 5–30 s. A number of attempts have been made to reduce this dead time by adapting sample tubes for on-line reagent addition, shortening sample lines, and using pressure boosts to speed sample to the laser beam (Kachel et al., 1982; Omann et al., 1985; Pennings et al., 1987; Kelley, 1989; Lindberg et al., 1994). These instrument modifications have pushed the dead time down to several seconds, but in general complete mixing has not been validated at these short times. Indeed, if magnetic stirring is used the mixing time is several seconds and represents a lower limit for instrument dead time. Recently, a coaxial flow cell design was employed to mix sample immediately upstream of the laser beam (Scampavia et al., 1995). Mixing times as short as 55 ms were reported, but this design does not allow one to watch the sample age in a continuous kinetic assay. Rather, a series of independent measurements at different time points would be required.

Fast mixing and measurement are routinely achieved by stopped-flow mixing devices used to measure enzyme and chemical kinetics. These instruments use stepper motor or ram-driven syringes to push sample through a mixing tee to a detection device, usually a fluorometer or spectrophotometer, and have dead times of milliseconds. We have adapted a stopped-flow mixing device to a flow cytometer, resulting in an instrument that provides subsecond kinetic resolution (Nolan et al., 1995). In principle, there is no reason why a flow cytometer could not replace a fluorometer or spectrophotometer as the detection device on a stopped-flow mixer. In practice, the 1000-fold discrepancy between sample flow rates provided by the mixer and the acceptable sample input rate for the cytometer represents a major obstacle to interfacing these technologies. Although the rapid kinetic mixer can mix and deliver sample at a rate in the range of microliters per millisecond, accurate cytometric measurements require sample flow rates of 2 μl/s or less.

This flow rate problem has been circumvented by incorporating a system of fast valves between the mixing device and the cytometer (Fig. 5), and modifying the mixer to provide slow sample flow for measurement, in addition to fast sample flow for mixing. The resulting instrument has a dead time of 300 ms between sample mixing and measurement, and sample is completely mixed within this time. Because mixing occurs almost instantaneously in the mixing tee, the cell population response to stimulant is much more synchronous than when mixing takes place over a period of several seconds as with magnetic stirring. Further developments have resulted in the ability to execute more complex, two-step mixing experiments (Nolan and Sklar, unpublished). This is important for applications such as determining receptor con-

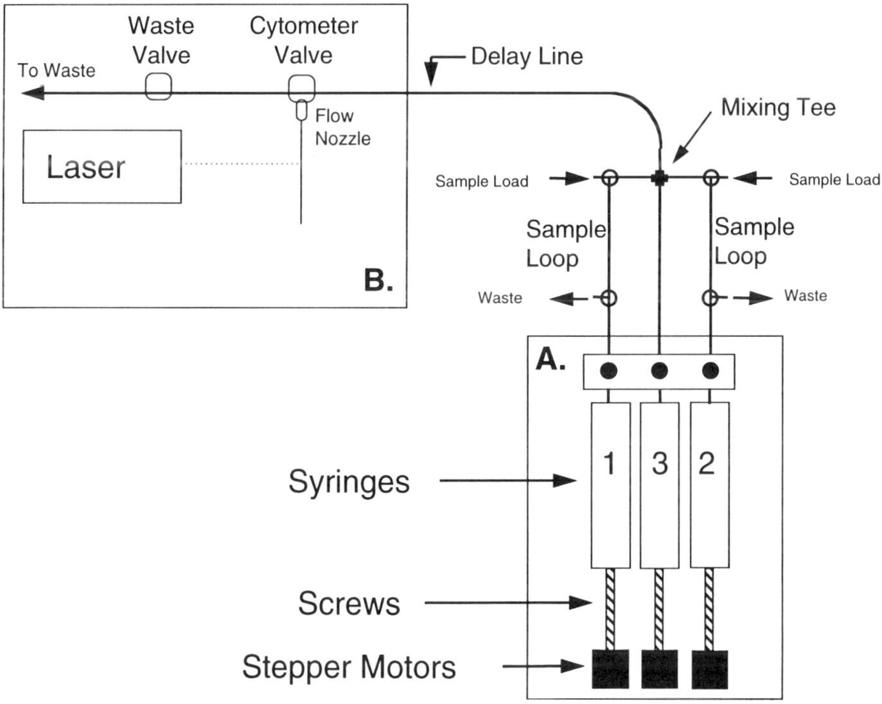

Fig. 5. Schematic diagram of the Rapid Kinetic Flow Cytometer. Flow cytometry with subsecond kinetic resolution is accomplished by connecting a stopped-flow mixing device (A) to a flow cytometer (B) via a series of sample loops, mixing tees, and delay lines. Computer-controlled valves, syringes, and time ramps allow fluorescence to be measured within 300 ms of sample mixing. Reprinted from Nolan et al. (1995).

formational changes as a function of time after ligand addition. In such an experiment, cells are first mixed with labeled ligand and allowed to bind. After some amount of time (<1 s to several minutes) the sample is mixed with excess unlabeled ligand to initiate dissociation of labeled ligand. This type of experiment is a key to understanding the rapid receptor dynamics that can occur in the intact cell. Another type of experiment that may be facilitated by this approach is the analysis of low-affinity ligands that must be used at high concentration. Normally, these ligands might wash away in a heterogeneous assay where the free and bound ligands are physically separated. In the rapid-mix cytometer, it may be possible to dilute free ligand so rapidly that bound ligand will not wash away within the observation time, thus improving the sensitivity of the detection.

Cuvette Methods for Measuring Ligand–Receptor Interactions

Spectrofluorometric analysis complements flow cytometric measurement of ligand–receptor interactions in cell suspensions. With optimal excitation and emission conditions, the fluorescence from 1 n*M* FITC ligand in a cuvette is readily detected

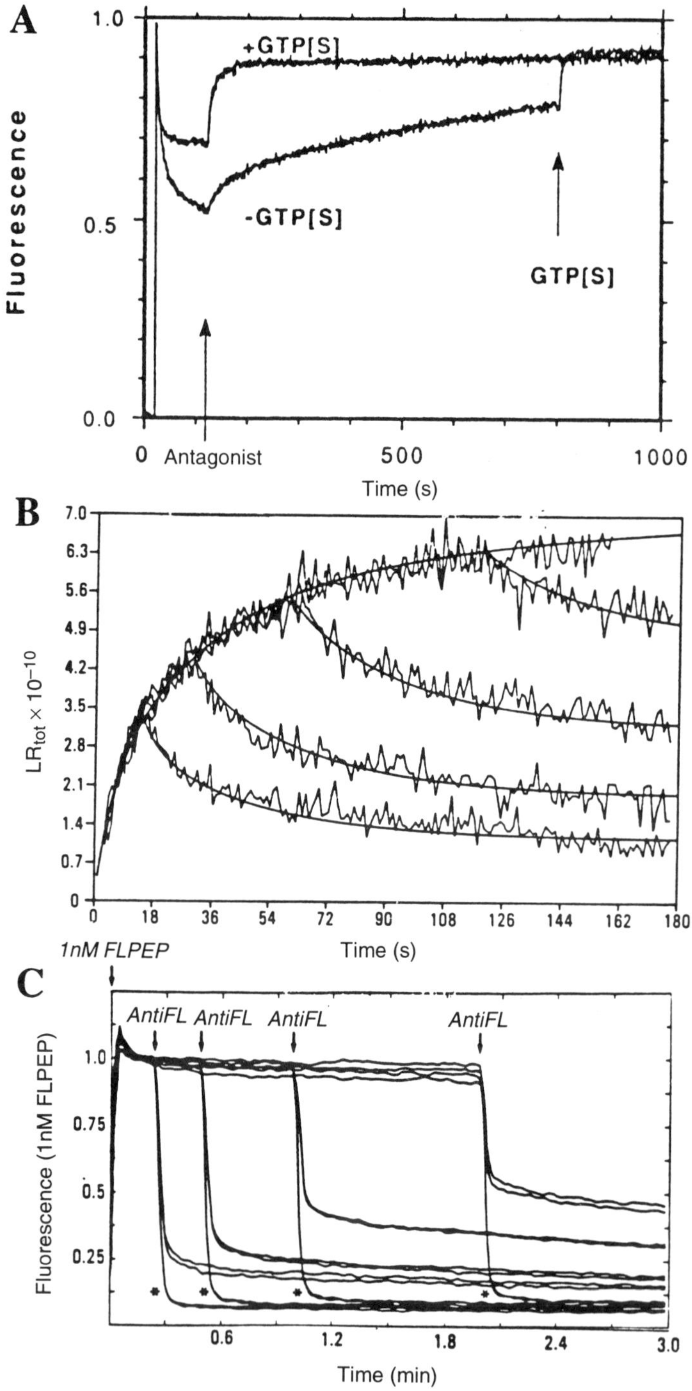
A
1.0
0.5
0.0
Fluorescence
+GTP[S]
-GTP[S]
GTP[S]
0
Antagonist
500
1000
Time (s)
B
7.0
6.3
5.6
4.9
4.2
3.5
2.8
2.1
1.4
0.7
0
$LR_{tot} \times 10^{-10}$
0
18
36
54
72
90
108
126
144
162
180
Time (s)
C
1nM FLPEP
AntiFL
AntiFL
AntiFL
AntiFL
1.0
0.75
0.5
0.25
Fluorescence (1nM FLPEP)
0.6
1.2
1.8
2.4
3.0
Time (min)

against the background scatter and autofluorescence from cells suspended at a concentration of 1×10^7 cells/ml (Sklar et al., 1984). For cells with 60,000 receptors per cell this represents a receptor concentration of approximately 1 n*M*. In general, in a measurement in suspension, the receptor concentration must be similar in magnitude to or greater than the K_d for ligand binding. Under these conditions, ligand binds to the receptor in quantities similar to the receptor concentration. Whereas flow cytometry intrinsically discriminates free and bound ligand, in suspension the discrimination depends upon the ligand's exhibiting distinct spectral features when free or bound (Sklar, 1987). For FITC ligands, whose fluorescence is often quenched upon receptor binding, the fluorescence decrease provides continuous monitoring of ligand–receptor binding (Fig. 6A). Another means of continuous discrimination of free from bound ligand is fluorescence polarization (Fig. 6B). Small ligands that tumble rapidly in solution have restricted mobility when bound to a receptor. The discrimination depends on the relative lengths of fluorescence lifetime and the tumbling time. The longer the lifetime, the more massive the free ligand whose polarization can be discriminated from bound ligand. For fluorescein, whose fluorescence lifetime is ~4 ns, the polarization method is practical only for ligands up to about 2 kDa. A third method involves the use of a high-affinity antibody to fluorescein. The antibody rapidly binds free ligand and quenches its fluorescence whereas the receptor-bound ligand is recognized more slowly, if at all, owing to its decreased diffusion and accessibility compared to free ligand. The antibody method does not allow continuous measurement of ligand binding, but does enable continuous measurement of ligand dissociation (Fig. 6C). None of these methods has yet been applied to a direct analysis of antibody–cell binding interactions because neither the change in intensity upon binding nor the change in polarization (free antibody is too large to depolarize fluorescein fluorescence) is adequate for detection.

The methodologies cited are all limited to receptor and ligand concentrations comparable to the K_d, which results in an appreciable fraction of the ligand being bound. Because receptor binding depletes free ligand, uncertainty in the free ligand concentration complicates the ligand–receptor analysis. However, as the ligand concentration increases above the available receptor, the fraction of bound ligand eventually becomes small and the signal-to-noise ratio becomes poor. As the K_d for a particular ligand–receptor interaction decreases, higher receptor concentrations are needed, which can increase the cost or setup time of the assay. On the positive side, however, the kinetic resolution of cuvette methods is superior to conventional flow cytom-

---

Fig. 6. Strategies for measuring receptor–ligand interactions in cell suspensions. (A) When ligand fluo rescence is quenched upon receptor binding, ligand association (20–110 s) and dissociation (110+ s) can be continuously monitored. Data on FITC–formyl peptide binding to digitonin-permeabilized neutrophils reprinted from Posner et al. (1994). (B) When ligand binding to receptor results in a significant decrease in rotational mobility (i.e., for small ligands), fluorescence anisotropy can be used to measure ligand association and dissociation in response to agonist (added at 15, 30, 60, and 120 s). Data on intact neutrophils reprinted from Sklar (1987). (C) Anti-fluorescein antibody can be used to measure the amount of ligand bound at discrete time points, as well as the ligand dissociation kinetics. Samples "blocked" with excess unlabeled peptide confirm specific binding. Data on intact neutrophils reprinted from Sklar et al. (1985).

etry, with time resolution of seconds by magnetic-stirrer mixing and millisecond resolution by stopped-flow mixing. For this reason, much of our early work on real-time analysis of receptor–ligand interactions was performed on suspended cells and membranes (Sklar and Finney, 1982; Sklar et al., 1985). The improved kinetic capabilities offered by stopped-flow flow cytometry will provide a new realm of analysis. For example, it will be possible to perform kinetic analysis at saturating concentrations of ligand, which is not possible in the cuvette. The cuvette methods will continue to provide independent measurements of the dynamics of ligand–receptor interactions, particularly at early times, for use in validating the performance of new cytometric technologies.

Summary

Measurement of the binding of fluorescent ligands to cell surface receptors by flow cytometry is an extremely useful alternative to radioligand binding measurements. Flow cytometry provides continuous discrimination between free and bound ligand, allowing for homogeneous assays of ligand binding. Recent developments in cytometric kinetic capabilities can take advantage of continuous kinetic assays to measure binding with subsecond resolution. Multiparameter measurements of individual cells allow simultaneous measurements of ligand binding and cell response. These advantages are currently being exploited to study the dynamics of receptor-mediated signal transduction in the human neutrophil.

ANTIBODY BINDING

Although fluorescent antibodies are commonly used to detect cell surface receptors, there remain a number of important issues about their quantitative use. Several recent reviews cover the practical aspects of using flow cytometry to analyze antibody binding to cell surface markers (Kantor and Roederer, 1995; Lanier and Recktenwald, 1993), as has a comparison between equilibrium binding measurements by radioligands and flow cytometry (Brotherick et al., 1994). In this section, we briefly highlight antibody binding in the context of the analytical approaches discussed above. Two questions that come up repeatedly are the following: (1) is divalent IgG bound in a stoichiometry of one IgG per two surface epitopes?, and (2) is the fluorescence intensity in flow cytometry linear with binding?

We have examined antibody binding to neutrophil adhesion molecules and characterized the equilibrium and kinetic features of IgG and Fab binding to a cell (Figs. 7 and 8). We used an approach in which the flow cytometer was calibrated in fluorescein equivalents per channel. The antibodies were calibrated in terms of fluorescing equivalents per antibody using "Simply Cellular" and FITC bead standards in tandem. Even though the f/p labeling ratio was ~10:1, we estimated an effective f/p of only 1.65 for the antibody. We performed the same tests for the Fab, and assuming two Fab sites for each IgG on beads, estimated an effective f/p ratio of 0.65.

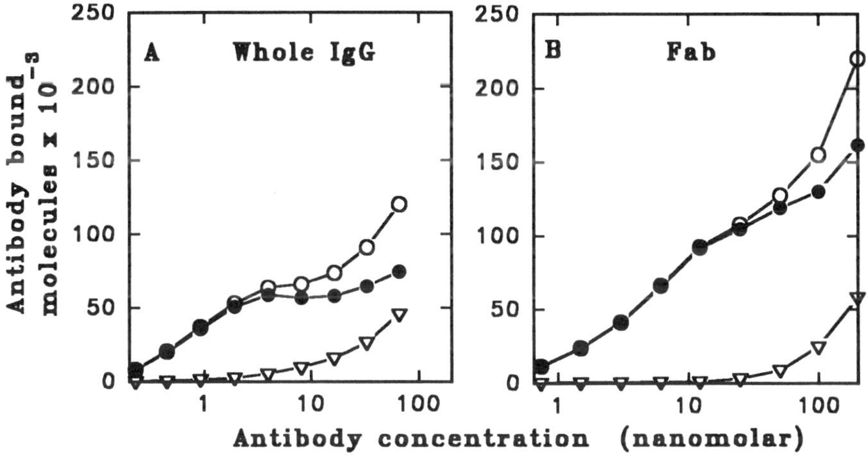

Fig. 7. Equilibrium binding of IgG and Fab fragments of anti-CD18 mAb IB4 to human neutrophils. Cells were incubated on ice with various amounts of FITC–IgG (A) or FITC–Fab (B), then analyzed by flow cytometry. Nonspecific binding (∇) measured in the presence of a 200-fold excess of unlabeled antibody. Specific binding (●) calculated by subtracting nonspecific binding from total binding (○).

Our experience with antibodies suggests that antibody binding, as with receptor–ligand binding, can be measured directly without washing excess antibody, at least over the range of concentrations associated with high-affinity binding. Based on the calibration assumption above, it appeared that there were approximately two high-affinity Fabs bound per each IgG. The affinity of binding is qualitatively reflected in the comparison of the dissociation kinetics of IgG as compared to Fab. However, in our studies, there were also a number of issues that were not resolved. There was low-affinity binding observed with both IgG and Fab that could not be attributed to monovalent association of IgG. Moreover, the association and dissociation rates were not precisely consistent with the K_d as shown in Equation 5. Thus we measured a K_d for the binding of Fab of ~6 n*M*, $k_{on} \sim 3.6 \times 10^5\ M^{-1}\ s^{-1}$ and $k_{off} \sim 2.9 \times 10^{-4}\ s^{-1}$.

The reasons for these discrepancies have not been investigated. However, as we have indicated above, the evaluation of the rate parameters is sensitive to the uncertainties of the concentrations of the antibody and the epitope, which depend directly on the calibration assumptions. One of the lingering concerns for antibody studies is that self-quenching of antibodies at high antibody surface density or with clustered antigens leads to a nonlinear relationship between antibody binding and fluorescence intensity that obscures the quantitative analysis of antibody binding. There is now hope that this issue may be resolved by phase-sensitive flow cytometry. In this technique, the fluorescence lifetime of the bound antibodies is measured as a function of the antibody surface density (Deka et al., 1996). Because fluorescence quenching results in reduced fluorescence lifetime, it may soon be possible to use the lifetime analysis to detect and correct for the quenching.

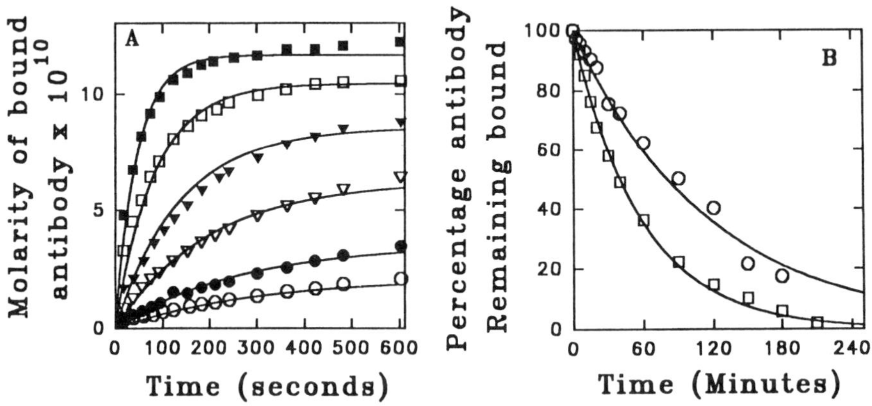

Fig. 8. Kinetics of antibody interaction with neutrophils. (A) IB4 Fab fragment binding. Fluorescent IB4 Fab fragments were added to cells at 4°C at various antibody concentrations (1.5 n*M*, ○; 3.1 n*M*, ●; 6.25 n*M*, △; 12.5 n*M*, ▼; 25 n*M*, □; and 50 n*M*, ■). Solid lines represent the best computer fit to a compartmental model of antibody–antigen interaction. (B) Cells were labeled for 30 min on ice with saturating concentrations of antibody. Dissociation was initiated by the addition of excess unlabeled antibody. Solid lines are exponential fits to the data from IgG (○) and Fab fragments (□).

LIGAND–RECEPTOR DYNAMICS IN THE NEUTROPHIL

Biological Responses and Receptor Occupation

In the neutrophil, binding of formyl peptide chemoattractant to its receptor triggers several rapid responses, including calcium elevation, degranulation, superoxide production, F-actin polymerization, pseudopod formation, and chemotaxis. These responses are initiated within seconds of ligand addition and can be halted within seconds by the addition of receptor antagonists. The kinetics of these phenomena reflect the dynamics of the assembly and disassembly of macromolecular complexes of ligand, receptor, G protein, and other components of intracellular signaling pathways. It is the sequence and dynamics of these macromolecular assembles that govern the magnitude and lifetime of cell response. Because the cell responses are regulated on a time scale of seconds, these molecular interactions must therefore occur on a subsecond time scale. The purpose of this section is to provide a specific example of how flow cytometric and suspension measurements of formyl peptide binding have been used to gain insight into cell and receptor biology. These data have been used to generate a model of the receptor-related events that regulate cell activation.

In early cytometric studies, we took advantage of the long known but poorly appreciated ability of flow cytometry to distinguish bound ligand to study fluoresceinated formyl peptide binding to intact cells (Sklar and Finney, 1982). We detected ~50,000 formyl peptide receptors per cell with a K_d of ~0.4 n*M* when analyzed by a single site model at 4°C. At 37°C, the analysis was complicated by internalization of the receptor–ligand complex, a process that is significant at 30 s after ligand addition (Finney and Sklar, 1983). Kinetic analysis by cytometry and radioligand dissociation revealed complex behavior: at early times after binding a fraction of surface bound ligand dissociated rapidly, whereas at later times the fraction of slow ligand dissociation increased (Zigmond and Sullivan, 1979; Vitkauskas et al., 1980; Sklar et al., 1984). These observations suggested that the formyl peptide receptor exists in at least two states on the surface of the cell, one low-affinity (fast dissociating) and one high-affinity (slow dissociating), and that these states interconverted in a time- and temperature-dependent manner (Sklar et al., 1984).

Subsequent studies of cell responses (Sklar et al., 1984, 1985; Sklar and Oades, 1985) noted that cell activation was triggered within seconds by occupancy of only a fraction of the total receptors present. Inhibition of new receptor occupation by addition of a receptor antagonist resulted in the rapid ($t_{1/2}$ ~30 s) decay of the cell response. It was also apparent that both the interconversion between the low- and high-affinity states and the ligand–receptor internalization were too slow to be involved in cell activation. Moreover, the slowly dissociating receptors appeared to be desensitized because they could be found on the cell surface at times when the cells were no longer responding to the stimulus. We therefore concluded that the receptors were only transiently capable of generating signals and that new receptor occupation, rather than ongoing occupation of old receptors, is necessary to maintain a response. To explain the rapid kinetics of cell response, the actual signal transduction must occur very rapidly and precede later receptor processing events.

Ternary Complex Dynamics in Permeabilized Cells

Further mechanistic insight has come from experiments using permeabilized cells. Treatment of neutrophils with digitonin results in a preparation of cells that retain ligand-binding activity and whose cytoplasmic compartment is contiguous with the extracellular buffer. Thus membrane-bound components remain concentrated on the bilayer surface and are readily accessible to exogenously added agents, whereas cytoplasmic components are greatly diluted, halting further receptor processing events such as desensitization and internalization.

Equilibrium cytometric binding studies on permeabilized neutrophils (Fig. 9A) demonstrated high-affinity binding, which could be converted to a low-affinity binding by the addition of excess nonhydrolyzable guanine nucleotide, GTPγS (Sklar, 1987; Fay et al., 1991). A family of guanine nucleotide-binding proteins, or G proteins, has been identified; they are involved in signal transduction by hormone receptors (Dohlman et al., 1991) and photoreceptors (Stryer, 1986), as well as the formyl peptide receptor (Koo et al., 1982). The kinetics of ligand association and dissociation in the presence or absence of GTPγS were measured by flow cytometry (Fig. 9B–D) and spectrofluorometry (Fig. 6A). The data were analyzed in terms of the ternary complex model (Fig. 10). Computer fitting indicates that the association rate constants are indistinguishable (~3×10^7 $M^{-1}s^{-1}$; Fay et al., 1991), whereas the dissociation rate constants at 37°C for LR (Step 1) are 0.1 s^{-1} as compared to 0.001 s^{-1} for LRG (Step 3). Thus in the presence of GTPγS, ligand dissociation occurs approximately 100 times faster than in its absence (Fig. 6A, 9D). Thus the change in ligand–receptor affinity by GTPγS in permeabilized cells is due to a change in k_{off}.

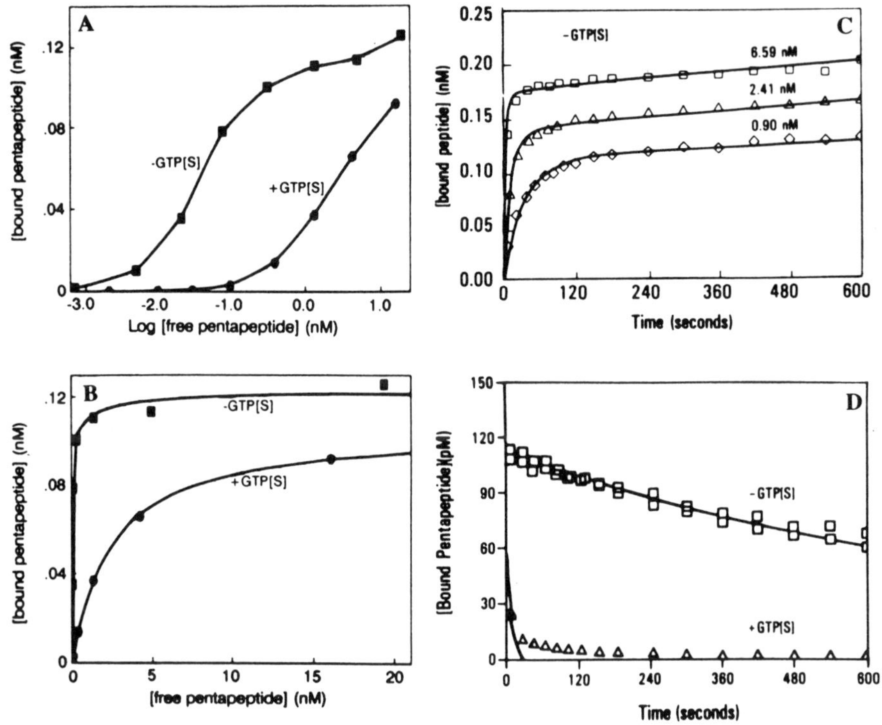

Fig. 9. (A) Equilibrium binding of FITC–formyl peptide to permeabilized neutrophils in the absence and presence of GTPγS. (B) Ligand binding kinetics in the presence and absence of GTPγS. Reprinted from Fay et al. (1991).

The interconversion between the low-affinity LR and the high-affinity LRG receptor states (step 2, forward), is predicted by the computer model to be slow. A fraction of the receptors couples to G protein very rapidly after ligand binding, so rapidly as to appear precoupled by these methods, while the remainder couples to G protein slowly ($k_{+2} = 0.01\ s^{-1}$, $t_{1/2} = 100$ s). We envision as a mechanistic consequence of this observation that there are separate pools of receptor in the cells. Those receptors that are not precoupled to the G proteins can not participate in rapid activation of the cells because they find G protein slowly. The other class of receptors rapidly form LRG ternary complex, which activates the cell more or less instantaneously in the presence of saturating GTP or GTPγS, which induce rapid dissociation into LR + activated G (Neubig and Sklar, 1993; Posner et al., 1994).

A Model for Intact Cell Dynamics and Cell Response

Ternary complex dynamics, as studied in the permeabilized cell, reflect the dynamics of the first two steps in G protein-mediated signal transduction: ligand–receptor binding and G protein activation in the presence of guanine nucleotide. According to

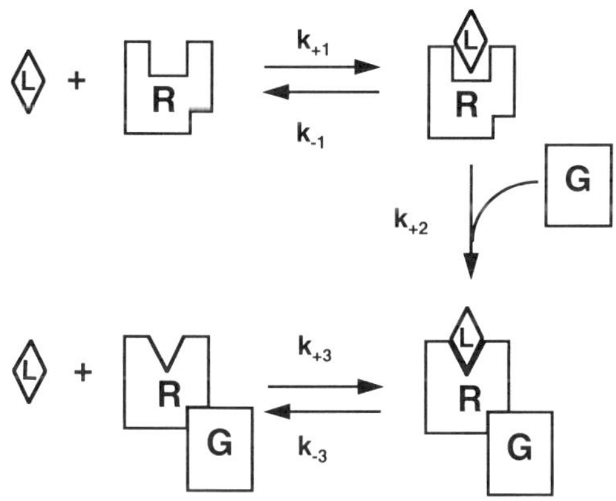

Fig. 10. Ternary complex model of ligand–receptor–G protein interactions.

the results outlined above, these steps occur very rapidly. In the intact cell, the ternary complex dynamics are connected with the rest of the cellular metabolism. This has several important effects. First, guanine nucleotide levels in the intact cell are high enough (Sklar, 1987) that ternary complex disassembly is expected to be very rapid. Thus any LRG formed upon addition of ligand should be instantly (<100 ms) converted to LR, the low-affinity uncoupled form of the ligand–receptor complex. Second, receptors can be sequestered and desensitized, presumably by mechanisms involving receptor phosphorylation (Tardif et al., 1993; Ali et al., 1993), cytoskeletal association (Klotz and Jesaitis, 1994), or association with a member of the arrestin family of receptor-sequestering proteins (Palczewski and Benovic, 1991). Desensitized receptors appear to have high affinity for ligand, but are sequestered from the signaling pathway. Third, ligand–receptor complexes can be internalized by endocytosis. Fourth, neutrophils can upregulate the number of surface receptors (Norgauer et al., 1991) or recycle internalized ligand in response to ligand binding. Finally, the effector systems that link the ternary complex to cell responses may feed back upon the ligand–receptor events. These factors make for a much more complex system than the permeabilized cell, but also provide the opportunity to investigate the biochemical events that follow ternary complex activation in the signal transduction pathway and relate them to the cell physiology. The general features of this scheme have recently been validated by continuous flow cytometric measurements (Hoffman et al., 1996).

Neutrophil responses to formyl peptide are initiated within seconds of ligand addition, reach maximal rates within tens of seconds, and decay slowly over a period of minutes. When binding is interrupted by the addition of a receptor antagonist, the cell responses decay within seconds. As described above, flow cytometry and cuvette studies of the kinetics of ligand binding to intact cells are consistent with an evolving distribution among the receptor forms. At early times after ligand addition, a rapidly dissociating fraction of receptors is detected together with a slowly dissoci-

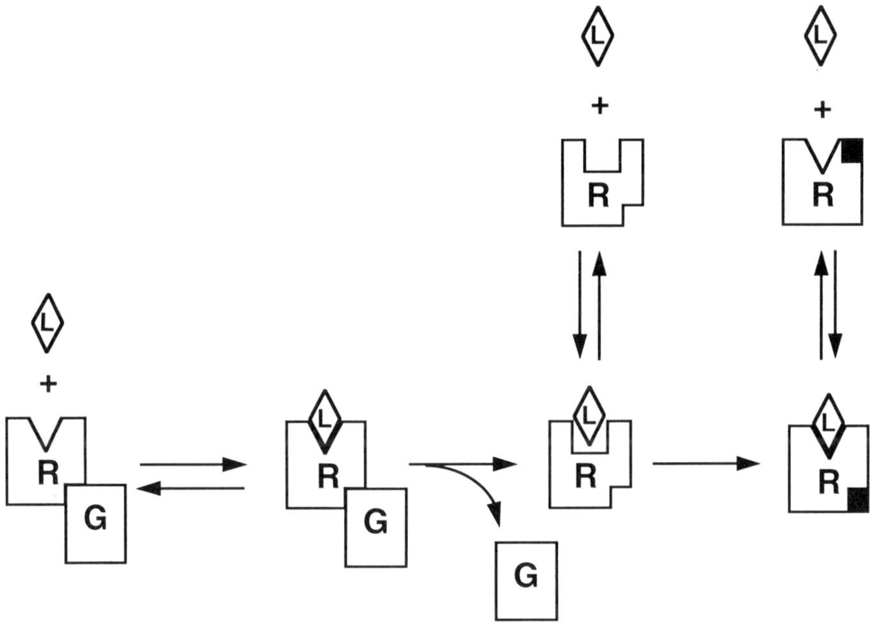

Fig. 11. Model of receptor dynamics in the intact neutrophil.

ating fraction (Fig. 6B). At later times (>1 min), the slowly dissociating form that predominates is not associated with G protein (Sklar et al., 1989). Taken together with the observed ternary complex dynamics in permeabilized cells, these data support the following model for signal transduction in the neutrophil (Fig. 11). When ligand binds, a pool of receptors precoupled to G proteins is rapidly activated (<100 ms) in the presence of GTP and immediately disassembles into LR. The total cellular LR reflects both complexes that have already activated G protein and complexes that have yet to find G protein. We hypothesize that the slowly dissociating receptor in the intact cell represents the sequestered receptor ("LRX"), a process that occurs within seconds. Because finding G appears to be slow compared to sequestration, the receptor may activate only a single G protein. As long as new ligand binding continues and precoupled RG complexes are available, LRG will generate new signal. When new binding is halted by addition of a receptor antagonist, LRG is rapidly replaced, first by LR and then by LRX.

Refinement of this model requires a detailed comparison between ligand binding kinetics, receptor biochemistry, and cell response. One new tool is rapid kinetic flow cytometry, which enables the receptors to be rapidly saturated with ligand, with receptors in both the LR and LRG forms. We expect that when LRG is formed in the presence of GTP, it is replaced by LR, and if there were no receptor processing, we would detect only LR. However, the rate of appearance of a high-affinity receptor should measure the newly desensitized receptor form. It is our hope to use receptor phosphorylation mutants to characterize the role of phosphorylation in the ligand binding kinetics and cell response and provide a rationale for the eventual biochemical isolation and characterization of the macromolecular complexes involved in signal transduction.

PROSPECTS

Flow cytometry represents a powerful and versatile tool for examining ligand–receptor interactions because of its intrinsic ability to discriminate fluorescent molecules on the cell surface from free fluorescent molecules in the volume surrounding the cells. This technique is likely to find application in ligand binding, ligand–receptor dynamics, and receptor processing. Moreover, it appears that the number of fluorescent ligands available from commercial sources will increase and that these ligands will in turn be used in the sorting of transfected cells expressing receptors, the characterization of the physiology of receptor pathways, and the development of antagonists for receptors in the pharmaceutical industry.

Kinetic analysis lends itself to the study of molecular mechanisms. Although we have focused on the ability of flow cytometry to study ligand–receptor complexes on cells, the example cited, the binding of fluorescent antibodies to cell surfaces, should make it obvious that binding or assembly interactions of other kinds are also readily accessible. Furthermore, we are not limited to applications involving living cells. Although the permeabilized cell has been cited as one example, fixed cells are often used with antibodies. More important is the possibility of assembling molecular complexes on surfaces. Cell-sized latex beads displaying antibodies are already used for calibration standards. Beads could similarly display purified receptors, other proteins or peptides, DNA, RNA, or oligonucleotides (Nolan et al., 1996). These tools will therefore provide a basis for molecular cytometry and provide tools for the analysis of mechanisms of binding and enzymatic activity.

REFERENCES

Ali H, Richardson RM, Tomhave ED, Didsbury JR, Snyderman R (1993): Differences in phosphorylation of formyl peptide and C5a chemoattractant receptors correlate with differences in desensitization. J Biol Chem 268:24247–24254.

Baindur N, Triggle DJ (1994): Concepts and progress in the development and utilization of receptor-specific ligands. Med Res Rev 14:591–664.

Brotherick I, Lennard TWJ, Wilkinson SE, Cook S, Angus B, Shenton BK (1994): Flow cytometric method for the measurement of epidermal growth-factor receptor and comparison with the radio-ligand binding assay. Cytometry 16:262–269.

Chang K-J, Jacobs S, Cuatrecasas P (1975): Quantitative aspects of hormone-receptor interactions of high affinity. Effect of receptor concentration and measurement of dissociation constants of labeled and unlabeled hormone. Biochim Biophys Acta 406:294–303.

Chatelier RC, Ashcroft RG, Lloyd CJ, Nice EC, Whitehead RH, Sawyer WH, Burgess AW (1986): Binding of fluoresceinated epidermal growth factor to A431 cell subpopulations studied using a model independent analysis of flow cytometric fluorescence data. EMBO J 5:1181–1186.

Deka C, Lehnert BE, Lehnert VM, Jones GM, Sklar LA, Steinkamp JA (1996): Analysis of fluorescence lifetime quenching of FITC-conjugated antibodies on cells by phase-sensitive flow cytometry. Cytometry 25:271–279.

Dohlman HG, Thorner J, Caron MG, Lefkowitz RJ (1991): Model systems for the study of seven-transmembrane-segment receptors. Annu Rev Biochem 60:535–588.

Erickson J, Goldstein B, Holowka D, Baird B (1987): The effect of receptor density on the forward rate constant for binding of ligand to cell surface receptors. Biophys J 52:657–662.

Fay SP, Posner RG, Swann WN, Sklar LA (1991): Real-time analysis of the assembly of ligand, receptor, and G protein by quantitative fluorescence flow cytometry. Biochemistry 30:5066–5075.

Finney DA, Sklar LA (1983): Ligand/receptor internalization: a kinetic flow cytometric analysis of the internalization of N-formyl peptides by human neutrophils. Cytometry 4:54–60.

Haugland R (1992): *Handbook of Fluorescent Probes and Research Chemicals.* Eugene, OR: Molecular Probes, Inc.

Hoffman JF, Linderman JJ, Omann GM (1996): Receptor up-regulation, internalization, and interconverting receptor states. Critical components of a quantitative description of *N*-formyl peptide-receptor dynamics in the neutrophil. J Biol Chem 271:18394–18404.

Hoyt DW, Harkins RN, Debanne MT, O'Connor-McCourt M, Sykes BD (1994): Interaction of transforming growth factor alpha with the epidermal growth factor receptor: binding kinetics and differential mobility within the bound TGF-alpha. Biochemistry 33:15283–15292.

Johnson ML (1992): Why, when and how biochemists should use least squares. Anal Biochem 206:215–226.

Kachel V, Glossner E, Schneider H (1982): A new flow cytometric transducer for fast sample throughput and time resolved kinetic studies of biological cells and other particles. Cytometry 3:202–212.

Kantor A, Roederer M (1996): FACS analysis of leukocytes. In *Handbook of Experimental Immunology,* vol. 2, ch. 49.

Kelley KA (1989): Sample station modification providing on-line reagent addition and reduced sample transit time for flow cytometers. Cytometry 10:796–800.

Ketelslegers J-M, Pirens G, Maghuin-Rogister G, Hennen G, Frere J-M (1984): The choice of erroneous models in hormone-receptor interactions: a consequence of illegitimate utilization of Scatchard graphs. Biochem Pharmacol 33:707–710.

Klotz KN, Jesaitis AJ (1994): Neutrophil chemoattractant receptors and the membrane skeleton. Bioessays 16:193–198.

Koo C, Lefkowitz RJ, Snyderman R (1982): Guanine nucleotides modulate the binding affinity of the oligopeptide chemoattractant receptor on human polymorphonuclear leukocytes. J Clin Invest 72:748–753.

Lanier LL, Recktenwald DJ (1993): Multicolor immunofluorescence and flow cytometry. Methods: A Companion to Methods in Enzymology 2:192–199.

Lauffenberger DA, Linderman JJ (1993): *Receptors: Models for Binding, Trafficking, and Signaling.* New York: Oxford University Press.

Lindberg W, Scampavia LD, Ruzicka J, Christian GD (1994): Fast kinetic measurements and on-line dilution using flow cytometry. Cytometry 16:324–330.

MacArthur MW, Driscoll PC, Thornton JM (1994): NMR and crystallography—complementary approaches to structure determination. Trends Biotechnol 12:168–172.

Martin JC, Schwartzenruber DE (1980): Time: a new parameter for flow cytometry. Science 207:199–201.

Mercola DA, Morris JWS, Arquilla ER (1972): Use of resononce interaction in the study of chain folding of insulin in solution. Biochemistry 11:3860–3874.

Model MA, Omann GM (1995): Ligand-receptor interaction rates in the presence of convective mass transport. Biophys J 69:1712–1720.

Motulsky HJ, Ransnas LA (1987): Fitting curves to data using nonlinear regression: a practical and nonmathematical review. FASEB J 1:365–374.

Murphy RF (1988): Ligand binding, endocytosis, and processing. In: Melamed M, Lindmo T, Mendelsohn M (eds). *Flow Cytometry and Sorting*. New York: Wiley-Liss Inc., pp 355–366.

Murphy RF, Powers S, Bisaccia F, Edelson FL, Cantor, CR (1984): Endosome pH measured by dual fluorescence flow cytometry: rapid acidification of insulin to pH 6. J Cell Biol 98:1757–1762.

Murphy RF, Powers S, Verderame M, Cantor CR, Pollack R (1982): Flow cytometric analysis of insulin binding and internalization by Swiss 3T3 cells. Cytometry 2:402–406.

Neubig R, Sklar LA (1993): Subsecond modulation of formyl peptide-linked guanine nucleotide-binding proteins by guanosine 5′-O-(3-thio) triphosphate in permeabilized neutrophils. Mol Pharmacol 45: 65–73.

Nolan JP, Posner RG, Martin JC, Habbersett RC, and Sklar LA (1995): A rapid kinetic flow cytometer with sub-second resolution. Cytometry 21:223–229.

Nolan JP, Shen B, Park MS, Sklar LA (1996): Kinetic analysis of human flap endonuclease-1 by flow cytometry. Biochemistry 35:11668–11676.

Norgauer J, Eberle M, Fay SP, Lemke HD, Sklar LA (1991): Kinetics of *N*-formyl peptide receptor upregulation during stimulation in human neutrophils. J Immunol 146:975–980.

Omann GM, Coppersmith W, Finney DA, Sklar LA (1985): A convenient on-line device for reagent addition, sample mixing, and temperature control of cell suspensions in flow cytometry. Cytometry 6: 69–73.

Palczewski K, Benovic JL (1991): G-protein-coupled receptor kinases. Trends Biochem Sci 16:387–391.

Pennings A, Speth P, Wessels H, Haanen C (1987): Improved flow cytometry of cellular RNA and DNA by on-line reagent addition. Cytometry 8:335–338.

Phizicky EM, Fields S (1995): Protein-protein interactions: methods for detection and analysis. Microbiol Rev 59:94–123.

Posner RG, Fay SP, Domalewski MD, and Sklar LA (1994): Continuous spectrofluorometric analysis of formyl peptide receptor ternary complex interactions. Mol Pharmacol 45:65–73.

Roederer M, Murphy RF (1986): Cell-by-cell autofluorescence correction for low signal to noise systems: application to EGR endocytosis for 3T3 fibroblasts. Cytometry 7:558–565.

Scampavia LD, Blankenstein G, Ruzicka J, Christian GD (1995): A coaxial jet mixer for rapid kinetic analysis in flow injection cytometry. Anal Chem 67:2743–2749.

Sklar LA (1987): Real-time spectroscopic analysis of ligand-receptor dynamics. Annu Rev Biophys Biophys Chem 16:479–506.

Sklar LA, Finney DA (1982): Analysis of ligand-receptor interactions with the fluorescence activated cell sorter. Cytometry 3:161–165.

Sklar LA, Finney DA, Oades ZG, Jesaitis AJ, Painter RG, Cochrane CG (1984): The dynamics of ligand-receptor interactions. Real-time analysis of association, dissociation, and internalization of an *N*-formyl peptide and its receptors on the human neutrophil. J Biol Chem 259:5661–5669.

Sklar LA, Hyslop PA, Oades ZG, Omann GM, Jesaitis AJ, Painter RG, Cochrane CG (1985): Signal transduction and ligand-receptor interactions in the human neutrophil. Transient responses and occupancy-response relationships at the formyl peptide receptor. J Biol Chem 260:11461–11467.

Sklar LA, Mueller H, Omann G, Oades Z (1989): Three states for the formyl peptide receptor on intact cells. J Biol Chem 264:8483–8486.

Sklar LA, Oades ZG (1985): Signal transduction and ligand-receptor interactions in the human neutrophil. Ca^{++} modulation and restoration. J Biol Chem 260:11468–11475.

Soper SA, Nutter HL, Keller RA, Davis LM, Shera EB (1993): The photophysical constants of several fluorescent dyes pertaining to ultrasensitive fluorescence spectroscopy. Photochem Photobiol 57:972–977.

Stryer L (1986): Cyclic GMP cascade of vision. Annu Rev Neurosci 9:87–119.

Tardif M, Mery L, Brouchon L, Boulay F (1993): Agonist-dependent phosphorylation of *N*-formylpeptide and activation peptide from the fifth component of C (C5a) chemoattractant receptors in differentiated HL-60 cells. J Immunol 150:3534.

Taylor SI (1975): Binding of hormones to receptors. An alternative explanation of nonlinear Scatchard plots. Biochemistry 14:2357–2361.

Turcatti G, Vogel H, Chollet A (1995): Probing the binding domain of the NK2 receptor with fluorescent ligands: evidence that hexapeptide agonists and antagonists bind differently. Biochemistry 34: 3972–3980.

Vitkauskas G, Showell HJ, Becker EL (1980): Specific binding of synthetic chemotactic peptides to rabbit peritoneal neutrophils: effect on dissociability of bound peptide, receptor activity and subsequent biological responsiveness (deactivation). Mol Immunol 17:171–180.

Weiland GA, Molinoff PB (1981): Quantitative analysis of drug-receptor interactions: I. Determination of kinetic and equilibrium properties. Life Sci 29:313–330.

Zigmond SH, Sullivan SJ (1979): Sensory adaptation of leukocytes to chemotactic peptides. J Cell Biol 82:517–527.

Zoller MJ (1991): New molecular biology methods for protein engineering. Curr Opin Biotechnol 2:526–531.

3

Pathophysiology of Human Fc-gamma Receptors in Health and Disease

Bruce H. Davis
William Beaumont Hospital, Royal Oak, Michigan

Paul K. Wallace
Dartmouth Medical School, Lebanon, New Hampshire

Paul M. Guyre
Dartmouth Medical School, Lebanon, New Hampshire

Myeloid cells play an important role in the physiological well-being of an individual, whether it be human or animal, being the major cellular component of the acute inflammatory process and playing a significant role in all forms of chronic and granulomatous inflammatory responses. As such, granulocytes and monocytes are involved in nearly every disease process. Among the growing number of myeloid cell surface proteins with functional significance elucidated in the past few decades are the family of proteins acting as receptors for immunoglobulins, which serve as a functional communication or interactive bridge between the humoral and cellular arms of the immune system. The purpose of this chapter is to provide an overview of the myeloid IgG Fc receptors (FcγR) from the perspective of both their role in the cell and molecular biology of myeloid cell responses and their potential diagnostic and therapeutic roles.

Phagocyte Function: A Guide for Research and Clinical Evaluation, Edited by J. Paul Robinson and George F. Babcock. ISBN 0-471-12364-1

Fc RECEPTORS FOR IgG (FcγR)

The IgG Fc receptors (FcγR) are a closely related family of leukocyte receptors with differing affinities for the Fc domain of IgG (Fanger et al., 1989; Morganelli et al., 1992; Ravetch and Kinet, 1991; van de Winkel, 1995; Wallace et al., 1994). Cross-linking FcγR, not simply ligand binding, results in the transduction of signals leading to activation and release of enzymes, cytokines, and inflammatory mediators by macrophages, neutrophils (polymorphonuclear leukocytes, PMNs), and natural killer (NK) cells. FcγR are involved in the clearance of immune complexes, phagocytosis of antibody-coated pathogens, enhancement of antigen presentation, and antibody-dependent cellular cytotoxicity (ADCC), and may play a role in atherogenesis (Graziano and Fanger, 1987a,b; Shen et al., 1989). The prime functions of IgG Fc receptors (FcγR), for which the mechanisms are relatively well understood, involve FcγR potentiation of the effector limb of the immune response as the result of bridging the interaction between specific myeloid cells and antibody-coated pathogens. Because many of the effector actions triggered by FcγR involve the release of toxic or inflammatory substances, it is necessary that the body strictly regulate their actions. This is accomplished by controlling the number of FcγR-bearing cells, their expression of FcγR, and the function of FcγR on these cells. Thus factors regulating the expression or function of FcγR have the potential to affect a wide array of physiological responses, ranging from initiation of the immune response through the elimination of IgG–antigen complexes. Recent evidence has shed new light on other FcγR functions of physiological significance (Chang et al., 1985; Gosselin et al., 1992; Manca et al., 1991). These functions involve the inductive arm of immunity, and include FcγR-directed capture and uptake of antigens by antigen-presenting cells for subsequent processing and presentation to T cells.

Human leukocyte expression of FcγR has been determined using monoclonal antibodies (mAb). Three different classes of human FcγR have been defined, FcγRI (CD64), FcγRII (CD32), and FcγRIII (CD16), encoded by at least eight distinct genes, all mapping to chromosome 1 (Sandor and Lynch, 1993; Takai et al., 1994; Unkeless, 1989; Unkeless et al., 1988; van de Winkel et al., 1995; van de Winkel and Anderson, 1991). The three classes of FcγR are structurally related but functionally distinct members of the Ig supergene family and are expressed primarily, but not exclusively except for CD64, on hematopoietic cells. These receptors have differing affinities for particular IgG subclasses and have particular patterns of expression on different cell types. Their functions vary widely depending on the cell type on which they are expressed, its degree of differentiation or maturation, and its state of activation.

FcγRI (CD64) is a 72-kDa glycoprotein that binds with high affinity ($K_A \sim 10^{-9}$ M) to monomeric human IgG1 and IgG3, mouse IgG2a and IgG3, and IgG immune complexes (Anderson et al., 1986, 1990; Fanger et al., 1989; Guyre et al., 1989; Kurlander and Batker, 1982). Thus in vivo, and in the presence of human serum in vitro, the ligand binding site is normally occupied by IgG. Three highly homologous genes coding for FcγRI have been identified (van de Winkel and Capel, 1993). Two genes, hFcγRIA and hFcγRIB, encode for membrane-anchored forms of FcγRI; the third, hFcγRIC, has a stop codon in an extracellular domain. FcγRI is a type 1 integral membrane protein with three Ig-like domains in its 292-amino acid extracytoplasmic portion, a transmembrane region of 21 amino acids, and a short charged cytoplasmic tail of either 61 (FcγRIA) or 31 (FcγRIB) amino acids. FcγRI is a potent cytotoxic trigger molecule whose expression is exclusively restricted to mononuclear phagocytes and activated PMNs (Fanger et al., 1989; Looney et al., 1986).

FcγRII (CD32) is not a single receptor, but a family of closely related 40-kDa glycoproteins that bind IgG with low affinity. It is the most broadly expressed FcγR, and is found on monocytes, macrophages, PMNs, platelets, LGL/NK cells, and B lymphocytes (Fanger et al., 1989; Looney et al., 1986; Unkeless et al., 1988; Vaughn et al., 1985; Zipf et al., 1983). Six isoforms have been identified (hFcγRIIa1, −a2, hFcγRIIb1, −b2, −b3, and hFcγRIIc), which are encoded by three separate genes (hFcγRIIA, −B, and −C). They all have similar extracellular regions of 180 amino acids containing two Ig-like domains, with the most significant differences located in the transmembrane and intracellular regions (Qiu et al., 1990; van de Winkel and Capel, 1993; van de Winkel and Anderson, 1991). FcγRII has a low affinity for IgG ($K_A < 10^{-7}$ *M*) and does not bind monomeric immunoglobulin, but reacts with IgG immune complexes and opsonized particles, and in vivo is important for immune complex clearance (Looney et al., 1986). It is a trigger molecule on monocytes, macrophages, PMNs, and eosinophils that mediates lysis and/or phagocytosis of tumor cell targets (Anderson et al., 1990; Fanger et al., 1989; Shen et al., 1989). Expression of CD32 varies from 30,000 to 60,000 sites per cell on all nucleated cells and its expression is relatively unaltered by cytokines. In comparative studies of FcγRI and FcγRII, triggering through CD32 was found to require targets to be coated with 10 times more antibody than required for triggering through FcγRI (van de Winkel et al., 1989), suggesting that CD64 may be a more efficient trigger molecule.

FcγRIII (CD16) is a 50 to 80-kDa glycoprotein with low affinity for monomeric IgG. It is present predominantly on PMNs, LGL/NK cells (Fleit et al., 1982), and a small subset (5–15%) of freshly isolated monocytes. As monocytes mature into macrophages, expression of FcγRIII increases, so that by day 3 of culture nearly all monocyte-derived macrophages express CD16 (Klaassen et al., 1990). FcγRIII has a moderate affinity for IgG ($K_A \sim 3 \times 10^{-7}$ *M*) and like CD32, preferentially binds immune complexes and opsonized particles. Two genes coding for CD16 have been identified, hFcγRIIIa and hFcγRIIIb, which encode for proteins consisting of two Ig-like extracellular domains of approximately 190 amino acids (van de Winkel and Capel, 1993). CD16a is expressed on macrophages and LGL/NK cells, has conventional transmembrane and cytoplasmic domains (Huizinga et al., 1988; Selvaraj et al., 1988), and functions as a cytotoxic and phagocytic trigger molecule (Clarkson et al., 1986). CD16b, which is highly expressed on PMNs, is linked to the surface via a phosphatidylinositol glycan (PIG) linkage (Huizinga et al., 1988; Selvaraj et al., 1988). A single amino acid change at amino acid position 203 is responsible for these different membrane anchoring mechanisms (Kurosaki and Ravetch, 1989; Lanier et al., 1989). Cross-linking of PMN CD16b does not initiate either ADCC or phagocytosis of tumor targets, but appears to stimulate a limited respiratory burst and may participate in the capture of immune complexes (Crockett-Torabi and Fantone, 1990; Edberg et al., 1992; Erbe et al., 1990; Shen et al., 1989).

Soluble forms of FcγRIIIb, also called IgG-binding factors, have been demonstrated in culture supernatants of immune cells and in sera from mice and humans (van de Winkel and Capel, 1993). Normal human sera contain large amounts (≅ 5 n*M*) of soluble FcγRIIIb, which is presumably released from PMNs by proteolytic cleavage and is believed to correlate with the activation state and size of the circulating and tissue PMN population (Huizinga et al., 1990; van de Winkel and Capel, 1993). In particular, the amount of soluble CD16b in the sera of neutropenic patients has been inversely correlated with their risk for infection, with higher CD16b associated with a significantly lower risk of infection (Huizinga et al., 1990). FcγRIIIa is also found in the plasma of healthy individuals, but in very low amount compared to FcγRIIIb (de Haas et al., 1994). Activation of LGL/NK cells results in the release of CD16a with concomitant decrease in ADCC activity. Elevated levels of CD16a have been detected in the plasma of some patients with a LGL/NK cell lymphocytosis and with rheumatoid arthritis (de Haas et al., 1994).

FcγR POLYMORPHISMS

Several polymorphisms of FcγR have been identified and provide important insights into the function and molecular nature of FcγR (reviewed by van de Winkel and Capel, 1993; van de Winkel and Anderson, 1991). These polymorphisms dictate differential recognition and phagocytosis of immune complexes or killing of antibody-coated targets. At least two allotypic forms of CD32 are known to exist (Anderson et al., 1987). This polymorphism was originally discovered in T cell mitogenesis studies in which monocytes from approximately 70% of Caucasians supported murine IgG1 anti-CD3-induced proliferation (high responders, HR) and the other 30% did so very poorly (low responders, LR). This functional polymorphism resides in a single Arg-His amino acid change at position 131, conferring on individuals with the H131 (LR) allotype a much higher affinity for murine IgG1 antibodies than those with the R131 (HR) form (van de Winkel and Anderson, 1991). Of particular clinical significance, phagocytes from H131 homozygous individuals also have a higher affinity for human IgG2 and therefore more readily ingest human IgG2-opsonized particles than do phagocytes from people with the R131 phenotype (Salmon et al., 1991, 1992; Sanders et al., 1995). Because FcγRIIa is the sole FcγR with affinity for hIgG2, the principal IgG subclass produced in response to certain antigens [e.g., bacterial polysaccharides (Salmon et al., 1992)], the R131 allotype may be less able to bind to and phagocytose certain immune complexes than the H131 allotype. This reasoning is supported by at least two studies indicating that children homozygous for the LR allotype had an increased susceptibility to infections with encapsulated bacteria (Bredius et al., 1994; Sanders et al., 1994).

CD16b also exists as a bi-allelic polymorphism of five nucleotides in the third exon generating two forms of PMN-specific CD16 antigens, NA1 and NA2, which differ in glycosylation at amino acid positions 65 and 82 (Huizinga et al., 1990; Ory et al., 1989). These polymorphisms can be identified using either serologic or molecular methods (Bux et al., 1993; Stein et al., 1994). The NA antigens constitute a clinically important neutrophil alloantigen system because antibodies to these antigens are associated with alloimmune neonatal neutropenia and transfusion-related lung injury. Neutrophils from NA1 homozygous individuals exhibit more efficient phagocytosis and CD16-mediated adherence than cells from NA2 homozygous donors, particularly when IgG opsonization is suboptimal or complement is absent (Bredius et al., 1994; Salmon et al., 1990). A third polymorphism involving CD16a expression on monocytes and LGL/NK cells has been identified in which approximately 50% of individuals express twice as many FcγRIIIA on their LGL/NK cells as do the remaining individuals (Vance et al., 1993). Cells with the higher FcγRIII expression performed ADCC of erythroid cell targets more efficiently, which may reflect an inherent difference in the ability of individuals to mediate cellular effector functions in vivo via CD16.

Individuals whose PMNs do not express CD16b have been described in the Caucasian population with a frequency of approximately 1 in 1000 (de Haas et al., 1995), although the frequency may vary in other ethnic groups (Muniz-Diaz, 1995). The FcγRIIIb deficiency is due to the complete absence of both FcγRIIIb alleles. In all 21 individuals examined, the FcγRIIIb deletion is accompanied by deletion of the two FcγRIIC genes, suggesting the defect may arise from an unequal crossover event during meiosis (de Haas et al., 1995). Loss of CD16b is associated with no obvious clinical ramifications, other than the predisposition for isoimmune neutropenia in the offspring of affected females. A family in Belgium has also been described containing four individuals with no detectable expression of CD64 on their monocytes even after culture with IFN-γ (van de Winkel et al., 1995). Again, all individuals are healthy and show no signs of illness, presumably due to redundancy within the FcγR.

Thus stable genetic polymorphisms exist in FcγR expression or function that may be important in the handling of immune complexes and in phagocytosis. Yet owing to the redundancy within the FcγR system, and the co-expression of different FcγR classes on myeloid cells, most combinations of polymorphisms remain silently expressed and lack any clinical manifestations. However, under certain circumstances, specific combinations of polymorphisms could significantly influence the immune responses, particularly to bacterial infections.

FcγR SUBUNITS AND SIGNALING

In addition to the ligand-binding α chain, CD64 and CD16a exist as hetero-oligomeric complexes with various disulfide-linked subunits associated with them (Ravetch, 1994; Ravetch and Kinet, 1991; van de Winkel and Capel, 1993). Both α chains associate with two 7.8-kDa γ chains to form a γ-γ dimer that is essential both for proper cell surface assembly and for mediating signaling within the cell. On NK cells the γ chain and its homologous family member, the ζ chain, originally defined as part of the TCR-CD3 complex, associate with FcγRIIIa as either homodimeric (γ-γ, ζ-ζ) or heterodimeric (γ-ζ) complexes. The FcγRIIIa complex also contains a β-chain subunit originally described in association with FcεR (CD23), which based on hydropathicity data, crosses the plasma membrane four times and has both amino and carboxy termini protruding into the cytoplasm (Metzger et al., 1989).

Cross-linking any of the FcγR results in a transduction of signals leading to activation of tyrosine kinases; elevation of intracellular calcium; release of inflammatory mediators such as leukotrienes, superoxide oxygen and nitrogen radicals, prostaglandins, and hydrolases; and the transcription of cytokine genes (Ravetch, 1994; Ravetch and Kinet, 1991; van de Winkel and Capel, 1993). The tyrosine-signaling motifs (TAM) are contained within the cytoplasmic tail of CD32 and within the γ/ζ chain subunits of CD64 and CD16a (Duchemin et al., 1994; Greenberg et al., 1994). These TAM broadly follow the pattern of a tandem YXXL amino acid sequence that was first identified in subunits of the T cell receptor ζ chain. The general scheme that has been deduced for TCR signaling (and B cell signaling) turns out to be similar for the FcγR. In that scheme, the TAM associates with an inactive tyrosine kinase of the Src family. Cross-linking of the extracellular domain of the α chain results in the activation of the kinase and phosphorylation of the TAM tyrosines. The phosphorylated TAM is then able to interact with another class of tyrosine kinases, exemplified by Syk and ZAP-70, through SH2 interactions with the TAM. The complex thus formed goes on to phosphorylate intracellular substrates, such as phospholipase C-γ 1 and phosphatidylinositol, triggering activation that culminates in the transcription of cytokine genes and release of inflammatory mediators.

The ligand-binding promiscuity of these receptors illustrates a general feature of this class of immune receptors in which the response of a particular cell to cross-linking by immune complexes is governed not solely by ligand specificity, as with T cell and B cell receptors, but by the unique transmembrane and cytoplasmic domains of the particular FcγR expressed on those cells. Thus common ligand-binding domains are coupled to distinct intracellular domains that thereby transduce different response signals to a single stimulus. This is best exemplified by the CD32 genes, which all encode for highly homologous extracellular portions, but have strong structural divergences in the cytoplasmic domains between different leukocytes (Brooks et al., 1989; Indik et al., 1995; Stuart et al., 1989; Van den Herik-Oudijk et al., 1995). Human PMNs, monocytes, and platelets express FcγRIIa whereas B cells express FcγRIIb. Both extracellular domains are 96% homologous and bind IgGs in an identical manner, yet rather than delivering an activation signal, the cross-linking of FcγRIIb on B cells results in a signal aborting B cell activation, proliferation, and antibody secretion (D'Ambrosio et al., 1995; Ravetch, 1994). One important difference between human FcγRIIa and human FcγRIIb in this context is that only the former includes a TAM in its tail. Though B cell CD32b contains a single YXXL motif in its cytoplasmic tail, this is insufficient to constitute the recognized signaling TAM. A recent study has produced mutations of human FcγRIIb in which additional YXXL motifs have been inserted and according to positioning, one such mutant is able to phagocytose as efficiently as FcγRIIa (Indik et al., 1994).

CYTOKINE REGULATION OF FcγR EXPRESSION

Because effector actions triggered by FcγR often involve the release of toxic or inflammatory substances, it is necessary that the body regulate these actions, activating them when needed and keeping them dormant when and where they are not required. This is accomplished by regulating both the number of FcγR-bearing cells and the expression and function of FcγR on these cells. The most important modulators of FcγR expression and function are hormones and immune cytokines, including the colony-stimulating factors (CSFs), interleukins (ILs), interferons (IFNs), and tumor necrosis factors (TNFs) (Capsoni et al., 1991; Guyre et al., 1988). Some of these cytokines (in particular IL-1 and TNF), when produced in response to infection or trauma, act on the hypothalamus and pituitary to induce the production of the adrenal glucocorticoid hormone cortisol. This newly released cortisol, in turn, feeds back to suppress the inflammatory response in part by suppressing further cytokine production (Arya et al., 1984; Snyder and Unanue, 1982). Indeed, the most striking example of the immunosuppressive and anti-inflammatory effects of glucocorticoids is their inhibition of the production of IFN-γ, IL-1, IL-2, and TNF-α (Romagnani, 1994a,b).

Our knowledge of cytokine regulation of FcγR on many cell types, although incomplete, underscores the potential for cytokine–FcγR interactions to modulate markedly the inflammatory response. It should be recognized, however, that most significant effects have been reported in cell culture, and it is not always clear that comparable effects occur in vivo, either physiologically or in response to therapeutic administration of the cytokine. Two cytokines for which there is substantial in vivo experience are IFN-γ and G-CSF. As documented in the sections that follow, these cytokines significantly increase FcγR expression and enhance FcγR-dependent function both in vitro and in vivo. Future studies are required to establish the relative in vivo activities and interactions of other cytokines.

MONOCYTES AND MACROPHAGES

Of the three types of FcγR, CD64 is the most thoroughly studied, with respect to modulation by the various cytokines that regulate FcγRI expression. In particular, the effects of IFN-γ on CD64 expression are best understood. Freshly isolated monocytes have from 15,000 to 40,000 CD64 sites per cell, whereas resting macrophages have more than 50,000 sites per cell (Anderson et al., 1990; Kurlander and Batker, 1982). Stimulation of monocytes with IFN-γ results in increased levels of mRNA for both FcγRIa and FcγRIb (Ernst et al., 1992; Pan et al., 1990), as well as a 5- to 10-fold increase in FcγRIa surface expression (Benech et al., 1992; Pan et al., 1990; Porges et al., 1992). These IFN-γ-induced events are mediated through the immediate transcriptional induction of many genes, including CD64, and the delayed induction of others, such as the MHC class II genes. Binding of IFN-γ to its cell surface receptor results in the tyrosine phosphorylation of a 91-kDa cytoplasmic DNA-binding protein which then assembles on a DNA site referred to as the IFN-γ response region (GRR) on phosphorylation (Farrar and Schreiber, 1993; Pearse et al., 1993). This 39-nucleotide sequence has been identified in the promoter for FcγRIA and is both necessary and sufficient for transcriptional activation by IFN-γ. In addition to the FcγRIa alpha chain, which contains the high affinity binding site for IgG, IFN-γ also markedly increases expression of the other key accessory molecules, including the γ subunit (Durden et al., 1994), and tyrosine kinases (Durden et al., 1995; Durden and Liu, 1995). Although FcγRIb mRNA is also upregulated following IFN-γ stimulation, the question of whether this protein is also increased in response to IFN-γ remains to be determined, for there are no FcγRIb-specific monoclonal antibodies yet available. Thus IFN-γ has multiple effects on myeloid cells that combine to increase markedly both the expression and functional potential of CD64.

Other cytokines are known to modulate FcγR expression on monocytes and macrophages. Tumor necrosis factor-alpha, IL-1β, IL-4, and IL-13 are generally inhibitory for expression of FcγR. Interleukin-4 and IL-13 decrease FcγRI, FcγRII, and FcγRIII expression on cultured monocytes (de Waal et al., 1993; te Velde et al., 1992), and such inhibitory effects, with respect to CD64, have been reported following IL-4 administration to patients (Wong et al., 1992). Interleukin-12 effects are those of increased CD64 expression, but they are secondary to its effect on T and NK cells to increase production and release of IFN-γ (Trinchieri, 1995; Trinchieri and Gerosa, 1996). Selectively enhanced expression of CD16 has been shown in response to TGF-β (Welch et al., 1990), and M-CSF (Young et al., 1990); CD32 is the only FcγR significantly increased by GM-CSF. Interleukin-10 selectively increases cell surface levels of CD64, but has no effect on expression of other FcγR (Capsoni et al., 1995; te Velde et al., 1992).

GRANULOCYTES

Resting PMNs constitutively express 30,000–60,000 CD32 sites per cell and >100,000 CD16 sites per cell (Huizinga et al., 1989). Although fewer than 500 CD64 sites per cell are found on freshly isolated granulocytes from healthy individuals, stimulation with either IFN-γ or G-CSF results in a rapid upregulation of FcγRI within 3–6 h (Pan et al., 1990; Perussia et al., 1983, 1987; Petroni et al., 1988). This IFN-γ-specific increase in FcγRI has also been observed in healthy volunteers and patients receiving IFN-γ therapy (Davis et al., 1995; Goulding et al., 1992; Shiff et al., 1997). As described below, PMN CD64 expression may therefore be useful as a monitor for IFN-γ therapy and/or bacterial infection in a variety of clinical settings (Davis et al., 1995; Goulding et al., 1992; Shiff et al., 1997). Similarly, G-CSF also upregulates FcγRI expression on neutrophils both in vitro and in vivo, which affects both immature myeloid cells and mature circulating PMNs (Gericke et al., 1995; Kerst et al., 1993a,b; Repp et al., 1991; Valerius et al., 1993).

In contrast to regulation of CD64 expression, cytokine regulation of CD32 and CD16 expression on neutrophils remains less clear. One report has shown downregulation of neutrophil FcγRIII by TNF-α, but there have been conflicting reports regarding the capacity of G-CSF, GM-CSF, and IFN-γ to upregulate CD16 and CD32 expression on neutrophils (Buckle and Hogg, 1989; Buckle et al., 1990; Davis et al., 1995; Graziano et al., 1987a; Maurer et al., 1991; Mendel et al., 1988; Reumaux et al., 1995; van Pelt et al., 1996).

CYTOKINE MODULATION OF FcγR FUNCTION

Classically, FcγR functions were defined by antibodies to opsonize targets, then measuring the consequences of the interaction of this antibody-coated target with FcγR-bearing cells. Effector cell functions measured include phagocytosis, ADCC, reactive oxygen intermediates (ROI) production, and cytokine release (Fanger et al., 1989; Sandor and Lynch, 1993; van de Winkel and Capel, 1993; van de Winkel and Anderson, 1991). Cytokines can dramatically affect these functions by modulating either the expression of FcγR or the particular function. Therefore, a cytokine can modulate the ability of the cell to mediate a particular function through FcγR without altering the number of FcγR expressed on the cell surface. Defining the role and function of individual FcγR and the influence of cytokines on mononuclear cells and PMNs is complicated by the expression of multiple FcγR classes and multiple cytokine receptors on individual cells, as well as the relative rather than absolute specificity of different IgG isotypes. Such an approach provides information on the role of FcγR acting in concert, but the data are difficult to interpret for individual classes of FcγR because each class recognizes several IgG isotypes to a greater or lesser extent once they are presented as a multivalent array (van de Winkel and Anderson, 1990).

ANTIBODY-DEPENDENT CELLULAR CYTOTOXICITY

Several model systems have been devised to evaluate the function of individual classes of FcγR when more than one class of FcγR is expressed on an effector cell. In one system, hybridoma cell lines bearing surface antibody directed to various myeloid cell surface molecules, including FcγR, were used as self-directed targets to define the triggering capabilities of these molecules (Graziano et al., 1987a,b; Shen et al., 1989). This approach established that the function of each receptor is determined by the nature of the effector cell on which it is expressed and the state of activation or differentiation of that cell. On human monocytes, FcγRI, FcγRII, and FcγRIII mediated killing of hybridoma cells whereas other surface molecules did not (Graziano et al., 1987a,b; Shen et al., 1989), suggesting that FcγR may be the only molecules capable of mediating antibody-dependent cellular cytotoxicity (ADCC) of these targets.

The capacity of monocytes to mediate ADCC through FcγR does not require prior cytokine activation. Freshly isolated monocytes are able to lyse tumor cells and erythrocytes effectively through either CD32 or CD64. Monocytes are so effective at mediating tumor cell killing that overnight (18 h) stimulation with a number of different cytokines, including IL-1, IL-2, IL-3, IL-4, IL-6, M-CSF, G-CSF, GM-CSF, and even IFN-γ, resulted in no significant increase in monocyte cytolytic activity (Erbe et al., 1990). However, at limiting antibody concentration, IFN-γ stimulation enhanced both the number of CD64 on monocytes and the ability of monocytes to perform FcγRI-mediated, but not FcγRII-mediated, ADCC tumor cell killing (Shen et al., 1989). More complicated single-cell studies addressing the role of various cytokines on monocyte cytolytic activity revealed that both IFN-γ and GM-CSF, but not IL-2, IL-3, IL-4, IL-6, or TNF-α, are able to enhance ADCC of erythrocytes through CD32 or CD64 (Connor et al., 1990). These studies confirmed further that IFN-γ increases not only the percentage of monocytes mediating ADCC but also the capacity of an individual monocyte to kill (perhaps due in part to upregulation of CD64), whereas GM-CSF affects only the former. Moreover, these studies suggest that the variation in FcγR-mediated ADCC by individual monocytes is in part the result of different maturational and/or activation states.

As the monocyte matures in culture, its cytolytic potential begins to decrease. Although this decrease in cytolytic activity may vary in some cases, monocytes cultured for 6 days usually lose most, if not all, measurable CD32- and CD64-mediated cytolytic potential (Erbe et al., 1990). If monocytes are cultured for a short term (0–6 days) in the presence of certain cytokines, the cytolytic activity of monocyte-derived macrophages is maintained. In particular, IFN-γ, TNF-α, GM-CSF, and IL-10 all maintained CD64- and CD32-mediated cytolytic activity of monocyte-derived macrophages (Erbe et al., 1990; te Velde et al., 1992), whereas IL-4 incubation resulted in decreased ADCC compared to the control (Liesveld et al., 1991; te Velde et al., 1990). Similarly, cytolytic activity was maintained when macrophages were cultured for longer periods of time (at least 9 days) in the presence of IL-3 or M-CSF (Young et al., 1990). Thus cytokines can play an important role in the modulation of FcγR-dependent cytolytic activity of the mononuclear phagocyte despite the inherent capacity of unstimulated monocytes to perform ADCC.

In contrast, unstimulated PMNs did not lyse self-directed hybridoma cells through FcγRI, FcγRII, or FcγRIII (Graziano et al., 1987a,b; Shen et al., 1989). However, when PMNs were cultured with IFN-γ, CD32 or CD64, but not CD16, became a functional cytotoxic trigger molecule on PMNs. Killing through CD64 correlated with the induction of this receptor on PMNs by IFN-γ (Petroni et al., 1988; Shen et al., 1987; Viackus et al., 1990). IFN-γ did not increase FcγRII expression on PMNs, but apparently induced changes that potentiated lytic activity or altered signal transduction (Cemerlic et al., 1991). Neutrophil CD16 did not mediate cytotoxicity toward hybridoma cells even after culture with IFN-γ (Graziano et al., 1987a). In contrast, other model systems of PMN ADCC have demonstrated a role for CD32 and CD16 (Bruchelt et al., 1989; Gavioli et al., 1991; Kushner and Cheung, 1992). This model-dependent ability of PMNs to mediate ADCC of tumor cells via CD16 contrasts with their efficient CD16-mediated lysis of erythrocytes, supporting the conclusion that different effector mechanisms are responsible for ADCC of tumor cells and erythrocytes (Shen et al., 1989).

Another cytokine that primes PMN FcγR-mediated cell killing is G-CSF. As discussed earlier, both in vitro and in vivo stimulation of PMNs by G-CSF increase the expression of FcγRI, but not of FcγRII or FcγRIII. G-CSF, like IFN-γ and GM-CSF, effectively primes the PMN for ADCC. Although short-term culture of PMNs with G-CSF resulted in effective tumor cell killing only through CD32 (Erbe et al., 1990), long-term (1–3 days) G-CSF therapy results in the upregulation of CD64 and in vitro cytolysis of Daudi lymphoma cells through either CD32 or CD64 (Cemerlic et al.,

1991; Valerius et al., 1993). Thus PMN FcγRI-mediated cytotoxicity from patients treated with G-CSF was significantly increased compared with that observed by PMNs following short term in vitro G-CSF priming (Repp et al., 1991)

Other cytokines, including IL-6, IL-8, and TNF-α, possess the ability to prime neutrophils for ADCC activity. IL-6 enhances PMN-mediated ADCC via CD64 by increasing secretory and primary degranulation, a property similarly shared with IL-8 (Djeu et al., 1990; Welch et al., 1990). In addition, TNF-α has been shown to prime PMNs for enhanced ADCC (Barbara et al., 1994).

PHAGOCYTOSIS

The phagocytic process and the corresponding role for FcγR have been characterized with opsonized targets. From these studies has emerged the zipper hypothesis, in which unoccupied FcγR are recruited to the periphery of the plasma membrane region already bound to the target as the membrane advances around the particle (Griffin et al., 1975), or polarized to the advancing membrane during cell locomotion (Davis et al., 1982; Zhou et al., 1992). Opsonized particles must be coated circumferentially with IgG ligand in order for ingestion to occur and if opsonin is deposited only on one pole, attachment to the macrophage membrane is seen without subsequent ingestion. As engulfment proceeds these lamellipodia form a continuous cup-like structure that ensheaths the particle, forming a phagosome. CD64-mediated phagocytosis results in the release of large amounts of oxygen and arachidonic acid metabolites, which may in part explain why phagocytosis of tumor cells three to four times the size of a macrophage can proceed by a "chipping away" at the target, accompanied by ingestion of smaller target pieces. Particle opsonization with complement facilities FcγR-mediated phagocytosis, but opsonization with complement alone may not be sufficient for phagocytosis to occur. Complement-coated red blood cells adhere without ingestion to unactivated macrophages; however, complement receptors on activated macrophages are constitutively competent for ingestion (Bianco et al., 1975; Griffin and Mullinax, 1990). Following phagocytosis by macrophages or PMNs the pH of the phagosome becomes first alkaline and then acidic. The basic milieu is optimal for defensins, small arginine- and cysteine-rich peptides, and other basic proteins; whereas the acidic pH is optimal for lysosomal enzymes that are released in the phagosome by phagolysosome fusion.

To investigate the structural determinants necessary for phagocytosis, Hutchinson and co-workers used a transient expression system for CD64 and various chimeras in COS-7 cells, a simian kidney fibroblast cell line (Hutchinson et al., 1995). Though FcγRI was able to rosette opsonized sheep red blood cells (SRBC) when expressed in COS cells, it did not internalize them. Chimeras in which the extracellular domain of FcγRI was attached to the transmembrane and cytoplasmic domains of either γ or FcγRIIa conferred upon COS cells the ability to phagocytose opsonized SRBC efficiently. In further studies, they established that internalization was ligand independent, suggesting that TAM phosporylation was the phagocytic signal and that the process depended on aggregation of these signaling motifs. The multiplicity of functional isoform combinations for FcγR-mediated phagocytosis and the co-dependence for TAM phosphorylation have recently been reviewed (Indik et al., 1995).

Many cytokines have been examined for their ability to modulate FcγR-mediated phagocytosis. Interferon-gamma treatment of monocytes, for example, decreased CD32-mediated phagocytosis (Backman and Guyre, 1994; Jungi et al., 1991) while increasing CD64 expression and FcγRI-mediated phagocytosis in short-term culture (Ruiz et al., 1992). Long-term IFN-γ treatment of monocyte-derived macrophages, however, resulted in almost complete loss of phagocytic ability through CD64. Interestingly, this decrease in phagocytic ability correlated with an increase in extracellular lytic activity (Durie et al., 1994). Monocyte-derived macrophages differentiated in M-CSF for 7 days were 10 to 100 times more efficient at antibody-dependent phagocytosis of tumor targets than cells cultured without cytokine, yet addition of IFN-γ resulted in a pronounced dose-dependent shift from the phagocytic form of cytotoxicity toward an extracellular form (Munn et al., 1993; Van Schie, 1992). In order to influence FcγR-mediated cytotoxicity, IFN-γ had to be present prior to the onset of differentiation; if it was added later it had progressively less effect. Neutrophil phagocytosis through CD64 is negibligible in the healthy human, but IFN-γ administration to normal adults results in a significant increase of PMN phagocytosis of *S. aureus* opsonized with heat-inactivated serum, which could further be correlated with the magnitude of CD64 expression (Schiff et al., 1997).

Tumor necrosis factor-alpha possesses a transient effect on monocyte phagocytic activity of IgG-opsonized beads (Liao and Simon, 1994), causing an initial downregulation followed by a reappearance of phagocytic activity to control levels within 24 h post-stimulation. This loss of phagocytic activity correlated with CD16 expression levels, suggesting CD16-dependent loss in phagocytic activity. In contrast, IL-1β downregulated both opsonized and nonopsonized phagocytosis by monocyte-derived macrophages, regardless of CD16 expression levels (Liao and Simon, 1994; Liao et al., 1994). This decreased phagocytic ability following IL-1β stimulation correlated with a decrease in cellular F-actin, implying an FcγR-independent loss in phagocytosis.

Other cytokines that have been demonstrated to alter FcγR-mediated phagocytosis by monocytes, macrophages, and PMNs include TNF-α, GM-CSF, IL-4, IL-6, IL-10, and IL-12. GM-CSF increased monocyte CD32 expression and induced higher phagocytic properties as well as increased cytotoxicity (Eischen et al., 1991; Thomassen et al., 1989); incubation of PMNs with GM-CSF increased CD16 expression and increased neutrophil phagocytosis of opsonized erythrocytes through FcγRIII (Capsoni et al., 1991). Interleukin-4 reduced the phagocytic activity by monocytes and macrophages, probably resulting from downregulation of all three classes of FcγR (Ho et al., 1992). In contrast, IL-4 increased FcγR-dependent PMN phagocytosis (Bober et al., 1995). Interleukin-10 upregulated FcγR-mediated phagocytosis by both monocytes and monocyte-derived macrophages (Capsoni et al., 1995). Tumor necrosis factor-alpha and IL-6 appear to have a synergistic effect in increasing PMN phagocytic activity of opsonized bacteria (Mullen et al., 1995).

SUPEROXIDE GENERATION

The two FcγR constitutively expressed by PMNs, CD16 and CD32, both serve as effective trigger molecules for ROI production upon cross-linking of either of these receptors (Crockett-Torabi, 1990; Zhou et al., 1994). Although freshly isolated PMNs failed to respond to FcγRI-specific ligands, there is a potent superoxide response to CD64 cross-linking in IFN-γ-treated PMNs (Akerley et al., 1991). Interferon-gamma not only increases FcγRI expression, but also increases expression of the NADPH oxidase component gp91-phox and prolongs the viability of PMNs (Amezaga et al., 1992; Klebanoff et al., 1992). Thus enhanced superoxide production observed following CD64-specific triggering is most likely a reflection of this dual effect of IFN-γ (Akerley et al., 1991). More detailed studies of the mechanism of FcγR-associated induction of superoxide anion production, both in cell lines and in blood monocytes, have been carried out (Pfefferkorn and Fanger, 1989a,b; Pfefferkorn et al., 1990). At subsaturating concentrations of FcγRI–cross-linking ligands, superoxide generation was found to be continuous over long periods of time and depended on the continuous de novo formation of cross-linked and activated CD64, with each FcγRI entering the pool of activated receptors making a unitary contribution to this signal. Cessation of oxidase activity occurred prior to endocytic uptake of cross-linked FcγRI clusters. This indicates that the activated state of surface cross-linked FcγRI is brief and that occupation of the receptors by cross-linking ligand does not sustain the activated state of the receptor.

FcγR-ENHANCED ANTIGEN PRESENTATION

Following up on a large number of studies showing antibody-enhanced antigen presentation in vitro (Cheng, 1985) and in vivo (Manca et al., 1991), Gosselin et al. (1992) showed that human monocyte IgG1-enhanced presentation of tetanus toxoid (TT) is mediated by CD64 but not CD32, even though monocytes bear twice as many CD32 receptors. Although targeting antigen to a wide variety of antigen-presenting cell (APC) molecules can enhance immune activation, FcγR are the most likely to play a physiological role once antigen-specific IgG has been produced. This influence could be very significant, for Sallusto and Lanzavecchia (1994) estimated that immune complex targeting to dendritic cells made them more efficient than antigen-specific B cells. It is noteworthy that antigen presentation by dendritic cells and macrophages can preferentially stimulate Th1 responses (Gajewski et al., 1991; Macatonia et al., 1995), suggesting that specific targeting of antigen to CD64, which is significantly expressed only on macrophages and dendritic cells, may prove to be particularly effective for selective stimulation of Th1 responses in vivo. This concept has been further validated using a novel human CD64-expressing transgenic mouse model and further indicating a potential role of anti-CD64 monoclonals as vaccine adjuvants (Heijnan and van de Winkel, 1995; Heijnan et al., 1996).

Another positive feature of targeting antigen for presentation by the macrophages is indicated by recent reports of class I-restricted T cell activation following phagocytic uptake of antigens by macrophages. For example, Pfeifer et al. (1993) demonstrated that phagocytic processing of an ovalbumin (OVA) cytotoxic T lymphocyte (CTL) epitope expressed within fusion proteins in *E. coli* elicited class I-restricted T-cell activation. When soluble OVA was attached to microspheres, it was presented in a class I-restricted manner by thymic and splenic FcR-positive cells, probably macrophages (Grant and Rock, 1992). Mice immunized with OVA targeted through this phagocytic pathway generated OVA-specific CTL that were sufficient to protect against subsequent challenge with OVA-transduced melanomas (Falo et al., 1995). Taken together, these findings suggest that targeting antigens to FcγR on macrophages or dendritic cells may lead to efficient class I- and class II-restricted presentation, resulting in Th1 and CTL responses capable of protecting against a variety of pathogens and tumors.

The effect of cytokines on FcγR-mediated enhancement of antigen presentation remains relatively undefined. Although initially it would seem that cytokine upregulation of FcγR on antigen-presenting cells would result in enhanced T-cell activation, the work of Nadler et al. (1994) suggests that cytokines affect multiple mechanisms of the antigen presentation pathway, and that the role of FcγR internalization of antigen is only one of these aspects. The ability of monocytes and the myeloid cell line THP-1 to internalize, process, and present peptides through CD32 and CD64 following IFN-γ stimulation demonstrated that IFN-γ can have dramatic effects on this process. Although increased CD64 in response to IFN-γ can increase the amount of antigen bound to the surface and internalized by the APC, these studies demonstrated a decrease in the efficiency of the APC to process antigen into peptides following IFN-γ stimulation. This decrease in peptide synthesis by these monocyte-like THP-1 cells was both dose dependent and time dependent, suggesting that IFN-γ can play a major role in antigen processing and presentation following internalization.

ROLE OF FcγR IN CLINICAL MEDICINE

As reviewed above, the leukocyte expression of FcγR is modulated by a variety of myeloid- and inflammatory-related cytokines, as well as influenced by several known genetic polymorphisms. The relationship of leukocyte FcγR expression and the presence of or susceptibility to disease, particularly of an inflammatory nature, is beginning to be elucidated. One of the major problems with the current laboratory tests for detecting the acute inflammatory response related to infection or tissue damage is the lack of specificity (Ardron et al., 1994). However, the pathophysiology of myeloid CD64 expression would appear to be ideally suited for a role as a diagnostic indicator of an active acute inflammatory response. Unlike many other activation-related antigenic changes on neutrophils and other myeloid cells, such as CD11b, CD18, CD16, CD45RA, and CD62L, neutrophil CD64 has negligible expression in the healthy individual and shows no evidence of sensitivity to blood sample manipulation (Davis, 1996; Davis et al., 1995). CD64 expression upregulation by cytokine occurs on a relatively short time scale of 4–6 h for cell surface expression and 1–3 h for detectable mRNA increases by Northern blot analysis (Cassatella et al., 1995; Davis et al., 1995; Pan et al., 1990; Petroni et al., 1988; Schiff et al., 1997). This upregulation of PMN CD64 appears to correlate directly with enhanced bactericidal (Schiff et al., 1997) and antifungal (Roilides et al., 1995) activity. Thus, given that normal neutrophil expression of CD64 is fewer than 500 sites per cell and that detectable upregulation can be found within 4–6 h after an acute tissue injury, dissemination of infection, or administration of recombinant IFN-γ or G-CSF becomes detectable, quantitative PMN CD64 measurements offer the potential to be a useful diagnostic parameter of a systemic acute inflammatory response.

Flow cytometric quantitative techniques have been the methodology employed in the majority of studies looking at CD64 expression in myeloid cells. However, studies using alternate methodologies such as RIA have also been reported (Moxey-Mims, 1993). Flow cytometric studies are based on either direct or indirect monoclonal antibody detection methods made quantitative through the use of standardized fluorescence microbeads (Schwartz et al., 1996). The methodology most refined with regard to clinical applications in mind is the work of Davis et al., which has validated a whole blood lysis technique using directly conjugated anti-CD64 (Davis, 1996; Davis et al., 1995; Pizza et al., 1995, 1996; Schiff et al., 1997). The current methodology requires less than 20 μl of whole blood, thus being suitable for testing of patients of all ages. The assay precision has been validated on replicates as having a coefficient of variation of less than 5% (Davis et al., 1995), making it quite competitive with other clinical laboratory assays. Of equal practical importance for a clinical diagnostic test, the quantitative PMN CD64 results are stable at least 6 h at room temperature and up to 72 h for anticoagulated blood held at refrigerated temperatures. The results can be quantified in terms of mean equivalent soluble fluorescence units (MESF) derived from the difference between binding of CD64 monoclonal antibody and isotype control antibody run under parallel conditions and calibrated through the use of Quantum 26 FITC beads (Flow Cytometry Standards, San Juan, Puerto Rico) and Quick-Cal for WinList (Verity Software House, Topsham, Maine). Alternatively, with accurate fluorochrome to protein (F/P) ratio determinations of monoclonal antibody reagents, the MESF units can be converted to actual average antibody binding or antigen sites (if the monoclonal reagent is monomeric) per cell.

Studies with quantitative PMN CD64 measurements to date have indicated a high degree of specificity for detection of systemic infection (Davis et al., 1995; Guyre et al., 1990; Herra et al., 1996; Simms et al., 1989). Interestingly, along with the sensitivity for detecting the acute inflammatory response in conjunction with infection there is only a weak correlation with neutrophilia, C-reactive protein, sedimentation rate, and/or the presence of bands (Davis, 1996; Davis et al., 1995), suggesting that additional diagnostic information is afforded by CD64 measurements compared to standard laboratory testing modalities. Clinical decisions regarding the likelihood of infection or sepsis are typically made on the basis of leukocyte counts, the presence of myeloid immature forms in the peripheral blood (bands or "left-shift"), and clinical findings, such as fever. Such indications are typically problematic and lack specificity, particularly in infants (Chiu et al., 1994) and the elderly (Pfitzenmeyer et al., 1995). Furthermore, recent evidence indicates that leukocyte counts may have no significant influence on clinical decisions about whether patients with abdominal pain receive surgery (Vermeulen et al., 1995) and that band counts provide little or no diagnostic information as to the probability of infection in a given patient (Ardron et al., 1994; Bentley, 1988). Given that there are multiple clinical explanations for elevated PMN counts, the potential for PMN CD64 expression to provide increased diagnostic sensitivity and specificity of infection, along with its ability to be performed in a clinical setting in a matter of hours, in contrast to the days required for culture confirmation of infection, would indeed find clinical utility. Although clinical blood samples show a weak tendency for increased PMN CD64 expression in parallel with increasing neutrophil counts or band counts (Davis et al., 1995), clinical samples are consistently identified in which PMN CD64 expression and leukocyte enumeration are discordant. Clinical samples with elevated absolute neutrophil counts, yet exhibiting normal CD64 expression, are commonly seen in three clinical situations, namely glucocorticoid or steroid therapy, the early post-surgical period, and chronic myeloproliferative disorders, conditions in which treating physicians have learned to expect elevated PMN counts without the necessity for further therapeutic action or concern for infection. Observations to date of more than 100 cases of chronic myelogenous leukemia, chronic myeloproliferative disorders, and myelodysplastic syndromes have shown no increased CD64 expression on mature myeloid cells, except in patients with documented evidence of infection or other recognizable or acute inflammatory process (Felzmann et al., 1993; Westwood et al., 1995). Thus one practical clinical application of myeloid CD64 expression could also be to distinguish acute inflammatory responses from hematologic malignancies resulting in neutrophilia. The corollary situation, where elevated PMN CD64 is observed in parallel with normal or decreased PMN counts, is observed most commonly in clinical situations in which patients have received myelosuppressive drugs, such as cancer chemotherapy, and yet have clinical symptoms or laboratory evidence of infection. Although additional studies are required to validate the diagnostic utility of quantitative PMN CD64 measurements, recent studies indicate the level of PMN CD64 expression correlates with both the clinical probability of infection (Davis, 1996; Leino et al., 1997) and the type of infection, with gram-negative bacteria inducing higher PMN CD64 expression than gram-positive bacteria (Herra et al., 1996). Additionally, the magnitude of PMN CD64 expression appears lower in localized, compared to systemic, infections (Davis, 1996; Herra et al., 1996), suggesting the potential for such a diagnostic test in the triage of patients having clinical symptoms suggestive of infection or as an adjunct test prior to or with blood cultures.

Bacterial infection or sepsis is also accompanied by changes in CD16 expression on both PMNs and monocytes. Aside from the numerous changes that occur in cytokine levels during sepsis (Jansen et al., 1996; Lin et al., 1994; Mesters et al., 1996; Mullen et al., 1995), an increasing list of antigenic changes on leukocytes has been described. Neutrophil activation-associated changes with sepsis include an upregulation of CD45RA, CD35, and CD11b (Guillou, 1994; Patterson et al., 1993; Rosenbloom et al., 1995), which is likely due in large part to degranulation of lysomal granule contents. These PMN antigenic changes with sepsis and septic shock are paralleled by a decrease in the expression of CD16 (Herra et al., 1996; Rosenbloom et al., 1995), the magnitude of which has been reported as correlated with the time-honored morphological indicator of neutrophil activation commonly referred to as "toxic" changes (Kabutomori et al., 1993). The mechanism for release of CD16 from the PMN into the plasma is presumably due to PMN activation by a variety of ligands, as occurs in vitro with chemotactic factors and other stimulants (Huizinga et al., 1988, 1990, 1994; Tosi et al., 1992). Similar loss of PMN CD16 in parallel with upregulation of CD11b has been reported with severe thermal injury in burn patients (Babcock et al., 1990; Vindenes and Bjerknes, 1994) and in HIV-infected individuals (Boros et al., 1990b). However, the diagnostic utility of PMN CD16 determinations appears limited, because the loss is typically seen only with severe infections or shock, clinical conditions that are not subtle. Furthermore, there is a wide range of expression density on PMN in normal individuals, thus limiting the predictive value in clinical samples where the normal base line or healthy PMN CD16 expression level is unknown (Davis et al., 1995; Herra et al., 1996).

Monocyte antigen changes in infection or sepsis have similarly been reported, which include alteration in CD16 and CD64 expression, as recently reviewed (Wallace et al., 1994). Just as increased HLA-DR expression on monocytes has been reported to correlate with sepsis development (Guillou, 1994), the expansion of the normally minor population of CD16+ monocytes in blood has been reported (Capsoni et al., 1994; Fingerle et al., 1993; Herra et al., 1996). However, the clinical significance of the CD16+ monocyte subset remains to be clearly defined, because their increase has also been reported in patients with monocytosis due to a variety of causes, cancer patients without infection, and patients treated with recombinant human M-CSF (Saleh et al., 1995). A reduction of all FcγR receptors in monocyte-derived macrophages obtained from HIV-infected individuals compared to controls leads to the hypothesis that this change may in part contribute to the propensity for opportunistic infections (Kent et al., 1994). Because similar reduction of macrophage FcγR expression or function has been reported with end-stage renal disease (Ruiz et al., 1990) and alcoholic cirrhosis (Gomez et al., 1994) patients, both clinical groups with an increased susceptibility for bacterial infections, this change may be secondary to these chronic diseases, rather than suggestive of a role in pathogenesis. Some authors have proposed that the combination of PMN CD64 measurements with determination of the percent of CD16+ monocytes could be used for rapid indication of bacterial infections (Herra et al., 1996). Monocyte CD64 increases with infection, just as in the PMNs the inflammatory-related cytokines such as IFN-γ and G-CSF induce rapid upregulation of this FcγR (Davis et al., 1995; Guyre et al., 1990; Herra et al., 1996). However, the diagnostic sensitivity for infection screening using the monocyte compared to PMN CD64 expression appears less robust, in large part owing to the inter-individual variation of the normal constitutive expression of CD64 on the monocyte population (Davis et al., 1995; Herra et al., 1996). Additional studies are required to determine whether these infection-related monocyte FcγR changes remain solely of research interest, providing insight into the immunologic defense mechanisms, or whether they find a role in diagnostic medicine.

Although leukocyte CD32 expression does not become modulated on either the PMN or monocyte during infections or other disease states, the above-described polymorphism for the R131/H131 allotypes of CD32 appears to provide at least partial insight into susceptibility to infection with encapsulated organisms and immune complex diseases. A disproportionately lower frequency of children with the H131/H131 phenotype, which is associated with higher IgG2 opsonization and phagocytic activity, has been reported in recurrent bacterial respiratory infection (Sanders et al., 1994) and meningococcal septic shock (Bredius et al., 1994). These findings have resulted in speculation that expanded characterization of FcγR polymorphism, of both the CD32 H131/R131 and the CD16 NA1/NA2 allotypes, might represent a means of identifying populations at risk for certain infections. The CD32 polymorphism has also been implicated in the genetic susceptibility for systemic lupus erythematosus, for there is a nearly twofold increase in the frequency of R131/R131 phenotype in these patients compared to ethnic-matched controls (Salmon et al., 1996).

Another acquired disease state associated with reduced PMN CD16, aside from septic shock or similar extreme clinical states of PMN activation, is paroxysmal nocturnal hemoglobinuria (PMH), an acquired clonal disorder affecting all hematopoietic lines, but clinically manifested primarily by periods of intravascular hemolysis and thrombotic events; the molecular abnormality has been defined as a somatic mutation in the PIG-A gene encoding a protein involved in the transfer of *N*-acetylglucosamine to phosphatidylinositol (Hillmen et al., 1995; Rosse and Ware, 1995). This defect affects all protein with a phosphatidylinositol (PI) linkage, including CD16b (Selvaraj et al., 1988). Thus contemporary diagnostic confirmation of PNH has evolved from the laborious Ham's test of in vitro increased complement-dependent lysis of PMH red cells to a flow cytometric assessment of PMN CD16, along with CD55 and CD59 expression either on red cells or neutrophils (Hall and Rosse, 1996; Iwamoto et al., 1995; van der Schoot, 1990). An apparent diagnostic caveat, as well as reason to perform simultaneous flow cytometric assessment of several types of PI-linked proteins, is a report that G-CSF therapy may result in the emergence of a PMN population with near-normal CD16 expression (Ninomiya et al., 1993). Paroxysmol nocturnal hemoglobinuria is another clinical state in which there is decreased expression of a FcγR form, yet there is no significant problem with infections in such patients, further indicating an apparent redundancy of protection afforded through multiple FcγR classes on leukocytes.

Although not a disease, the neonatal period of human development has been associated with numerous alterations in myeloid cell structure and function (Bektas et al., 1990). Among the differences observed between adult and neonatal PMN to explain the reduced function responses by these cells and increased susceptibility to infection is the expression of FcγR. Several studies have reported lower CD16 expression on the neonatal PMN (Carr and Davies, 1990; Maeda et al., 1996; Smith et al., 1990; Takahashi et al., 1993), thus adding to the previously documented decrease in stimulation-induced upregulation of CD11b (Bruce et al., 1987). Additionally, it is apparent from these studies that neonatal or cord blood PMN may express either higher (Maeda et al., 1996) or similar (Takahashi et al., 1993) levels of CD64 relative to healthy adults. What is not clear from these recent studies is the clinical situation from which these neonatal blood samples were derived. These changes in PMN CD16 and CD64 levels could simply reflect PMN activation secondary to the birth process or associated infection, which as in adults might explain the decrease of PMN CD16 and increase in CD64. Clearly additional studies with quantitative, standardized measurements in well defined clinical situations are required before it can be confidently concluded that the pathophysiology of neonatal myeloid FcγR differs significantly from that of adults.

A variety of myeloid FcγR changes have been reported in diseases having in common an autoimmune component, leading to speculation about the role of myeloid FcγR in either pathogenesis or diagnostic monitoring of disease activity. Studies on PMNs from patients with rheumatoid arthritis have demonstrated increased CD64 expression on cells isolated from inflamed joints, but not the peripheral blood (Felzmann et al., 1991; Watson et al., 1993); this presumably reflects the localized recruitment and activation of the PMNs. Neutrophil signaling through CD32 and CD16 is defective in rheumatoid arthritis (Goulding and Guyre, 1992; Robinson et al., 1994), although how this is related to reports of high titers of autoantibodies to CD32 and CD16 in such patients (Boros et al., 1990a, 1993, 1994) remains to be elucidated, particularly because several autoimmune diseases including leukocytoclastic vasculitis (Hundt et al., 1993), systemic vasculitis and glomerulonephritis (Mulder et al., 1994), systemic lupus erythematosus, scleroderma, and Raynaud's syndrome (Boros et al., 1993, 1994) have similarly been reported with autoantibodies to CD16 and/or CD32. A common theme with these inflammatory-mediated diseases may be immune complex activation of PMNs through CD16 and CD32 and subsequent release of such FcγR from cells into a soluble plasma form (Hutin et al., 1994), followed by autoantibody development. The role of myeloid FcγR measurements as a diagnostic indicator of disease activity in inflammatory-mediated autoimmune diseases has only begun to be investigated. Two intriguing observations along these lines are recent reports of correlations of cell-free CD16 serum levels and disease activity in systemic lupus erythematosus (Hutin et al., 1994) and PMN CD64 expression in the Crohn's disease form of inflammatory bowel disease (Kayaba et al., 1993). Additional studies with larger patient numbers and correlation with standard laboratory measurements of disease activity in the various autoimmune diseases with a significant acute inflammatory component are required to define better the diagnostic and prognostic significance of various myeloid FcγR measurements, but it would seem a potentially fruitful area of continued research.

The final promising clinical application of the FcγR field is its role in therapeutics, particularly through the use of bispecific monoclonal antibody (BsAb) constructs (Fanger et al., 1990; Garrido et al., 1990; Wallace et al., 1994). The operative principle is that FcγR-bearing cells could be targeted to a specific cell population, such as cancer cells (Wallace et al., 1994) or HIV-infected cells (Howell et al., 1994), and through ADCC result in further disease regression. The BsAbs, made through the covalent linking of monoclonal antibodies with different specificities, represent a form of immunotherapy whereby potentially both complement and cell-mediated forms of cytotoxicity could be directed against a targeted cell population. Such therapeutic attempts to utilize the host myeloid cells, monocytes and PMNs alike, as a form of adjuvant therapy are now in clinical trials. Although a complete review of this subject is beyond the scope of this chapter, it is clear from recent reports that this use of FcγR-directed reagents and FcγR-bearing cells is encouraging. Bispecific monoclonal antibodies directed against both CD16 and CD64 have been reported to show efficacy in metastatic colon, ovarian, and breast cancer (Repp et al., 1995; Valone et al., 1995; Weiner et al., 1995). In vitro work has also suggested a potential for therapeutic efficacy of CD64 × CD19 or CD64 × CD37 constructs in lymphoma (Ely et al., 1996). The availability of human CD64 transgenic mice promises to provide an animal model for further design and testing of BsAb therapy (Heijnen and van de Winkel, 1995); this could help address issues of therapeutic relevance, such as the role of G-CSF or IFN-γ in upregulating monocyte and PMN CD64 prior to BsAb administration and thereby involving both myeloid cell types for maximal leukocyte cytotoxicity. If the use of FcγR BsAb proves to be efficacious in the final analysis, this would truly represent a successful transition of work from the research bench to the hospital bedside.

REFERENCES

Akerley WL, Guyre PM, Davis BH (1991): Neutrophil activation through high-affinity Fc gamma receptor using a monomeric antibody with unique properties. Blood 77:607–615.

Amezaga MA, Bazzoni F, Sorio C, Rossi F, Cassatella MA (1992): Evidence for the involvement of distinct signal transduction pathways in the regulation of constitutive and interferon gamma-dependent gene expression of NADPH oxidase components (gp91-phox, p47-phox, and p22-phox) and high-affinity receptor for IgG (Fc gamma R-I) in human polymorphonuclear leukocytes. Blood 79:735–744.

Anderson CL, Guyre PM, Whitin JC, Ryan DH, Looney RJ, Fanger MW (1986): Monoclonal antibodies to Fc receptors for IgG on human mononuclear phagocytes: antibody characterization and induction of superoxide production in a monocyte cell line. J Biol Chem 261:12856–12864.

Anderson CL, Looney RJ, Culp DJ, et al. (1990): Alveolar and peritoneal macrophages bear three distinct classes of Fc receptors for IgG. J Immunol 145:196–201.

Anderson CL, Ryan DH, Looney RJ, Leary PC (1987): Structural polymorphism of the human monocyte 40 kilodalton Fc receptor for IgG. J Immunol 138:2254–2256.

Anderson CL, Shen L, Eicher DM, Wewers MD, Gill JK (1990): Phagocytosis mediated by three distinct Fc gamma receptor classes on human leukocytes. J Exp Med 171:1333–1345.

Ardron MJ, Westengard JC, Dutcher TF (1994): Band neutrophil counts are unnecessary for the diagnosis of infection in patients with normal leukocyte counts. Am J Clin Pathol 102:646–649.

Arya SK, Wong-Staal F, Gallo RC (1984): Dexamethasone-mediated inhibition of human T cell growth factor and γ-interferon messenger RNA. J Immunol 133:273–276.

Babcock GF, Alexander JW, Warden GD (1990): Flow cytometric analysis of neutrophil subsets in thermally injured patients developing infection. Clin Immunol Immunopathol 54:117–125.

Backman KA, Guyre PM (1994): Gamma-interferon inhibits Fc receptor II-mediated phagocytosis of tumor cells by human macrophages. Cancer Res 54:2456–2461.

Barbara JA, Smith WB, Gamble JR, et al. (1994): Dissociation of TNF-alpha cytotoxic and proinflammatory activities by p55 receptor- and p75 receptor-selective TNF-alpha mutants. EMBO J 13:843–850.

Bektas S, Goetze B, Speer CP (1990): Decreased adherence, chemotaxis, and phagocytic activities of neutrophils for preterm neonates. Acta Paediatr Scand 79:1031–1038.

Benech PD, Sastry K, Iyer RR, Eichbaum QG, Raveh DP, Ezekowitz RA (1992): Definition of interferon gamma-response elements in a novel human Fc gamma receptor gene (Fc gamma RIb) and characterization of the gene structure. J Exp Med 176:1115–1123.

Bentley SA (1988): Alternatives to the neutrophil band count. Arch Pathol Lab Med 112:883–884.

Bianco G, Griffin F, Silverstein SC (1975): Studies of the macrophage complement receptor: alteration of receptor function upon macrophage activation. J Exp Med 141:1278–1290.

Bober LA, Waters TA, Pugliese-Sivo CC, Sullivan LM, Narula SK, Grace MJ (1995): IL-4 induces neutrophilic maturation of HL-60 cells and activation of human peripheral blood neutrophils. Clin Exp Immunol 99:129–136.

Boros P, Chen JM, Bona C, Unkeless JC (1990a): Autoimmune mice make anti-Fc gamma receptor antibodies. J Exp Med 171:1581–1595.

Boros P, Gardos E, Bekesi GJ, Unkeless JC (1990b): Change in expression of Fc gamma RIII (CD16) on neutrophils from human immunodeficiency virus-infected individuals. Clin Immunol Immunopathol 54:281–289.

Boros P, Muryoi T, Spiera H, Bona C, Unkeless JC (1993): Autoantibodies directed against different classes of Fc gamma R are found in sera of autoimmune patients. J Immunol 150:2018–2024.

Boros P, Odin JA, Chen J, Unkeless JC (1994): Specificity and class distribution of Fc gamma R-specific autoantibodies in patients with autoimmune disease. J Immunol 152:302–306.

Bredius RG, Derkx BH, Fijen CA, et al. (1994): Fc gamma receptor IIa (CD32) polymorphism in fulminant meningococcal septic shock in children. J Infect Dis 170:848–853.

Bredius RG, Fijen CA, de Haas M, et al. (1994): Role of neutrophil Fc gamma RIIa (CD32) and Fc gamma RIIIb (CD16) polymorphic forms in phagocytosis of human IgG1- and IgG3-opsonized bacteria and erythrocytes. Immunol 83:624–630.

Brooks DG, Qiu WQ, Luster AD, Ravetch JV (1989): Structure and expression of human IgG FcRII(CD32). Functional heterogeneity is encoded by the alternatively spliced products of multiple genes. J Exp Med 170:1369–1385.

Bruce MC, Baley JE, Medvik KA, Berger M (1987): Impaired surface membrane expression of C3bi but not C3b receptors on neonatal neutrophils. Pediatr Res 21:306–311.

Bruchelt G, Handgretinger R, Fierlbeck G (1989): Lysis of neuroblastoma cells by the ADCC-reaction: granulocytes of patients with chronic granulomatous disease are more effective than those of healthy donors. Immunol Lett 22:217–220.

Buckle AM, Hogg N (1989): The effect of IFN-gamma and colony-stimulating factors on the expression of neutrophil cell membrane receptors. J Immunol 143:2295–2301.

Buckle AM, Jayaram Y, Hogg N (1990): Colony-stimulating factors and interferon-gamma differentially affect cell surface molecules shared by monocytes and neutrophils. Clin Exp Immunol 81:339–345.

Bux J, Kober B, Kiefel V, Mueller-Eckhardt C (1993): Analysis of granulocyte-reactive antibodies using an immunoassay based upon monoclonal antibody-specific immobilization of granulocyte antigens. Transfusion Med 3:157–162.

Capsoni F, Bonara P, Minonzio F, et al. (1991): The effect of cytokines on human neutrophil Fc receptor-mediated phagocytosis. J Clin Lab Immunol 34:115–124.

Capsoni F, Minonzio F, Ongari AM (1994): Fc receptors expression and function in monoclear phagocytes from AIDS patients: modulation by IFN-γ. Scand J Immunol 39:45–50.

Capsoni F, Minonzio F, Ongari AM (1995): IL-10 up-regulates human monocyte phagocytosis in the presence of IL-4 and IFN-g. J Leukoc Biol 58:351–358.

Carr R, Davies JM (1990): Abnormal FcRIII expression by neutrophils from very preterm neonates. Blood 76:607–611.

Cassatella MA, Flynn RM, Amezaga MA, Bazzoni F, Vicentini F, Trinchieri G (1990): Interferon gamma induces in human neutrophils and macrophages expression of the mRNA for the high affinity receptor for monomeric IgG (Fc gamma R-I or CD64). Biochem Biophys Res Commun 170:582–588.

Cemerlic D, Dadey B, Han T, Vaickus L (1991): Cytokine influence on killing of fresh chronic lymphocytic leukemia cells by human leukocytes. Blood 77:2707–2715.

Chang TW (1985): Regulation of immune response by antibodies: the importance of antibody and monocyte Fc receptor interaction in T cell activation. Immunol Today 6:245–259.

Chiu C-H, Lin T-Y, Bullard MJ (1994): Application of criteria identifying febrile outpatient neonates at low risk for bacterial infections. Pediatr Infect Dis J 13:946–949.

Clarkson SB, Kimberly RP, Valinsky JE, et al. (1986): Blockade of clearance of immune complexes by an anti-Fc gamma receptor monoclonal antibody. J Exp Med 164:474–489.

Connor RI, Shen L, Fanger MW (1990): Evaluation of the antibody-dependent cytotoxic capabilities of individual human monocytes. Role of Fc gamma RI and Fc gamma RII and the effects of cytokines at the single cell level. J Immunol 145:1483–1489.

Crockett-Torabi E, Fantone JC (1990): Soluble and insoluble immune complexes activate human neutrophil NADPH oxidase by distinct Fc gamma receptor-specific mechanisms. J Immunol 145:3026–3032.

D'Ambrosio D, Hippen KL, Minskoff SA, et al. (1995): Recruitment and activation of PTP1C in negative regulation of antigen receptor signaling by FcγRIIB1. Science 268:293–296.

Davis BH: Quantitative neutrophil CD64 expression (1996): Promising diagnostic indicator of infection or systemic acute inflammatory response. Clin Immunol Newsl 16:121–130.

Davis BH, Bigelow NC, Curnutte JT, Ornvold K: Neutrophil CD64 expression (1995): Potential diagnostic indicator of acute inflammation and therapeutic monitor of interferon-γ therapy. Lab Hematol 1:3–12.

Davis BH, Walter RJ, Pearson CB, Becker EL, Oliver JM (1982): Membrane activity and topography of f-Met-Leu-Phe-treated polymorphonuclear leukocytes. Acute and sustained responses to chemotactic peptide. Am J Pathol 108:206–216.

de Haas M, Kleijer M, Minchinton RM, Roos D, von dem Borne A (1994): Soluble Fc gamma RIIIa is present in plasma and is derived from natural killer cells. J Immunol 152:900–907.

de Haas M, Kleijer M, van Zwieten R, Roos D, von dem Borne AE (1995): Neutrophil Fc gamma RIIIb deficiency, nature, and clinical consequences: a study of 21 individuals from 14 families. Blood 86:2403–2413.

de Waal MR, Figdor CG, Huijbens R, et al. (1993): Effects of IL-13 on phenotype, cytokine production, and cytotoxic function of human monocytes. Comparison with IL-4 and modulation by IFN-gamma or IL-10. J Immunol 151:6370–6381.

Djeu JY, Matsushima K, Oppenheim JJ, Shiotsuki K, Blanchard DK (1990): Functional activation of human neutrophils by recombinant monocyte-derived neutrophil chemotactic factor/IL-8. J Immunol 144:2205–2210.

Duchemin AM, Ernst LK, Anderson CL (1994): Clustering of the high affinity Fc receptor for immunoglobulin G (Fc gamma RI) results in phosphorylation of its associated gamma-chain. J Biol Chem 269:12111–12117.

Durden DL, Kim HM, Calore B, Liu Y (1995): The Fc gamma RI receptor signals through the activation of hck and MAP kinase. J Immunol 154:4039–4047.

Durden DL, Liu YB (1994): Protein-tyrosine kinase p72syk in Fc gamma RI receptor signaling. Blood 84:2102–2108.

Durden DL, Rosen H, Cooper JA (1994): Serine/threonine phosphorylation of the gamma-subunit after activation of the high-affinity Fc receptor for immunoglobulin G. Biochem J 299:569–577.

Durie FH, Foy TM, Masters SR, Laman JD, Noelle RJ (1994): The role of CD40 in the regulation of humoral and cell-mediated immunity. Immunol Today 15:406–411.

Edberg JC, Salmon JE, Kimberly RP (1992): Functional capacity of Fc gamma receptor III (CD16) on human neutrophils. Mol Immunol 11:239–251.

Eischen A, Vincent F, Bergerat JP, et al. (1991): Long term cultures of human monocytes in vitro. Impact of GM-CSF on survival and differentiation. J Immunol Meth 143:209–221.

Ely P, Wallace PK, Given AL, Graziano RF, Guyre PM, Fanger MW (1996): Bispecific-armed, interferon-γ primed macrophage-mediated phagocytosis of malignant non-Hodgkin's lymphoma. Blood 87:3813–3821.

Erbe DV, Collins JE, Shen L, Graziano RF, Fanger MW (1990): The effect of cytokines on the expression and function of Fc receptors for IgG on human myeloid cells. Mol Immunol 27:57–67.

Ernst LK, van de Winkel JGJ, Chiu IM, Anderson CL (1992): Three genes for the human high affinity Fc receptor for IgG (Fc gamma RI) encode four distinct transcription products. J Biol Chem 267:15692–15700.

Falo J, Kovacsovics-Bankowski M, Thompson K, Rock KL (1995): Targeting antigen into the phagocytic pathway in vivo induces protective tumour immunity. Nature Med 1:649–653.

Fanger MW, Segal DM, Wunderlich JR (1990): Going both ways: bispecific antibodies and targeted cellular cytotoxicity. FASEB J 4:2846–2849.

Fanger MW, Shen L, Graziano RF, Guyre PM (1989): Cytotoxicity mediated by human Fc receptors for IgG. Immunol Today 10:92.

Farrar MA, Schreiber RD (1993): The molecular cell biology of interferon-gamma and its receptor. Annu Rev Immunol 11:571–611.

Felzmann T, Gadd S, Majdic O, et al. (1991): Analysis of function-associated receptor molecules on peripheral blood and synovial fluid granulocytes from patients with rheumatoid and reactive arthritis. J Clin Immunol 11:205–212.

Felzmann T, Gisslinger H, Krieger O, Majdic O, Ludwig H, Koller U (1993): Immunophenotypic characterization of myelomonocytic cells in patients with myelodysplastic syndrome. Br J Haematol 84:428–435.

Fingerle G, Pforte A, Passlick B, Strobel M, Ziegler-Heitbrock HW (1993): The novel subset of CD14+/CD16+ blood moncytes is expanded in sepsis patients. Blood 82:3170–3176.

Fleit HB, Wright SD, Unkeless JC (1982): Human neutrophil Fc gamma receptor distribution and structure. Proc Natl Acad Sci USA 79:3275–3279.

Gajewski TF, Pinnas M, Wong T, Fitch FW (1991): Murine Th1 and Th2 clones proliferate optimally in response to distinct antigen-presenting cell populations. J Immunol 146:1750–1758.

Garrido MA, Valdayo MJ, Winkler DF, et al. (1990): Refocusing the immune system to react with human tumors by targeting human lymphocytes with bispecific antibodies. Dev Biol Stand 71:33–42.

Gavioli R, Spisani S, Giuliani AL, Cosulich E, Risso A, Traniello S (1991): CD16 and CR3 receptors distinguish between the two mechanisms of tumour cytotoxicity in neutrophils. Br J Haematol 79:170–176.

Gericke GH, Ericson SG, Pan L, Mills LE, Guyre PM, Ely P (1995): Mature polymorphonuclear leukocytes express high-affinity receptors for IgG (Fc gamma RI) after stimulation with granulocyte colony-stimulating factor (G-CSF). J Leukoc Biol 57:455–461.

Gomez F, Ruiz P, Schreiber AD (1994): Impaired function of macrophage Fcγ receptors and bacterial infection in alcoholic cirrhosis. N Engl J Med 331:1122–1128.

Gosselin EJ, Wardwell K, Gosselin DR, Alter N, Fisher JL, Guyre PM (1992): Enhanced antigen presentation using human Fc gamma receptor (monocyte/macrophage)-specific immunogens. J Immunol 149:3477–3481.

Goulding NJ, Guyre PM (1992): Impairment of neutrophil Fc gamma receptor mediated transmembrane signalling in active rheumatoid arthritis. Ann Rheum Dis 51:594–599.

Goulding NJ, Knight SM, Godolphin JL, Guyre PM (1992): Increase in neutrophil Fc gamma receptor I expression following interferon gamma treatment in rheumatoid arthritis. Ann Rheum Dis 51:465–468.

Grant EP, Rock KL (1992): MHC class I-restricted presentation of exogenous antigen by thymic antigen-presenting cells in vitro and in vivo. J Immunol 148:13–18.

Graziano RF, Fanger MW (1987a): Fc gamma RI and Fc gamma RII on monocytes and granulocytes are cytotoxic trigger molecules for tumor cells. J Immunol 139:3536–3541.

Graziano RF, Fanger MW (1987b): Human monocyte-mediated cytotoxicity: the use of Ig-bearing hybridomas as target cells to detect trigger molecules on the monocyte cell surface. J Immunol 138:945–950.

Greenberg S, Chang P, Silverstein SC (1994): Tyrosine phosphorylation of the gamma subunit of Fc gamma receptors, p72syk, and paxillin during Fc receptor-mediated phagocytosis in macrophages. J Biol Chem 269:3897–3902.

Griffin F, Griffin JA, Leider JE, Silverstein SC (1975): Studies on the mechanism of phagocytosis. I. Requirement for circumferential attachment of particle-bound ligands to specific receptors on the macrophage plasma membrane. J Exp Med 142:1263–1282.

Griffin FMJ, Mullinax PJ (1990): High concentrations of bacterial lipopolysaccharide, but not microbial infection-induced inflammation, activate macrophage C3 receptors for phagocytosis. J Immunol 145:697–701.

Guillou PJ (1994): Biologic variation in the development of sepsis after surgery or trauma. Lancet 342:217–220.

Guyre PM, Campbell AS, Kniffin WD, Fanger MW (1990): Monocytes and polymorphonuclear neutrophils of patients with streptococcal pharyngitis express increased numbers of type I IgG Fc receptors. J Clin Invest 86:1892–1896.

Guyre PM, Girard MT, Morganelli PM, Manganiello PD (1988): Glucocorticoid effects on the production and actions of immune cytokines. J Steroid Biochem 30:89–93.

Guyre PM, Graziano RF, Vance BA, Morganelli PM, Fanger MW (1989): Monoclonal antibodies that bind to distinct epitopes on Fc gamma RI are able to trigger receptor function. J Immunol 143: 1650–1655.

Hall S, Rosse W (1996): The use of monoclonal antibodies and flow cytometry in the diagnosis of paroxysmal nocturnal hemoglobinuria. Blood 87:5332–5340.

Heijnen IA, van de Winkel J (1995): A human Fc gamma RI/CD64 transgenic model for in vivo analysis of (bispecific) antibody therapeutics. J Hematother 4:351–356.

Heijnen IA, van Vugt M, Fanger NA, et al. (1996): Antigen targeting to myeloid-specific human Fc gamma RI/CD64 triggers enhanced antibody responses in transgenic mice. J Clin Invest 97:331–338.

Herra CM, Keane CT, Whelan A (1996): Increased expression of Fcγ receptors on neutrophils and monocytes may reflect ongoing bacterial infection. J Med Microbiol 44:135–140.

Hillmen P, Lewis SM, Bessler M, Luzzatto L, Dacie JV (1995): Natural history of paroxysmal nocturnal hemoglobinuria. N Engl J Med 333:1253–1258.

Ho JL, He SH, Rios MJ, Wick EA (1992): Interleukin-4 inhibits human macrophage activation by tumor necrosis factor, granulocyte-monocyte colony-stimulating factor, and interleukin-3 for anti-leishmanial activity and oxidative burst capacity. J Infect Dis 165:344–351.

Howell AL, Guyre PM, You K, Fanger MW (1994): Targeting HIV-1 to Fc gamma R on human phagocytes via bispecific antibodies reduces infectivity of HIV-1 to T cells. J Leukoc Biol 55:385–391.

Huizinga TW, de Haas M, Kleijer M, Nuijens JH, Roos D, von dem Borne AE (1990): Soluble Fc gamma receptor III in human plasma originates from release by neutrophils. J Clin Invest 86:416–423.

Huizinga TW, de Haas M, van Oers MH, et al. (1994): The plasma concentration of soluble Fc-gamma RIII is related to production of neutrophils. Br J Hematol 87:459–463.

Huizinga TW, Kleijer M, Tetteroo PA, Roos D, von dem Borne A (1990): Biallelic neutrophil Na-antigen system is associated with a polymorphism on the phospho-inositol-linked Fc gamma receptor III (CD16). Blood 75:213–217.

Huizinga TW, van der Shoot C, Jost C, et al. (1988): The PI-linked receptor FcRIII is released on stimulation of neutrophils. Nature 333:667–669.

Huizinga TW, van Kemenade F, Koenderman L, et al. (1989): The 40-kDa Fc gamma receptor (FcRII) on human neutrophils is essential for the IgG-induced respiratory burst and IgG-induced phagocytosis. J Immunol 142:2365–2369.

Hundt M, Zielinska SM, Schmidt RE (1993): Fc gamma receptor activation of neutrophils in cryoglobulin-induced leukocytoclastic vasculitis. Arthritis Rheum 36:974–982.

Hutchinson MJ, Harrison PT, Floto RA, Allen JM (1995): Fc gamma receptor-mediated phagocytosis requires tyrosine kinase activity and is ligand independent. Eur J Immunol 25:481–487.

Hutin P, Lamour A, Pennec YL, et al. (1994): Cell-free Fc-gamma receptor III in sera from patients with systemic lupus erythematosus: correlation with clinical and biological features. Int Arch Allergy Immunol 103:23–27.

Indik Z, Park J, Hunter S, Schreiber A (1995): The molecular dissection of Fcγ receptor mediated phagocytosis. Blood 86:4389–4399.

Indik ZK, Pan XQ, Huang MM, McKenzie SE, Levinson AI, Schreiber AD (1994): Insertion of cytoplasmic tyrosine sequences into the nonphagocytic receptor Fc gamma RIIB establishes phagocytic function. Blood 83:2072–2080.

Iwamoto N, Kawaguchi T, Nagakura S, et al. (1995): Markedly high population of affected reticulocytes negative for decay-accelerating factor and CD59 in paroxysmal nocturnal hemoglobinuria. Blood 85:2228–2232.

Jansen P, van der Pouw Kraan T, de Jong I, et al. (1996): Release of interleukin-12 in experimental Escherichia coli septic shock in baboons: Relation to plasma levels of interleukin-10 and interferon-γ. Blood 87:5144–5151.

Jungi TW, Brcic M, Leutwyler C, Pfister H, Spycher MO (1991): Interferon-gamma treatment impairs Fc receptor type II-mediated phagocytosis of human macrophages by a post-receptor-binding mechanism. Immunology 74:439–445.

Kabutomori O, Iwatani Y, Fushimi R, Amino N (1993): Decrease in the density of IgG-Fc receptor III (CD16) on "toxic" neutrophils. Acta Haematol 89:163–164.

Kayaba Y, Hiwatashi N, Nagatomi R, Toyota T, Majima T (1993): Peripheral granulocytes of patients with Crohn's disease express increased numbers of functional IgG Fc receptor I. J Clin Lab Immunol 40:5–18.

Kent SJ, Stent G, Sonza S, Hunter SD, Crowe SM (1994): HIV-1 infection of monocyte-derived macrophages reduces Fc and complement receptor expression. Clin Exp Immunol 95:450–454.

Kerst JM, de Haas M, van der Schoot C, et al. (1993a): Recombinant granulocyte colony-stimulating factor administration to healthy volunteers: induction of immunophenotypically and functionally altered neutrophils via an effect on myeloid progenitor cells. Blood 82:3265–3272.

Kerst JM, van de Winkel JGJ, Evans AH, et al. (1993b): Granulocyte colony-stimulating factor induces hFc gamma RI (CD64 antigen)-positive neutrophils via an effect on myeloid precursor cells. Blood 81:1457–1464.

Klaassen RJ, Ouwehand WH, Huizinga TW, Engelfriet CP, von dem Borne A (1990): The Fc-receptor III of cultured human monocytes. Structural similarity with FcRIII of natural killer cells and role in the extracellular lysis of sensitized erythrocytes. J Immunol 144:599–606.

Klebanoff SJ, Olszowski S, Van Voorhis WC, Ledbetter JA, Waltersdorph AM, Schlechte KG (1992): Effects of gamma-interferon on human neutrophils: protection from deterioration on storage. Blood 80:225–234.

Kurlander R, Batker J (1982): The binding of human immunoglobulin G1 monomer and small, covalently cross-linked polymers of immunoglobulin G1 to human peripheral blood monocytes and polymorphonuclear leukocytes. J Clin Invest 69:1–8.

Kurosaki T, Ravetch JV (1989): A single amino acid in the glycosyl phosphatidylinositol attachment domain determines the membrane topology of Fc gamma RIII. Nature 342:805–807.

Kushner BH, Cheung NK (1992): Absolute requirement of CD11/CD18 adhesion molecules, FcRII and the phosphatidylinositol-linked FcRIII for monoclonal antibody-mediated neutrophil antihuman tumor cytotoxicity. Blood 79:1484–1490.

Lanier LL, Cwirla S, Yu G, Testi R, Phillips JH (1989): Membrane anchoring of a human IgG Fc receptor (CD16) determined by a single amino acid. Science 246:1611–1613.

Leino L, Sorvajarvi K, Katajisto J, et al. (1997): Febrile infection changes the expression of IgG Fc receptors and complement receptors in human neutrophils in vivo. Clin Exp Immunol 107:37, 1997.

Liao G, Simon SR (1994): Temporal down-regulation of Fc gamma RIII expression and Fc gamma receptor-mediated phagocytosis in human monocyte-derived macrophages induced by TNF-alpha and IL-1 beta. J Leukoc Biol 55:702–710.

Liao G, Simone J, Simon SR (1994): Paracrine downregulation of Fc gamma RIII in human monocyte-derived macrophages induced by phagocytosis of nonopsonized particles. Blood 83:2294–2304.

Liesveld JL, Frediani KE, Winslow JM, Duerst R, Abboud CN (1991): Cytokine effects and role of adhesive proteins and Fc receptors in human macrophage-mediated antibody dependent cellular cytotoxicity. J Cell Biochem 45:381–390.

Lin RY, Astiz ME, Saxon JC, Saha DC, Rackow EC (1994): Relationships between plasma cytokine concentrations and leukocyte functional antigen expression in patients with sepsis. Crit Care Med 22:1595–1602.

Looney RJ, Abraham GN, Anderson CL (1986): Human monocytes and U937 cells bear two distinct Fc receptors for IgG. J Immunol 136:1641–1647.

Macatonia SE, Hosken NA, Litton M, et al. (1995): Dendritic cells produce IL-12 and direct the development of Th1 cells from naive CD4+ T cells. J Immunol 154:5071–5079.

Maeda M, van Schie R, Yuksel B, et al. (1996): Differential expression of Fc receptors for IgG by monocytes and granulocytes from neonates and adults. Clin Exp Immunol 103:343–347.

Manca F, Fenoglio D, Li Pira G, Kunkl A, Celada F (1991): Effect of antigen/antibody ratio on macrophage uptake, processing, and presentation to T cells of antigen complexed with polyclonal antibodies. J Exp Med 173:37–48.

Maurer D, Fischer GF, Felzmann T, et al. (1991): Ratio of complement receptor over Fc-receptor III expression: a sensitive parameter to monitor granulocyte-macrophage colony-stimulating factor effects on neutrophils. Ann Hematol 62:135–140.

Mendel DB, Shen L, Guyre PM (1988): The effect of tumor necrosis factor on three IgG Fc receptors of human neutrophils. Clin Res 35:460A.

Mesters R, Mannucci P, Coppola R, Keller T, Ostermann H, Kienast J (1996): Factor VIIa and Antithrombin III activity during severe sepsis and septic shock in neutropenic patients. Blood 88:881–886.

Metzger H, Kinet JP, Blank U, Miller L, Ra C (1989): The receptor with high affinity for IgE. Ciba Found Symp 147:93–113.

Morganelli PM, Kitzmiller TJ, Hemmer R, Fanger MW (1992): Redirected targeting of LDL to human monocyte Fc gamma receptors with bispecific antibodies. Arterioscler Thromb 12:1131–1138.

Moxey-Mims MM, Frank MM, Lin EY, Francis C, Gaither TA (1993): Increased expression of Fc gamma RI on isolated PMN from individuals of African descent. Clin Immunol Immunopathol 69:117–121.

Mulder AH, Heeringa P, Brouwer E, Limburg PC, Kallenberg CG (1994): Activation of granulocytes by anti-neutrophil cytoplasmic antibodies (ANCA): a Fc gamma RII-dependent process. Clin Exp Immunol 98:270–278.

Mullen PG, Windsor AC, Walsh CJ, Fowler A, Sugerman HJ (1995): Tumor necrosis factor-alpha and interleukin-6 selectively regulate neutrophil function in vitro. J Surg Res 58:124–130.

Muniz-Diaz E, Madoz P, de la Calle Martin O, Puig L (1995): The polymorphonuclear neutrophil Fc gamma RIIIb deficiency is more frequent than hitherto assumed [letter]. Blood 86:3999.

Munn DH, Armstrong E (1993): Cytokine regulation of human monocyte differentiation in vitro: the tumor cytotoxic phenotype induced by macrophage colony stimulating factor is developmentally regulated by gamma-interferon. Cancer Res 53:2603–2613.

Nadler SG, Rankin BM, Moran-Davis P, Cleaveland JS, Kiener PA (1994): Effect of interferon-gamma on antigen processing in human monocytes. Eur J Immunol 24:3124–3130.

Ninomiya H, Muraki Y, Shibuya K, Nagasawa T, Abe T (1993): Induction of Fc gamma R-III (CD16) expression of neutrophils affected by paroxysmal nocturnal haemoglobinuria by administration of granulocyte colony-stimulating factor. Br J Haematol 84:497–503.

Ory PA, Goldstein IM, Kwoh EE, Clarkson SB (1989): Characterization of polymorphic forms of Fc receptor III on human neutrophils. J Clin Invest 83:1676–1681.

Pan LY, Mendel DB, Zurlo J, Guyre PM (1990): Regulation of the steady state level of Fc gamma RI mRNA by IFN-gamma and dexamethasone in human monocytes, neutrophils, and U-937 cells. J Immunol 145:267–275.

Patterson WP, Caldwell CW, Yesus YW (1993): In vivo upregulation of CD45RA in neutrophils of acutely infected patients. Clin Immunol Immunopathol 68:35–40.

Pearse RN, Feinman R, Shuai K, Darnell JEJ, Ravetch JV (1993): Interferon gamma-induced transcription of the high-affinity Fc receptor for IgG requires assembly of a complex that includes the 91-kDa subunit of transcription factor ISGF3. Proc Natl Acad Sci USA 90:4314–4318.

Perussia B, Daytonh E, Lazarus R, Fanning V, Trinchieri G (1983): Immune interferon induces the receptor for monomeric IgG1 on human moncytic and myeloid cells. J Exp Med 158:1092–1113.

Perussia B, Kobayashi M, Rossi ME, Anegon I, Trinchieri G (1987). Immune interferon enhances functional properties of human granulocytes: role of Fc receptors and effect of lymphotoxin, tumor necrosis factor, and granulocyte-macrophage colony-stimulating factor. J Immunol 138:765–774.

Petroni KC, Shen L, Guyre PM (1988): Modulation of human polymorphonuclear leukocyte IgG Fc receptors and Fc receptor-mediated functions by IFN-gamma and glucocorticoids. J Immunol 140:3467–3472.

Pfefferkorn L, Fanger M (1989a): Cross-linking of the high affinity Fc receptor for human immunoglobulin G1 triggers transient activation of NADPH oxidase activity. J Biol Chem 264:14112.

Pfefferkorn LC, Fanger MW (1989b): Transient activation of the NADPH oxidase through Fc gamma RI. Oxidase reactivation precedes internalization of cross-linked receptors. J Immunol 143:2640.

Pfefferkorn LC, Guyre PM, Fanger MW (1990): Functional comparison of the inductions of NADPH oxidase activity and Fc gamma RI in IFN gamma-treated U937 cells. Mol Immunol 27:263–272.

Pfeifer JD, Wick MJ, Roberts RL, Findlay K, Normark SJ, Harding CV (1993): Phagocytic processing of bacterial antigens for class I MHC presentation to T cells. Nature 361:359–362.

Pfitzenmeyer P, Decrey H, Auckenthaler R, Michel JP (1995): Predicting bacteremia in older patients. J Am Geriatr Soc 43:230–235.

Pizza F, Davis B, Hendrickson S, et al. (1996): Adaptation to eccentric exercise: effect on CD64 and CD11b/CD18 expression. J Appl Physiol 80:47–55.

Pizza F, Mitchell J, Davis B, Starling R, Holtz R, Bigelow N (1995): Exercise-induced muscle damage: effect on circulating leukocyte and lymphocyte subsets. Med Sci Sports Exercise 27:363–370.

Porges AJ, Redecha PB, Doebele R, Pan LC, Salmon JE, Kimberly RP (1992): Novel Fc gamma receptor I family gene products in human mononuclear cells. J Clin Invest 90:2102–2109.

Qiu WQ, de Bruin BD, Brownstein BH, Pearse R, Ravetch JV (1990): Organization of the human and mouse low-affinity Fc gamma R genes: duplication and recombination. Science 248:732–735.

Ravetch JV (1994): Fc receptors: rubor redux. Cell 78:553–560.

Ravetch JV, Kinet JP (1991): Fc receptors. Annu Rev Immunol 9:457–492.

Repp R, Valerius T, Sendler A, et al. (1991): Neutrophils express the high affinity receptor for IgG (Fc gamma RI, CD64) after in vivo application of recombinant human granulocyte colony-stimulating factor. Blood 78:885–889.

Repp R, Valerius T, Wieland G, et al. (1995): G-CSF-stimulated PMN in immunotherapy of breast cancer with a bispecific antibody to Fc gamma RI and to HER-2/neu (MDX-210). J Hematother 4:415–421.

Reumaux D, Vossebeld PJ, Roos D, Verhoeven AJ (1995): Effect of tumor necrosis factor-induced integrin activation on Fc gamma receptor II-mediated signal transduction: relevance for activation of neutrophils by anti-proteinase 3 or anti-myeloperoxidase antibodies. Blood 86:3189–3195.

Robinson JJ, Watson F, Bucknall RC, Edwards SW (1994): Role of Fc gamma receptors in the activation of neutrophils by soluble and insoluble immunoglobulin aggregates isolated from the synovial fluid of patients with rheumatoid arthritis. Ann Rheum Dis 53:515–520.

Roilides E, Holmes A, Blake C, Pizzo P, Walsh TJ (1995): Effects of granulocyte colony-stimulating factor and interferon-γ on antifungal activity of human polymorphonuclear neutrophils against pseudohyphae of different medically important Candida species. J Leukoc Biol 57:651–656.

Romagnani S (1994a): Lymphokine production by human T cells in disease states. Annu Rev Immunol 12:227–257.

Romagnani S (1994b): Regulation of the development of type 2 T-helper cells in allergy. Curr Opin Immunol 6:838–846.

Rosenbloom AJ, Pinsky MR, Bryant JL, Shin A, Tran T, Whiteside T (1995): Leukocyte activation in the peripheral blood of patients with cirrhosis of the liver and SIRS. J Am Med Assoc 274:58–65.

Rosse WF, Ware RE (1995): The molecular basis of paroxysmal nocturnal hemoglobinuria. Blood 86:3277–3286.

Ruiz P, Gomez F, Lopez R, Chien P, Rossman MD, Schreiber AD (1992): Granulocyte Fc gamma receptor recognition of cell bound and aggregated IgG: effect of gamma-interferon. Am J Hematol 39:257–263.

Ruiz P, Gomez F, Schreiber AD (1990): Impaired function of macrophage Fc gamma receptors in end-stage renal disease. N Engl J Med 322:717–722.

Saleh MN, Goldman S, LoBuglio AF, et al. (1995): CD16+ monocytes in patients with cancer: Spontaneous elevation and pharmacologic induction by recombinant human macrophage colony-stimulating factor. Blood 85:2910–2917.

Sallusto F, Lanzavecchia A (1994): Efficient presentation of soluble antigen by cultured human dendritic cells is maintained by granulocyte/macrophage colony-stimulating factor plus interleukin 4 and down-regulated by tumor necrosis factor alpha. J Exp Med 179:1109–1118.

Salmon JE (1996): Role of allelic polymorphisms of Fcγ receptor IIa in systemic lupus erythematosus. Clin Immunol Newsl 16:130–134.

Salmon JE, Edberg JC, Brogle NL, Kimberly RP (1991): Functional consequence of allelic variants of Fc gamma receptors (Fc gamma R) on human phagocytes. FASEB J 5:A1718.

Salmon JE, Edberg JC, Brogle NL, Kimberly RP (1992): Allelic polymorphisms of human Fc gamma receptor IIA and Fc gamma receptor IIIB. Independent mechanisms for differences in human phagocyte function. J Clin Invest 89:1274–1281.

Salmon JE, Edberg JC, Kimberly RP (1990): Fc gamma receptor III on human neutrophils. Allelic variants have functionally distinct capacities. J Clin Invest 85:1287–1295.

Sanders LA, Feldman RG, Voorhorst OM, et al. (1995): Human immunoglobulin G (IgG) Fc receptor IIA (CD32) polymorphism and IgG2-mediated bacterial phagocytosis by neutrophils. Infect Immun 63:73–81.

Sanders LA, van de Winkel J, Rijkers GT, et al. (1994): Fc gamma receptor IIa (CD32) heterogeneity in patients with recurrent bacterial respiratory tract infections. J Infect Dis 170:854–861.

Sandor M, Lynch RG (1993): The biology and pathology of Fc receptors. J Clin Immunol 13:237–246.

Schiff D, Rae J, Martin T, Davis B, Curnutte J (1997): Increased phagocyte CD64 expression and improved Fc-receptor mediated phagocytosis following in vivo recombinant human interferon-γ treatment of normal human subjects. Blood 90:3187–3194.

Schwartz A, Fernandez-Repollet E, Vogt R, Gratama J (1996): Standardizing flow cytometry: construction of a standardizing fluorescence calibration plot using matching spectral calibrators. Cytometry 26:22–31.

Selvaraj P, Rosse WF, Silber R, Springer TA (1988): The major Fc receptor in blood has a phosphatidylinositol anchor and is deficient in paroxysmal nocturnal haemoglobinuria. Nature 333:565–567.

Shen L, Graziano RF, Fanger MW (1989): The functional properties of Fc gamma RI, II and III on human myeloid cells: a comparative study of killing of erythrocytes and tumor cells mediated through the different Fc receptors. Mol Immunol 26:959–969.

Shen L, Guyre PM, Fanger MW (1987): PMN triggered through the high affinity Fc receptor for monomeric IgG. J Immunol 139:534.

Simms HH, Frank MM, Quinn TC, Holland S, Gaither TA (1989): Studies on phagocytosis in patients with acute bacterial infections. J Clin Invest 83:252–260.

Smith JB, Campbell DE, Ludomirsky A, et al. (1990): Expression of the complement receptors CR1 and CR3 and the type III Fc gamma receptor on neutrophils from newborn infants and from fetuses with Rh disease. Pediatr Res 28:120–126.

Snyder DS, Unanue ER (1982): Corticosteroids inhibit murine macrophage Ia expression and interleukin-1 production. J Immunol 129:1803–1805.

Stein EL, Bux J, Santoso S, Mueller EC (1994): Typing of the granulocyte-specific NA antigens by restriction fragment length polymorphism analysis. Br J Haematol 87:428–430.

Stuart SG, Simister NE, Clarkson SB, Kacinski BM, Shapiro M, Mellman I (1989): Human IgG Fc receptor (hFcRII; CD32) exists as multiple isoforms in macrophages, lymphocytes and IgG-transporting placental epithelium. EMBO J 8:3657–3666.

Takahashi N, Nishida H, Kuratsuji T (1993): Fc gamma RI and Fc gamma RIII on polymorphonuclear leucocytes in cord blood. Arch Dis Child 70:F31.

Takai S, Kasama M, Yamada K, et al. (1994): Human high-affinity Fc gamma RI (CD64) gene mapped to chromosome 1q21.2-q21.3 by fluorescence in situ hybridization. Hum Genet 93:13–15.

te Velde A, de Waal Malefijt R, Huijbens RJ, de Vries J, Figdor CG (1992): IL-10 stimulates monocyte Fc gamma R surface expression and cytotoxic activity. Distinct regulation of antibody-dependent cellular cytotoxicity by IFN-gamma, IL-4, and IL-10. J Immunol 149:4048–4052.

te Velde A, Huijbens RJ, de Vries J, Figdor CG (1990): IL-4 decreases Fc gamma R membrane expression and Fc gamma R-mediated cytotoxic activity of human monocytes. J Immunol 144:3046–3051.

Thomassen MJ, Barna BP, Rankin D, Wiedemann HP, Ahmad M (1989): Differential effect of recombinant granulocyte macrophage colony-stimulating factor on human monocytes and alveolar macrophages. Cancer Res 49:4086–4089.

Tosi MF, Zakem H (1992): Surface expression of Fc gamma receptor III (CD16) on chemoattractant-stimulated neutrophils is determined by both surface shedding and translocation from intracellular storage compartments. J Clin Invest 90:462–470.

Trinchieri G (1995): Interleukin-12 and interferon-γ. Do they always go together? Am J Pathol 147:1534–1538.

Trinchieri G, Gerosa F (1996): Immunoregulation by interleukin-12. J Leuk Biol 59:505–511.

Unkeless J (1989): Function and heterogeneity of human Fc receptors for immunoglobulin G. J Clin Invest 83:355–361.

Unkeless JC, Scigliano E, Freedman VH (1988): Structure and function of human and murine receptors for IgG. Annu Rev Immunol 6:251–281.

Valerius T, Repp R, de Wit TP, et al. (1993): Involvement of the high-affinity receptor for IgG (Fc gamma RI; CD64) in enhanced tumor cell cytotoxicity of neutrophils during granulocyte colony-stimulating factor therapy. Blood 82:931–939.

Valone FH, Kaufman PA, Guyre PM, et al. (1995): Phase Ia/Ib trial of bispecific antibody MDX210 (anti-HER-2/neu x anti-Fc gamma RI) in patients with advanced breast or ovarian cancer that over-express the proto-oncogen HER-2/neu. J Clin Oncol 13:2281–2288.

van de Winkel J, Anderson CL (1990): Biological functioning of human IgG Fc receptors. Res Immunol 141:64–67.

van de Winkel J, Anderson CL (1991): Biology of human immunoglobulin G Fc receptors. J Leukoc Biol 49:511–524.

van de Winkel J, Boonen GJ, Janssen P, Vlug A, Hogg N, Tax W (1989): Activity of two types of Fc receptors, Fc gamma RI and Fc gamma RII, in human monocyte cytotoxicity to sensitized erythrocytes. Scand J Immunol 29:23–31.

van de Winkel J, Capel P (1993): Human IgG Fc receptor heterogeneity: molecular aspects and clinical implications. Immunol Today 14:215–221.

van de Winkel J, de Wit T, Ernst LK, Capel PJ, Ceuppens JL (1995): Molecular basis for a familial defect in phagocyte expression of IgG receptor I (CD64). J Immunol 154:2896–2903.

Van den Herik-Oudijk IE, ter Bekke MW, Tempelman MJ, Capel PJ, Van de Winkel JG (1995): Functional differences between two Fc receptor ITAM signaling motifs. Blood 86:3302–3307.

van der Schoot CE, Huizenga TW, Veer-Korthof ET, Wijmans R, Pinkster J, von dem Borne AEG (1990): Deficiency of glycosyl-phosphatidylinositol-linked membrane glycoproteins of leukocytes in paryoxsymal nocturnal hemoglobinuria, description of a new diagnostic cytofluorometric assay. Blood 76:1853–1859.

van Pelt L, Huisman M, Weening R, von demBorne A, Roos D, van Oers R (1996): A single dose of GM-CSF induces systemic interleukin-8 release and neutrophil activation in healthy volunteers. Blood 87:5305–5313.

Van Schie RC, Verstraten RG, van de Winkel JG, Tax WJ, de Mulder HM (1992): Effect of recombinant IFN-γ on the mechanism of human macrophage IgG FcRI-mediated cytotoxicity. J Immunol 148:169–176.

Vance BA, Huizinga TW, Wardwell K, Guyre PM (1993): Binding of monomeric human IgG defines an expression polymorphism of Fc gamma RIII on large granular lymphocyte/natural killer cells. J Immunol 151:6429–6439.

Vaughn M, Taylor M, Mohanakumar T (1985): Characterization of human IgG Fc receptors. J Immunol 135:4059–4065.

Vermeulen B, Morabia A, Unger P-F (1995): Influence of white cell count on surgical decision making in patients with abdominal pain in the right lower quandrant. Eur J Surg 161:483–486.

Viackus L, W B, Cemerlic D, Foon K (1990): Interferon gamma augments Lym-1 dependent, granulocyte-mediated tumor cell lysis. Blood 75:2408–2416.

Vindenes H, Bjerknes R (1994): Activation of polymorphonuclear neutrophilic granulocytes following burn injury: alteration of Fc-receptor and complement-receptor expression and of opsonophagocytosis. J Trauma 36:161–167.

Wallace P, Howell A, Fanger M (1994): Role of Fcγ receptors in cancer and infectious disease. J Leukoc Biol 55:816–826.

Watson F, Robinson JJ, Phelan M, Bucknall RC, Edwards SW (1993): Receptor expression in synovial fluid neutrophils from patients with rheumatoid arthritis. Ann Rheum Dis 52:354–359.

Weiner LM, Clark JI, Davey M, et al. (1995): Phase I trial of 2B1, a bispecific monoclonal antibody targeting c-erbB-2 and Fc gamma RIII. Cancer Res 55:4586–4593.

Welch GR, Wong HL, Wahl SM (1990): Selective induction of Fc gamma RIII on human monocytes by transforming growth factor-beta. J Immunol 144:3444–3448.

Westwood NB, Copson ER, Page LA, Mire SA, Brown KA, Pearson TC (1995): Activated phenotype in neutrophils and monocytes from patients with primary proliferative polycythaemia. J Clin Pathol 48:525–530.

Wong HL, Lotze MT, Wahl LM, Wahl SM (1992): Administration of recombinant IL-4 to humans regulates gene expression, phenotype, and function in circulating monocytes. J Immunol 148:2118–2125.

Young DA, Lowe LD, Clark SC (1990): Comparison of the effects of IL-3, granulocyte-macrophage colony-stimulating factor, and macrophage colony-stimulating factor in supporting monocyte differentiation in culture. Analysis of macrophage antibody-dependent cellular cytotoxicity. J Immunol 145:607–615.

Zhou MJ, Brown EJ (1994): CR3 (Mac-1, alpha M beta 2, CD11b/CD18) and Fc gamma RIII cooperate in generation of a neutrophil respiratory burst: requirement for Fc gamma RIII and tyrosine phosphorylation. J Cell Biol 125:1407–1416.

Zhou MJ, Poo H, Todd RF III, Petty HR (1992): Surface-bound immune complexes trigger transmembrane proximity between complement receptor type 3 and the neutrophil's cortical microfilaments. J Immunol 148:3550–3553.

Zipf TF, Lauzon GJ, Longenecker BM (1983): A monoclonal antibody detecting a 39,000 M.W. molecule that is present on B lymphocytes and chronic lymphocytic leukemia cells but is rare on acute lymphoblastic leukemia blasts. J Immunol 131:3064–3072.

4

Early Events in Receptor-Mediated Neutrophil Signal Transduction

Elizabeth R. Simons
Boston University School of Medicine, Boston, Massachusetts

Although the nature of the specific agonist receptors on phagocytic cells has now been delineated, the mechanisms and pathways by which these receptors mediate and modulate the functional responses of these cells to specific, and different, agonists have not been easy to elucidate. Thus, as is discussed in some detail below, even though the important domains in these receptors (e.g., the FcR class of agonist receptors) have been identified, and some of their modes of action delineated (e.g., the involvement of specific phospholipases), the actual sequence of events leading to expression of the cells' function (e.g., phagocytosis, degranulation, and oxidative product formation) has been elusive. In particular, one of the limitations of the bulk techniques that are generally available is an inability to determine whether a specific result or response is attributable to all of the cells or to a subpopulation of these cells and if the latter, to characterize the responding versus nonresponding cells. This limitation, already identified when normal mature cells were being investigated, is even more serious in differentiating cells (in which responses have been shown to develop as the differentiation proceeds) and in cells from knockout animals or from transformed cultures (in which, frequently, only some of the cells have been altered while others remain the wild type). Realization of these problems has led to increasing emergence of techniques permitting the individual study of large numbers of single cells, ac-

Phagocyte Function: A Guide for Research and Clinical Evaluation, Edited by J. Paul Robinson and George F. Babcock. ISBN 0-471-12364-1

complished most readily by flow cytometry or less readily (because of the number of images required) by microscope-linked imaging. This chapter addresses one of these techniques, flow cytometry, its development into a kinetic tool for functional analysis of phagocytes, the limitations, and the current applications, keeping in mind that the latter are expanding very rapidly.

FLOW CYTOMETRY: INHERENT ADVANTAGES, DISADVANTAGES, AND LIMITATIONS

A flow cytometer interrogates fluorophore-labeled cells, suspended singly in a drop, as they pass in front of a laser beam, recording the ensuing fluorescence on as many photomultipliers as are in the optical path of the laser. This is an oversimplification of the instrument and its purpose; its detailed construction and description are summarized elsewhere (Shapiro, 1995). No further details are given here because it is the applications of flow cytometry, especially to kinetic studies of neutrophil and phagocyte precursor functions, that are the subject of this chapter. One of the main advantages of flow cytometry is its applicability to simultaneous evaluation of several functional responses of individual cells and coincidentally, of the detection of subpopulations that respond differently; because of the time scale involved, real time measurements involve relatively rapid responses, whereas slower ones are more likely to involve the end product of such responses. This technique, because it interrogates single cells, also permits correlation between several parameters, for example, that of receptor occupancy with the consequences, for that cell, of its receptor liganding. In this respect flow cytometry is far superior (albeit more cumbersome) than fluorimetry, which cannot yield single cell data because it records the average of all the cells in the light path, and therefore is unable to distinguish between, for example, a full response by half the cells and a half-maximal response by all the cells. Furthermore, leakage of probe or the presence of the detected entity both inside and outside the cells is a significant problem in fluorimetry (Davies et al., 1987, 1988) but immaterial in flow cytometry because the cells are surrounded mainly by sheath fluid. For all these reasons, our laboratory has developed and used multiparameter flow kinetic techniques for the determination of cellular signal transduction and consequent functional execution.

PROBE SELECTION

The use of multiple photomultipliers to record the fluorescence emission, using filters at several different wavelengths, resulting from a single excitation, permits the simultaneous observation of several parameters for the same cell. An overlap of these emissions can be a serious problem, but the availability of a large number of new probes (e.g., the expanding Molecular Probes Catalogue) has reduced this problem to the point where compensation, the bane of existence of the early flow cytometers, rarely has to be used to correct overlap problems.

Among the earliest real-time fluorescent indicators of cellular functions were those reporting stimulation-elicited changes in membrane potentials and in cytoplasmic cation (Ca^{2+}, H^+, Na^+, K^+) and anion (Cl^-) concentrations. More recently, as discussed below, indicators of phospholipase activity, of oxidative product formation, and of lytic enzyme release have been developed. Flow cytometry has made it possible to use all these indicators, together with fluorescently labeled ligands of cell surface proteins, to examine cell responses even when these are expressed by only a fraction of the cells in a given sample (i.e., a subpopulation) (Davies et al., 1988, 1989, 1990; Ryan, 1990; Brunkhorst et al., 1991, 1992; Omann and Porasik-Lowes, 1991; Elsner et al., 1992, 1993; Strohmeier, 1995a,b; Guillemot et al., 1996); no other technique, except laborious microscopy, permits such observations.

The choice of probes is dictated by the available excitation laser wavelengths. The most frequently used laser, an argon laser, emits at 488 nm, with a weaker band at 514 nm, and an ultraviolet band at 353 nm for a high-energy argon laser. Simultaneous excitation by uv (353 nm) and visible (488 nm) is also possible. Krypton and tunable dye lasers have also been used, and applications for multiple air-cooled lasers emitting at different frequencies, rather than water-cooled lasers, are being developed.

Surface Labeling (Direct and Indirect)

For identification of cell surface proteins, use of either directly labeled protein-specific monoclonal antibodies, or unlabeled antibodies followed by labeled polyclonal antibodies, is still the most common application in both clinical and research laboratories. Fluorophores such as fluorescein, rhodamine, phycoerythrin and their derivatives, highly efficient internal energy transfer ("piggyback" or "Duo") labels, and biotin-labeled ligands detected with avidin-labeled antibodies are the most frequently used single labels for identification of surface proteins. The fluorescent labels may reside directly on the ligand or on an antibody either to that ligand or to a secondary antibody that recognizes the primary one; the latter technique affords the largest amplification but requires careful controls. Because the cell surface proteins are in their native state, the specific antibodies must recognize that state and therefore antibodies used for Western blots are often unsatisfactory. In recent years, many of the antibodies and some ligands have become available as biotinylated derivatives whose fluorescence can then be developed by utilization of high affinity biotin–avidin complexes. However labeled, surface protein-specific ligands or antibodies permit evaluation of the extent of occupancy (and/or of blockage of a specific epitope) of that protein on a given cell.

If the possibility exists, it is usually more accurate to use a specific ligand than an antibody to that ligand's receptor. This is especially true of leukocytes, lymphocytes, and platelets because all three possess receptors for the Fc ends of immunoglobulins, especially IgG. Monocytes and macrophages have three classes of FcγR, with FcγRI recognizing all IgG with high affinity whereas FcγRII and FcγRIII bind and respond only to aggregated IgG or immune complexes; neutrophils express only FcγRII and III unless the cells have been treated with γ-interferon, and are fully activated only when they are cross-linked, for example, occupied by multivalent Fc ends of IgG

(Brunkhorst et al., 1991, 1992; Strohmeier et al., 1995a,b). This tends to complicate studies of lymphocytes, platelets, and leukocytes, because mere identification of these cells with specific ligand-recognizing antibodies (such as CD4, CD8, CD11, and CD18) can at least prime if not actually activate them. It also makes deduction of the specific mechanisms of activation attributable to each receptor subclass difficult unless one utilizes the Fab or Fab_2 fragments of the desired antibodies to block each subclass in turn (Brunkhorst et al., 1991, 1992; Strohmeier, 1995a,b).

Probes for Activation-correlatable Cytoplasmic Changes

The measurement of cytoplasmic ion concentrations has been helped immeasurably by Tsien's design of a series of probes (Grynkiewicz et al., 1985; Tsien, 1989) that are non-indicating but membrane permeating when esterified (i.e., acetates or acetoxymethylates), but become fluorescent in the cytoplasm after esterolysis by ubiquitous mammalian cell cytoplasmic esterases, yielding an indicating free acid but also lowering the pH_i. Although it has been demonstrated that most of the de-esterification occurs, and therefore the probe is retained, in the cytoplasm (Thomas et al., 1979; Simons et al., 1985), some leakage to the exterior in certain classes of cells [e.g., platelets (Davies et al., 1987)] does occur. Recently, Molecular Probes has developed new forms of these indicators, whose de-esterified form can bind to cytoplasmic reduced glutathione through formation of covalent disulfide bonds and which have reduced leakage. The utilization of some of the cytoplasmic probes, including precautions, has also recently been reviewed (Simons and Davies, 1994).

For any fluorescence-based (or absorbance-based) technique, one assumes that the Beer–Lambert law is obeyed so that the fluorescence emitted is linearly proportional to the concentration of the entity being measured and to the light path through the cell. This also means that one assumes that the cells being measured are all loaded equally with probe, are of equal size, and have the same surface area and distribution of receptors, and that this holds for a given class of cells not only for every experiment but for every day and every donor. For most cultured cells, of uniform size, content, and stage of differentiation, these assumptions appear to hold but for differentiating cells such as DAMI™ (unpublished data) or HL-60 (Bernardo et al., 1990) cells, which are mixtures at various stages of differentiation (i.e., nonsynchronous), and for freshly isolated human cells, which are rarely adequately uniform, these assumptions are probably not fully valid. There is hence a great advantage in using the ratiometric probes (i.e., probes whose absorbance or emission spectra exhibit a wavelength shift as well as a change in intensity upon binding their specific ligand) or in setting up an artificial ratio, frequently by using separate noninterfering probes.

If one wishes to correlate the extent of surface protein occupancy and the specific protein occupied, whether by a ligand or by an antibody to a cell surface receptor, with the functional consequences of that occupancy, specific indicators of these consequences are the choice. For virtually all mammalian cells, the initial responses to specific functional receptor (as distinguished from specific binding site) occupancy include a change in transmembrane potential, a rise in the cytoplasmic concentrations of Ca^{2+} ($[Ca^{2+}]_i$) and H^+ (pH_i), generally followed by a decrease in H^+ (increase in pH_i) due to the opening of the relatively ubiquitous Na^+/H^+ antiport. As mentioned above, excellent and highly specific probes for all these entities as well as for cations such as Mg^{2+} and for at least one anion, Cl^-, are now commercially available.

Probes for Transmembrane Potential Changes

The early cellular responses to specific stimuli include membrane potential changes consisting of a very rapid and small (hence sometimes difficult to detect) hyperpolarization, followed by a larger (therefore often obscuring) and slower depolarization, both attributable to changes in cytoplasmic K^+/Na^+ (Korchak and Weissmann, 1980; Whitin et al., 1980; Simchowitz et al., 1982; Nasmith and Grinstein, 1986; Luscinskas et al., 1988; Demaurex et al., 1993) and probably Cl^-/HCO_3^- concentrations (Korchak et al., 1982). The indicators of choice (Waggoner, 1988) have been the cyanines (lipophilic cations, $diXC_{n(m)}$) whose chemical structure, as shown below, varies depending upon the substituent X (S, O, I, or C) on the heterocyclic ring, the number of conjugated carbon atoms (m) in the link between the rings, and the number of carbon atoms (n) in the heterocyclic ring's "tail." These indicators distribute between the exterior and interior (including organelles and granules) of the cells, depending upon the membrane potential (i.e., as the latter increases, more probe enters the cell).

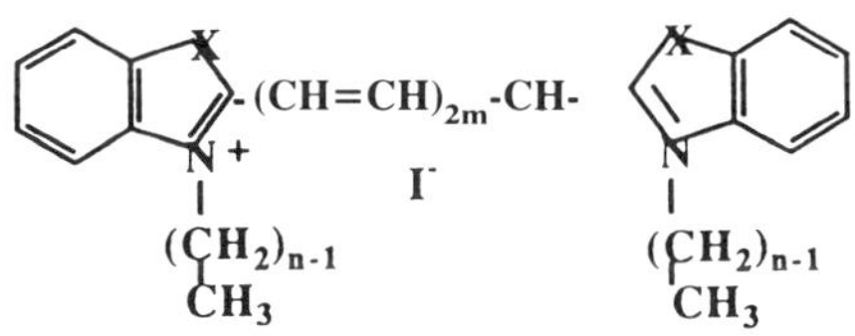

Oxonols, which are negatively charged, have also been used, though the mechanism by which they are measures of membrane potentials is less clear and the values obtained sometimes differ from those of the cyanines, probably due to differential entry into the granules. However, it has been shown (Seligmann and Gallin, 1983; Fletcher and Seligmann, 1985; Lazzari et al., 1990) that for early events, these two classes of indicators yield comparable results, probably because changes in the mitochondria (strongly labeled with cyanines, not at all with oxonols) and release of granules (and the probe they contain) are slow compared to the trans-plasma membrane potential changes. The utilization of these probes, including precautions, has recently been reviewed (Simons, 1993; Simons and Davies, 1994).

Quantitation and Calibration

Where quantitation of either binding or activation-initiated events is desired, calibration is necessary. For calculation of the numbers of binding sites or receptors which are occupied, calibration is most readily accomplished with fluorescence intensity standards (e.g., beads) or when these are not available, by using standards with a known numbers of binding sites (obtained, for example, with isotopically labeled analogues). To determine the actual concentration of specific ions reflected by a given fluorescence, calibration curves are most readily obtainable by fluorimetry, because the gains and hence the ratio are fixed and therefore reproducible day to day. Calibration is more difficult for the flow cytometer, because the gains on its photomultiplier amplifiers are set separately, yielding different values and hence ratios day to day. In addition, a fluorimeter reports an average value over all the cells in suspension, whereas a flow cytometer reports this value for each cell, making it highly preferable to use a ratiometric probe such as those discussed above (Thomas et al., 1979; Grynkiewicz et al., 1985). If the probe's K_d is known, and a linear relationship between the ratio and the ion's concentration can be assumed, the corresponding ion concentration can be calculated from the observed ratio and the ratios at maximal and at minimal concentration of the ion by the equation (Grynkiewicz et al., 1985) $[\text{ion}] = K_d(\text{ratio}_{obs} - \text{ratio}_{min})/(\text{ratio}_{max} - \text{ratio}_{obs})$. Like all such equations, it becomes inapplicable as ratio_{obs} approaches ratio_{max} because the denominator then approaches zero. For Indo 1, a Ca^{2+} indicator, we have developed an exponential equation (Schonhorn and Simons, 1989; Davies et al., 1997) that is linear throughout the range of ratios observed, even those at maximal $[Ca^{2+}]_i$.

Probes for Phospholipase Activation

Another stimulus response of virtually all cells is an initiation of phospholipid metabolism by one or more phospholipases. Fluorescence-based assays are being developed for these enzymes (Hendrickson, 1994). Phospholipase A2 activity can be detected in real time in a flow cytometer (Meshulam et al., 1992), using a probe specifically designed for this purpose, bis(BODIPY™)phosphatidylcholine. The fluorescence of the two BODIPY groups, one each in positions 1 and 2 of the phospholipid, is internally quenched in the intact phospholipid, rendering the probe nearly nonfluorescent when incorporated into the inner leaflet of a cell's plasma membrane via liposomes (Meshulam et al., 1992). Upon activation of phospholipase A, but not of phospholipases C or D, both the lysophospholipid and the cleaved fatty acid are highly fluorescent. This has permitted use of the probe to investigate the mechanism of phospholipase A activation resulting from cellular phospholipid metabolism stimulation. The activity of phospholipase D can be measured by fluorimetric high-pressure liquid chromatography detection of phosphatidic acid (in presence of low concentration of ethanol), but not yet by flow cytometry (Gewirtz and Simons, 1996).

Probes for Oxidative Products

Activation of phagocytes leads to initiation of an oxidative burst and to the formation of a number of oxidative species (superoxide, peroxide, O· and OH· free radicals, and in the presence of myeloperoxidase and Cl^-, OCl^-). Starting with the development of an oxidizable fluorescein derivative that could be introduced into the cytoplasm as an ester (Bass et al., 1983), it has been possible to use real-time flow cytometry to follow the oxidative burst, described in more detail in Chapter 9. A cytoplasmic probe can detect only the small fraction of oxidative products that, after excretion from a cell, manages to rediffuse into the cytoplasm and to escape capture by catalases and peroxidases. Probes that reside in the phagocytic vacuole, such as reduced dihydrochlorofluorescein-labeled immune complexes, which can therefore detect a much larger fraction of secreted products, were hence developed (Ryan et al., 1990) and are now in extensive use for flow cytometry (Bass et al., 1983; Ryan et al., 1990; Robinson et al., 1988; Brunkhorst et al., 1991, 1992; Haugland and Johnson, 1993; Hardy, 1994; Holevinsky and Nelson, 1995; Strohmeier et al., 1995a,b; Guillemot et al., 1996). A line of oxidizable rhodamine-labeled entities has now been added to the fluorescein-labeled ones. Because there is always a small amount of already oxidized probe, binding of the labeled entities can be evaluated either at 4°C, a temperature at which cells produce no oxidative products, or at 37°C immediately upon binding, for the oxidative burst is usually a late (>30 s) consequence of phagocyte activation.

Probes for Secreted Granule Products

As studies of cellular functions have progressed, it has become clearer that responses to stimulation are often expressed by only a subpopulation of cells. This has made it imperative to develop probes for exocytosed products that are applicable to flow cytometric measurement, for example, probes for products trapped in a phagocytic vacuole and hence still tightly associated with the cell, or probes that are cell-linked and can report the presence of a product that has itself been secreted into the external medium. In addition to the oxidative burst indicators described above, a new line of enzyme substrates linked to a phagocytosable entity has been developed. A stimulus-linked fluorescent methylcoumarin peptide probe for human neutrophil elastase, analogous to the oxidative burst-indicating immune complex, now exists (Gewirtz and Simons, 1996).

NEUTROPHIL SIGNAL TRANSDUCTION AND DEGRANULATION

One of the earliest applications of multiparameter flow cytometry kinetics was the study of neutrophil membrane potential (ψ) and cytoplasmic calcium concentration ($[Ca^{2+}]_{in}$) responses to chemotactic (Sklar, 1985; Lazzari et al., 1986, 1990; Sullivan et al., 1988; Elsner et al., 1992, 1993) and phagocytic (Brunkhorst et al., 1991, 1992; Davies et al., 1992; Strohmeier et al., 1995a,b) stimuli, as well as to the development of such functional responses as myeloid precursors mature into functional neutrophils (Sullivan et al., 1988; Bernardo et al., 1990) or macrophages (Bernardo et al., 1993).

Early Events in Neutrophil Activation

Early studies of neutrophil activation (Korchak and Weissmann, 1980; Korchak et al., 1982) demonstrated a short, small hyperpolarization followed, after a lag period, by a much larger depolarization of the cell's transmembrane potential. It has since been shown (Lazzari et al., 1990) that this hyperpolarization is (*a*) small (approx. −4 mV), (*b*) detectable most readily at very low doses of chemotactic peptides where the depolarization is barely discernible, (*c*) specific (e.g., not mimicked by *t*-boc-chemotactic peptide), (*d*) chemotactic peptide dose-dependent, (*e*) independent of cytoplasmic [Ca^{2+}] changes, (*f*) mimicked by increased K^+ permeability of the plasma membrane, (*g*) independent of the Na^+/H^+ antiport, and (*h*) independent of Na^+/K^+ ATPase functions.

The large depolarization that follows this hyperpolarization was one of the first stimulus responses observed via fluorescent probes in real time (Korchak and Weissmann, 1980; Whitin et al., 1980; Seligmann and Gallin, 1983). It has proved easy to perform such measurements on the flow cytometer, and to correlate membrane potential changes with other stimulus responses such as $\Delta[Ca^{2+}]$ in the same cells (Lazzari et al., 1986, 1990; Ryan et al., 1990; Brunkhorst et al., 1991, 1992; Meshulam et al., 1992). In this connection, flow cytometry permits the determination of whether a given response is exhibited by all the cells or by a subpopulation of the cells, and to verify whether the presence of such subpopulations is attributable to the number of occupied receptors, that is, to study the up- or downregulation of receptors and their correlation with subpopulation responses.

For neutrophils, stimuli can be roughly divided into chemotactic versus phagocytic agonists, with the former, such as f-Met-Leu-Phe, acting through the G-protein-linked fMLP-receptor, and the latter acting either through Fc (aggregated IgG, immune complexes, IgG-opsonized particles) or C3b (complement-opsonized particles) receptors. The mechanisms by which activation is modulated differ for the three classes of receptors. Via simultaneous multiparameter analysis, cell by cell, of several stimulus responses, flow cytometry has markedly furthered the understanding of independence versus interdependence of stimulus-induced events such as changes in membrane potential, cytoplasmic cation concentrations, phospholipase activities, and/or degranulation.

All three classes of stimuli have been found to bind uniformly to all the neutrophils in a given suspension, whether the dose of stimulus is saturating or subsaturating (Lazzari et al., 1990; Brunkhorst et al., 1991, 1992). Thus if there were two forms of any of the corresponding specific receptors (e.g., two conformations, one capable of transducing a signal, the other capable only after a stimulus-induced conformational change), the transformation from one to the other must be extremely rapid. With standard commercial flow cytometers, it has been difficult to determine times of full binding of certain stimuli (e.g., f-Met-Leu-Phe) and consequent initiation of stimulus responses if these are shorter than 10–15 s owing largely to the distance between the sample chamber and the flow nozzle and the need to open the flow chamber to add the agonist. Therefore, several laboratories have adapted their flow cytometers, following a design first used by Sklar (Sklar, 1985; Lazzari et al., 1986, 1990; Ryan et al., 1990; Brunkhorst et al., 1991, 1992; Meshulam et al., 1992) to permit agonist injection into a stirred sample (Sklar, 1985; Ryan et al., 1990; Elsner et al., 1992, 1993), thereby reducing the time before the first cells reach the nozzle to <2 s; an improved apparatus for this purpose has been described (Kelley, 1989) and is now commercially available. A detailed discussion of rapid flow kinetics systems can be found in Chapter 2 of this volume. It should be noted that the difficulty in observing cells in the first second after stimulation is attributable not only to instrumental limitations and to the distance between sample compartment and nozzle, but also to factors imposed by the cells' response to shear as they are pushed rapidly through the capillary tubing. For example, we have found that neutrophils can readily be forced through the nozzle at more than 4000/s without becoming activated or primed, but this is not true for platelets, which become primed or pre-activated, sometimes even forming microaggregates, when subjected to comparable treatment.

The rate and extent of agonist binding to neutrophils are readily measured in a flow cytometer. For neutrophils, as for platelets (Davies et al., 1988) and monocytes (Bernardo et al., 1990), binding appears to be uniform to all the cells in a given suspension for all doses of stimulus (Lazzari et al., 1986, 1990; Ryan et al., 1990; Brunkhorst et al., 1991, 1992; Elsner et al., 1992, 1993). That is, even for subsaturating concentrations of a given stimulus, flow cytometric evaluation demonstrates that there is no evidence of disproportionation or differential regulation of receptors, yielding several populations expressing different numbers of liganded receptors. For fMLP, this uniform binding is rapid (Lazzari et al., 1986, 1990; Bernardo et al., 1990; Brunkhorst et al., 1992; Elsner et al., 1992, 1993) and complete within <7 s (Elsner et al., 1992, 1993); for immune complexes, it is considerably slower, achieving maximal binding only after approximately 30 s, even in a stirred sample (Ryan et al., 1990; Brunkhorst et al., 1991, 1992; Strohmeier et al., 1995a,b).

Within the limits of these time delays for maximal binding, one can use the correlation between receptor occupancy and response that flow cytometry offers to investigate at least some of the components of the mechanism by which binding to a specific receptor is translated into initiation of signal transduction. As discussed above, the parameters of cell activation whose kinetics are currently measurable by flow cytometry with commercially available probes include $\Delta\psi$, $\Delta[Ca^{2+}]_i$, ΔpH_i, $\Delta(Na^+)_i$, $\Delta(K^+)_i$, oxidative burst generation, and phospholipase A activity.

Neutrophil Activation via Chemotactic Receptors

Stimulation of neutrophils via their chemotactic peptide receptors by a range of concentrations of f-Nleu-Leu-Phe-Nleu-Tyr-Lys (a potent chemotactic peptide analogue of fMLP that can be fluorescently labeled on the lysine residue) results in uniform binding to and uniform $\Delta\psi$ and $\Delta[Ca^{2+}]_i$ responses by all of the cells, for peptide concentrations ranging from 10^{-13} to 10^{-8} M, 10^{-9} M being a saturating dose of this stimulus (Lazzari et al., 1986, 1990; Ryan et al., 1990; Elsner et al., 1992, 1993). At doses below 10^{-11} M, the $\Delta[Ca^{2+}]_i$ transient cannot be detected, but a small hyperpolarization can; at higher peptide doses, both a hyperpolarization and a $\Delta[Ca^{2+}]_i$ transient appear simultaneously in all the neutrophils (Lazzari, 1986, 1990), attaining their maximal values within 6–7 s of peptide addition (Lazzari, 1986, 1990; Bernardo et al., 1990; Elsner et al., 1992, 1993) (Fig. 1). The magnitude of both changes depends on the dose of peptide used to elicit them, yielding a normal hyperbolic saturation curve. By using fluorescein-or rhodamine-labeled peptides one can follow their binding and determine that maximal binding is also achieved within the same time period (Bernardo et al., 1990; Lazzari et al., 1990; Elsner et al., 1992, 1993). Fur-

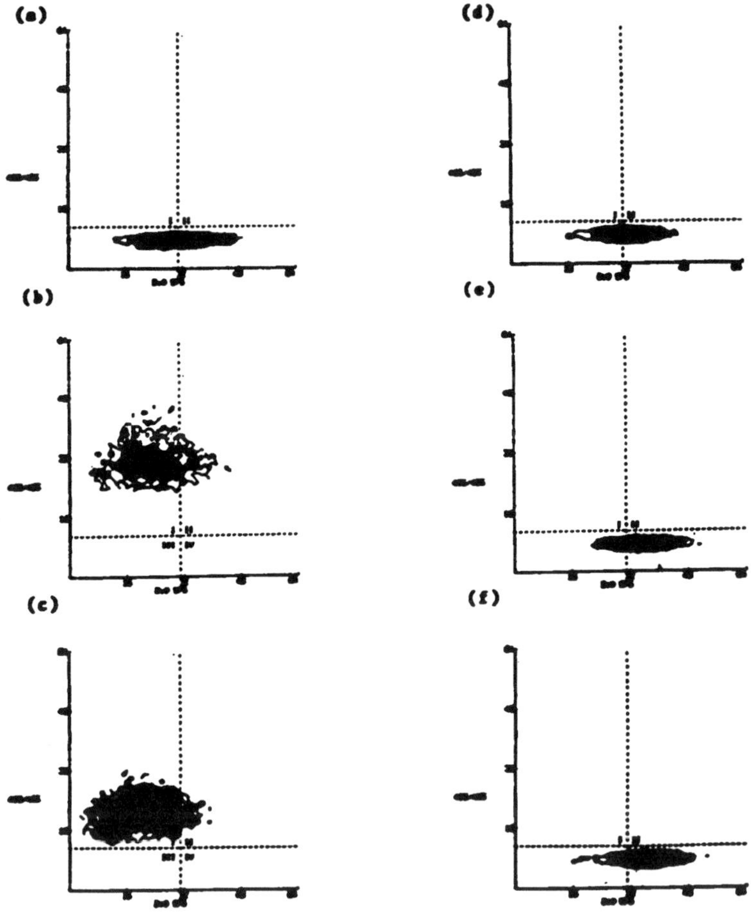

Fig. 1. Neutrophil fluorescence distribution changes in [Ca^{2+}] and membrane potential during fMLP stimulation. Population distributions displaying simultaneous membrane potential changes (DiO-C_3-(3) fluorescence) on the *horizontal axis* and cytosolic [Ca^{2+}] changes (indo-1 fluorescence ratio) on the *vertical axis*. The fluorescence intensity of each cell is indicated on the respective axis, from 0 to 64 relative fluorescence units. Each frame represents a total of 2000 cells. *a* and *d* depict the resting cell distribution, in *b* and *e* are cells 15 s after stimulation with 10^{-7} and 10^{-11} *M* fMLP, respectively, and *c* and *f* are cells 30 s after stimulation. From Lazzari et al, 1990

thermore, flow cytometric studies, using the previously mentioned bis(BODIPY™) phosphatidylcholine probe, also demonstrated unequivocally that phospholipase A was not activated within the first 2.5 min after fMLP addition (Meshulam et al., 1992). Additionally, as anticipated in view of the vulnerability to oxidative destruction of chemotactic peptides, stimulation with an oxidizable fluorescein-labeled peptide, even at 10^{-7} *M* (a saturating dose), failed to elicit an oxidative burst in the absence of cytochalasin.

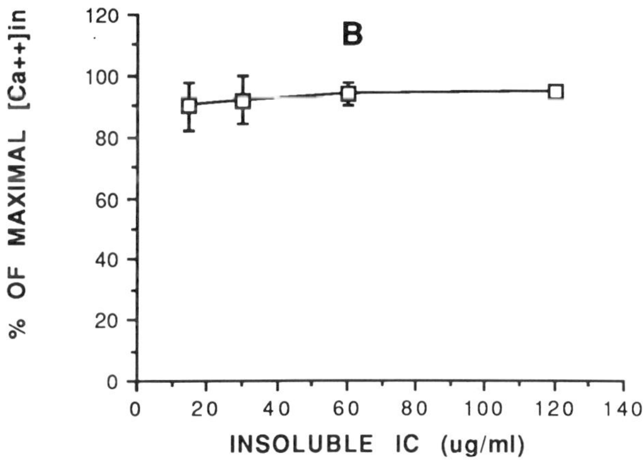

Fig. 2. The maximal $[Ca^{2+}]_{in}$ increase in $[Ca^{2+}]_{in}$-responding cells exposed to various doses of SIC and IIC. Data were calculated using maximal increases of $[Ca^{2+}]_{in}$ elicited by ≥750 μg/ml SIC or 120 μg/ml IIC as 100% response and were from more than three separate experiments. *, $p < .05$. From Brunkhorst et al, 1991

Neutrophil Stimulation via Fc Receptors

Neutrophil activation via its Fc receptors, that is, with multivalent immune complexes or aggregated IgG, exhibited uniform binding to and Δψ changes in all the cells, but a $[Ca^{2+}]$ transient in only a fraction of the same cells (Brunkhorst et al., 1991, 1992; Strohmeier et al., 1995a,b). The magnitude of the response was maximal in the

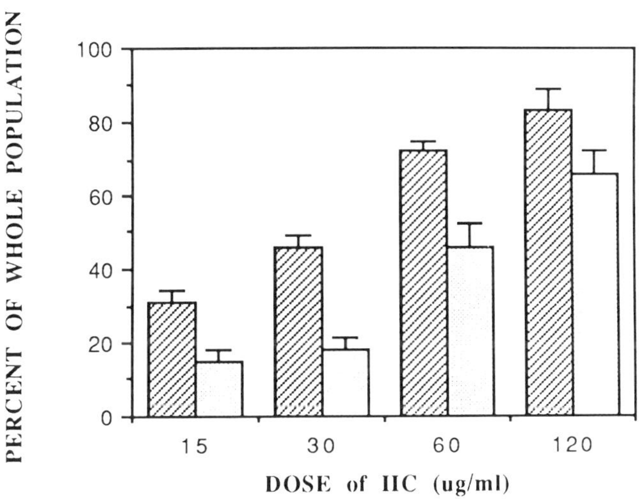

Fig. 3. Expression of increases in $[Ca^{2+}]_{in}$ within the F-IIIC-labeled neutrophil population. □, the percent of the whole population of cells that bound F-IIIC; □, the percent of the whole population that exhibited an increase in $[Ca^{2+}]_{in}$. All data represent mean ±SE and are from three or more separate experiments. From Ryan et al, 1991

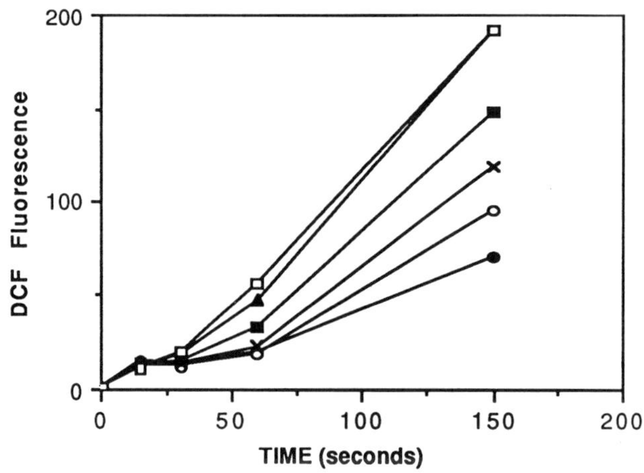

Fig. 4. Stimulus dose response for stirred samples of DCFH-IC-activated human neutrophils. Stirred cell suspensions of neutrophils were exposed to a range of complex concentrations (60–480 μg/ml) and monitored for 2.5 min. 60 (●), 120 (○), 180 (X), 240 (■), 360 (▲), 480 (□) μg/ml. From Brunkhorst et al, 1991

responding cells and nil in the nonresponding ones at any given subsaturating dose, that is, there was an all-or-none response (Figs. 2 and 3). However, a higher dose elicited this response in a larger proportion of the neutrophils, leading eventually to a hyperbolic relationship between the dose of immune complex used and the fraction of cells responding with a $\Delta[Ca^{2+}]$ transient. The relationship between immune complex dose and generation of an oxidative burst followed the same disproportionation mechanism (Fig. 4), but the change in membrane potential did not, $\Delta\Psi$ being exhibited to the same immune complex dose-dependent extent in all the cells (Ryan et al., 1990; Brunkhorst et al., 1991, 1992; Strohmeier et al., 1995a,b).

In contrast to fMLP, multivalent immune complexes do activate phopholipase A (Fig. 5), starting approximately 30 s after their addition to neutrophils, and only in those cells (Fig. 6) that have already exhibited the transient $\Delta[Ca^{2+}]_i$ response (Meshulam et al., 1992).

The flow cytometric studies described above have led to a much better understanding of the mechanisms of neutrophil activation via the fMLP versus Fc receptors and of the interdependence of the various events associated with signal transduction. Among the conclusions that have been reached are the following:

1. Activation via the fMLP receptor proceeds uniformly in all the cells, a hyperpolarization being detectable at a much lower stimulus dose than $\Delta[Ca^{2+}]_i$.
2. Activation of neutrophils via the fMLP receptor does not elicit an oxidative burst or initiate degranulation, nor does it activate a phospholipase A.
3. Activation via the Fc receptor leads to a disproportionation into both $\Delta[Ca^{2+}]$-responding and nonresponding subpopulations. Because the latter exhibit identical ligand (immune complex) binding, the difference between the sub-

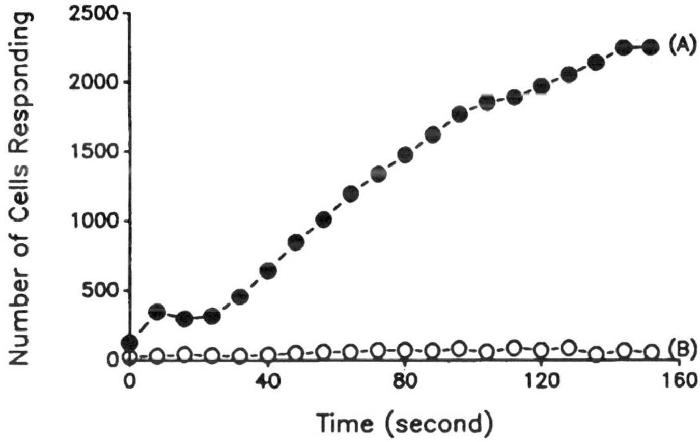

Fig. 5. Representative curves indicating the proportion of cells exhibiting PLA activation in fMLP versus IIC-stimulated neutrophils, as determined by flow cytometry. Neutrophils loaded with bis-BODIPY™-PC and Indo-1 were stimulated with 60 μg/ml IIC (●) or 100 n*M* fMLP (○) at t = 0. From Meshulam et al, 1992

populations cannot be attributed to extent of receptor occupancy and no cooperativity or receptor upregulation appears to occur.

4. Activation via cross-linked Fc receptors elicits an oxidative burst in the same subpopulation of neutrophils, but 15–20 s later, that exhibits a $\Delta[Ca^{2+}]_i$.
5. Although neutrophils do exhibit a stimulus-induced membrane potential change, it is not necessarily accompanied by a $\Delta[Ca^{2+}]_i$, indicating that these are independent and separable events.
6. Phospholipase A is not activated by chemotactic peptides, but is activated by immune complexes, acting via the Fc receptor, only in neutrophils whose cytoplasm has already undergone a $[Ca^{2+}]_i$ transient.
7. Efficient activation via the Fc receptors requires their cross-linking by multivalent stimuli; monovalent stimuli (i.e., those exhibiting a single Fc terminus) elicit a very small $\Delta[Ca^{2+}]_i$ transient and barely degranulate.

The specific class of Fc receptors involved in signal transduction and in degranulation control has recently been identified. Using specific antibodies to FcγRII (monoclonal antibody IV.3) and FcγRIII (monoclonal antibody 3G8) to block each class separately, it has now been demonstrated that FcγRIII glycosyl phosphatidyl inositol-anchored receptors, control most of the neutrophil responses to multivalent immune complexes whereas FcγRII functions mainly as a modulator of neutrophil activation (Brunkhorst et al., 1992; Strohmeier et al., 1995a,b) and a downregulator of $\Delta\ [Ca^{2+}]_{in}$ (Van den Herik-Oudijk et al., 1995). The difference between the anchoring of FcγRIII on the one hand, and the presence of a cytoplasmic tail and a number of phosphorylation sites on FcγRII and the fMLP receptors on the other, may yield a

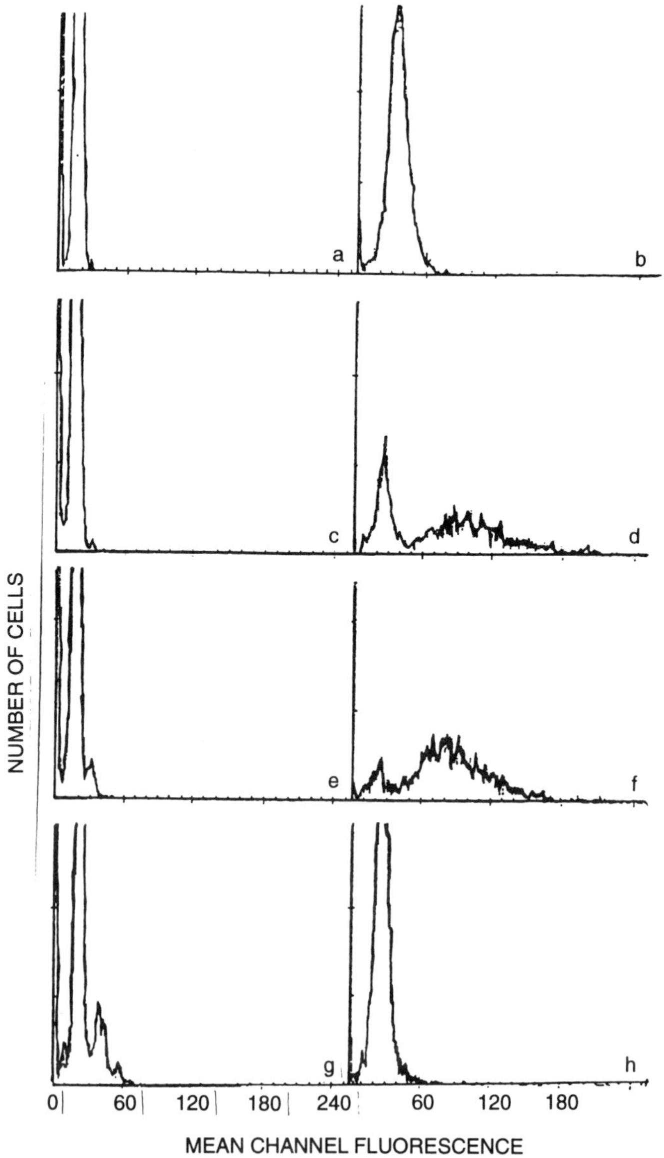

Fig. 6. Representative time course of the IIC-induced changes in bis-BODIPY™-PC fluorescence and in [Ca^{2+}] response, in the same cell, as measured by flow cytometry. 60 μg/ml IIC was injected into neutrophils loaded with both bis-BODIPY™-PC and Indo-1 in the flow cytometer. Measurements were made, 4000 cells per time point, at 4-s intervals. Both BODIPY fluorescence (λ excitation = 488, λ emission = 530 nm) and Indo-1 fluorescence ratio (λ excitation = 357; λ emission = 405 and 485 nm) were recorded for each cell. These fluorescences are shown in panels *a* and *b*, respectively, at 0 s, *c* and *d* at 32 s, *e* and *f* at 42 s (the time at which the ratio and therefore [Ca^{2+}]$_{in}$ is maximal), and *g* and *h* at 230 s (a time at which [Ca^{2+}]$_{in}$ has redistributed to a final concentration of 200 n*M*, but BODIPY fluorescence is still rising, that is, PLA is still active). From Meshulam et al, 1992

clue to the control of degranulation of neutrophils. There appears, futhermore, to be a difference in activation mechanisms when FcγRII- and FcγRIII-mediated activation is initiated through their role as Fcγ receptors rather than as antigens for the specific monoclonal antibodies, IV.3 and 3G8, respectively, that recognize them; for this reason, studies aimed at the elucidation of the specific role of each as an Fcγ receptor are generally performed by blocking the other antibody with non-Fc containing Fab or Fab_2 fragments.

ACQUISITION OF SIGNAL TRANSDUCTION CAPABILITIES

The hyperpolarization discussed above was one of the few fMLP responses detectable in undifferentiated human bone marrow precursor cells (Sullivan et al., 1988) that were incapable of exhibiting any $\Delta[Ca^{2+}]_{in}$, ΔpH_{in}, or $\Delta\psi$ responses. As these precursors differentiated and matured into neutrophils in the presence of GM–CSF, their responses were measured daily. Over a period of 7 days, these cells acquired the ability to mount a $\Delta[Ca^{2+}]_{in}$, ΔpH_{in}, and depolarization response, with the latter eventually almost obscuring the hyperpolarization noted earlier (Sullivan et al., 1988). Similar studies have now demonstrated that undifferentiated cells such as HL-60 cells (Bernardo et al., 1990) (Fig. 7), circulating monocytes (Bernardo et al., 1988, 1990, 1993) or naive lymphocytes (Macauley and Simons, 1991) cannot mount these responses but acquire the ability to do so as the cells mature into neutrophils, macrophages, or memory B-cells, respectively. It is not yet clear, in most of these cases, whether this phenomenon is due to incomplete assembly of existent signal transduction components in the right conformation, to the presence of these components in an inactive form (e.g., putative or unexpressed receptors), or to induction of synthesis of these components by the growth factors and cytokines present in the differentiation incubate. For HL-60 cells, it has been shown that at least part of the maturation process involves increased surface expression of the fMLP receptor (Bernardo et al., 1990; Laskin et al., 1991) although the extra receptors alone cannot account for the $\Delta\Psi$, $\Delta[Ca^{2+}]_i$, and oxidative burst responses noted in the mature but not the immature cells. It is therefore likely that maturation of these cells also involves the conversion of putative components into an actual functional signal transduction system. A more detailed discussion of the HL-60 model can be found in chapter 11.

Somewhat similar studies have been undertaken with mononuclear cells. As these mature into macrophages over a period of several days, they acquire the ability to respond with enhanced $\Delta\Psi$, $\Delta[Ca^{2+}]_i$, and oxidative burst, whereas, interestingly, their chemotactic response and their ΔpH_i response are invariant over the same time period (Bernardo et al., 1988, 1990, 1993). Furthermore, this ability to mature and to acquire the functional responses of a macrophage depends on adherence; circulating mononuclear cells (monocytes) isolated by adherence to plastic and incubated for several days while adhering, acquire signal transduction and degranulation responses, whereas cells isolated negatively (removal of all others with magnetic bead technology) and kept in suspension in Teflon beakers over the same time period do not mature and retain their minimal responsiveness (Bernardo et al., 1996).

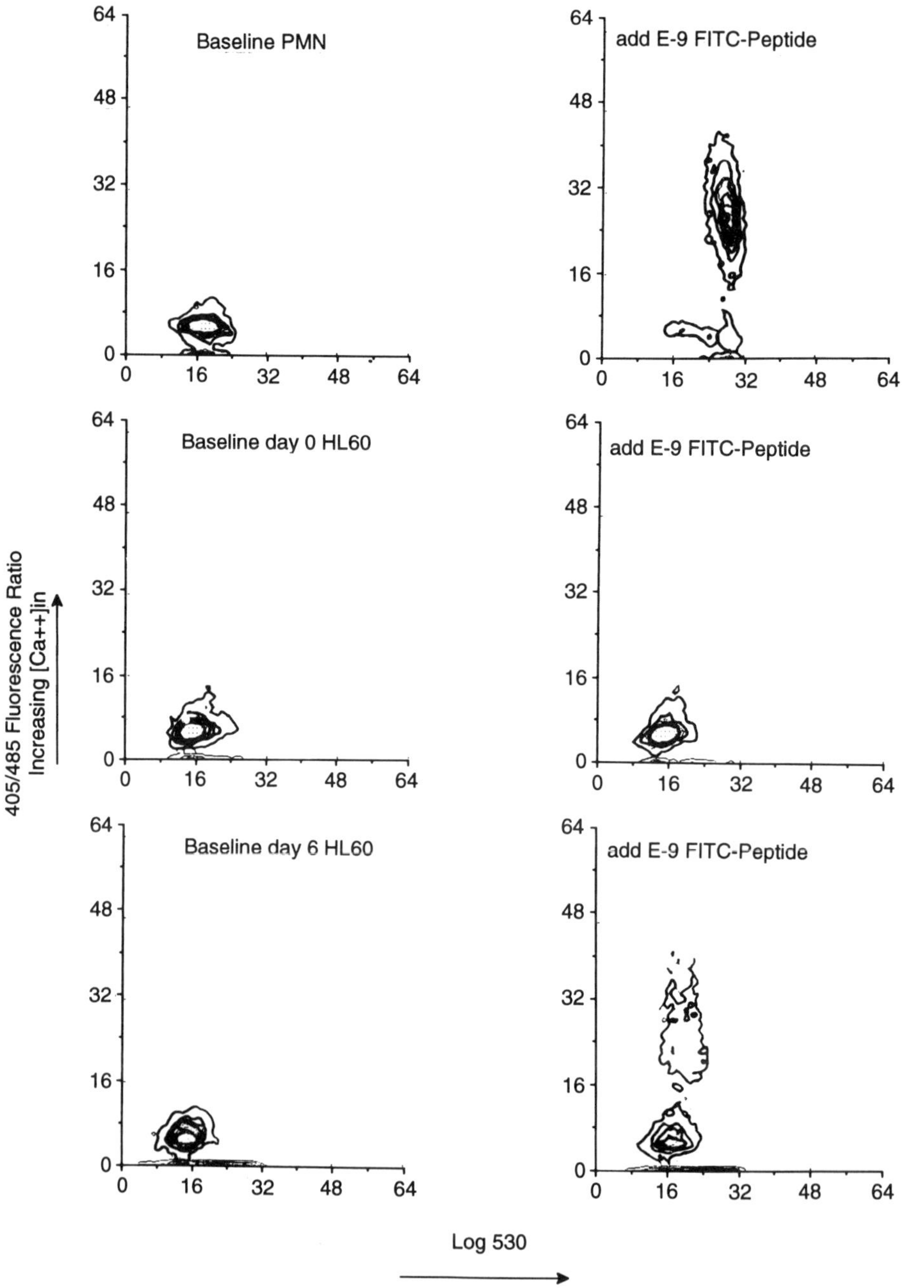

Fig. 7. Measurements of intracellular [Ca^{2+}] (Indo-1 fluorescence at 405/485 nm; ordinate) vs. log 530 nm FITC fluorescence emission (excitation wavelength 488 nm: abscissa) before (left column) and 8 s following stimulation with FITC-labeled chemotactic peptide (10^{-9} *M*) in neutrophils (top row) and in d 0 (middle row) and d 6 (bottom row) HL-60 granulocytes. From Bernardo et al, 1990

SUMMARY

The use of flow cytometric techniques has permitted significant advances in the elucidation of neutrophil activation mechanisms, to determine the role of specific receptors and the expression of functional responses in individual cells and to follow the acquisition of responses as precursor cells mature. These techniques are those of choice, where probes are available, because they permit determination of multiple facts about a given cell at the same time, and therefore permit correlation of these facts, for example, occupancy of a given receptor type on a single cell with the resultant responses of that individual cell. In addition to this obvious advantage, flow cytometric measurements permit investigation of a small subpopulation of cells that for any of a number of reasons cannot be isolated in a pure state. They are also proving to be useful in studies of transformed or transfected cells that do not express different surface markers from the wild type cell, but do exhibit different functions or responses. As more and more knockout techniques or other modifications that involve a cell's proteins but leave the surface intact are attempted, flow cytometric techniques are bound increasingly to be the methods of choice in analyzing the cell's functions.

ACKNOWLEDGMENTS

I am grateful to all of my associates, graduate students, and post-doctoral fellows without whose contributions to both the experiments and their interpretations this chapter could not have been written. Special thanks go to Dr. Theresa Davies not only for her scientific contributions but also for her critical reading of this manuscript and preparation of its bibliography. Support of those studies described here that were conducted in my laboratory was from N.I.H. grants HL-07501, DK-31056, and HL-19717.

REFERENCES

Bass D, Parce J, Dechatelet L, Szejda P, Seeds M, Thomas M (1983): Flow cytometric studies of oxidative product formation by neutrophils: a graded response to membrane stimulation. J Immunol. 130(4): 1910–1917.

Bernardo J, Billingslea A, Ortiz MF, Seetoo K, Macauley J, Simons ER (1997): Adherence-dependent calcium signaling in monocytes: Induction of a CD-14-high phenotype, formyl peptide responsive, subpopulation. J Leukoc Biol in press.

Bernardo J, Brennan L, Brink HF, Ortiz MF, Newburger PE, Simons ER (1993): Chemotactic peptide-induced cytoplasmic pH changes in incubated human monocytes. J Leukoc Biol 53:673–678.

Bernardo J, Brink HF, Simons ER (1988): Time dependence of transmembrane potential changes and intracellular calcium flux in stimulated human monocytes. J Cell. Physiol 134:131–136.

Bernardo J, Newburger PE, Brennan L, Brink HF, Bresnick SA, Weil G, Simons ER (1990): Simultaneous flow cytometric measurements of cytoplasmic Ca^{2+} and membrane potential changes upon fMLP exposure as HL-60 cells mature into granulocytes: using $[Ca^{2+}]_{in}$ as an indicator of granulocyte maturity. J Leukoc Biol 47:265–274.

Brunkhorst BA, Lazzari KG, Strohmeier G, Weil G, Simons ER (1991): Calcium changes in immune complex-stimulated human neutrophils: Stimultaneous measurement of receptor occupancy and activation reveals full population stimulus binding but subpopulation activation. J Biol Chem 266(20):13035–13043.

Brunkhorst BA, Strohmeier G, Lazzari K, et al. (1992): Differential roles of FcγRII and FcγRIII in immune complex stimulation of human neutrophils. J Biol Chem 267:20659–20666.

Davies TA, Bernardo J, Lazzari K, Brennan L, Simons ER (1992): Cytosolic calcium determination: a fluorometric technique. J Nutr Biochem 2:102–106.

Davies TA, Billingslea A, Johnson R, Greenberg S, Ortiz M, Long H, Sgro K, Tibbles H, Seetook, Rathbun W, Schonhorn J, Simons ER (1997): Stimulus Responses and amyloid precursor protein processing in DAMI megakaryocytes. J Lab Clin Med 130:21–32.

Davies TA, Drotts D, Weil GJ, Simons ER (1988): Flow cytometric measurement of cytoplasmic calcium changes in human platelets. Cytometry 9:138–142.

Davies TA, Drotts DL, Weil GJ, Simons ER (1989): Cytoplasmic Ca^{2+} is necessary for thrombin-induced platelet activation. J Biol Chem 264:19600–19606.

Davies TA, Dunn JM, Simons ER (1987): Evaluation of changes in cytoplasmic pH in thrombin-stimulated human platelets. Anal Biochem 167:118–123.

Davies TA, Weil GJ, Simons ER (1990): Simultaneous flow cytometic measurements of thrombin-induced cytosolic pH and Ca^{2+} fluxes in human platelets. J Biol Chem 265:11522–11526.

Demaurex N, Schrenzel J, Jaconi ME, Lew DP, Krause KH (1993): Proton channels, plasma membrane potential, and respiratory burst in human neutrophils. Eur J Haematol 51:309–312.

Elsner J, Kaever V, Emmendorffer A, Breidenbach T, Lohmann-Matthes ML, Roesler J (1992): Heterogeneity in the mobilization of cytoplasmic calcium by human polymorphonuclear leukocytes in response to fMLP, C5a and IL-8/NAP-1. J Leukoc Biol 51:77–83.

Elsner J, Norgauer J, Dobos GJ, et al. (1993): Flow cytometry reveals different lag times in rapid cyoplasmic calcium elevations in human neutrophils in response to N-formyl-peptide. J Cell Physiol 157:637–643.

Fletcher MP, Seligmann BE (1985): Monitoring human neutrophil granule secretion by flow cytometry: secretion and membrane potential changes assessed by light scatter and a fluorescent probe of membrane potential. J Leukoc Biol 37:431–447.

Gewirtz AT, Simons ER (1997): Phospholipase D mediates Fc receptor activation of neutrophils and provides signaling specificity between chemotactic and phagocytic signaling pathways. J Leukoc Biol 61:522–528.

Grynkiewicz G, Poenie M, Tsien RY (1985): A new generation of Ca^{2+} indicators with greatly improved fluorescence properties. J Biol Chem 260(6):3440–3450.

Guillemot JC, Kruskal BA, Adra CN, et al. (1996): Targeted disruption of guanosine diphosphate-dissociation inhibitor for Rho-related proteins, GDID4: normal hematopoietic differentiation but subtle defect in superoxide production by macrophages derived from in vitro embryonal stem cell differentiation. Blood 88(7):2722–2731.

Hardy MM (1994): Superoxide dismutase mimetics inhibit neutrophil-mediated human aortic endothelial cell lung injury in vitro. J Biol Chem 269:1835.

Haugland RP, Johnson ID (1993): Detecting enzymes in living cells using fluorogenic substrates. Fluorescence 3:119.

Hendrickson HS (1994): Fluorescence-based assays of lipases, phospholipases, and other lipolytic enzymes. Anal Biochem 219:1.

Holevinsky KO, Nelson DJ (1995): Simultaneous detection of free radical release and membrane current during phagocytosis. J Biol Chem 270:8328.

Kelley KA (1989): Sample station modification providing on-line reagent addition and reduced sample transit time for flow cytometers. Cytometry 10:796–800.

Korchak HM, Eissenstat BA, Smolen JE, Rutherford LE, Dunhamn PB, Weissman G (1982): Stimulus response coupling in the human neutrophil: the role of anion fluxes in degranulation. J Biol Chem 257:6919–6922.

Korchak HM, Weissmann G (1980): Stimulus response coupling in the human neutrophil: Transmembrane potential and the role of extracellular Na^+. Biochim Biophys Acta 601:180–194.

Laskin DL, Sirak AA, Robertson FM, Laskin JD (1991): Distinct patterns of chemotactic peptide-induced calcium mobilization in differentiated myeloid leukemia cells and peripheral blood neutrophils. J Leukoc Biol 49:369–379.

Lazzari KG, Proto P, Simons ER (1990): Neutrophil hyperpolarization in response to a chemotactic peptide. J Biol Chem 265(19):10959–10967.

Lazzari KG, Proto PJ, Simons ER (1986): Simultaneous measurement of stimulus-induced changes in cytoplasmic Ca^{2+} and in membrane potential of human neutrophils. J Biol Chem 261(21):9710–9713.

Luscinskas FW, Mark DE, Brunkhorst B, Lionetti FJ, Cragoe EJ Jr, Simons ER (1988): The role of transmembrane cationic gradients in immune complex stimulation of human polymorphnuclear leukocytes. J Cell Physiol 134:211–219.

Macauley JB, Simons ER (1991): Differences in calcium responsiveness in memory vs. naive CD4+ lymphocytes from human peripheral blood upon crosslinking of CD3 and CD4 surface antigens. J Cell Biol 115:272a.

Meshulam T, Herscovitz H, Casavant D, et al. (1992): Flow cytometry kinetic measurements of neutrophil phospholipase A activation. J Biol Chem 267:21465–21470.

Nasmith PE, Grinstein S (1986): Impairment of Na^+/H^+ exchange underlies inhibitory effect of Na^+-free media on leukocyte function. FEBS Lett. 202:79–85.

Omann GM, Porasik-Lowes MM (1991): Graded G-protein uncoupling by pertussis toxin treatment of human polymorphonuclear leukocytes. J Immunol 146:1303–1308.

Robinson JP, Bruner LH, Bassoe C-F, Hudson JL, Ward PA, Phan SH (1988): Measurement of intracellular fluorescence of human monocytes relative to oxidative metabolism. J Leukoc Biol 43:304–310.

Ryan TC, Weil GJ, Newburger PE, Haugland R, Simons ER (1990): Measurement of superoxide release in the phagovacuoles of immune complex-stimulated human neutrophils. J Immunol Meth 130:223–233.

Schonhorn JE, Simons ER (1989): The role of the Ca^{++} transient in neutrophil activation. J Cell Biol 109: 221a.

Seligmann BE, Gallin JI (1983): Comparison of indirect probes of membrane potentials utilized in studies of human neutrophils. J Cell Physiol 115:105–115.

Shapiro HM (1995): *Practical Flow Cytometry*. New York: Alan R. Liss, Inc.

Simchowitz L, Spilberg I, de Weer P (1982): Sodium and potassium fluxes and membrane potential of human neutrophils. J Gen Physiol 79:453–479.

Simons ER (1993): *Fluorescent and Luminescent Probes for Biological Activity*. San Diego: Academic Press.

Simons ER, Davies TA (1994): Identification and characterization of functional secretory cells: Advantages of multiparameter flow cytometry kinetics. In *Advances in Molecular and Cell Biology*. JAI Press Inc., pp. 151–170.

Sklar LA (1985): Signal transduction and ligand-receptor dynamics in the neutrophil. J Biol Chem 260(21): 11468–11475.

Strohmeier GR, Brunkhorst BA, Seetoo KF, Bernardo J, Weil GJ, Simons ER (1995a): Neutrophil functional responses depend on immune complex valency. J Leukoc Biol 58:403–414.

Strohmeier GR, Brunkhorst BA, Seetoo KF, Meshulum T, Bernardo J, Simons ER (1995b): Role of FcγR subclasses FcγRII and FcγRIII in the activation of human neutrophils by low and high valency immune complexes. J Leukoc Biol 58:415–422.

Sullivan R, Griffin JD, Wright J, et al. (1988): Effects of recombinant human granulocyte-macrophage colony-stimulating factor on intracellular pH in mature granulocytes. Blood 72:1665–1673.

Thomas JA, Buchsbaum RN, Zimniak A, Racker E (1979): Intracellular pH measurements in Ehrlich Ascites tumor cells utilizing spectroscopic probes generated in situ. Biochemistry 18:2210–2218.

Tsien RY (1989): Fluorescent probes of cell signaling. Annu Rev Neurosci 12:227–253.

Van den Herik-Oudijk IE, Capel PJ, van der Bruggen T, Van de Winkel JG (1995): Identification of signaling motifs within human Fc gamma RIIa and Fc gamma RIIb isoforms. Blood 85(8):2202–2211.

Waggoner AS (1988): *Mechanisms of Membrane Potential Probes*. New York: Rockefeller University Press.

Whitin JC, Clark RA, Simons ER, Cohen HJ (1980): Effects of myeloperoxidase system on fluorescent probes on granulocyte membrane potential, J Biol Chem 256:8904–8906.

5

Lipopolysaccharide-Binding Proteins

Robert C. Bass
University of Cincinnati College of Medicine and the Shriners Burns Institute, Cincinnati, Ohio

Robert J. Burnett, MD
University of Cincinnati College of Medicine and the Shriners Burns Institute, Cincinnati, Ohio

Alice N. Neely, PhD
University of Cincinnati College of Medicine and the Shriners Burns Institute, Cincinnati, Ohio

Joseph S. Solomkin, MD
University of Cincinnati College of Medicine and the Shriners Burns Institute, Cincinnati, Ohio

Gram-negative infections are associated with significant illness in humans. Such infections, particularly if bacteremic, result in a syndrome of fever, hypotension, coagulopathy, and organ injury (Bone, 1991). This clinical picture is now commonly called the sepsis syndrome (Bone et al., 1989). Statistics from the National Hospital Discharge Survey prepared by the Centers for Disease Control revealed that septicemia discharges totaled 2,570,000 from 1979 to 1987. During the reporting period, septicemia rates increased 139%, from 73.6 per 100,000 persons in 1979 to 175.9 in 1987 (Anonymous, 1990). The mortality associated with this discharge diagnosis was 25%.

Phagocyte Function: A Guide for Research and Clinical Evaluation, Edited by J. Paul Robinson and George F. Babcock. ISBN 0-471-12364-1

from 20 years ago have suggested that gram-negative septicemia alone has an incidence of 70,000 to 300,000 cases per year (McCabe, 1974; Wolf and Bennett, 1974). A current estimate of annual incidence would be 400,000 cases of septicemia, 200,000 bouts of septic shock, and 100,000 deaths from this disease (Parillo et al., 1990).

The pathogenesis of this syndrome has been the subject of intensive research efforts. The discovery of cytokine networks involving interstitial and alveolar macrophages and epidermal, mesenchymal, and endothelial cells has provided a novel regulatory system for the induction of acute inflammation (Barker et al., 1991; Elner et al., 1991; Goodman et al., 1992b; Kunkel et al., 1991a, 1991b; Rankin et al., 1990; Rolfe et al., 1991; Schroder et al., 1990; Sica et al., 1990; Standiford et al., 1990; Thornton et al., 1990). Gram-negative infection has provided an important clinical model for study of the consequences of the interplay between elements in these cytokine networks. Based on current notions of this system, therapeutic agents intended to improve the consequences of gram-negative infection have been developed and are in various phases of clinical testing.

Recent theories regarding the involvement of cytokines in septic shock have centered upon the role of elevated plasma levels of cytokines, most notably TNF-α, IL-1, and IL-6 (Calandra et al., 1991; Damas et al., 1992; Dofferhoff et al., 1992; Girardin et al., 1988, 1992; Hack et al., 1989, 1992; Hamilton et al., 1992; Waage et al., 1989). Various independent clinical studies have identified an association between bacteremia, endotoxemia, elevated cytokine levels, and clinical shock (Brandtzaeg et al., 1989; Danner et al., 1991; Levin et al., 1970; Shenep et al., 1988; Stumacher et al., 1973; Van Deventer et al., 1988; Wortel et al., 1992). The preclinical background includes studies showing elevations in plasma TNF-α, IL-6, and other cytokines and cytokine regulators in animals and human volunteers infused with endotoxin, and the consequences of intravenous infusion of cytokines in both animals and humans (Fong et al., 1989; Granowitz et al., 1991; Hesse et al., 1988; Martich et al., 1991; Michie et al., 1988; van der Poll et al., 1992; van Deventer et al., 1990; Van Zee et al., 1991). Furthermore, the efficacy of an anti-endotoxin monoclonal antibody is confined to patients with shock, endotoxemia, and elevated cytokine levels (Baumgartner et al., 1990; Ziegler et al., 1991).

Conversely, these data argue that an intravascular basis for the pathogenesis of late organ failure syndromes and the sepsis syndrome without shock is unlikely. Elevations in plasma endotoxin levels are not found in the great majority of such patients. It is probable that many forms of clinical sepsis evolve over a time interval sufficient to allow downregulation of responsiveness to endotoxin, as observed in animal studies of repetitive endotoxin injection (Mathison et al., 1990). These results are paralleled in studies of patients with infection that reveal an absence of plasma cytokine response except, as described above, in the presence of acute bacteremia or its likely surrogate, endotoxemia.

Certainly our insight into the role of cytokine-mediated processes in subacute events such as organ failure in intact hosts is limited. Current notions of whole organism responses to endotoxin are in large part derived from in vitro experiments examining responses of purified cell populations in isolation or acute infusion of pharmacological doses of endotoxins or mediators into otherwise healthy animal species

(Beutler et al., 1985; Hinshaw et al., 1990; Smedegard et al., 1989; Tracey, 1987a,b). In these experimental systems, important potential downregulatory systems are either not present or overcome by the supraphysiological doses employed. Complicating the understanding of the complex series of events resulting in clinical sepsis is the recognition that many biologically relevant mediators of inflammation have overlapping effects that depend on the state of the target cells as well as on the presence or absence of other cytokines (Elias et al., 1990).

One approach to understanding the involvement of cytokines and endotoxin in clinical disease states is to examine mechanisms that might produce compartmentalization of endotoxin. This approach appears particularly suited to determining the basis for the syndrome of multiple organ failure, a sequela of major infection. A variety of plasma and cellular export proteins are capable of binding endotoxin. These proteins serve multiple functions, including providing ligands for specific cell surface receptors. It is probable that cellular, organ, and organism response to endotoxin depends on the particular clearance mechanisms present in the individual compartment in which endotoxin is encountered. The purpose of this chapter is to describe currently recognized plasma protein and cellular mechanisms for binding endotoxin. The central thesis we hope to develop is that interactions of various cell populations with varying lipopolysaccharide (LPS)-binding protein-specific receptors may serve to define the individual response to endotoxin.

LIPOPOLYSACCHARIDE–PLASMA PROTEIN INTERACTIONS

Endotoxin is known to bind to a number of plasma proteins that regulate the interaction of endotoxin with the host. These plasma proteins can be divided into at least four classes. These include the high- and low-density lipoproteins, the lipopolysaccharide-binding protein (LBP)/bactericidal permeability-increasing protein (BPI) family of lipid-binding proteins, complement, and a novel class of proteases termed septins. These interactions appear to be necessary steps in the recognition of LPS by different cell populations, induction of specific cellular responses, and the clearance and detoxification of LPS.

BINDING OF LIPOPOLYSACCHARIDE TO LIPOPROTEINS

It has long been recognized that plasma interacts with LPS and decreases its toxicity (Rudbach and Johnson, 1964; Skarnes, 1968). The plasma components involved in this detoxification remained unknown until the late 1970s. At that time, Ulevitch and colleagues characterized the biophysical changes that occurred to LPS in serum and noted a decrease in both the buoyant density and the endotoxic activities of LPS (Ulevitch and Johnston, 1978; Ulevitch et al., 1979). In the absence of serum, LPS preparations are highly aggregated with buoyant densities approximating 1.4 g/cm^3. However, in experiments in which LPS from either *Escherichia coli* 0111:B4 or *Salmonella minnesota* Re595, which lacks the polysaccharide chain, was added to plasma or sera, a marked reduction in the buoyant density of the parent LPS to <1.2 g/cm^3 occurred (Ulevitch and Johnston, 1978). Delipidation of the plasma or sera prevented this density shift, which could be restored only by addition of physiological amounts of high-density lipoprotein (HDL), but not low-density or very-low-density lipoproteins (LDL, VLDL). The ability of normal plasma to inhibit the febrile response, neutropenia, and thrombocytopenia induced by LPS was also lost after delipidation but was restored by addition of HDL (Ulevitch et al., 1979). Following immunoprecipitation of Re595–serum mixtures, amino acid analysis of the LPS–protein complex revealed apolipoprotein A-I, the major protein component of HDL (Porat et al., 1992).

Rudbach and Johnson had originally proposed disaggregation of the LPS to explain the size reduction with concomitant LPS detoxification by sera. Cesium chloride density centrifugation of LPS plus delipidated sera indicated that LPS had undergone a change in particle size in the delipidated sera. Lipopolysaccharide therefore undergoes modification, probably disaggregation, in serum prior to interaction with HDL (Ulevitch et al., 1979). The recently isolated plasma protein lipopolysaccharide binding protein (LBP) delays binding of LPS to HDL (Tobias et al., 1985; Tobias and Ulevitch, 1983) and may represent the disaggregating factor in plasma. Additional plasma factors in animals made tolerant to endotoxin may act to hasten lipoprotein binding and detoxification (Moreau and Skarnes, 1973). Warren et al. (1988) suggested that a macrophage factor may be secreted in response to LPS that acts to increase lipoprotein binding to LPS.

The interaction of HDL with LPS has a number of biologic effects. LPS-HDL is less active than LPS alone in the *Limulus* amebocyte lysate (LAL) assay (Warren et al., 1985), and less active than LPS in the induction of neutropenia, thrombocytopenia, complement activation, fever, hypotension, and death in adrenalectomized mice (Ulevitch and Johnston, 1978; Ulevitch et al., 1979). However, the complexed LPS can still induce lethal hypotensive shock and activate B cells and macrophages (Ulevitch and Johnston, 1978). Lipopolysaccharide–HDL primes neutrophils for oxidative responses at a slower rate than does LPS-LBP, but eventually higher levels of priming are reached than with LPS alone (Vosbeck et al., 1990). The complex has been shown to be 10–100 times less active in inducing TNF production by monocytes than the parent LPS (Mathison et al., 1988), and is less active in stimulating IL-1 synthesis in THP-1 cells and human monocytes (Baumberger et al., 1991). In addition, complexation of LPS with HDL results in a decrease in circulating half-life for rough but not smooth LPS. (Injected smooth LPS complexes rapidly with HDL and therefore preincubation does not shorten its already brief half-life.) Clearance of a bolus of radiolabeled Re595 LPS injected into the circulation showed $t_{1/2}$ = 2 min for uncomplexed LPS compared to $t_{1/2}$ = 12 h for complexed LPS (Mathison and Ulevitch, 1979). High-density lipoprotein complexing of the LPS also changes the cellular targeting of the LPS. Specifically, uncomplexed LPS is taken up by tissues rich in macrophages, whereas LPS-HDL complexes are taken up more by tissues metabolizing cholesterol, that is, the adrenal and the liver (Konig et al., 1988; Mathison and Ulevitch, 1979; Munford et al., 1982). Nonspecific binding to red blood cells and leukocytes is also prevented (Freudenberg and Galanos, 1988).

LIPOPOLYSACCHARIDE BINDING TO LOW-DENSITY AND VERY-LOW-DENSITY LIPOPROTEINS

Though most studies have focused on the effects of HDL binding to LPS, there is increasing evidence that LDL and VLDL can act to modify the effects of LPS. Van Lenten et al. (1986), using radio-iodinated LPS, demonstrated that all major lipoprotein classes bind LPS in proportion to their cholesterol concentration, and that this association is quite stable and mediated via a lipid–lipid interaction. Low-density lipoprotein, VLDL, and chylomicrons as well as HDL protect against endotoxin-induced death in mice (Harris et al., 1990). Studies by Navab and others have shown that LDL as well as HDL markedly reduces the cytoxic effects of LPS in bovine and rat aortic endothelial cell culture systems (Morel et al., 1986; Navab et al., 1988). The complexes are transported across the endothelium in a biologically active form where they may be able to initiate subendothelial inflammation (Navab et al., 1988).

During the acute phase response, levels of HDL decrease whereas levels of LDL and VLDL increase (Cabana et al., 1989; Evans and Williamson, 1991; Flegel et al., 1989). Flegel et al. have observed that HDL possesses less LPS-inactivating capacity than LDL. They propose that a relative increase in LDL and a decrease in HDL during an acute phase response would act to increase the LPS-inactivating capacity of the plasma. Thus the lipoproteins may act as a buffer system that handles small amounts of LPS without massive activation of an inflammatory response, but still allows for sufficient activation of the immune system in the event of a larger influx of LPS (Flegel et al., 1989).

Recently Emancipator et al. (1992) demonstrated that purified HDL, LDL, and the apolipoprotein AI inactivated LPS as assayed by the LAL assay. They also found that apo AI and apo B, in the absence of lipid, reduced the activity of endotoxin in the LAL and that apo AI, in the absence of lipid, reduced the lapine pyrogenic response. They concluded that a protein–endotoxin interaction is responsible for the detoxification process. Their findings that lipoprotein or apolipoproteins could inactivate endotoxin in the absence of other plasma factors is supported by Munford et al. (1981) but is in contrast to the earlier findings of Ulevitch (Porat et al., 1992; Ulevitch and Johnston, 1978). Emancipator proposed that because they utilized physiological doses of LPS, in contrast to the microgram quantities employed in earlier work, the excess of protein relative to LPS compensated for the slow reaction rate of LPS–lipoprotein in the absence of disaggregating agents.

OXIDIZED LOW-DENSITY LIPOPROTEIN

Investigations of the role of LDL in atherogenesis have implicated LDL derivatives in production of cholesterol-laden macrophages ("foam cells") and in the generation of early atherogenic inflammation. Although the interaction between endotoxin and LDL derivatives has not been characterized, we believe this phenomenon likely to be relevant to the host's response to endotoxin. Modification of LDL by acetylation, conjugation with malondialdehyde, or in vivo or metal-catalyzed oxidation leads to its uptake through scavenger receptors located predominantly on macrophages (Heinecke et al., 1987; Henriksen et al., 1981; Navab et al., 1988). Oxidation by endothelial cells or macrophages appears to depend on lipoperoxides generated intracellularly through 15-lipoxygenase activity and then transferred to the LDL (Parthasarathy et al., 1989; Rankin et al., 1991). Reactive oxygen species secreted into the medium may also be involved in lipid peroxidation of LDL (Heinecke et al., 1987). Neutrophils are prominent sources of oxide radicals, and it is probable that acute inflammatory microenvironments support LDL oxidation, although this remains unexplored. There is some evidence that superoxide anion is involved, but may act indirectly (Cathcart et al., 1989). Copper or iron ions are absolute requirements and generate peroxy radicals from lipoperoxides. Ceruloplasmin, a copper chelator, and transferrin, an iron chelator, are acute phase reactants and may serve to affect these reactions. This material has gained relevance since the recent observation, discussed below, that LPS internalized through scavenger receptors does not appear to initiate a cytokine response.

In vivo formation of oxidized lipoproteins and fatty acids has recently been demonstrated following cytokine or endotoxin infusion. Mouse recombinant tumor necrosis factor-alpha (mr-TNF-α) significantly decreased the H-1 methyl and methylene lipid line widths by nuclear magnetic resonance (NMR), and the ^{13}C spectra indicated a decrease in the relative concentration of polyunsaturated fatty acids (McDonagh et al., 1992). The same changes were directly confirmed by gas chromatographic analysis, showing decreases in the amount of linoleic and arachidonic acids and other polyunsaturated fatty acids relative to monounsaturated fatty acids and in the ratio of polyunsaturated to monounsaturated fatty acids. Serial plasma samples from volunteers receiving an infusion of endotoxin showed similar changes in their ^{13}C NMR spectroscopy at times when peak TNF-α values were measured. In addition, in these samples the ^{13}C NMR spectra showed direct evidence of lipid peroxidation products. Other cytokines (human recombinant interleukin-1 alpha, hr-IL-2, mouse recombinant interferon-gamma) did not produce these effects.

Oxidized low-density lipoprotein formation can induce formation of pro-inflammatory mediators, including monohydroxy fatty acid derivatives and oxysterols (Bhadra et al., 1991; Wang et al., 1992). Lipid peroxidation results in a chain reaction amplifying the number of lipoperoxides and resulting in double bond rearrangement (Witztum and Steinberg, 1991). Phospholipase A_2 activity, intrinsic to apolipoprotein B, results in conversion of lecithin to lysolecithin and propagation of peroxides (Parthasarathy and Barnett, 1990). Fatty acid fragmentation occurs with formation of highly reactive intermediates that can complex with apolipoprotein B and with phospholipids (Jurgens et al., 1993). One such linoleate reaction product, 9-hydroxyoctadecadienoic acid and its cholesteryl ester, is able to induce IL-1β production by human macrophages (Ku et al., 1992). Oxidized LDL is cytotoxic to endothelial cells.

Because lipoprotein binding inhibits LPS activity in the LAL assay, it is possible that considerable transvascular trafficking of LPS may occur that is not currently recognized. This question has been directly examined in fulminant meningococcal disease by high-speed centrifugation of plasma from patients with high levels of LAL activity (Brandtzaeg et al., 1992). After 15 min of centrifugation, 84 $\pm$ 2% (mean $\pm$ SE) of the recovered LPS was found in the lower one-third of the centrifuged volume, whereas 6 ± 1% remained in the upper one-third volume, indicating that meningococcal endotoxin circulates as complexes with high sedimentation coefficients. Bacterial outer membrane fragments were detected in the bottom fractions of three patient plasmas examined by means of electron microscopy. The plasma samples selected for analysis, with high LAL activity, may not accurately reflect events in less fulminant clinical disease where bacterial fragments may not be present.

LIPOPOLYSACCHARIDE AND ACUTE PHASE SERA: LIPOPOLYSACCHARIDE BINDING PROTEIN

Recent DNA analysis techniques have revealed a family of proteins with sequence homology that bind LPS and are involved in the host's response to endotoxemia (Tobias, 1988). The family includes lipopolysaccharide binding protein (LBP), bactericidal/permeability-increasing protein (BPI), and cholesterol esterase transfer protein (CETP). Cholesterol esterase transfer protein shares a lipid binding domain with LBP and BPI (Au-Young and Fielding, 1992) but has not been examined in relation to the processing of LPS. The conserved sequence in these proteins may represent a binding site for nonpolar lipids in otherwise unrelated proteins (Au-Young and Fielding, 1992; Ooi et al., 1991; Tobias et al., 1989).

Tobias et al. found that LPS reacts differently with acute phase sera (APS) than with normal sera (NS). In APS, LPS from *S. minnesota* Re595 forms a protein complex with density of 1.3 g/cm^3, designated complex 1.3. Lipopolysaccharide binding to HDL is significantly slower (20 to 40-fold for rabbits; 10-fold for humans) in APS than in NS. When Re595-LPS/serum mixtures are dialyzed against a low salt buffer, Re595-LPS precipitates in a less soluble form from APS than from NS. The precipitate from the Re595-LPS-APS mixtures contains a protein (molecular weight = 60,000) that does not precipitate from Re595-LPS–NS mixtures or from NS or APS alone (Tobias et al., 1985; Tobias and Ulevitch, 1983). This protein has been isolated and designated LPS-binding protein (LBP) (Schumann et al., 1990; Tobias et al., 1988). That LBP is a component of complex 1.3 is supported by two experiments. First, adding purified LBP to NS allows the formation of complex 1.3, and second, immunoprecipitating LBP and APS renders it incapable of forming complex 1.3.

LBP is a 60-kDa glycoprotein present in normal plasma at <0.5 μg/ml, but 24 h after induction of an acute phase response, 50 μg/ml LBP can be detected (Tobias et al., 1986). As with other acute phase reactants, LBP is synthesized mainly in hepatocytes (Ramadori et al., 1990). Recent work by Su et al., however, has demonstrated LBP mRNA in extrahepatic sites (Schmouder et al., 1995). These experiments used a rat model of the acute phase response with injection of either turpentine into the hindlimb or LPS into the peritoneum. Northern blot analysis revealed basal expression of LBP mRNA in hepatocytes that is markedly upregulated with injection of either turpentine or LPS. Also detected were lower levels of base-line mRNA expression in lung and kidney tissue that was upregulated with the inflammatory response. In additional studies, a sterile polyurethane sponge was implanted into the subcutaneous tissue of rats for 7 days. The capsule and inflammatory tissue surrounding the sponge was harvested and examined for LBP mRNA. This tissue revealed expression of LBP message; however, no LBP bioactivity was detectable in serum-free culture. These studies suggest that LBP may play a role in extrahepatic, extravascular sites; however, no direct correlation between message expression and protein production has been made. Opal et al. have shown that abscess fluid contains LBP levels comparable to that of plasma (Corbett et al., 1994).

LBP can be measured in plasma by several methods. The most obvious method involves a sandwich ELISA using two monoclonal antibodies generated to LBP (Thomas et al., 1994); however, this method requires access to, or generation of, these antibodies. Heumann et al. (1994) described two alternate methods, one of which does not require an anti-LBP monoclonal antibody. The first involves monophosphoryl lipid A-coated plates reacted with serial dilutions of LBP of a known concentration followed by incubation with an ^{125}I-labeled anti-LBP IgG. This radioimmunoassay standard curve can then be used to test samples with unknown concentrations of LBP. Alternatively, because LBP facilitates the binding of LPS to mononuclear cells, FITC-labeled LPS can be used in a flow cytometry assay to measure LBP concentrations without the use of a specific anti-LBP antibody. Monocytes are incubated with increasing concentrations of purified LBP and labeled LPS, and fluorescence is measured by flow cytometry. This curve was found to be linear in the concentration range between 0.1 and 5 μg/ml of LBP (Heumann et al., 1994).

Lipopolysaccharide-binding protein binds to lipid A and shows high-affinity binding to LPS of both rough and smooth bacteria (Tobias et al., 1988, 1989). Recognition of the LBS-LBP complex is by a specific receptor on the phagocyte membrane, CD14 (Wright et al., 1989b). Cross-linking experiments by Ulevitch et al. suggest that the monocyte LPS receptor consists of two components, CD14 and a yet unidentified lipid A-binding protein (Ulevitch and Tobias, 1988). Macrophage responsiveness to nanogram quantities of LPS is enhanced by LBP with a resultant increase in the rate and amount of TNF production (Mathison et al., 1992; Schumann et al., 1990; Wright et al., 1989b) presumably responsible for the initiation of the inflammatory cascade. Removal of LBP from the serum of rabbits suppresses LPS-induced TNF production in rabbit whole blood (Mathison et al., 1992). Priming of neutrophils for increased oxidative responses and upregulation of fMLP and CD18 receptors in response to LPS is accelerated by the presence of serum and specifically LBP (Aida and Pabst, 1990; Lynn et al., 1991; Vosbeck et al., 1990). These responses and increased adherence can be blocked by monoclonal antibody to CD14 (Vosbeck et al., 1990; Worthen et al., 1992). Truncation of the LBP molecule at Ile-197 yields a protein that is capable of binding LPS but does not complex with soluble or membrane-bound CD14 (Han et al., 1994). This suggests that the amino terminal portion of the molecule is responsible for the LPS-binding properties, whereas the carboxy terminus allows for the formation of complexes with CD14. This information is significant in the development of a therapeutic LBP fragment that binds LPS but does not initiate the inflammatory response.

The ability of LBP to sensitize monocytes, macrophages, and neutrophils to nanogram quantities of LPS and its rise in concentration during the acute phase response suggest that it plays a role in augmenting the host response to gram-negative infection. The initial disaggregation step of LPS in plasma would allow LPS to bind either LBP or HDL. Under normal circumstances HDL binding would be expected to predominate and inhibit responses to LPS. However, under conditions of injury or infection, increased concentrations of LBP would predipose the host to the formation of LPS LBP complexes and increased activation of the inflammatory response (Ulevitch and Tobias, 1988). Wurful et al. have demonstrated a neutralizing/potentiating effect of plasma in an assay measuring the stimulation of human neutrophils to adhere to a fibrinogen coated surface (Walz et al., 1993). They demonstrated that LPS incubated with plasma for short time periods (up to 10 min) and then incubated with neutrophils resulted in a stimulatory effect. Longer LPS/plasma incubations resulted in greatly reduced stimulatory effect with >80% lost at 120 min. The full effect was restored with the addition of LPS but not LBP, indicating that decreased stimulation over time is due to a decrease in LPS availability. In the same study, HDL particles were reconstituted from highly purified apoA-I, phosphatidylcholine, and cholesterol (R-HDL), which were shown to possess little of the LPS-neutralizing ability of serum HDL that has been previously reported (Munford et al., 1981; Rogivue et al., 1995; Skarnes, 1968; Ulevitch et al., 1979). However, the addition of rLBP in substoichiometric amounts resulted in a rapid increase in the amount of LPS associated with the R-HDL fraction. This suggests a catalytic role for LBP in the transfer of LPS to HDL. It was also shown that >99% of LBP detectable by ELISA is removed from plasma with its passage over an anti-apoA-I column. This finding suggests that LBP circulates in association with apoA-I in a position that allows for the transfer of LPS to either HDL or CD14, resulting in a neutralizing or potentiating effect. Such an equilibrium is probably responsible for the eventual host response, or lack thereof, to LPS that reaches the vascular compartment. In the tissue space, or extravascular compartment, evidence shows that LBP may be in competition with another LPS-binding protein, bactericidal/permeability-increasing protein (BPI), to determine cellular response.

BACTERICIDAL/PERMEABILITY-INCREASING PROTEIN

Bactericidal/permeability-increasing protein is a 55-kDa neutrophil azurophilic granule protein that has sequence homology and immuno-crossreactivity with LBP and binds to the lipid A portion of LPS (Gazzano-Santoro et al., 1992; Gray et al., 1989; Koch et al., 1994; Marra et al., 1990; Weersink et al., 1993; Weiss and Olsson, 1987). Unlike LBP, which facilitates the potent cellular stimulatory effect of LPS (Mathison et al., 1992; Vosbeck et al., 1990; Wright et al., 1990b), BPI does not stimulate target cells when bound to endotoxin (Dentener et al., 1993; Heumann et al., 1993; Koch et al., 1994; Marra et al., 1990; Meszaros et al., 1993; Ooi et al., 1991; Weiss et al., 1992). This anti-inflammatory effect has created interest in BPI as a potential therapeutic agent in gram-negative infections (Calvano et al., 1994; Donnelly et al., 1994; Fisher et al., 1994; Kelly et al., 1993; Kohn et al., 1993).

Bactericidal/permeability-increasing protein has been shown to be specifically cytotoxic to gram-negative bacteria in vitro as well as in whole blood (Car et al., 1994; Donnelly et al., 1993; Mannion et al., 1989; Weiss et al., 1992). The mechanism by which BPI is cytotoxic to gram-negative but not gram-positive bacteria is thought to occur in two stages (Mannion et al., 1990). The first stage involves BPI binding to bacterial LPS, resulting in growth arrest, permeabilization, and selective activation of bacterial enzymes. The second phase involves changes in the cytoplasmic membrane resulting in cell death (Mannion et al., 1990; Walz, 1992). The presence of LBP potentiates the bactericidal activity of BPI by nearly 10,000 fold (Walz et al., 1991).

Several forms of the protein, including native BPI, recombinant BPI, and the 23-kDa *N*-terminal fragment, have been shown to bind lipid A and neutralize the toxic effects of LPS in vitro (Marra et al., 1990; Meszaros et al., 1993; Ooi et al., 1987), most commonly measured as a suppression of LPS-induced mononuclear cell TNF-α production. Further evidence of the anti-endotoxin activities of BPI is the finding that BPI blocks the LPS-stimulated upregulation of CR1 and CR3 on neutrophils (Marra et al., 1990). Additionally, BPI has been found to inhibit LPS activity in the LAL assay, LPS priming of neutrophils for release of LTB4, and LPS-mediated production of TNF in whole blood (Ooi et al., 1991). Marra et al. (1990) demonstrated that BPI expression on the neutrophil surface is increased by neutrophil treatment with LPS, fMLP, or TNF-α. More recently, animal studies have shown a protective effect of BPI in lethal endotoxemia as well as prevention of endotoxin-induced fever and shock when given intravenously prior to, or up to 1 h after, LPS administration (Fisher et al., 1994; Kohn et al., 1993; Walz, 1992). In addition, early TNF-α levels after endotoxemia were significantly lower in BPI-treated animals than in controls (Walz, 1992). Similar protection was reported from a study of an animal model of gram-negative pneumonia (Kelly et al., 1993).

Because cellular response to LPS is remarkably different in the presence of BPI versus LBP, much work has been done to determine the interactions between LPS, LBP, BPI, and mononuclear cells. Before individual data about these interactions can be discussed, however, two fundamental questions with regard to location and concentration must be addressed in order to give physiological significance to such discussion. Lipopolysaccharide-binding protein serum concentrations have been established both in the normal state and in the acute phase response with levels rising from 1–10 to 50–100 μg/ml (Tobias et al., 1996). Serum BPI levels rarely exceed picogram quantities even in endotoxemia and sepsis (Calvano et al., 1994). Therefore, in the absence of exogenously administered BPI, it seems unlikely that BPI plays a major role in the processing of intravascular LPS. It is more likely that the host response to circulating endotoxin is determined by whether LBP transfers LPS to HDL or to CD14, as discussed previously. In extravascular sites, where neutrophils are known to degranulate, BPI concentrations may reach levels that equal or exceed LBP concentrations. Measurements of abscess fluid have shown this to be true with BPI concentrations of 1–17 μg/ml and LBP concentrations reaching only 1–2 μg/ml (Corbett et al., 1994). The importance of the extravascular processing of LPS should not be underestimated. Most gram-negative infections begin in extravascular compartments such as the peritoneal cavity, lungs, or soft tissues, and the local host response determines both adjacent organ injury as well as systemic response. We speculate that bacteremia or endotoxemia may occur when the BPI processing of LPS is overwhelmed in a tissue-space infection. The resultant host response to bloodborne LPS is well documented.

Wilde et al. have done extensive work comparing the binding affinities of LBP and BPI for LPS (Koch, 1994). They compared the ability of LBP and BPI to inhibit the binding of biotinylated BPI to immobilized LPS. Inhibition by BPI was 100 times greater than that by LBP. Fluorescein isothiocyanate-labeled LPS and flow cytometry were utilized to determine BPI inhibition of LBP-mediated LPS binding to mononuclear cells. These data revealed that 1 μg/ml of BPI was capable of significantly reducing the LBP-mediated binding of LPS over a range of 1–10 μg/ml of LBP. Additionally, soluble BPI effectively abolished the TNF-α response to LBP/LPS stimulation. Similar results were seen with the 23-kDa *N*-terminal fragment of BPI ($rBPI_{23}$) (Chang et al., 1994). These are strong arguements for BPI's role in the neutralization of LPS in tissue-space gram-negative infections.

Our laboratory has confirmed the findings of Wilde et al. using tritiated LPS and the mononuclear cell line THP-1. Indeed, soluble BPI is capable of inhibiting the LBP-mediated binding of ^{3}H-LPS to these cells and abrogating subsequent TNF-α secretion. We also confirmed the ability of soluble BPI to mediate significant uptake of labeled LPS in the absence of LBP or serum, suggesting the uptake of BPI/LPS complexes by mononuclear cells. Interestingly, cells treated with BPI and then washed also bound more labeled LPS than cells not pretreated. Tumor necrosis factor-alpha response was also significantly reduced in cells pretreated with BPI in a dose-dependent fashion (Burnett et al., 1995). These data suggest that mononuclear cells exposed to BPI can take up LPS in a fashion that does not stimulate an inflammatory response, possibly as a disposal mechanism for LPS. Additionally, cells exposed to BPI are then rendered resistant to subsequent LBP/LPS stimulation, supporting an anti-inflammatory effect of BPI other than mere soluble phase competition. We have shown that BPI treatment of mononuclear cells does not affect the number of CD14 receptors expressed, nor does BPI remain on the cell surface (Burnett et al., 1995). Internalization of BPI by THP-1 cells was demonstrated by Western blotting to be rapid, dose dependent, nonsaturable, and not inhibited by the addition of excess unlabeled BPI. Cross-linking studies did not define a specific protein entity associated with the internalized BPI. These data suggest uptake of BPI by a nonreceptor-mediated mechanism consistent with pinocytosis (Burnett et al., 1996). We have yet to determine the process by which the internalization of BPI renders mononuclear cells resistant to subsequent LBP/LPS stimulation.

The importance of BPI in host response to gram-negative infection is evolving. Current evidence supports a role for BPI in the killing of gram-negative organisms as well as in the extravascular noninflammatory processing of LPS. Bacterial/permeability-increasing protein is considered to be a potential therapeutic agent in the setting of bacteremia or endotoxemia.

SEPTINS

Wright et al. (1992) recently described a novel set of plasma factors that opsonize LPS for recognition by the CD14 receptor on neutrophils. The factor, which they have termed "septin," allows subnanogram quantities of LPS to induce stimulation of the adhesive capacity of neutrophils and production of TNF by monocytes. Septin activity requires two distinct protein species and appears to function as a plasma protease and a substrate. The opsonic activity of plasma is not blocked by addition of neutralizing antibody against LBP nor by depletion of the complement or clotting cascades. The receptor for this activity appears to be CD14. Wright has proposed that septin may be a protease cascade related to that seen in *Limulus* amebocytes. Gelation of *Limulus* plasma proteins is initiated by LPS-binding factor C (Minetti et al., 1991; Muta et al., 1990). The enzyme has a mosaic structure with complement-, epidermal growth factor-, and lectin-like domains (Muta et al., 1991). This results in the autolytic cleavage of factor C, activating proteolytic activity against an additional zymogen, factor B. Two additional proteolytic steps result in the cleavage of a fibrin-like molecule that makes up the clot.

COMPLEMENT

In normal plasma, LPS is able to activate complement by mechanisms that do not require the presence of specific anti-LPS antibody (Morrison, 1977). There is evidence that the lipid A region of LPS mediates activation of the classical pathway via binding of C1q, whereas the presence of polysaccharide on the LPS obligates activation via the alternate pathway (Cooper and Morrison, 1978; Loos et al., 1990; Morrison and Kline, 1977; Pangburn, 1989; Vukajlovich et al., 1987). Activation of either pathway by LPS appears to be a mutually exclusive event (Vukajlovich et al., 1987). We have some evidence that complement may be a co-factor for maximally efficient LPS-mediated priming of neutrophils for oxidative responses.

It remains unclear at present whether complement activation products generated by endotoxin interactions affect clinical responses to LPS. Complement activation results in the production of the anaphylatoxins C3a and C5a, which promote neutrophil activation and enhancement of aggregation, adherence, and migration. Prior exposure to C5a enhances monocyte cytokine responses to LPS through initiation of messenger transcription, with the translation signal being provided by LPS (Morin et al., 1991; Schindler et al., 1990). In patients at risk for the adult respiratory distress syndrome, both endotoxemia and elevated C3adesArg levels correlated with the appearance of the pulmonary lesion (Parsons et al., 1989). In an animal model of the adult respiratory distress syndrome, a synergistic interaction between trace amounts of lipopolysaccharide and complement-mediated neutrophil activation was noted (Worthen et al., 1987). Volunteers infused with reference endotoxin did not have elevations in plasma C3adesArg or C5adesArg, in contradistinction to patients with intra-abdominal and other forms of gram-negative infection (Solomkin et al., 1981; van Deventer et al., 1990). Tissue injury, as occurs in burns, appears sufficient to induce complement activation, although there is evidence that endotoxin may be involved in this process (Davis et al., 1987; Moore et al., 1986; Solomkin, 1984).

LEUKOCYTE RECEPTORS FOR LIPOPOLYSACCHARIDE

Induction of cell responses by LPS depends upon internalization of the lipid A moiety rather than activation of receptor-coupled transduction mechanisms. This conclusion is based upon the finding that lipid IV_A and deacylated LPS are able to block induction of macrophage-like cell responses to endotoxin without inhibiting receptor-mediated uptake (Kitchens et al., 1992; Kovach et al., 1990). One possibility is that a lipid A partial structure is able to bind to and activate protein kinase C (Wrightman and Raetz, 1984). Interactions between lipid A structures and platelet protein kinase C have been characterized (Grabarek et al., 1990). Lipid A is metabolized within leukocytes by cleavage of acyloxyacyl bonds and by release of myristic acid residues (Erwin and Munford, 1990; Hagen et al., 1991; Munford and Hall, 1989; Northoff et al., 1987). Myristic acid derivatives are potent protein kinase C activators. Certain effects of LPS, such as cytokine secretion, arachidonic acid metabolism, downregulation of TNF-α receptors, expression of cryptic receptors on neutrophils, and activation of the neutrophil NADPH oxidase, are protein kinase C-dependent phenomena (Downey et al., 1992; Humes, 1988; Nauseef et al., 1991; Porteu and Nathan, 1990).

THE CD18 FAMILY OF RECEPTORS

Lipopolysaccharide and intact *E. coli* specifically bind to a family of adhesion-promoting receptors on the surface of phagocytes (Wright and Detmers, 1988). Lymphocyte function-associated antigen-1 (LFA-1), CR-3, and p150,95 are structurally homologous, dimeric glycoproteins consisting of identical 95-kDa β1 chains (CD18) and structurally and antigenically distinct 150- to 190-kDa α1 chains (CD11a, CD11b, CD11c) (Sanchez-Madrid et al., 1983). They have been shown to be members of the larger family of cell surface adhesion-promoting proteins known as integrins (Hynes, 1987).

The lymphocyte function-associated antigen (CD11a/CD18) found on monocytes, macrophages, neutrophils, T and B lymphocytes, and natural killer (NK) cells functions in T and NK cell target binding as well as macrophage-tumor binding. CR3 (CD11b/CD18) functions as the C3bi complement receptor on monocytes, macrophages, neutrophils, and NK cells. Finally, p150,95 (CD11c/CD18), found on monocytes, macrophages, and neutrophils, has yet to have a specific function ascribed to it (Wright and Detmers, 1988).

Wright and Jong (1986) demonstrated that macrophage-associated LFA-1, CR3, and p150,95 were each capable of recognizing *E. coli* in the absence of IgG, complement, or bacterial adhesins. These experiments were performed using rosetting of sheep erythrocytes coated with *E. coli.* Removal of all three receptors with monoclonal antibodies to the common β1 subunit, CD18, markedly inhibited binding of *E. coli,* whereas blockage of any one or two receptors with antibody directed against the specific α1 chain caused negligible inhibition of binding. It was thus concluded that each of the three families of receptors was capable of independently binding *E. coli* (Wright and Jong, 1986). The moiety recognized was determined to be the lipid A portion of LPS. Adhesion of macrophages to LPS-coated surfaces markedly diminished the ability of macrophages to bind *E. coli,* and treatment of LPS with polymyxin B, a drug that binds to lipid A, blocked macrophage recognition of LPS-coated erythrocytes (Wright and Jong, 1986). Further studies by Wright et al. (1989a) using monoclonal antibodies directed against different epitopes of CR3 demonstrated that LPS and C3bi bind to different sites on CR3.

Cells bearing CR3, LFA-1, and/or p150,95 are known to respond to low concentrations of LPS: macrophages secrete IL-1 and TNF-α; neutrophils are primed for increased release of toxic oxygen intermediates; and B lymphocytes undergo blast transformation (Anderson et al., 1977; Guthrie et al., 1984). However, it has yet to be proved that the CD11/CD18 family of receptors mediate these biologic responses, and more recent data suggest that they do not. Although neutrophils from patients deficient in the CD18 family of receptors are unable to bind LPS-coated erythrocytes (ELPS) or *E. coli,* they are able to respond normally to LPS in vitro (Wright et al., 1990a). Wright proposed that the CD18 family of receptors may be sufficient but not necessary for cells to respond to LPS.

CD14

The purification and characterization of LBP led to the discovery that the phagocytic cell receptor CD14 was capable of binding LBP complexed with LPS on the surface of bacteria or inserted in the membranes of erythrocytes (Wright et al., 1991). CD14 is a 55-kDa glycoprotein monocyte differentiation antigen that is attached to the cell membrane via phosphatidyinositol–glycan anchors (Haziot et al., 1988). It is strongly expressed on monocytes and macrophages, and in lesser amounts of neutrophils (Simmons et al., 1989; Wright et al., 1990b). This receptor was functionally characterized by examining adherence of LPS-coated erythrocytes (ELPS) to leukocytes. Treatment of macrophages with monoclonal antibodies directed against CR1, CR3, FcγR1, FcγRII, and FcγRIII did not affcct binding of LBP-coated particles (Wright et al., 1990b). Leukocytes from CD18-deficient patients were also able to bind LBP (Wright et al., 1990a). Depletion of macrophage apical surface receptors with adherent antibody to CD14 did, however, suppress binding of ELPS (Wright et al., 1990b). That CD14 was the postulated novel LPS receptor was supported by the findings that (1) removal of 70% of cell-surface CD14 by treatment with phosphatidylinositol-specific phospholipase C resulted in an 80% reduction in subsequent binding of ELPS, and (2) monoclonal antibody against CD14 inhibited monocyte TNF-α synthesis in response to 0.01–1 ng/ml LPS and blocked LPS-mediated neutrophil priming (Worthen et al., 1992; Wright et al., 1990a,b). More recently, cross-linking experiments using LPS that contained a radio-iodinated, photoreactive phenyl azide directly demonstrated the iodination of CD14 on the surface of mononuclear cells (Tobias et al., 1993). Deletional mutation studies revealed that the *N*-terminal 65 amino acids are responsible for the serum-dependent binding of LPS to CD14 (Lukacs et al., 1995). CD14 truncated from the *C*-terminus to 152 amino acids maintains all the bioactivity of holoCD14 (Strieter et al., 1992).

Neutrophils have a spare CD14 receptor population and substantially increase CD14 expression following stimulation with secretagogues. Treatment of neutrophils with TNF-α, G-CSF, GM-CSF, or fMLP results in a rapid (20 min) doubling of neutrophil surface expression of CD14 as assessed by immunofluorescent staining with anti-CD14 monoclonal antibody (Wright et al., 1991). CD14 is necessary for the LBP-potentiated binding by neutrophils of erythrocytes coated with LPS (Wright et al., 1991). Pretreatment of neutrophils with the above-listed cytokines resulted in doubling of the binding of ELPS-LBP complexes, in parallel with the upregulation of CD14 (Wright et al., 1991). Thus CD14 acts not only as a signaling receptor but also as an opsonic receptor and enhances the ability of neutrophils to recognize LPS, and may be important in the clearance of aggregated or surface-bound LPS from the bloodstream. Neutrophils and macrophages possess acyloxyacyl hydrolase, which is capable of detoxifying LPS (Erwin and Munford, 1990; Munford and Hall, 1986). Neutrophilia may therefore represent a means of detoxifying intravascular lipopolysaccharide. However, recent work by Haziot et al. has demonstrated that neutrophils are capable of synthesizing CD14, which can be isolated as both the GPI-anchored form and a smaller soluble form (Winn et al., 1993). In addition, CD14 ligation by LBP/LPS complexes has been shown to stimulate TNF production by neutrophils (Haziot et al., 1993). It would therefore seem that the role of neutrophil CD14 as an inflammatory or anti-inflammatory mediator remains to be determined.

The CD14 receptor is solubilized during acute illness and may serve to restrict the activation of CD14-bearing cells by LBP-LPS complexes (Bazil and Strominger, 1991; Kruger et al., 1991; Maliszewski, 1991; Schutt et al., 1991). Most of the total CD14 content in blood is plasma-, not membrane-associated. Recent experiments have shown that cells that do not express membrane-bound CD14 become LPS responsive when exposed to solubilized CD14/LPS/LBP or septin complexes (Frey et al., 1992). Cell types studied in these experiments included bovine pulmonary arterial endothelial cells and human umbilical vein endothelial cells. Lipopolysaccharide-induced cytotoxicity (bovine cells) and expression of endothelial leukocyte adhesion molecule-1 (human cells) were shown to depend on the presence of CD14 in the incubation mixture. In these experiments, LBP was found to be unnecessary, although LBP contamination could not be ruled out. Similar work by Arditi et al. has shown that other serum factors are not required for the effects noted in bovine endothelial cells and human umbilical vein endothelial cells (Goodman et al., 1992a). Soluble CD14 thus probably plays a significant role in the systemic inflammatory response to LPS by mediating endothelial cell changes associated with endotoxemia.

MACROPHAGE SCAVENGER RECEPTORS

More recent attention has been focused upon the fate of oxidized LDL/LPS complexes. Goldstein et al. (1979) first proposed modification of LDL as a prerequisite for macrophage uptake and accumulation of cholesterol. This provided a mechanism for generation of "foam cells," a characteristic finding in early atheroma formation. Lysine residues in apolipoprotein B are required for LDL/LDL receptor binding, and as these are derivatized LDL receptor binding decreases (Haberland et al., 1982). An acetylLDL receptor was subsequently identified and termed the "scavenger" receptor because of a very broad ligand specificity (Brown and Goldstein, 1983). This receptor binds acetylLDL, oxyLDL, maleylated bovine serum albumin, certain polyribonucleotides, some polysaccharides, certain phospholipids, and bacterial liposaccharide (Hampton et al., 1991a). The scavenger receptor recognizes a modified fragment of apolipoprotein B (Parthasarathy et al., 1987). Oxidized low-density lipoprotein and its surrogates bind to and are internalized by macrophages through this receptor.

Hampton and co-workers (1991a) found that macrophage scavenger receptors on RAW 264.7 cells bound lipid IV_A and mediated its metabolism to the less active monophosphoryl form. Label binding was inhibited by the scavenger receptor ligands acetylated LDL, maleylated bovine serum albumin, and polyinosinic acid. Studies in mice demonstrated that up to 79% of injected labeled lipid IV_A accumulated in the liver within 10 min of injection, and co-injection of unlabeled scavenger receptor ligands blocked hepatic accumulation of the labeled ligand. Scavenger receptor binding of LPS or lipid IV_A was not involved in activation of macrophages in vitro. The authors concluded that though the scavenger receptor apparently has no role in activation of murine macrophages, it may be important in the hepatic clearance of LPS. Relatively small amounts of lipid IV_A were infused, perhaps more closely duplicating the clinical situation.

This theme was explored by Fong et al. (1991), who found that LDL oxidation inhibited IL-1β mRNA synthesis in response to LPS in a mouse macrophage system. These investigators, however, reported that this effect was not mediated through scavenger receptors in that oxidized LDL also inhibited scavenger receptor-negative THP-1 cell IL-1β mRNA expression. They suggested that an oxyLDL derivative such as 9-HODE may be involved. The ability of LDL receptors to internalize LPS/LDL complexes has not been directly studied, and it is possible that this hypothesized circumstance might serve similarly to suppress LPS responses. In support of this, experiments with a human monocyte–macrophage culture system for studying the suppression of scavenger receptor activity by LPS showed that only LPS bound to LDL mimicked the suppressive activity seen when LPS was added to cells in serum (VanLenten et al., 1986).

It is important to note that other studies suggest that uptake of oxyLDL itself may be a pro-inflammatory stimulus. Such uptake is associated with generation of MCP-1 (Yla Herttuala et al., 1991). This has been studied primarily in relation to atherogenesis and has not been examined in relation to LPS. Considerable interplay may occur at the level of the macrophage surface, involving oxidation of LDL with oxyLDL and other derivative generation, binding of LPS directly to scavenger receptors, and possibly binding of LPS to oxyLDL and subsequent internalization through scavenger and other as yet undefined receptors.

This pathway for uptake of LPS is subject to regulation. Scavenger receptor ligands for RAW 264.7 cells attenuate the metabolism of lipid IV_A to the 4′-monophosphoryl lipid IV_A (Hampton and Raetz, 1991b). TGF-β profoundly inhibits scavenger receptor activity, in part through a twofold loss of receptor mRNA in treated cells (Bottalico et al., 1991). In human monocytes, LPS similarly inhibits scavenger receptor mRNA formation (Van Lenten and Fogelman, 1992). Addition of anti-TNF-α prevented this suppression, whereas exogenous TNF-α mimicked the LPS effect.

High-density lipoprotein receptors are present on macrophages and lymphocytes (Rahim et al., 1991; Takahashi et al., 1989; Xu et al., 1992). Macrophages internalize HDL receptors and it is therefore probable that LPS bound to HDL is brought into the cell through this mechanism. Cross-linked and perhaps oxyHDL are also internalized through scavenger receptors (Miyazaki et al., 1991). Why LPS internalized through lipoprotein receptor pathways does not induce cell responses is unclear. Whether intracellular processing depends on the specific uptake receptor is unknown but may explain the paradoxical cell responses seen.

OTHER LIPOPOLYSACCHARIDE RECEPTORS

Morrison et al. have identified a third LPS receptor on murine splenic lymphoid cells, using a radio-iodinated, disulfide-reducible, photoactivatable LPS derivative (ASD-LPS). Based on evidence that LPS can act directly on B and T cells, these investigators screened murine splenic cell lysates with ASD-LPS and detected an 80-kDa LBP on the surface of murine T and B cells and adherent macrophages (Lei and Morrison, 1988b). Of note, the 80-kDa protein was also found on splenocytes from the LPS nonresponder C3HeB/FeJ strain of mice, and no differences were detectable between the receptor proteins of responder and nonresponder strains as assessed by two-dimensional gel electrophoresis (Lei and Morrison, 1988b). The putative receptor is membrane localized and displays linear binding of LPS at low concentrations and saturability at high concentrations. Binding of the probe is blocked with both unlabeled LPS and polysaccharide-free lipid A, suggesting that it binds the biologically active lipid A portion of LPS (Lei and Morrison, 1988a).

Other potential cell surface LPS receptors have been described, but are less well characterized. Hampton et al. (1991a) identified 31- and 95-kDa proteins on the surface of the macrophage-like cell line RAW 264.7 that bound the lipid A precursor lipid IV_A. Binding was specific, saturable, and inhibitable with excess unlabeled lipid IV_A. The authors felt that the 31-kDa protein is a histone, and they raised the possibility that the 95-kDa protein is actually the 95-kDa β1 chain, CD18, found in LFA-1, CR3, and p150,95.

CONCLUSIONS

Endotoxin is clearly an important inducer of clinically significant local and systemic inflammation. Elaborate mechanisms have evolved to induce a host response and in parallel, to regulate the response. Regulatory responses occur at several levels, including binding of LPS to plasma proteins, targeted uptake into cell populations that do or do not respond with large-scale mediator release, intracellular metabolism to various components, and cellular sensitization or desensitization. We have attempted to highlight in this chapter newly described systems that may serve to alter clinically deleterious responses through alterations in receptor targeting.

The characterization of LPS/lipoprotein/receptor system interactions appears to be a promising research area. The novel functioning of soluble CD14 receptors similarly represents a recently described means of altering host responses. The exact targeting and receptor mediation of LPS/BPI complexes may also provide therapeutically relevant information. The development of LPS preparations with high specific activities was a technical requirement for studies examining the possibility that the route of LPS internalization may control the cell response to it.

REFERENCES

Aida Y, Pabst MJ (1990): Priming of neutrophils by lipopolysaccharide for enhanced release of superoxide. Requirement for plasma but not for tumor necrosis factor-alpha. J Immunol 145:3017–3025.

Anderson JA, Coutinho A, Lernhardt W, Melchers F (1977): Clonal growth and maturation to immunoglobulin secretion in vitro of every growth-inducible B lymphocyte. Cell 10:27.

Anonymous (1990): Increase in national hospital discharge survey rates for septicemia—United States, 1979–1987. MMWR 39:31–34.

Au-Young J, Fielding CJ (1992): Synthesis and secretion of wild-type and mutant human plasma cholesteryl ester transfer protein in baculovirus-transfected insect cells: The carboxyl-terminal region is required for both lipoprotein binding and catalysis of transfer. Proc Natl Acad Sci USA 89:4094–4098.

Barker JN, Jones ML, Mitra RS, et al. (1991): Modulation of keratinocyte-derived interleukin-8 which is chemotactic for neutrophils and T lymphocytes. Am J Pathol 139:869–876.

Baumberger C, Ulevitch RJ, Dayer JM (1991): Modulation of endotoxic activity of lipopolysaccharide by high-density lipoprotein. Pathobiology 59:378–383.

Baumgartner JD, Heumann D, Gerain J, Weinbreck P, Grau GE, et al. (1990): Association between protective efficacy of anti-lipopolysaccharide (LPS) antibodies and suppression of LPS-induced tumor necrosis factor alpha and interleukin 6. Comparison of O side chain-specific antibodies with core LPS antibodies. J Exp Med 171:889–896.

Bazil V, Strominger JL (1991): Shedding as a mechanism of down-modulation of CD14 on stimulated human monocytes. J Immunol 147:1567–1574.

Beutler B, Milsark IW, Cerami AC (1985): Passive immunization against cachectin/tumor necrosis factor protects mice from lethal effect of endotoxin. Science 229:869–871.

Bhadra S, Arshad MA, Rymaszewski Z, et al. (1991): Oxidation of cholesterol moiety of low density lipoprotein in the presence of human endothelial cells or Cu+2 ions: identification of major products and their effects. Biochem Biophys Res Commun 176:431–440.

Bone RC, Fisher CJ, Clemmer T (1989): Sepsis syndrome: a valid clinical entity. Crit Care Med 17:389–393.

Bone RC (1991): Sepsis syndrome. New insights into its pathogenesis and treatment. Infect Dis Clin North Am 5:793–805.

Bottalico LA, Wager RE, Agellon LB, Assoian RK, Tabas I (1991): Transforming growth factor-beta 1 inhibits scavenger receptor activity in THP-1 human macrophages. J Biol Chem 266:22866–22871.

Brandtzaeg P, Bryn K, Kierulf P, et al. (1992): Meningococcal endotoxin in lethal septic shock plasma studied by gas chromatography, mass-spectrometry, ultracentrifugation, and electron microscopy. J Clin Invest 89:816–823.

Brandtzaeg P, Kierulf P, Gaustad P, et al. (1989): Plasma endotoxin as a predictor of multiple organ failure and death in systemic meningococcal disease. J Infect Dis 159:195–204.

Brown MS, Goldstein JL (1983): Annu Rev Biochem 52:223–261.

Burnett RJ, Lyden CA, Tindal CJ, Cave CM, Marra MN, Solomkin JS (1996): Mononuclear cell line THP-1 internalizes bactericidal/permeability-increasing protein by a non-receptor-mediated mechanism consistent with pinocytosis. Arch Surg 131:200–205.

Burnett RJ, Lyden CA, Tindal CJ, Marra MN, Neely AN, Solomkin JS (1995): Effects of bactericidal/permeability-increasing protein upon mononuclear cell interactions with endotoxin: evidence for cell surface binding. J Endotox Res 2:349–357.

Cabana VG, Siegel JN, Sabesin SM (1989): Effects of the acute phase response on the concentration and density distribution of plasma lipids and apolipoproteins. J Lipid Res 30:39–49.

Calandra T, Gerain J, Heumann D, Baumgartner JD, Glauser MP (1991): High circulating levels of interleukin-6 in patients with septic shock: evolution during sepsis, prognostic value, and interplay with other cytokines. The Swiss-Dutch J5 Immunoglobulin Study Group. Am J Med 91:23–29.

Calvano SE, Thompson WA, Marra MN, et al. (1994): Changes in polymorphonuclear leukocyte surface and plasma bactericidal/permeability-increasing protein and plasma lipopolysaccharide binding protein during endotoxemia or sepsis. Arch Surg 129:220–226.

Car BD, Meloni F, Luisetti M, et al. (1994): Elevated IL-8 and MCP-1 in the bronchoalveolar lavage fluid of patients with idiopathic pulmonary fibrosis and pulmonary sarcoidosis. Am J Respir Crit Care Med 149:655–659.

Cathcart MK, McNally AK, Morel DW, Chisold GM (1989): Superoxide anion participation in human monocyte-mediated oxidation of low-density lipoprotein and conversion of low-density lipoprotein to a cytotoxin. J Immunol 142:1963–1969.

Chang MS, McNinch J, Basu R, Simonet S (1994): Cloning and characterization of the human neutrophil-activating peptide (ENA-78) gene. J Biol Chem 269:25277–25282.

Cooper NR, Morrison DC (1978): Binding and activation of the first component of human complement by the lipid A region of lipopolysaccharides. J Immunol 120:1862–1868.

Corbett MS, Schmitt I, Riess O, Walz A (1994): Characterization of the gene for human neutrophil-activating peptide 78 (ENA-78). Biochem Biophys Res Commun 205:612–617.

Damas P, Ledoux D, Nys M, et al. (1992): Cytokine serum level during severe sepsis in human IL-6 as a marker of severity. Ann Surg 215:356–362.

Danner RL, Elin RJ, Hosseini JM, et al. (1991): Endotoxemia in human septic shock. Chest 99:169–175.

Davis CF, Moore FDJ, Rodrick ML, Fearon DT, Mannick JA (1987): Neutrophil activation after burn injury: contributions of the classic complement pathway and of endotoxin. Surgery 102:477–484.

Dentener MA, Von Asmuth EJ, Francot GJ, Marra MN, Buurman WA (1993): Antagonistic effects of lipopolysaccharide binding protein and bactericidal/permeability-increasing protein on lipopolysaccharide-induced cytokine release by mononuclear phagocytes. Competition for binding to lipopolysaccharide. J Immunol 151:4258–4265.

Dofferhoff AS, Bom VJ, de Vries Hospers HG, et al. (1992): Patterns of cytokines, plasma endotoxin, plasminogen activator inhibitor, and acute-phase proteins during the treatment of severe sepsis in humans. Crit Care Med 20:185–192.

Donnelly SC, Strieter RM, Kunkel SL, et al. (1993): Interleukin-8 and development of adult respiratory distress syndrome in at-risk patient groups [see comments]. Lancet 341:643–647.

Donnelly SC, Strieter RM, Kunkel SL, et al. (1994): Chemotactic cytokines in the established adult respiratory distress syndrome and at-risk patients. Chest 105:98S–99S.

Downey GP, Chan CK, Lea P, Takai A, Grinstein S (1992): Phorbol ester-induced actin assembly in neutrophils: role of protein kinase C. J Cell Biol 116:695–706.

Elias JA, Freundlich B, Kern JA, Rosenbloom J (1990): Cytokine networks in the regulation of inflammation and fibrosis in the lung. Chest 97:1439–1445.

Elner VM, Strieter RM, Pavilack MA, et al. (1991): Human corneal interleukin-8, IL-1 and TNF-induced gene expression and secretion. Am J Pathol 139:977–988.

Emancipator K, Csako G, Elin RJ (1992): In vitro inactivation of bacterial endotoxin by human lipoproteins and apolipoproteins. Infect Immun 60:596–601.

Erwin AL, Munford RS (1990): Deacylation of structurally diverse lipopolysaccharides by human acyloxyacyl hydrolase. J Biol Chem 265:16444–16449.

Evans RD, Williamson DH (1991): Signals, mechanisms, and function of the acute lipid response to sepsis. Biochem Cell Biol 69:320–321.

Fisher CJ Jr, Marra MN, Palardy JE, et al. (1994): Human neutrophil bactericidal/permeability-increasing protein reduces mortality rate from endotoxin challenge: a placebo-controlled study [see comments]. Crit Care Med 22:553–558.

Flegel WA, Wolpl A, Mannel DN, Northoff H (1989): Inhibition of endotoxin-induced activation of human monocytes by human lipoproteins. Infect Immun 57:2237–2245.

Fong LG, Fong TA, Cooper AD (1991): Inhibition of lipopolysaccharide-induced interleukin-1 beta mRNA expression in mouse macrophages by oxidized low density lipoprotein. J Lipid Res 32:1899–1910.

Fong Y, Moldawer LL, Marano M, et al. (1989): Endotoxemia elicits increased circulating beta 2-IFN/IL-6 in man. J Immunol 142:2321–2324.

Freudenberg MA, Galanos C (1988): The metabolic fate of endotoxins. Prog Clin Biol Res 272:63–75.

Frey EA, Miller DS, Jahr TG, Sundan A, Bazil V, et al. (1992): Soluble CD14 participates in the response of cells to lipopolysaccharide. J Exp Med 176:1665–1671.

Gazzano-Santoro H, Parent JB, Grinna L, et al. (1992): High-affinity binding of the bactericidal/permeability-increasing protein and a recombinant amino-terminal fragment to the lipid A region of lipopolysaccharide. Infect Immun 60:4754–4761.

Girardin E, Grau GE, Dayer JM, Roux Lombard P, Lambert PH (1988): Tumor necrosis factor and interleukin-1 in the serum of children with severe infectious purpura. N Engl J Med 319:397–400.

Girardin E, Roux Lombard P, Grau GE, et al. (1992): Imbalance between tumour necrosis factor-alpha and soluble TNF receptor concentrations in severe meningococcaemia. The J5 Study Group. Immunology 76:20–23.

Goldstein JL, Ho YK, Basu SK, Brown MS (1979): Binding site on macrophages that mediates uptake and degradation of acetylated low density lipoprotein, producing massive cholesterol deposition. Proc Natl Acad Sci USA 76:333–337.

Goodman RB, Foster DC, Mathewes SL, et al. (1992a): Molecular cloning of porcine alveolar macrophage-derived neutrophil chemotactic factors I and II; identification of porcine IL-8 and another intercrine-alpha protein. Biochemistry 31:10483–10490.

Goodman RB, Wood RG, Martin TR, Hanson Painton O, Kinasewitz GT (1992b): Cytokine-stimulated human mesothelial cells produce chemotactic activity for neutrophils including NAP-1/IL-8. J Immunol 148:457–465.

Grabarek J, Her GR, Reinhold VN, Hawiger J (1990): Endotoxic lipid A interaction with human platelets. Structure-function analysis of lipid A homologs obtained from Salmonella minnesota Re595 lipopolysaccharide. J Biol Chem 265:8117–8121.

Granowitz EV, Santos AA, Poutsiaka DD, et al. (1991): Production of interleukin-1-receptor antagonist during experimental endotoxaemia. Lancet 338:1423–1424.

Gray PW, Flaggs G, Leong SR, et al. (1989): Cloning of the cDNA of a human neutrophil bactericidal protein. Structural and functional correlations. J Biol Chem 264:9505–9509.

Guthrie LA, McPhail LC, Henson PM, Johnston RBJ (1984): Priming of neutrophils for enhanced release of oxygen metabolites by bacterial lipopolysaccharide. J Exp Med 160:1656–1671.

Haberland ME, Fogelman AM, Edwards PA (1982): Specificity of receptor-mediated recognition of malondialdehyde-modified low density lipoproteins. Proc Natl Acad Sci USA 79:1712–1716.

Hack CE, De Groot ER, Felt Bersma RJ, et al. (1989): Increased plasma levels of interleukin-6 in sepsis. Blood 74:1704–1710.

Hack CE, Hart M, van Schijndel RJ, et al. (1992): Interleukin-8 in sepsis: relation to shock and inflammatory mediators. Infect Immun 60:2835–2842.

Hagen FS, Grant FJ, Kuijper JL, et al. (1991): Expression and characterization of recombinant human acyloxyacyl hydrolase, a leukocyte enzyme that deacylates bacterial lipopolysaccharides. Biochemistry 30:8415–8423.

Hamilton G, Hofbauer S, Hamilton B (1992): Endotoxin, TNF-alpha, interleukin-6 and parameters of the cellular immune system in patients with intraabdominal sepsis. Scand J Infect Dis 24:361–368.

Hampton RY, Golenbock DT, Penman M, Krieger M, Raetz CRH (1991a): Recognition and plasma clearance of endotoxin by scavenger receptors. Nature 352:342–344.

Hampton RY, Raetz CR (1991b): Macrophage catabolism of lipid A is regulated by endotoxin stimulation. J Biol Chem 266:19499–19509.

Han J, Mathison JC, Ulevitch RJ, Tobias PS (1994): Lipopolysaccharide (LPS) binding protein, truncated at Ile-197, binds LPS but does not transfer LPS to CD14. J Biol Chem 269:8172–8175.

Harris HW, Grunfeld C, Feingold KR, Rapp JH (1990): Human very low density lipoproteins and chylomicrons can protect against endotoxin-induced death in mice. J Clin Invest 86:696–702.

Haziot A, Chen S, Ferrero E, et al. (1988): The monocyte differentiation antigen, CD14, is anchored to the cell membrane by a phosphatidyl-inositol linkage. J Immunol 141:547.

Haziot A, Tsuberi B, Goyert SM (1993): Neutrophil CD14: biochemical properties and role in the secretion of tumor necrosis factor-α in response to lipopolysaccharide. J Immunol 150:5556–5565.

Heinecke JW, Rosen H, Chait A (1987): Iron and copper promote modification of low-density lipoprotein by human arterial smooth muscle cells in culture. J Clin Invest 74:1890–1894.

Henriksen TE, Mahoney EM, Steinberg D (1981): Enhanced macrophage degradation of low density lipoprotein previously incubated with cultured endothelial cells: recognition by receptors for acetylated low density lipoproteins. Proc Natl Acad Sci USA 78:6499–6503.

Hesse DG, Tracey KJ, Fong Y, et al. (1988): Cytokine appearance in human endotoxemia and primate bacteremia. Surg Gynecol Obstet 166:147–153.

Heumann D, Gallay P, Betz-Corradin S, et al. (1993): Competition between bactericidal/permeability-increasing protein and lipopolysaccharide-binding protein for lipopolysaccharide binding to monocytes. J Infect Dis 167:1351–1357.

Heumann D, Gallay P, Le Roy D, Landmann R, Glauser MP (1994): Radioimmunoassay versus flow cytometric assay to quantify LPS-binding protein (LBP) concentrations in human plasma. J Immunol Meth 171:169–176.

Hinshaw LB, Tekamp Olson P, Chang AC, et al. (1990): Survival of primates in LD100 septic shock following therapy with antibody to tumor necrosis factor (TNF alpha). Circ Shock 30:279–292.

Humes JL (1988): Regulation of leukotriene formation in inflammatory cells. Ann NY Acad Sci 524:252–259.

Hynes RO (1987): Integrins: a family of cell surface receptors. Cell 45:549–554.

Jurgens G, Hoff HF, Chisolm GMI, Esterbauer H (1993): Modification of human serum low density lipoprotein by oxidation: characterization and pathophysiological implications. Chem Phys Lipids.

Kelly CJ, Cech AC, Argenteanu M, et al. (1993): Role of bactericidal permeability-increasing protein in the treatment of gram-negative pneumonia. Surgery 114:140–146.

Kitchens RL, Ulevitch RJ, Munford RS (1992): Lipopolysaccharide (LPS) partial structures inhibit responses to LPS in a human macrophage cell line without inhibiting LPS uptake by a CD14-mediated pathway. J Exp Med 176:485–494.

Koch AE, Kunkel SL, Harlow LA, et al. (1994): Epithelial neutrophil activating peptide-78: a novel chemotactic cytokine for neutrophils in arthritis. J Clin Invest 94:1012–1018.

Kohn FR, Ammons WS, Horwitz A, et al. (1993): Protective effect of a recombinant amino-terminal fragment of bactericidal/permeability-increasing protein in experimental endotoxemia. J Infect Dis 168: 1307–1310.

Konig V, Hopf U, Moller B, et al. (1988): The significance of high-density lipoproteins (HDL) in the clearance of intravenously administered bacterial lipopolysaccharides (LPS) in mice. Hepatogastroenterology 35:111–115.

Kovach NL, Yee E, Munford RS, Raetz CR, Harlan JM (1990): Lipid IVA inhibits synthesis and release of tumor necrosis factor induced by lipopolysaccharide in human whole blood ex vivo. J Exp Med 172:77–84.

Kruger C, Schutt C, Obertacke U, et al. (1991): Serum CD14 levels in polytraumatized and severely burned patients. Clin Exp Immunol 85:297–301.

Ku G, Thomas CE, Akeson AL, Jackson RL (1992): Induction of interleukin 1 beta expression from human peripheral blood monocyte derived macrophages by 9-hydroxyoctadecadienoic acid. J Biol Chem 267:14183–14188.

Kunkel SL, Standiford T, Kasahara K, Strieter RM (1991a): Interleukin-8 (IL-8): the major neutrophil chemotactic factor in the lung. Exp Lung Res 17:17–23.

Kunkel SL, Strieter RM, Chensue SW, Campbell DA, Remick DG (1991b): The role of TNF in diverse pathologic processes. Biotherapy 3:135–141.

Lei MG, Morrison DC (1988a): Specific endotoxic lipopolysaccharide-binding proteins on murine splenocytes. II. Membrane localization and binding characteristics. J Immunol 141:1006–1011.

Lei MG, Morrison DC (1988b): Specific endotoxic lipopolysaccharide-binding proteins on murine splenocytes. I. Detection of lipopolysaccharide-binding sites on splenocytes and splenocyte subpopulations. J Immunol 141:996–1005.

Levin J, Poore TE, Zauber NP, Oser RS (1970): Detection of endotoxin in the blood of patients with sepsis due to gram-negative bacteria. N Engl J Med 283:1313–1316.

Loos M, Euteneuer B, Clas F (1990): Interaction of bacterial endotoxin (LPS) with fluid phase and macrophage membrane associated C1Q. The FC-recognizing component of the complement system. Adv Exp Med Biol 256:301–317.

Lukacs NW, Kunkel SL, Allen R, et al. (1995): Stimulus and cell-specific expression of C-X-C and C-C chemokines by pulmonary stromal cell populations. Am J Physiol 268:L856–L861.

Lynn WA, Raetz CR, Qureshi N, Golenbock DT (1991): Lipopolysaccharide-induced stimulation of CD11b/CD18 expression on neutrophils. Evidence of specific receptor-based response and inhibition by lipid A-based antagonists. J Immunol 147:3072–3079.

Maliszewski CR (1991): CD14 and immune response to lipopolysaccharide. Science 252:1321.

Mannion BA, Kalatzis ES, Weiss J, Elsbach P (1989): Preferential binding of the neutrophil cytoplasmic granule-derived bactericidal/permeability increasing protein to target bacteria. Implications and use as a means of purification. J Immunol 142:2807–2812.

Mannion BA, Weiss J, Elsbach P (1990): Separation of sublethal and lethal effects of the bactericidal/permeability increasing protein on Escherichia coli. J Clin Invest 85:853–860.

Marra MN, Wilde CG, Griffith JE, Snable JL, Scott RW (1990): Bactericidal/permeability-increasing protein has endotoxin-neutralizing activity. J Immunol 144:662–666.

Martich GD, Danner RL, Ceska M, Suffredini AF (1991): Detection of interleukin 8 and tumor necrosis factor in normal humans after intravenous endotoxin: the effect of anti-inflammatory agents. J Exp Med 173:1021–1024.

Mathison JC, Tobias PS, Wolfson E, Ulevitch RJ (1992): Plasma lipopolysaccharide (LPS)-binding protein. A key component in macrophage recognition of gram-negative LPS. J Immunol 149:200–206.

Mathison JC, Ulevitch RJ (1979): The clearance, distribution and cellular localization of lipopolysaccharide (LPS) in rabbits. J Immunol 123:2133–2143.

Mathison JC, Virca GD, Wolfson E, et al. (1990): Adaptation to bacterial lipopolysaccharide controls lipopolysaccharide-induced tumor necrosis factor production in rabbit macrophages. J Clin Invest 85:1108–1118.

Mathison JC, Wolfson E, Ulevitch RJ (1988): Participation of tumor necrosis factor in the mediation of Gram-negative bacterial lipopolysaccharide-induced injury in rabbits. J Clin Invest 81:1925–1937.

McCabe WR (1974): Gram-negative bacteremia. Adv Intern Med 19:135–158.

McDonagh J, Fossel ET, Kradin RL, et al. (1992): Effects of tumor necrosis factor-alpha on peroxidation of plasma lipoprotein lipids in experimental animals and patients. Blood 80:3217–3226.

Meszaros K, Parent JB, Gazzano Santoro H, et al. (1993): A recombinant amino terminal fragment of bactericidal/permeability-increasing protein inhibits the induction of leukocyte responses by LPS. J Leukoc Biol 54:558–563.

Michie HR, Manogue KR, Spriggs DR, et al. (1988): Detection of circulating tumor necrosis factor after endotoxin administration. N Engl J Med 318:1481–1486.

Minetti CA, Lin YA, Cislo T, Liu TY (1991): Purification and characterization of an endotoxin-binding protein with protease inhibitory activity from Limulus amebocytes. J Biol Chem 266:20773–20780.

Miyazaki A, Rahim AT, Araki S, Morino Y, Horiuchi S (1991): Chemical cross-linking alters high-density lipoprotein to be recognized by a scavenger receptor in rat peritoneal macrophages. Biochim Biophys Acta 1082:143–151.

Moore FDJ, Davis C, Rodrick M, Mannick JA, Fearon DT (1986): Neutrophil activation in thermal injury as assessed by increased expression of complement receptors. N Engl J Med 314:948–953.

Moreau SC, Skarnes RC (1973): Host resistance to bacterial endotoxemia: mechanisms in endotoxin tolerant animals. J Infect Dis 128:S122–S133.

Morel DW, DiCorleto PE, Chisolm GM (1986): Modulation of endotoxin-induced endothelial cell toxicity by low density lipoprotein. Lab Invest 55:419–426.

Morin M, Schindler R, Wakabayashi G, et al. (1991): Picogram concentrations of endotoxin stimulate synthesis of IL-1 beta and TNF alpha by human peripheral blood mononuclear cells exposed to recombinant human C5a. Eur Cytokine Netw 2:27–30.

Morrison DC, Kline LF (1977): Activation of the classical and properdin pathways of complement by bacterial lipopolysaccharides (LPS). J Immunol 118:362–368.

Munford RS, Hall CL, Dietschy JM (1981): Binding of *Salmonella typhimurium* lipopolysaccharides to rat high-density lipoproteins. Infect Immun 34:835–843.

Munford RS, Hall CL (1986): Detoxification of bacterial lipopolysaccharides (endotoxins) by a human neutrophil enzyme. Science 234:203.

Munford RS, Hall CL (1989): Purification of acyloxyacyl hydrolase, a leukocyte enzyme that removes secondary acyl chains from bacterial lipopolysaccharides. J Biol Chem 264:15613–15619.

Munford RS, Hall CL, Lipton JM, Dietschy JM (1982): Biological activity, lipoprotein-binding behavior and in vivo disposition of extracted and native forms of S. typhimurium lipopolysaccharides. J Clin Invest 70:877–888.

Muta T, Hashimoto R, Miyata T, et al. (1990): Proclotting enzyme from horseshoe crab hemocytes. cDNA cloning, disulfide locations, and subcellular localization. J Biol Chem 265:22426–22433.

Muta T, Miyata T, Misumi Y, et al. (1991): Limulus factor C. An endotoxin-sensitive serine protease zymogen with a mosaic structure of complement-like, epidermal growth factor-like, and lectin-like domains. J Biol Chem 266:6554–6561.

Nauseef WM, Volpp BD, McCormick S, Leidal KG, Clark RA (1991): Assembly of the neutrophil respiratory burst oxidase. Protein kinase C promotes cytoskeletal and membrane association of cytosolic oxidase components. J Biol Chem 266:5911–5917.

Navab M, Hough GP, Van Lenten BJ, Berliner JA, Fogelman AM (1988): Low density lipoproteins transfer bacterial lipopolysaccharides across endothelial monolayers in a biologically active form. J Clin Invest 81:601–605.

Northoff H, Gluck D, Wolpl A, Kubanek B, Galanos C (1987): Lipopolysaccharide-induced elaboration of Interleukin 1 by human monocytes: use for detection of lipopolysaccharide in serum and the influence of serum-lipopolysaccharide interactions. Rev Infect Dis 9 (suppl 5):S599–601.

Ooi CE, Weiss J, Doerfler ME, Elsbach P (1991): Endotoxin-neutralizing properties of the 25 kD N-terminal fragment and a newly isolate 30 kD C-terminal fragment of the 55-60 kD bactericidal/permeability-increasing protein of human neutrophils. J Exp Med 174:649–655.

Ooi CE, Weiss J, Elsbach P, Frangione B, Mannion B (1987): A 25-kDa NH2-terminal fragment carries all the antibacterial activities of the human neutrophil 60-kDa bactericidal/permeability-increasing protein. J Biol Chem 262(31):14891–14894.

Pangburn MK (1989): Analysis of recognition in the alternative pathway of complement. Effect of polysaccharide size. J Immunol 142:2766–2770.

Parillo JE, Parker MM, Natanson C, et al. (1990): Septic shock in humans. Advances in the understanding of pathogenesis, cardiovascular dysfunction, and therapy. Ann Intern Med 113:227–242.

Parsons PE, Worthen GS, Moore EE, Tate RM, Henson PM (1989): The association of circulating endotoxin with the development of the adult respiratory distress syndrome. Am Rev Respir Dis 140:294–301.

Parthasarathy S, Barnett J (1990): Phospholipase A2 activity of low density lipoprotein: evidence for an intrinsic phospholipase A2 activity of apoprotein B-100. Proc Natl Acad Sci USA 87:9741–9745.

Parthasarathy S, Fong LG, Otero D, Steinberg D (1987): Recognition of solubilized apoproteins from delipidated, oxidized low density lipoprotein (LDL) by the acetyl-LDL receptor. Proc Natl Acad Sci USA 84:537–540.

Parthasarathy S, Wieland E, Steinberg D (1989): A role for endothelial cell lipoxygenase in the oxidative modification of low density lipoprotein. Proc Natl Acad Sci USA 86:1046–1050.

Porat R, Yanoov M, Johns MA, Shibolet S, Michalevicz R (1992): Effects of endotoxin-associated protein on hematopoiesis. Infect Immun 60:1756–1760.

Porteu F, Nathan C (1990): Shedding of tumor necrosis factor receptors by activated human neutrophils. J Exp Med 172:599–607.

Rahim AT, Miyazaki A, Morino Y, Horiuchi S (1991): Biochemical demonstration of endocytosis and subsequent resecretion of high-density lipoprotein by rat peritoneal macrophages. Biochim Biophys Acta 1082:195–203.

Ramadori G, Meyer zum Buschenfelde KH, Tobias PS, Mathison JC, Ulevitch RJ (1990): Biosynthesis of lipopolysaccharide-binding protein in rabbit hepatocytes. Pathobiology 58:89–94.

Rankin JA, Sylvester I, Smith S, Yoshimura T, Leonard EJ (1990): Macrophages cultured in vitro release leukotriene B4 and neutrophil attractant/activation protein (interleukin 8) sequentially in response to stimulation with lipopolysaccharide and zymosan. J Clin Invest 86:1556–1564.

Rankin SM, Parthasarathy A, Steinberg D (1991): Evidence for a dominant role of lipoxygenase(s) in the oxidation of LDL by mouse peritoneal macrophages. J Lipid Res 32:449–456.

Rogivue C, Car BD, Allmann-Iselin I, Zwahlen RD, Walz A (1995): Bovine melanoma growth stimulatory activity: a new monocyte-macrophage-derived cytokine of the IL-8 family. Partial structure, function, and expression in acute pulmonary inflammation. Lab Invest 72:689–695.

Rolfe MW, Kunkel SL, Standiford TJ, et al. (1991): Pulmonary fibroblast expression of interleukin-8: a model for alveolar macrophage-derived cytokine networking. Am J Respir Cell Mol Biol 5:493–501.

Rudbach JA, Johnson AG (1964): Restoration of endotoxin activity following alteration by plasma. Nature 202:811–812.

Sanchez-Madrid F, Nagy JA, Robbins E, Simon P, Springer TA (1983): A human leukocyte differentiation antigen family with distinct alpha-subunits and a common beta-subunit: the lymphocyte function-associated antigen (LFA-1), the C3bi complement receptor (OKM1/Mac-1), and the p150,95 molecule. J Exp Med 158:1785–1803.

Schindler R, Gelfand JA, Dinarello CA (1990): Recombinant C5a stimulates transcription rather than translation of interleukin-1 (IL-1) and tumor necrosis factor: translational signal provided by lipopolysaccharide or IL-1 itself. Blood 76:1631–1638.

Schmouder RL, Streiter RM, Walz A, Kunkel SL (1995): Epithelial-derived neutrophil-activating factor-78 production in human renal tubule epithelial cells and in renal allograft rejection. Transplantation 59: 118–124.

Schroder JM, Sticherling M, Henneicke H, Preissner WC, Christophers E (1990): IL-1α or tumor necrosis factor-α stimulate release of three nap-1/IL-8-related neutrophil chemotactic proteins in human dermal fibroblasts. J Immunol 144:2223–2232.

Schumann RR, Leong SR, Flaggs GW, et al. (1990): Structure and function of lipopolysaccharide binding protein. Science 249:1429–1431.

Schutt C, Schilling T, Kruger C (1991): sCD14 prevents endotoxin inducible oxidative burst response of human monocytes. Allerg Immunol (Leipz) 37:159–164.

Shenep JL, Flynn PM, Barrett FF, Stidham GL, Westenkirchner DF (1988): Serial quantitation of endotoxemia and bacteremia during therapy for Gram-negative bacterial sepsis. J Infect Dis 157(3): 565–568.

Sica A, Matsushima K, Van Damme J, et al. (1990): IL-1 transcriptionally activates the neutrophil chemotactic factor/IL-8 gene in endothelial cells. Immunology 69:548–553.

Simmons DL, Tan S, Tenen DG, Nicholson-Weller A, Seed B (1989): Monocyte antigen CD14 is a phospholipid-anchored membrane protein. Blood 73(1):284–289.

Skarnes RC (1968): In vivo interaction of endotoxin with a plasma lipoprotein having esterase activity. J Bacteriol 95:2031–2034.

Smedegard G, Cui L, Hugli TE (1989): Endotoxin-induced shock in the rat. A role for C5a. Am J Pathol 135:489–497.

Solomkin JS, Jenkins MK, Nelson RD, Chenoweth D, Simmons RL (1981): Neutrophil dysfunction in sepsis. II Evidence for the role of complement activation products in cellular deactivation. Surgery 90:319–326.

Solomkin JS, Nelson RD, Chenoweth DE, Solem LD, Simmons RL (1984): Regulation of neutrophil migratory function in burn injury by complement activation productions. Ann Surg 200:742–746.

Standiford TJ, Kunkel SL, Basha MA, et al. (1990): Interleukin-8 gene expression by a pulmonary epithelial cell line. A model for cytokine networks in the lung. J Clin Invest 86:1945–1953.

Strieter RM, Kunkel SL, Burdick MD, Lincoln PM, Walz A (1992): The detection of a novel neutrophil-activating peptide (ENA-78) using a sensitive ELISA. Immunol Invest 21:589–596.

Stumacher RJ, Kovnat MJ, McCabe WR (1973): Limitations of the usefulness of the limulus assay for endotoxin. N Engl J Med 283:1261–1264.

Takahashi K, Fukuda S, Naito M, et al. (1989): Endocytic pathway of high density lipoprotein via trans-Golgi system in rat resident peritoneal macrophages. Lab Invest 61:270–277.

Thomas KM, Taylor L, Prado G, et al. (1994): Functional and ligand binding specificity of the rabbit neutrophil IL-8 receptor. J Immunol 152:2496–2500.

Thornton AJ, Strieter RA, Lindley I, Baggiolini M, Kunkel SL (1990): Cytokine-induced gene expression of a neutrophil chemotactic factor/IL-8 in human hepatocytes. J Immunol 144:2609–2613.

Tobias PS, Mathison JC, Ulevitch RJ (1988): A family of lipopolysaccharide binding proteins involved in responses to gram-negative sepsis. J Biol Chem 263:13479–13481.

Tobias PS, McAdam KP, Soldau K, Ulevitch RJ (1985): Control of lipopolysaccharide-high-density lipoprotein interactions by an acute-phase reactant in human serum. Infect Immun 50:73–76.

Tobias PS, Soldau K, Kline L, et al. (1993): Cross-linking of lipopolysaccharide (LPS) to CD14 on THP-1 cells mediated by LPS-binding protein. J Immunol 150:3011–3021.

Tobias PS, Soldau K, Ulevitch RJ (1986): Isolation of a lipopolysaccharide-binding acute phase reactant from rabbit serum. J Exp Med 164:777–793.

Tobias PS, Soldau K, Ulevitch RJ (1989): Identification of a lipid A binding site in the acute phase reactant lipopolysaccharide binding protein. J Biol Chem 264:10867–10871.

Tobias PS, Ulevitch RJ (1983): Control of lipopolysaccharide-high density lipoprotein binding by acute phase protein(s). J Immunol 131:1913–1916.

Tracey KJ, Fong Y, Hesse DG, et al. (1987a): Anti-cachectin/TNF monoclonal antibodies prevent septic shock during lethal bacteraemia. Nature 330:662–664.

Tracey KJ, Lowry SF, Fahey TJI, Albert JD (1987b): Cachectin/Tumor Necrosis Factor induces lethal shock and stress hormone responses in the dog. Surg Gynecol Obstet 164:415–422.

Ulevitch RJ, Johnston AR (1978): The modification of biophysical and endotoxic properties of bacterial lipopolysaccharides by serum. J Clin Invest 62:1313–1324.

Ulevitch RJ, Johnston AR, Weinstein DB (1979): New function for high density lipoproteins. Their participation in intravascular reactions of bacterial lipopolysaccharides. J Clin Invest 64:1516–1524.

Ulevitch RJ, Mathison JC, Lee JD, et al. (1992): A new model for receptor mediated stimulation of macrophages (m) by bacterial lipopolysaccharide (lps) [abstract]. J Cell Biochem Suppl 16c:154

Ulevitch RJ, Tobias PS, (1988): Interactions of bacterial lipopolysaccharide with serum proteins. Prog Clin Biol Res 272:309–318.

van der Poll T, van Deventer SJ, Hack CE, et al. (1992): Effects on leukocytes after injection of tumor necrosis factor into healthy humans. Blood 79:693–698.

van Deventer SJ, Buller HR, ten Cate JW, et al. (1990): Experimental endotoxemia in humans: analysis of cytokine release and coagulation, fibrinolytic, and complement pathways. Blood 76:2520–2526.

Van Deventer SJH, Buller HRtW, Sturk A, Pauw W (1988): Endotoxemia: an early predictor of septicaemia in febrile patients. Lancet 1:605–609.

Van Lenten BJ, Fogelman AM (1992): Lipopolysaccharide-induced inhibition of scavenger receptor expression in human monocyte-macrophages is mediated through tumor necrosis factor-alpha. J Immunol 148:112–116.

VanLenten BJ, Fogelman AM, Haberland ME, Edwards PA (1986): The role of lipoproteins and receptor-mediated endocytosis in the transport of bacterial lipopolysaccharide. Proc Natl Acad Sci USA 83: 2704–2708.

Van Zee KJ, DeForge LE, Fischer E, et al. (1991): IL-8 in septic shock, endotoxemia, and after IL-1 administration. J Immunol 146:3478–3482.

Vosbeck K, Tobias P, Mueller H, et al. (1990): Priming of polymorphonuclear granulocytes by lipopolysaccharides and its complexes with lipopolysaccharide binding protein and high density lipoprotein. J Leukoc Biol 47:97–104.

Vukajlovich SW, Hoffman J, Morrison DC (1987): Activation of human serum complement by bacterial lipopolysaccharides: structural requirements for antibody independent activation of the classical and alternative pathways. Mol Immunol 24:319–331.

Waage A, Brandtzaeg P, Halstensen A, Kierulf P, Espevik T (1989): The complex pattern of cytokines in serum from patients with meningococcal septic shock. Association between interleukin 6, interleukin 1, and fatal outcome. J Exp Med 169:333–338.

Walz A (1992): Generation and properties of neutrophil-activating peptide 2. [review]. Cytokines 4:77–95.

Walz A, Burgener R, Car B, et al. (1991): Structure and neutrophil-activating properties of a novel inflammatory peptide (ENA-78) with homology to interleukin 8. J Exp Med 174:1355–1362.

Walz A, Strieter RM, Schnyder S (1993): Neutrophil-activating peptide ENA-78. [review]. Adv Exp Med Biol 351:129–137.

Wang T, Yu WG, Powell WS (1992): Formation of monohydroxy derivatives of arachidonic acid, linoleic acid, and oleic acid during oxidation of low density lipoprotein by copper ions and endothelial cells. J Lipid Res 33:525–537.

Warren HS, Novitsky TJ, Ketchum PA, et al. (1985): Neutralization of bacterial lipolysaccharides by human plasma. J Clin Microbiol 22:590–595.

Warren HS, Riveau GR, de Deckker FA, Chedid LA (1988): Control of endotoxin activity and interleukin-1 production through regulation of lipopolysaccharide-lipoprotein binding by a macrophage factor. Infect Immun 56:204–212.

Weersink AJL, vanKessel KPM, vandenTol ME, et al. (1993): Human granulocytes express a 55-kDa lipopolysaccharide-binding protein on the cell surface that is identical to the bactericidal/permeability-increasing protein. J Immunol 150:253–263.

Weiss J, Elsbach P, Shu C, et al. (1992): Human bactericidal/permeability-increasing protein and a recombinant NH2-terminal fragment cause killing of serum-resistant gram-negative bacteria in whole blood and inhibit tumor necrosis factor release induced by the bacteria. J Clin Invest 90:1122–1130.

Weiss J, Olsson I (1987): Cellular and subcellular localization of the bactericidal/permeability-increasing protein of neutrophils. Blood 69:652–659.

Wilson F, Parrillo JEA, eds. (1984): *Major Issues in Critical Care Medicine.* Baltimore: Williams & Wilkins. Surgical intensive care units, pp 17–33.

Winn RK, Mileski WJ, Kovach NL, et al. (1993): Role of protein synthesis and CD11/CD18 adhesion complex in neutrophil emigration into the lung. Exp Lung Res 19:221–235.

Witztum JL, Steinberg D (1991): Role of oxidized low density lipoprotein in atherogenesis. J Clin Invest 88:1785–1792.

Wolf SM, Bennett JV (1974): Gram-negative-rod bacteremia. N Engl J Med 291:733.

Wortel CH, Von der Möhlen MAM, Van Deventer SJH, et al. (1992): Effectiveness of a human monoclonal anti-endotoxin antibody (HA-1A) in gram-negative sepsis: Relationship to endotoxin and cytokine levels. J Infect Dis 166:1367–1374.

Worthen GS, Avdi N, Vukajlovich S, Tobias PS (1992): Neutrophil adherence induced by lipopolysaccharide in vitro. Role of plasma component interaction with lipopolysaccharide. J Clin Invest 90:2526–2535.

Worthen GS, Haslett C, Rees AJ, et al. (1987): Neutrophil-mediated pulmonary vascular injury. Synergistic effect of trace amounts of lipopolysaccharide and neutrophil stimuli on vascular permeability and neutrophil sequestration in the lung. Am Rev Respir Dis 136:19–28.

Wright SD, Detmers PA (1988): Adhesion-promoting receptors on phagocytes. J Cell Sci 9(Suppl):99–120.

Wright SD, Detmers PA, Aida Y, et al. (1990a): CD18-deficient cells respond to lipopolysaccharide in vitro. J Immunol 144:2566–2571.

Wright SD, Jong MTC (1986): Adhesion-promoting receptors on human macrophage recognize Escherichia coli by binding to lipopolysaccharide. J Exp Med 164:1876–1888.

Wright SD, Levin SM, Jong MTC, Chad Z, Kabbash LG (1989a): CR3 (CD11b/CD18) expresses one binding site for Arg-Gly-Asp-containing peptides and a second site for bacterial lipopolysaccharide. J Exp Med 169:175–183.

Wright SD, Ramos RA, Hermanowski Vosatka A, Rockwell P, Detmers PA (1991): Activation of the adhesive capacity of CR3 on neutrophils by endotoxin: dependence on lipopolysaccharide binding protein and CD14. J Exp Med 173:1281–1286.

Wright SD, Ramos RA, Patel M, Miller DS (1992): Septin: a factor in plasma that opsonizes lipopolysaccharide-bearing particles for recognition by CD14 on phagocytes. J Exp Med 176:719–727.

Wright SD, Ramos RA, Tobias PS, Ulevitch RJ, Mathison JC (1990b): CD14, a receptor for complexes of lipopolysaccharide (LPS) and LPS binding protein. Science 249:1431–1433.

Wright SD, Tobias PS, Ulevitch RJ, Ramos RA (1989b): Lipopolysaccharide (LPS) binding protein opsonizes LPS-bearing particles for recognition by a novel receptor on macrophages. J Exp Med 170: 1231–1241.

Wrightman PD, Raetz CRH (1984): The activation of protein kinase C by biologically active lipid moieties of lipopolysaccharides. J Biol Chem 259:10048–10052.

Xu Q, Jurgens G, Huber LA, et al. (1992): Lipid utilization by human lymphocytes is correlated with high-density-lipoprotein binding site activity. Biochem J 285:105–112.

Yla Herttuala S, Lipton BA, Rosenfeld ME, et al. (1991): Expression of monocyte chemoattractant protein 1 in macrophage-rich areas of human and rabbit atherosclerotic lesions. Proc Natl Acad Sci USA 88:5252–5256.

Ziegler EJ, Fisher CJ, Sprung CL, et al. (1991): Treatment of gram-negative bacteremia and septic shock with HA-1A human monoclonal antibody against endotoxin. A randomized, double-blind, placebo-controlled trial. The HA-1A Sepsis Study Group. N Engl J Med 324:429–436.

6

Lipopolysaccharide-Binding Proteins and Receptors

Julie A. Gegner
The University of San Diego, San Diego, California

Peter S. Tobias
The Scripps Research Institute, La Jolla, California

Invasion of gram-negative bacteria into the blood can initiate septic shock, which is defined progressively by high fever, hypotension, disseminated intravascular coagulation, multiple-organ failure, and death. The disease septic shock rarely makes its way into conversations outside the medical and scientific communities. This may be because it is often, but not exclusively, associated with other primary traumas, infections, and autoimmune disorders. It is a major medical problem with more than 100,000 cases occurring annually in the United States alone. Mortality rates remain high at 20–50% and although research efforts in this field are intense, no effective treatment has been found (Glauser et al., 1991; Bone, 1991).

Recent advances have identified the early molecular events triggered by gram-negative bacteria. A complex web of interacting serum proteins and membrane surface receptors is emerging; those interactions discussed in this chapter are summarized in Figure 1. Central to our discussion is lipopolysaccharide (LPS), the endotoxic moiety shed from the outer membrane of gram-negative bacteria. Lipopolysaccharide is a unique and highly conserved molecule that initiates a cascade of cellular events. The

Phagocyte Function: A Guide for Research and Clinical Evaluation, Edited by J. Paul Robinson and George F. Babcock. ISBN 0-471-12364-1

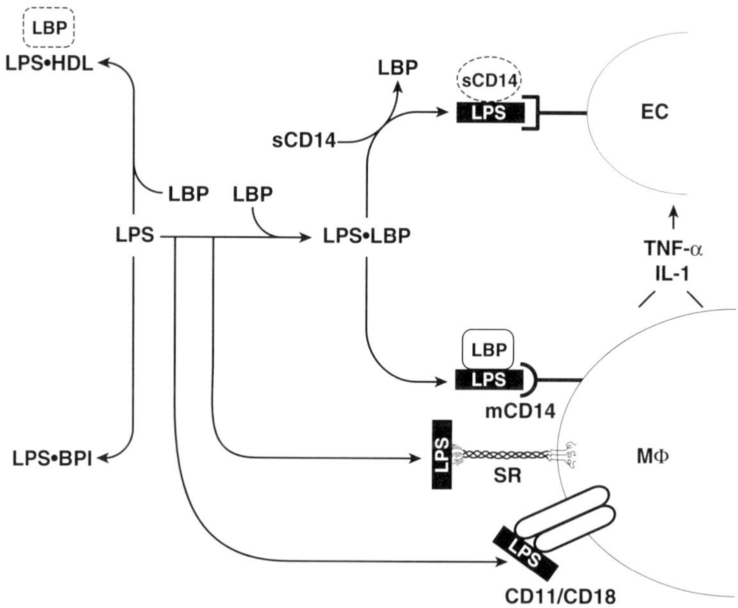

Fig. 1. Lipopolysaccharide (LPS) pathways. Lipopolysaccharide in blood is bound by the LPS-binding protein (LBP) and delivered to membrane CD14 (mCD14) on myeloid cells (MΦ). This initiates cellular activation events leading to cytokine release. The cytokines such as tumor necrosis factor-α (TNF-α) and interleukin 1 (IL-1) subsequently activate endothelial and epithelial cells (EC). In a second pathway, LBP delivers LPS to soluble CD14 (sCD14). The LPS–sCD14 complex can activate EC directly. The dotted line around sCD14 illustrates that interaction between sCD14 and EC has been proposed but not actually demonstrated. Lipopolysaccharide also binds to CD11a-c/CD18. LPS initiates nuclear transcription factor NF-κB activation through CD11b/CD18 although the biological relevance of this activation pathway has not been established. Lipopolysaccharide is inactivated by several mechanisms. Lipopolysaccharide–LBP complexes bound to mCD14 can internalize without cellular activation. Scavenger receptors (SRs) also internalize LPS without triggering cellular activation. In a serum reaction catalyzed by LBP, LPS is neutralized by associating with high-density lipoproteins (HDL). The dotted line around LBP indicates that the nature of the HDL–LBP interaction is not fully defined. Bactericidal/permeability-increasing protein (BPI) binds LPS in the outer membrane of gram-negative bacteria and initiates events leading to cell death. The variables controlling the distribution of LPS among these molecules have not been identified.

major cellular activation pathway identified involves LPS binding to the myeloid cell surface receptor membrane CD14 (mCD14). Lipopolysaccharide binding to mCD14 leads to gene transcription and cytokine release. Most notably, levels of tumor necrosis factor-α (TNF-α), interleukin-1 (IL-1), IL-6, and IL-8 increase. Levels of other mediators, such as nitric oxide (NO), superoxide (O_2^-), and hydrogen peroxide (H_2O_2), similarly increase as do the levels of the lipids prostaglandin E_2, thromboxane A_2, and platelet activating factor. Cells lacking mCD14, such as endothelial and epithelial cells (EC), are activated either indirectly by the cytokines released from stimulated mCD14-bearing cells or directly by LPS bound to the soluble form of CD14 (sCD14) found in serum. Adhesive proteins on EC surfaces are upregulated in response to LPS. Increased expression of E-selectin, VCAM, and ICAM-1 is observed, as is increased release of IL-6 and IL-8. An alternative LPS activation pathway may signal through the leukocyte integrin CD11b/CD18. The biological significance of the CD11b/CD18 LPS receptor system has yet to be identified.

Lipopolysaccharide interactions with both soluble and membrane forms of CD14 are facilitated by a serum component termed LPS binding protein (LBP). Lipopolysaccharide-binding protein binds and dissociates LPS aggregates, then catalytically transfers LPS to sCD14 or forms a stoichiometric, ternary LBP–LPS–mCD14 complex on the cell surface. Lipopolysaccharide association with mCD14 initiates the cellular activation events described above, or in a process distinct from cellular activation, the LBP–LPS complex internalizes. Internalization can be separated from cellular activation by monoclonal antibodies (mAbs) specific for CD14 and LBP. Another LPS clearance route utilizes macrophage scavenger receptors. Although the scavenger receptors bind and internalize LPS, they do not trigger the cellular activation events described above. The functional consequences of which receptor internalizes LPS are not known. A third path of LPS clearance from serum is its association with high-density lipoproteins (HDL) in a reaction again catalyzed by LBP. Substoichiometric amounts of LBP loosely associate with HDL, enabling LPS to bind. Another LBP function observed in vitro involves the bactericidal/permeability-increasing protein (BPI), an LBP homologue important in the killing and clearance of bacteria. It binds LPS and disrupts the bacterial membrane. It is highly bactericidal whereas LBP is not. The in vitro results suggest that LBP can potentiate the bactericidal activity of BPI. The mechanism of this phenomenon is not understood.

To develop effective treatments against the pathogenesis of septic shock, it is critical to understand how the functions of proteins that bind LPS are orchestrated by the immune system. Low levels of LPS are presumably without detrimental consequences. However, when an organism is stressed by a prior trauma or higher levels of LPS are present, hyperreactivity occurs and gram-negative septic shock can result. Although we are beginning to identify the critical components in the LPS signal transduction pathway, there are few scientific data to describe how their activities are coordinated. One of the earliest observations is the dramatic rise in LBP serum levels in response to inflammatory stimuli. Serum levels of sCD14 also increase. Modulation of LBP and sCD14 serum levels suggests one way in which the response to LPS may be regulated. It is becoming increasingly clear from the number of proteins that bind LPS identified in serum and on the cell surface that LPS distribution among these molecules is a key determinant in the type of immune response elicited.

LIPOPOLYSACCHARIDES: THE LIGAND

Basic Structure and Solubility Properties of Lipopolysaccharide

The bioactive moiety that triggers the cascade of immune responses associated with gram-negative sepsis is lipid A. Lipid A is found in the outer layer of the outer membrane of gram-negative bacteria and is the membrane anchor of a larger structure known as lipopolysaccharide (LPS) or endotoxin (Fig. 2). Lipopolysaccharide is an amphipathic molecule composed of hydrophilic polysaccharides and the lipophilic

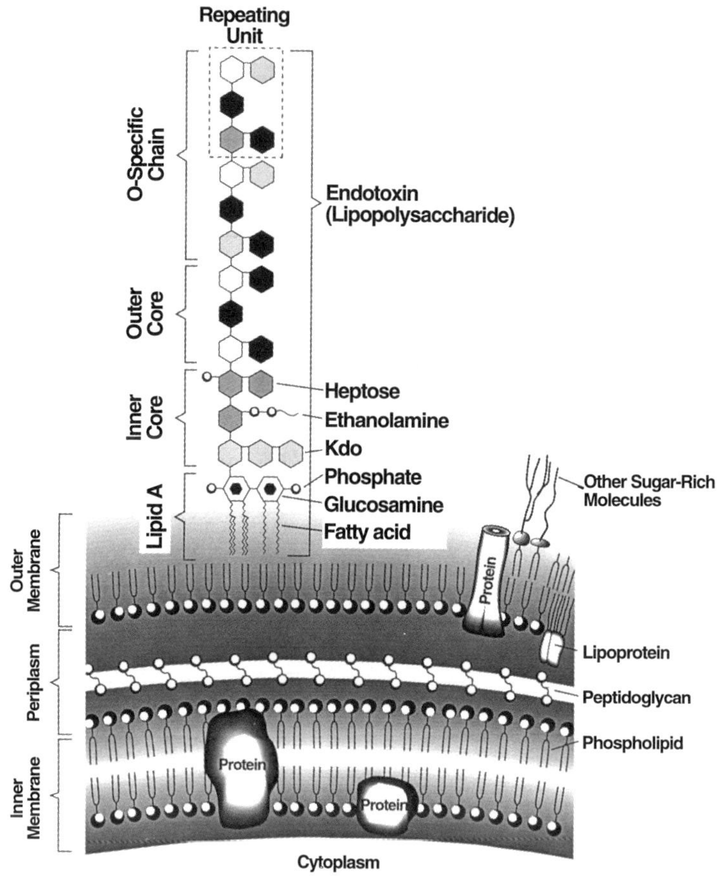

Fig. 2. Lipopolysaccharide core structure and location in the bacterial membrane. Lipopolysaccharide is an identifying feature of gram-negative bacteria and is an abundant component of the outer membrane. The schematic cross-section illustrates how the hydrophobic lipid A moiety of LPS is inserted into the bacterial membrane. The *O*-specific sugar chain, which contains up to five repeating sugar residues, varies from one bacterial species to another and is attached to lipid A by a two-part core. The LPS molecule is not drawn to scale and is adapted from Rietschel and Brade (1992).

lipid A (Raetz, 1990; Rietschel, 1984; Rietschel et al., 1994). It contains up to four structural domains that become progressively more variable. The first domain, lipid A, is a highly conserved phosphoglycolipid that contains D-glucopyranosidic hexosamine residues present as a β(1,6)-linked disaccharide. The lipid A backbone is typically phosphorylated at positions 1 and 4′, and is substituted with acylated, ester-, and amide-linked (*R*)-3-hydroxy fatty acids. Variations in lipid A structure result from the type of hexosamine present, the degree of phosphorylation, the presence of phosphate substituents, and most notably for endotoxicity, the nature, chain length, number, and location of the fatty acyl groups. Lipid A is glycosylated with certain unique sugars such as the eight-carbon sugar 3-deoxy-D-mannooctulosonic acid (ketodeoxyoctulosonate, KDO). Attached to the KDO are ethanolamine and heptose. The KDO together with ethanolamine and heptose is referred to as the inner core. An outer core made up of hexoses such as glucose, galactose, and *N*-acetyl-D-glucosamine, is attached to the inner core. The outer core is followed by a repeating oligosaccharide polymer known as the *O*-antigen. The enormous structural variability in the *O*-antigen determines the serological specificity of the LPS and may have evolved to help bacteria avoid detection by the immune system. Although lipid A alone is sufficient to trigger an endotoxic response, it is important to note that the other domains of the LPS molecule and perhaps other bacterial components may contribute to endotoxemia in vivo (Danner et al., 1990).

Mutant strains of bacteria have been characterized that produce incomplete forms of LPS (Raetz, 1984). Historically these mutants proved to be invaluable tools in elucidating the details of LPS biosynthesis. The nomenclature used in describing different forms of LPS refers to the type of bacterial strain from which it is isolated. For example, smooth LPS is purified from strains that produce all four domains whereas rough LPS is purified from mutants that produce LPS molecules that lack the *O*-antigen. Deep rough mutants (Re mutants) are isolated from heptose-deficient strains and consist of KDO sugars and lipid A. Lipid IV_A is a disaccharide precursor of lipid A that accumulates in KDO-deficient mutants. Lipid IV_A is not glycosylated and has no secondary fatty acyl groups (Raetz et al., 1985).

Besides the structural diversity of LPS among bacterial species and strains, LPS microheterogeneity is found within each organism and hence each LPS preparation (Caroff et al., 1993; Nowotny, 1984). The smooth LPS preparations demonstrate the largest degree of molecular polymorphism, primarily in size, and ReLPS, generally, the least. LPS typically is quantified either by an assay to detect 2-keto-3-deoxyoctanoic acid (such as KDO) (Cynkin and Ashwell, 1960) or by assays of bioactivity. The most common bioassay is the chromogenic assay referred to as the *Limulus* amebocyte lysate (LAL) assay for endotoxin. *Limulus* amebocyte lysate, obtained from the horseshoe crab *Limulus polyphemus,* contains a coagulation system that is activated by LPS (Cooper et al., 1972). Sometimes the KDO and bioactivity assays can give different results. Although the KDO content is related to the lipid A content of LPS, heterogeneity in the LPS preparation allows for only approximate calculations of lipid A molarity. A disadvantage with the bioassay is that it excludes LPS associated with neutralizing molecules such as polymyxin B and HDL unless prior processing is included. An alternative method used with chemically defined preparations, such as lipid IV_A, is mass spectroscopy (Raetz et al., 1985). Mass spectroscopy is not practical for heterogeneous LPS preparations.

Another difficulty that arises when working with the LPS ligand is its limited solubility in aqueous solutions (Galanos and Luderitz, 1984). Solubility depends on a number of variables such as polysaccharide content, phosphate content (negative charge), and salt form. For example, smooth LPS generally is more soluble than rough LPS because of its higher polysaccharide content. Negatively charged phosphate and carboxyl groups improve solubility whereas the presence of divalent cations such as Ca^{2+} and Mg^{2+} cause LPS to form turbid solutions. Solubility improves with Na^+, K^+, NH_4^+, pyridinium, and ethanolammonium salts and even more with triethylammonium salt. Although clear solutions can often be produced by sonication, even a clear solution contains highly aggregated LPS. Monomeric LPS exists in equilibrium with aggregated LPS, varying in concentration as a function of pH and ionic state. Monomer solubility has been determined to be about 10^{-8} *M* for ReLPS (Takayama et al., 1990; Din et al., 1993). Solubility studies suggest that LPS monomers are in equilibrium with polymorphic aggregates but not spherical micelles (Hofer et al., 1991). Dissolution of LPS aggregates into monomers has been proposed as the initial step in the activation of cells (Takayama et al., 1990; Din et al., 1993).

Crude preparations of LPS are often contaminated with low-molecular-weight cations that associate with the negatively charged ligand. Ethylenediaminetetraacetic acid (EDTA) is added to chelate the cations and facilitate dispersion of the LPS aggregates. Because LPS is a very sticky molecule, LPS preparations are typically stored dried or as concentrated stock solutions (1–5 mg/ml) to minimize loss due to nonspecific adhesion to tubes. To reduce aggregate size and facilitate dispersion of the LPS aggregates, LPS solutions are sonicated upon reconstitution. Lipopolysaccharide is often used in the presence of a protein carrier such as fatty acid-free bovine serum albumin (BSA) to minimize nonspecific binding to surfaces and facilitate dispersion (Galanos et al., 1971). Activity is retained when LPS is associated with BSA (Takayama et al., 1990). LPS also associates with albumin that is bound to the cell surface (Dziarski, 1994). The relevance of LPS binding to cell-bound albumin is not known.

Structure and Bioactivity

Lipid A has long been suspected of being the endotoxic component of LPS. Decisive evidence was obtained after the *Escherichia coli* lipid A structure was elucidated and chemically synthesized (Imoto et al., 1985). The biological properties of the synthetic lipid A were identical to those of the bacterial lipid A (Galanos et al., 1985). The structural requirements for lipid A endotoxicity are quite specific. This is clearly demonstrated when the endotoxic lipid A from *E. coli* is compared to the nontoxic version from *Rhodobacter sphaeroides* (Fig. 3). Lipid A from *E. coli* has six fatty acyl groups with the chain length of the hydroxy fatty acids being 14, whereas the *R. sphaeroides* version has only five fatty acyl groups with a hydroxy fatty acid chain length of 10 and one degree of unsaturation on the secondary acyl chain (Qureshi et al., 1988). A synthetic version of lipid A from *R. sphaeroides* also has been produced

Fig. 3. Lipopolysaccharide structures and taxol. (A) The deep rough *E. coli* ReLPS shown is composed of lipid A and two KDO sugars. In rough LPS mutants, R^1 is where heptose and phosphoethanolamine would add to complete the inner core. In LPS from smooth strains of bacteria, the outer core and *O*-antigen regions would follow the inner core. R^2 is a potential site of phosphoethanolamine attachment. The arrows indicate sites deacylated by acyloxyacyl hydrolase (AOAH). (B) The *R. sphaeroides* diphosphoryl lipid A (RsDPLA) structure differs from *E. coli* lipid A by its hydroxy fatty acid chain length (10 vs. 14), number of acyl chains (5 vs. 6), and fatty acid unsaturation (1 vs. 0). (C) The lipid A precursor lipid IV_A has only the glucosamine sugars and four fatty acyl groups. The 4′-phosphate group on the right is radiolabeled with [γ^{32}P]ATP by *E. coli* 4′-kinase to form 4′[^{32}P]lipid IV_A. (D) Although the structure of taxol is quite dissimilar from the bioactive moiety lipid A, it behaves as an LPS mimetic.

A. ReLPS

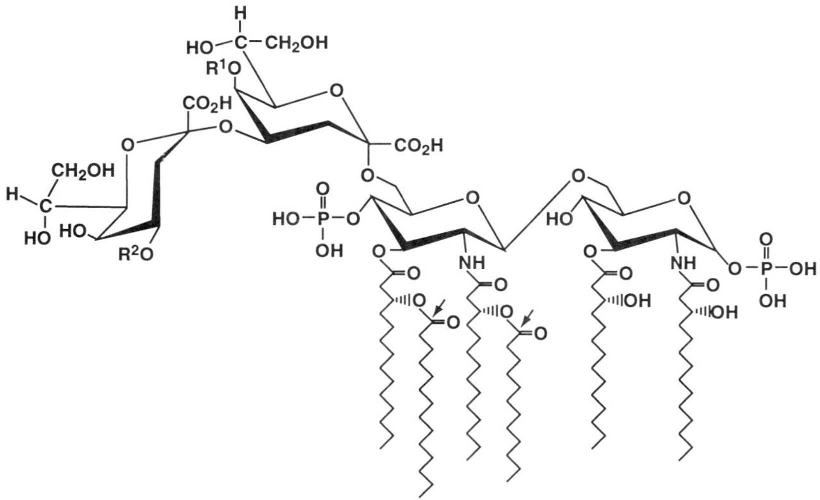

B. RsDPLA

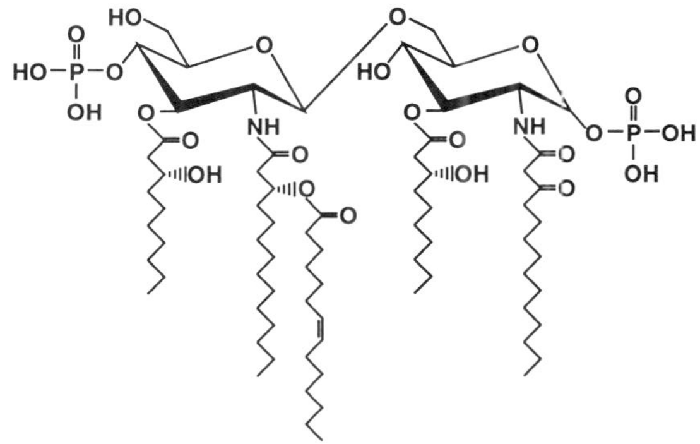

C. Lipid IV_A

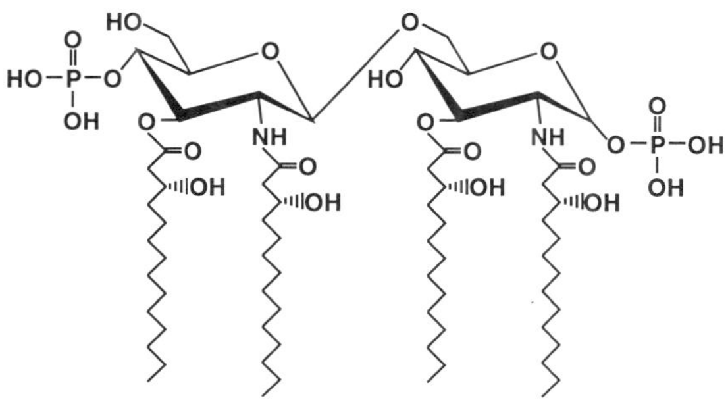

D. Taxol

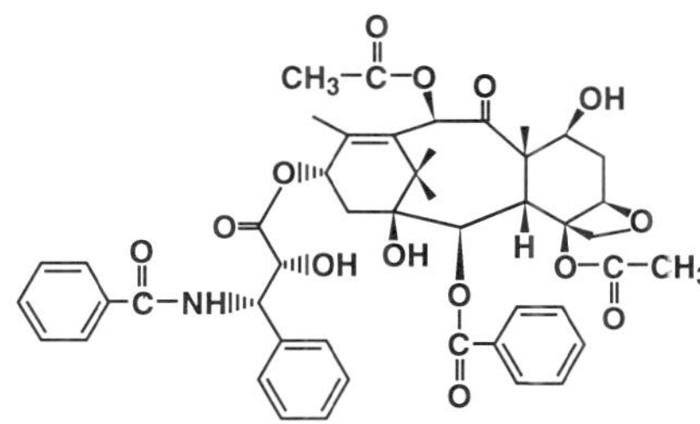

(Rose et al., 1995). In this case, the biological properties of the synthetic molecule were similar but not identical to the bacterial lipid A. Very high concentrations (10–100 μM) of the bacterial *R. sphaeroides* molecule can activate cells whereas the synthetic version lacks agonistic effects. This difference suggests that either the proposed structure is incorrect or the bacterial preparation is contaminated with an unidentified agonist.

The key property shared by the biological and synthetic *R. sphaeroides* lipid A molecules is their ability to inhibit the effects of endotoxic lipid A molecules (Rose et al., 1995; Takayama et al., 1989). The synthetic and biological molecules are equally potent at inhibiting TNF-α release and generation of NO in murine macrophages. *R. sphaeroides* diphosphoryl lipid A (RsDPLA) inhibits the endotoxic effects of ReLPS when added at 1–10 mol RsDPLA per mole of ReLPS. Lipid A antagonists such as RsDPLA have potential as therapeutic agents (Christ et al., 1995). One typical biological response to LPS initiated by RsDPLA is tolerance to subsequent LPS challenges. Pretreatment (1 h) of normal mice with either ReLPS or RsDPLA prevents the induction of TNF-α and protects the mice against a lethal ReLPS challenge (Zuckerman and Qureshi, 1992). Although the mechanism(s) of tolerance have not been fully elucidated, glucocorticoids appear to be involved. In adrenalectomized (adrex) mice, pretreatment with ReLPS or RsDPLA does not inhibit induction of TNF-α or protect the mice from a lethal ReLPS challenge, whereas addition of exogenous glucocorticoids restores the ability of adrex mice to become LPS tolerant (Zuckerman and Qureshi, 1992).

The endotoxicity of LPS is altered by acyloxyacyl hydrolase, which detoxifies LPS by removing the secondary fatty acids of lipid A (Munford and Hall, 1986). Not only is deacylated LPS (dLPS) nontoxic, but it also acts as an antagonist toward many of the known effects of LPS (Pohlman et al., 1987; Dal Nogare and Yarbrough, 1990). Binding studies have shown that the antagonistic effects of dLPS can not be explained by competitive binding to mCD14 (Kitchens et al., 1992). Deacylated LPS has similar antagonist effects upon endothelial cells and macrophages (Riedo et al., 1990), suggesting that the downstream signaling step blocked by dLPS in these cells may be similar. An alternative possibility is that rather than block a signal, dLPS initiates a negative signal to counteract the signal generated by LPS. Experiments designed to resolve these issues will provide insights into the mechanism(s) of LPS signal transduction.

The endotoxicity of lipid A depends not only on structure but also on the target species (Golenbock et al., 1991). Differential responses among hosts become apparent when the effects of the biosynthetic lipid A precursor lipid IV_A are evaluated (Fig. 3). The structure of lipid IV_A is essentially identical to that of dLPS except that lipid IV_A lacks KDO groups. Whereas lipid IV_A acts as an LPS antagonist toward human cells (Kovach et al., 1990; Lynn et al., 1991), it acts as an LPS-mimetic toward mouse cells (Hampton et al., 1991). The basis for the species difference is not known. Transfection of human CD14 into mouse fibroblast cells does not alter the responsiveness of the mouse cells to lipid IV_A, nor does transfection of murine CD14 into human cells alter the responsiveness of the human cells. Hamster cells show yet another species specificity toward the various lipid A molecules. ReLPS, lipid IV_A, and even RsD-

PLA activate hamster cells. As before, the species origin of CD14 had no effect on responsiveness (Delude et al., 1994). This suggests that the species origin of the mCD14 is not the determining factor and that a species-specific molecule distinct from mCD14 is the signal transducing target of these molecules. The biological relevance of the different species sensitivity to the lipid A-like molecules is not known.

C3H/HeJ mice are an interesting strain of mice that are genetically defective in their responsiveness to LPS (Sultzer, 1968). Although the chromosomal location of the defect has been identified, the defect has not. The defect does not affect the binding, internalization, or deacylation of LPS by C3H/HeJ macrophages (Munford and Hall, 1985). C3H/HeJ mice appear to respond normally to other activators such as TNF-α and IL-1 but not to the LPS mimetic taxol (Fig. 3) (Ding et al., 1990). Taxol was originally characterized as a microtubule stabilizer with anticancer activity (Schiff et al., 1979; Wani et al., 1971). The mimetic effects of taxol are mCD14 independent. This is illustrated by experiments done with the mouse macrophage-like cell line J7.DEF3, a mutant isolated from the parental J774.1 cells (Kirikae et al., 1995). J7.DEF3 cells are defective in mCD14 and hence LPS responsiveness but still respond normally to taxol. In addition, taxol-induced activation of J7.DEF3 cells is inhibited by the LPS antagonist RsDPLA. The antagonist effects of RsDPLA are specific for LPS and taxol and without effect toward zymosan, an anticomplementary factor from yeast, and heat-killed *Listeria monocytogenes,* a gram-positive mCD14-binding activator. Together these results suggest that LPS, taxol, and RsDPLA have similar targets downstream from mCD14.

Lipopolysaccharide Derivatives

The physical interactions of LPS with soluble and cell surface molecules have been evaluated using several modified versions of LPS (Fig. 4). One of the most useful is fluorescein isothiocyanate covalently attached to LPS (FITC-LPS). Fluorescein isothiocyanate attachment is assumed to be through the free and accessible amino groups on LPS. The substitution ratio for FITC-LPS is typically 0.2–0.3 mol FITC per mole of monomer ReLPS (Skelly et al., 1979). The FITC-LPS probe has been used in flow cytometry and microscopy experiments to detect and monitor LPS interactions with cells, and fluorimetry has been used to detect LPS interactions in solution.

Flow cytometry has been used to screen for cells that bind LPS, and the dependence of this binding on mCD14 has received a lot of attention. Expression of mCD14 either through cell differentiation or by transfection results in a dramatic increase in FITC-LPS binding (Lee et al., 1992; Tobias et al., 1993). Addition of anti-CD14 mAbs MY4, 3C10, 60b, MEM-18, or 28C5, which inhibit LPS activation of cells, inhibits binding of FITC-LPS to human monocytes. Release of mCD14 by cleavage of its GPI anchor (Haziot et al., 1988) with phosphatidylinositol-specific phospholipase C (PIPLC) reduces FITC-LPS binding as does addition of sCD14 to compete with mCD14 for LPS (Grunwald et al., 1993). LPS binding to mCD14 requires a source of LBP, either purified or in serum. When LBP is not added, FITC-LPS binding to mCD14 is not observed (Tobias et al., 1993; Wilde et al., 1994). In another flow cy-

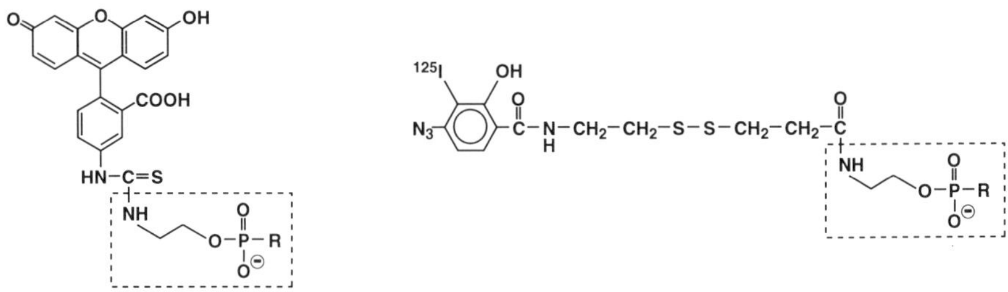

FITC-LPS **^{125}I-ASD-LPS**

Fig. 4. Chemically modified LPS. Both the FITC and ^{125}I-ASD substitutions are thought to attach randomly to accessible phosphoethanolamine groups on the LPS molecule. The dotted box encloses the phosphoethanolamine group of the LPS molecule. R refers to the remaining LPS structure.

tometry experiment, competition of LBP-meditated binding of FITC-LPS to mCD14 on monocytes by BPI was observed (Wilde et al., 1994). Binding of physiologically relevant concentrations of 10–20 ng/ml FITC-LPS can be detected by flow cytometry (Gallay et al., 1993; Bochsler et al., 1993).

Internalization of FITC-LPS can be observed with either flow cytometry or microscopy. In flow cytometry, fluorescence quenching of extracellular FITC-LPS by the nonpenetrating dye trypan blue has been used to differentiate between extracellular and internalized FITC-LPS (Weersink et al., 1990). Fluorescein isothiocyanate–lipopolysaccharide bound to neutrophils and monocytes at 4°C is almost completely quenched compared to only 30–40% of FITC-LPS bound at 37°C. Confocal microscopy also has been used to monitor FITC-LPS binding and internalization; binding to monocytes was observed within 1 min and internalization within 5 min (Gallay et al., 1993). The mechanism of LPS internalization awaits clarification.

Fluorimetery has been used to characterize the solution interactions of FITC-LPS with LBP and sCD14 (Tobias et al., 1995). As discussed previously, LPS is highly aggregated in aqueous solutions. The increase in fluorescence observed upon mixing FITC-LPS with LBP or LBP plus sCD14 is thought to come about from relief of self-quenching upon removal of FITC-LPS monomers from the aggregate. Kinetic analysis of the fluorescence change reveals that LBP acts as a catalyst in promoting the binding of LPS to sCD14. This is supported by electrophoretic mobility band-shift experiments using ^{3}H-labeled LPS (Hailman et al., 1994). In both sets of experiments, substoichiometric amounts of LBP were sufficient for maximal LPS binding to sCD14. The dissociation constants for the interaction of LPS with LBP and sCD14 were determined by this method and found to be 3.5×10^{-9} and 29×10^{-9} *M*, respectively (Tobias et al., 1995).

Another modified LPS is ^{125}I-labeled 2-(*p*-azidosalicylamido)-1,3′-dithiopropionamide-LPS (^{125}I-ASD-LPS) (Wollenweber and Morrison, 1985). ^{125}I-ASD-LPS has been conjugated with a bifunctional, photoreactive cross-linking reagent (SASD). Addition of a reducing agent cleaves the thiol bond of ^{125}I-ASD-LPS with the ^{125}I-moiety transferred to the target molecule; the SASD attaches primarily to phosphoethanolamine residues localized to the inner core of LPS. The estimated substitution ratio for ASD-LPS of 0.3 mol ASD per mole of monomer ReLPS is similar to the FITC-LPS substitution ratio. Typical iodination of the ASD-LPS is 0.003 mol ^{125}I per mole of monomer ReLPS. Iodine-125-labeled ASD–LPS has been used to demonstrate a direct interaction between LPS and mCD14 (Tobias et al., 1993). LPS cross-links to several unidentified proteins on the cell surface with reported SDS–PAGE mobilities of 38, 25, and 18 kDa (Lei et al., 1993; Kirkland et al., 1990;). ^{125}I-ASD-LPS also has been used to demonstrate the direct binding of soluble proteins that interact with LPS, such as LBP, truncated LBP$^{\Delta 1\text{-}197}$, and sCD14 (Tobias et al., 1986; Han et al., 1994; Pugin et al., 1993). Iodine-125-labeled ASD–LPS has been used to confirm the dissociation constants of LBP and sCD14 determined with FITC-LPS. When LBP and sCD14 were mixed in equal molar amounts with limiting ^{125}I-ASD-LPS, essentially all the LPS cross-linked to LBP. When sCD14 was added in approximately 10-fold excess to LBP, the majority of the LPS cross-linked to sCD14 (Tobias et al., 1995).

As previously mentioned, the LAL assay is used to quantify the bioactivity of LPS. The LAL fraction was probed with ^{125}I-ASD-LPS to identify proteins that bind LPS (Roth and Tobias, 1993). An 82-kDa protein was identified as a potential negative regulator of coagulation. A minor 50-kDa protein was observed to bind LPS and is a candidate for the LPS-sensitive coagulation protein in *L. polyphemus*.

The LPS precursor lipid IV_A has been modified with a 4′-kinase from *E. coli* membranes and [γ^{32}P]ATP to generate the radioactive ligand 4′[^{32}P]lipid IV_A (Raetz et al., 1985). The 4′[^{32}P]lipid IV_A has been used to detect LPS binding to scavenger receptors (Hampton et al., 1991) as well as to two other unidentified 31- and 95-kDa proteins (Hampton et al., 1988). The advantage of 4′[^{32}P]lipid IV_A over metabolically ^{3}H-labeled LPS is its higher specific activity. The specific activity of 4′[^{32}P]lipid IV_A is 3000 Ci/mmol. The specific activity of the commercially available ^{3}H-labeled LPS (from List Biological Laboratories) is 3 Ci/mmol. For further comparison purposes, the specific activity of ^{125}I-ASD–LPS is approximately 15 Ci/mmol (Wollenweber and Morrison, 1985). An obvious drawback to using 4′[^{32}P]lipid IV_A is its different agonist/antagonist effects among species.

SERUM LIPOPOLYSACCHARIDE-BINDING PROTEIN: THE CATALYST

Lipopolysaccharide-binding Protein

Lipopolysaccharide-binding protein (LBP) is a serum component first identified as an acute-phase reactant that modulates LPS binding to HDL (Tobias et al., 1986); it subsequently was found to enhance the responsiveness of myeloid, endothelial, and epithelial cells to LPS (Schumann et al., 1990; Pugin et al., 1993). Lipopolysaccharide binding protein is a major mediator of the lethal effects of LPS. Elimination of LBP in vivo by addition of anti-mouse LBP immune IgG neutralizes LPS endotoxicity (Gallay et al., 1994). From recent work, it has become increasingly apparent that LBP is multifunctional; it enhances LPS binding to mCD14 and subsequent cellular activation (Wright et al., 1990a), and mediates the mCD14-dependent internalization of LPS (Gegner et al., 1995). Lipopolysaccharide-binding protein catalyzes the transfer of LPS to sCD14 (Hailman et al., 1994; Tobias et al., 1995) and lipoproteins (Wurfel et al., 1994), and potentiates the bactericidal activity of BPI (Horwitz et al., 1995).

Lipopolysaccharide-binding protein binds the lipid A moiety of LPS and forms a high-affinity LBP-LPS complex with an apparent dissociation constant of 10^{-9} *M* (Tobias et al., 1989). Size analysis of LBP-LPS complexes by mobility in sucrose density gradients indicates that LBP-LPS complex size varies as a function of the molar ratio of LBP to LPS (Tobias et al., 1995). When LPS is in stoichiometric excess to LBP, LBP co-migrates with the large LPS aggregates, whereas when LBP is in excess, LPS shifts to the mobility of LBP. These data suggest that LBP binds to both LPS monomers and LPS aggregates. The reduced size of LPS in the presence of LBP also suggests that LBP facilitates the dissolution of LPS from LPS aggregates.

Lipopolysaccharide-binding protein is a 60-kDa serum glycoprotein present in normal human serum at 1–10 μg/ml. During an acute-phase response, LBP levels rise to 50–200 μg/ml (Tobias et al., 1992). Expression of LBP was first observed in hepatocytes (Ramadori et al., 1990) and subsequently in other tissues (Su et al., 1994). Regulation of LBP synthesis in human (Grube et al., 1994) and rat (Wang et al., 1995) hepatocytes is similar to that of other acute-phase reactants.

Soluble CD14

Activation of EC by LPS is greatly enhanced by sCD14 (Frey et al., 1992; Read et al., 1993; Pugin et al., 1993). Soluble CD14 is a 50–55 kDa LPS-binding glycoprotein found in serum and urine (Maliszewski et al., 1985; Bazil et al., 1986). Normal human serum levels are in the microgram range (Kruger et al., 1991). Soluble CD14 is either secreted or shed from the cell surface (Bazil and Strominger, 1991). Band shift experiments using nondenaturing gel electrophoresis indicate that although LPS binds sCD14 directly in the absence of other serum components, addition of LBP accelerates this process (Hailman et al., 1994). Lipopolysaccharide-binding protein does not interact with sCD14 in the absence of LPS nor is a ternary LBP-LPS-sCD14 complex observed. Fluoroscein isothiocyanate–lipopolysaccharide was used to measure the kinetics of the LPS-sCD14 interaction (Tobias et al., 1995). The rate limiting step in the formation of LPS-sCD14 complexes catalyzed by LBP appears to be the dissolution of LPS from LPS aggregates. This is further substantiated by the observation that unlike LBP, sCD14 binding to LPS aggregates is not observed.

As is apparent from the relative dissociation constants of 10^{-8} for sCD14 and 10^{-9} for LBP, LPS binding to sCD14 is favored only when sCD14 is present in excess to LBP. In normal human serum, LBP and sCD14 are present at comparable levels. During an acute-phase response, LBP levels generally rise more than levels of sCD14 (Tobias et al., 1992; Kruger et al., 1991). Even so, sufficient LPS is transferred to sCD14 to initiate EC activation in normal human serum at 1 ng/ml LPS (Pugin et al., 1993). In other experiments, addition of sCD14 (5–30 mg/ml) inhibited LPS activation of human monocytes (Schutt et al., 1992), indicating that sCD14 can compete with mCD14 for LPS. Addition of sCD14 also inhibits TNF-α release and prevents LPS-induced mortality in mice (Haziot et al., 1995). How LPS complexed to LBP is distributed between sCD14 and mCD14 is not entirely clear.

High-density Lipoproteins

Plasma lipoproteins, particularly HDL, have been shown to bind and neutralize LPS bioactivity (Ulevitch et al., 1979; Freudenberg et al., 1980). Neutralized LPS loses its ability to cause fever and death in experimental animals (Ulevitch and Johnston, 1978) and to give a positive response in the LAL assay (Johnston et al., 1977). Neutralized LPS in serum is not chemically modified (Rudbach and Johnson, 1964; Ulevitch and Johnston, 1978). Lipopolysaccharide neutralization may result from its association with the phospholipid surface of the lipoproteins (Parker et al., 1995). There is now evidence that neutralization of LPS through association with lipoproteins requires LBP. Lipopolysaccharide binding to HDL-like particles reconstituted from purified apolipoprotein A-1 (apoA-1), phospholipid, and free cholesterol was observed only on addition of LBP (Wurfel et al., 1994). When apoA-1-containing lipoproteins were isolated from normal human plasma by affinity immunosorption, essentially all the LBP co-purified with the particles. In a reciprocal experiment, apoA-1 was captured with anti-LBP antibodies to suggest further that LBP and apoA-1 are physically associated in the same particle. The interaction is apparently weak, for purification by ultracentrifugation separates LBP from the lipoprotein particles (Tobias et al., 1986). Substoichiometric amounts of LBP were sufficient for neutralization of LPS by the HDL-like particles. These observations suggest that LBP is an LPS transfer protein that facilitates the binding of LPS not only to sCD14 but also to lipoproteins.

Experimental evidence suggests serum components can both potentiate and neutralize the endotoxic effects of LPS. Short incubations of LPS with serum (<10 min) enhance LPS activity whereas its endotoxic effects are lost during longer incubations (Wurfel et al., 1994). Serum enhances the sensitivity of cells to LPS by several orders of magnitude. The enhancement effect of serum can be reconstituted by addition of purified LBP for mCD14-bearing cells or LBP plus sCD14 for EC cells. Conversely, LPS is neutralized by associating with lipoproteins such as HDL in a reaction catalyzed by LBP. Neutralization of LPS by lipoproteins is faster in normal than in acute-phase rabbit serum (Tobias et al., 1986). Because levels of LBP are elevated in acute-phase serum, additional experimentation is required to explain the relationship between the proposed catalytic transfer of LPS to HDL by LBP observed in vitro, and the slower LPS–lipoprotein association rate observed in vivo. Although LBP has an apparent dual role of both enabling and inhibiting responses to LPS, neutralization of LBP with mAbs in in vivo mouse experiments indicate LBP is important in LPS endotoxicity (Gallay et al., 1994). Additional insights into the functions of LBP most likely will result from studies involving LBP-deficient mice.

Bactericidal/permeability-increasing Protein

Bactericidal/permeability-increasing protein (BPI) and LBP have homologous amino acid sequences (Table 1) and share the ability to bind lipid A (Tobias et al., 1989; Wilde et al., 1994); BPI is found in azurophilic granules of polymorphonuclear leukocytes, on the surface of neutrophils, in plasma, and in inflammatory fluids (Weersink et al., 1993; Marra et al., 1990; Opal et al., 1994). Bactericidal/permeability-increasing protein is present at 0.2–2 ng/ml in normal plasma and at >200 ng/ml in plasma

TABLE 1. LBP Homologous Proteins[a]

Protein	Identity (%)	Reference
Bactericidal/permeability-increasing protein	45	Gray et al., 1989
Phospholipid transfer protein	25	Day et al., 1994
Cholesteryl ester transfer protein	23	Drayna et al., 1987
Rat olfactory mucosa proteins	25	Dear et al., 1991

[a]LPS-binding protein (LBP) shares sequence homology with several other proteins. LBP and bactericidal/permeability-increasing protein (BPI) both bind LPS and share the greatest homology in their LPS-binding domains. LBP, phospholipid transfer protein (PLTP), and cholesteryl ester transfer protein (CETP) are all lipid transfer proteins. The rat olfactory proteins are specifically expressed in the olfactory mucosa and may bind lipophilic molecules.

from septic patients (White et al., 1994; Calvano et al., 1994); it binds gram-negative bacteria and initiates cytotoxic events. An *N*-terminal fragment of BPI, $rBPI_{23}$, retains all the known functions of BPI (Ooi et al., 1987; Weiss et al., 1992).

Although the apparent binding affinity of BPI for LPS is 10- to 100-fold higher than that of LBP for LPS (Gazzano-Santoro et al., 1994; Wilde et al., 1994), LBP serum levels are 1000-fold higher than BPI levels. This suggests serum BPI does not effectively compete with LBP for LPS. One exception is that higher BPI levels relative to LBP are found in fluid from closed-space infections (Opal et al., 1994). Under such circumstances, BPI may effectively compete with LBP for LPS and block the inflammatory response. In vitro competition experiments between BPI (or $rBPI_{23}$) and LBP support this hypothesis (Marra et al., 1992; Heumann et al., 1993).

To determine if BPI in plasma functions in the presence of 1000-fold molar excess of LBP, the effects of LBP on BPI bactericidal activity were evaluated (Horwitz et al., 1995). Although LBP alone has little bactericidal activity, it was found to potentiate the bactericidal activity of BPI by 10,000-fold. The mechanism by which LBP potentiates the effects of BPI is not known. The *N*-terminal half fragment of LBP ($LBP^{\Delta 1\text{-}197}$) also was effective in potentiating the bactericidal activity of BPI. Recombinant $LBP^{\Delta 1\text{-}197}$ binds LPS but no longer can transfer LPS to either mCD14 or sCD14 (Han et al., 1994; Theofan et al., 1994). Thus BPI competition with LBP as a defense mechanism against LPS endotoxicity may not have biological relevance if the in vitro LBP potentiation of BPI activity occurs in vivo.

MEMBRANE LIPOPOLYSACCHARIDE-BINDING PROTEINS: THE RECEPTORS

Signal Transduction by mCD14 and CD11c/CD18

mCD14 was first identified as a differentiation antigen expressed by myeloid cells (Maliszewski et al., 1985). mCD14 specifically binds LPS in a reaction promoted by LBP. More importantly, mCD14 assists in the recognition of LPS necessary to trigger cellular activation (Wright et al., 1990a). This function was identified by the ability of specific antibodies against CD14 not only to block the binding of erythrocytes coated with LBP–LPS complexes to monolayers of macrophages, but also to inhibit TNF-α production in whole blood. The importance of mCD14 in LPS signal transduction has been evaluated in several other ways. For example, normally hyporesponsive cells, such as mouse pre-B 70Z/3 cells and Chinese hamster ovary (CHO) cells, become LPS sensitive after transfection with CD14 (Lee et al., 1992; Golenbock et al., 1993). In other experiments, whole blood was tested from patients suffering from paroxysmal nocturnal hemoglobinuria (PNH) (Duchow et al., 1993). The PNH blood cells lack glycophosphatidylinositol (GPI)-anchored proteins and thus lack mCD14, a GPI-anchored protein. The response of PNH blood to LPS was reduced 100- to 1000-fold over normal blood. In experiments designed to evaluate the effects of blocking LPS–CD14 interactions in vivo, it was observed that the addition of anti-CD14 mAbs protected mice (Gallay et al., 1994) and cynomolgus monkeys (Leturcq et al., 1994) from the lethal effects of LPS. Additional evidence for the in vivo role of CD14 in septic shock has been obtained from experiments with CD14-knockout mice (Haziot et al., 1994). These mice are said to be extremely resistant to the lethal effects of purified LPS as well as to injection of gram-negative bacteria. Taken together these results demonstrate the importance of mCD14 in LPS activation.

Cells can be loosely grouped into several categories based on their responsiveness to LPS. Monocytes, macrophages, CD14-transfected murine pre-B 70Z/3 cells, and CD14-transfected CHO (fibroblasts) cells respond to LPS primarily through mCD14. The serum-dependent responses to LPS evaluated in these cells (such as NF-κB activation) can be fully reconstituted by adding purified LBP. In 70Z/3 and CHO cells, soluble CD14 cannot substitute for the membrane-attached form (Lee et al., 1993; Delude et al., 1994). On the other hand, purified sCD14 enhances the sensitivity of EC to LPS by several orders of magnitude. Both sCD14 and mCD14 may function in LPS activation of polymorphonuclear leukocytes (PMNs, neutrophils). Polymorphonuclear leukocytes express low but detectable levels of mCD14 (Weingarten et al., 1993), but show sCD14-dependent LPS expression of E-selectin and PMN adhesion to fibrinogen (Hailman et al., 1994). Finally, in T-lymphocytes such as murine EL4 cells, no response to LPS has been observed even after transfection with CD14 (J.-D. Lee, unpublished results).

The GPI anchor of mCD14 is not involved in certain aspects of LPS cellular activation (Lee et al., 1993). In transfection experiments with 70Z/3 cells, the GPI anchor of mCD14 was replaced with either another GPI anchor (from decay-accelerating factor) or a transmembrane fragment (from tissue factor or murine class 1 molecule). The form of the CD14 membrane attachment did not affect LPS binding, NF-κB activation, or IgM expression. However, the GPI-anchored CD14 does appear to have some unique properties. When the GPI-tailed form of CD14 is cross-linked with antibodies, a Ca^{2+} burst is observed. This was observed in human monocytes and in CD14-transfected THP-1 cells (Beekhuizen et al., 1993; J. Pugin, unpublished results). Release of Ca^{2+} is not detected upon cross-linking of transmembrane-CD14 transfected into THP-1 cells (J. Pugin, unpublished results). Additionally, THP-1 cells transfected with the GPI-tailed form of CD14 show enhanced phagocytosis of gram-negative bacteria over either cells transfected with transmembrane-CD14 or control cells (D. Schiff, unpublished results).

Although CD14 is not homologous with any other known proteins, it does contain

TABLE 2. The Diversity of Ligands for CD14: A Pattern Recognition Receptor[a]

Ligand	Reference
Lipopolysaccharide	Wright et al., 1990
Polyuronic acid polymers	Espevik et al., 1993
Mycobacterial lipoarabinomannan	Zhang et al., 1993
Gram-positive cell walls	Pugin et al., 1994
Soluble peptidoglycan	Weidemann et al., 1994
Chitosans (GlcN/GlcNAc polymer)	Otterlei et al., 1994
Blastomyces WI-1 protein antigen	Newman et al., 1995
Streptococcal rhamnose/glucose polymer	Soell et al., 1995

[a]Receptors with broad binding specificity may function to discriminate between self and nonself (Krieger and Hertz, 1994). The role of CD14 in initiating cell activation by lipopolysaccharide is well established. CD14 also appears to serve as a pattern recognition molecule for a wide variety of microbial structures. These structures are highly conserved among pathogens. Recognition of infectious particles by macrophages leads to a rapid activation of nonspecific defenses typical of the innate immunity.

the structural motif known as leucine-rich repeats (LRRs) (Kobe and Deisenhofer, 1994; Shapiro et al., 1995). The number of repeats in known proteins varies from 1 to 30; CD14 contains 8 LRRs. Leucine-rich repeats are thought to be involved in protein–protein interactions. The determination of the crystal structure of the ribonuclease inhibitor (15 LRRs)-ribonuclease A complex provides some insights on such interactions (Kobe and Deisenhofer, 1995). The two unusual features of the ribonuclease A inhibitor structure are its nonglobular shape and the solvent-exposed parallel β-sheet. The nonglobular shape allows for an extensive binding area and thus the formation of many favorable contacts. The solvent-exposed parallel β-sheet is thought to provide conformational flexibility that enables an LRR protein to change shape relative to the requirements of the ligand. That CD14 may have conformational flexibility is suggested by its 40% β-sheet content (McGinley et al., 1995) and by the diversity of ligands it binds (Table 2). The lack of α-helix structure in CD14 as determined from the CD spectrum argues that its LRR structure may be rather different from the ribonuclease inhibitor structure. Proteolysis of sCD14 suggests it is composed of two tightly folded domains connected by a bridge that binds LPS (McGinley et al., 1995).

In addition to facilitating recognition of LPS in cellular activation, mCD14 may assist in the capture of a large number of LPS molecules (Gegner et al., 1995). When LBP is present in molar amounts equivalent to LPS, binding of LPS to mCD14 does not saturate. This is in sharp contrast to the substoichiometric amounts of LBP needed to catalyze fully the transfer of LPS to sCD14. The binding of LBP-LPS complexes to mCD14 on the cell surface is inferred from the ability of phosphatidylinositol-specific phospholipase C (PI-PLC) to cleave the GPI-anchor of mCD14 and release it along with the mCD14-associated LBP-LPS complexes from the cell surface. This binding does not appear to be involved in cellular activation. An anti-LBP mAb 18G4 inhibits this phenomenon with no observable effect on either NF-κB activation or

TNF-α release. The binding instead appears to lead to LBP-LPS complex internalization.

Deletion studies have shown that a large portion of CD14 is expendable. The *N*-terminal half of CD14 ($sCD14^{1-152}$) missing five LRRs was found to be biologically active (Juan et al., 1995b). The truncated sCD14 bound ^3H-labeled LPS, enabled human astroglioma U373 cells to secrete IL-6 in response to LPS, inhibited LPS-induced LAL response, and mediated LPS-induced adhesion of human neutrophils to fibrinogen-coated wells. In addition, LBP still catalyzed the binding of LPS to $sCD14^{1-152}$. What effect deleting the *C*-terminal half of CD14 has on the responses of EC and the functions of membrane CD14 awaits investigation. Other studies identified several regions within the *N*-terminal 65 amino acids critical for LPS binding to CD14. Deletion of amino acids 8–11, 21–24, 34–38, 58–62, and/or 57–64 abrogated LPS binding (Viriyakosol and Kirkland, 1995; Juan et al., 1995b; McGinley et al., 1995). This was further substantiated by the loss of the binding epitopes of anti-CD14 mAbs that inhibit LPS binding (Viriyakosol and Kirkland, 1995; Juan et al., 1995a).

Chinese hamster ovary cells transfected with CD11c/CD18 (p150, 95 or CR4) acquire serum-independent responsiveness to LPS (Ingalls and Golenbock, 1995). CD11c/CD18 is an obligate α/β heterodimeric adhesion molecule. Response to LPS in CD11c/CD18-transfected CHO cells (CHO/CD11c) was distinct from the response in CHO cells transfected with CD14 (CHO/CD14) in that (1) serum enhanced nuclear transcription factor NF-κB activation in CHO/CD14 cells but reduced it in CHO/CD11c cells, (2) NF-κB activation in CHO/CD11c cells was slower than that in CHO/CD14 cells, and (3) CHO/CD11c cells required approximately 10-fold higher levels of LPS for NF-κB activation. Lipopolysaccharide also interacts with the other members of this integrin family, CD11a/CD18 (LFA-1) and CD11b/CD18 (CR3) (Wright and Jong, 1986). The cations Mg^{2+} and Ca^{2+} are required for serum-independent binding of *E. coli* and LPS-coated erythrocytes to CD11/CD18 on monocytes (Wright and Jong, 1986). Binding of iC3b (the major complement C3 opsonin), NIF (a hookworm-derived neutrophil adhesion inhibitor), CD54 (intercellular adhesion molecule-1, ICAM-1), and fibrinogen to CD11b/CD18 are all metal ion-dependent (Lee et al., 1995). A metal coordination site within the A domain of CD11b is important in protein ligand binding (Lee et al., 1995). The metal dependence of the LPS–CD11/CD18 interaction suggests the A domain is involved in the LPS as well as the other ligand interactions with CD11b. However, the LPS binding site is distinct from the region of CD11/CD18 that recognizes the RGD amino acid sequence (Wright et al., 1989).

The functional significance of the serum-independent signaling by LPS through the CD11c/CD18 receptor is not understood. CD14-mediated signaling appears to be the primary means by which the inflammatory response is initiated. Monocytes and macrophages from CD18-deficient patients released normal amounts of TNF-α and IL-1β in response to LPS (Wright et al., 1990b), whereas blood lacking mCD14 from patients with paroxysmal nocturnal hemoglobinuria showed impaired responsiveness (Duchow et al., 1993). However, CD11/CD18 is critical to the function of polymorphonuclear cells and monocytes in inflammation and injury. Inherited deficiency of

CD18 impairs phagocyte chemotaxis, adhesion, transmigration across endothelium, and clearance of invading microorganisms through phagocytosis and cell-mediated killing (Arnaout et al., 1990).

CD14 and CD11a/CD18 have been implicated in mediating the homotypic adhesion of human monocytes (Lauener et al., 1990). Induction of Mg^{2+}-dependent, monocyte aggregation/adhesion by anti-CD14 mAb monovalent F(ab) fragments was observed. Aggregation by monovalent F(ab) fragments indicates CD14-mediated monocyte adhesion does not require FcγR engagement or cross-linking. Aggregation was inhibited by mAbs to CD11a, CD18, and ICAM-1. In other experiments, cross-linking of CD14 was implicated in CD11a/CD18-dependent adherence of monocytes to EC (Beekhuizen et al., 1991; Beekhuizen et al., 1993). Although these observations appear to link CD14 with CD11a/CD18, a role for LPS in this phenomenon remains undefined.

Clearance by mCD14 and Scavenger Receptors

Lipopolysaccharide is neutralized and cleared by a number of pathways. Estimations for LPS concentrations circulating in the blood of septic patients are 0.1–1 ng/ml (Danner et al., 1991). In rabbits, the time course for clearance of a bolus dose of LPS from the circulation in vivo is biphasic (Mathison and Ulevitch, 1979). Roughly half of a 250-μg bolus (1.5–2 μg/ml) of LPS injected into rabbits is cleared from the blood within 5 min; the half-life, depending on the type of LPS used, is about 2 min. The remaining LPS is cleared from the blood with a much slower half-life of 12 h. Appearance of a low-density LPS fraction, identified as LPS complexed to lipoproteins such as HDL, correlates with the decrease in clearance rate. The cellular targets of LPS in the two clearance phases are also different. The rapidly cleared LPS is taken up by tissues containing macrophages, primarily hepatic macrophages (Kupffer cells), whereas LPS-HDL complexes are taken up by tissues metabolizing cholesterol, notably the adrenal gland.

Lipopolysaccharide-binding protein and mCD14 both function in LPS clearance (Kitchens et al., 1992; Luchi and Munford, 1993). Lipopolysaccharide-binding protein complexes with LPS bind to mCD14 in large clusters and internalize together (Gegner et al., 1995). The majority of mCD14 either remains on the cell surface or recycles to the cell surface. Although transfection experiments with transmembrane-CD14 and GPI-CD14 demonstrate the importance of the GPI anchor in phagocytosis, it is not clear how CD14-mediated internalization of LPS occurs. Electron microscopy of mCD14 internalization after mAb binding revealed microclustering and formation of coated pits and receptor-bearing endosomes (Schutt et al., 1988). It is unclear if LBP–LPS–mCD14 complexes follow the same internalization pathway as cross-linked mCD14. Multimerization may regulate sequestration of GPI-anchored receptors. GPI-anchored folate receptors cluster (approximately 750 receptors/cluster) in caveolae upon cross-linking (Rothberg et al., 1990; Mayor et al., 1994).

Lipopolysaccharide is bound and internalized by macrophage scavenger receptors (SRs) (Hampton et al., 1991), which bind a number of ligands and like CD14, have been referred to as pattern recognition receptors (Abrams et al., 1992). They bind and internalize negatively charged ligands such as acetylated low-density lipoproteins, maleylated BSA, dextran sulfate, fucoidan, and polyinosinic acid. Surface receptors are structurally quite distinct from CD14. The macrophage SR-type I structure, deduced by complementary DNA cloning, is a trimeric membrane glycoprotein with a cysteine-rich *C*-terminal extracellular domain connected to the transmembrane domain by a collagen-like helix and an α-helical coiled coil (Kodama et al., 1990). MARCO, a new SR-like receptor structurally related to SR-type I, is expressed in a subset of macrophages in the marginal zone of the spleen and the medullary cord of the lymph nodes (Elomaa et al., 1995). COS cells transfected with MARCO bound FITC-labeled *E. coli* and *Staphylococcus aureus* but not Zymosan A. Whether or not the LPS moiety specifically was involved in the MARCO interaction with *E. coli* was not reported.

Scavenger-receptor-mediated lysosomal dephosphorylation of lipid IV_A is observed in murine RAW 264.7 macrophage-like cells (Hampton and Raetz, 1991). The resulting molecule, 4′-monophosphoryl lipid IV_A, is less endotoxic (Madonna et al., 1986). Another type of LPS processing that results in reduced LPS endotoxicity is deacylation by acyloxyacyl hydrolase (AOAH), found in phagocytic cells such as neutrophils, monocytes, macrophages, and vascular endothelial cells but not in fibroblasts (Munford and Hunter, 1992). It is a phospholipase A1 with optimal activity at acidic pH. Its activity is inhibited by agents that reduce lysosomal acidification such as ammonium chloride, methylamine, and chloroquine to suggest that at some point it is in an endosomal or lysosomal compartment. The AOAH precursor is proteolytically processed into two subunits (52-, 14-kDa) that are linked by a disulfide bond (Hagen et al., 1991). The smaller subunit has amino acid sequence similarity to sphingolipid activator proteins called saposins. The saposin domain influences the intracellular localization, stability, and catalytic activity of AOAH (Staab et al., 1994). Acyloxyacyl hydrolase also has lysophospholipase, diacylglycerollipase, and acyltransferase activities in vitro (Munford and Hunter, 1992). In support of these in vitro observations, radioactivity derived from LPS fatty acids has been found in cellular lipids (Hall and Munford, 1983; Munford and Hall, 1985), and association of AOAH with the plasma membrane may occur in stimulated cells (Munford and Hunter, 1992). The physiological role of AOAH and LPS deacylation remains undefined. Acyloxyacyl hydrolase may function in some aspect of the LPS immune response or it may simply catabolize LPS.

Thus LPS clearance is initiated in at least three ways. Rapid clearing by macrophages and Kupffer cells appears to be mediated by both mCD14 and scavenger receptors. How LPS is distributed between these two receptors is not known, but LBP clearly facilitates the LPS–mCD14 interaction. Although lipid IV_A–SR binding was observed in the absence of serum (Hampton et al., 1991), the role of either LBP or serum in the lipid IV_A-SR interaction has not been systematically investigated. Lipopolysaccharide also associates with lipoproteins such as HDL in an LBP-dependent manner. Again it is not understood how LPS is distributed between macrophage clearance and clearance mediated by lipoproteins. The physiological regulators of these various clearance pathways as well as the biological implications for the different LPS clearance routes remain important unanswered questions.

MISCELLANEOUS LIPOPOLYSACCHARIDE-BINDING MOLECULES

Polymyxin B

An LPS-binding peptide frequently used to remove traces of LPS from solutions is polymyxin B (Karplus et al., 1987). Polymyxin B belongs to a family of antibiotics that contain a cationic cyclic heptapeptide and a side chain that terminates with a short fatty acid chain such as methyloctanoic acid. The amphiphilic polymyxin B interacts with the lipid A region of the LPS molecule (Morrison and Jacobs, 1976). Addition of polymyxin B reduces both the in vitro and in vivo endotoxicity of LPS (Stokes et al., 1989).

Horseshoe Crab Lipopolysaccharide-binding Proteins

A number of LPS-binding proteins have been identified in the amebocyte (hemocyte) lysate from horseshoe crabs. The horseshoe crab (*Limulus polyphemus* and *Tachypleus tridentatus*) has a simple blood defense system against invading bacteria. The bacteria interact with the outer membrane of amebocytes, the only circulating blood cell in horseshoe crabs, causing them to aggregate. This is followed by rapid degranulation of the amebocytes and clot formation. Degranulation releases the coagulation proteins from their intracellular location. Coagulation is a proteolytic cascade activated by the interaction of LPS with the LPS-sensitive factor C of the coagulation system, which ultimately leads to the clotting of the blood and immobilization of the invading microorganism.

An 80-kDa protein from amebocyte membranes has been isolated and found to enhance *Limulus* lysate gelation (Liang et al., 1980). This protein has no enzymatic activity but has been proposed as an LPS receptor which stimulates the amebocyte to release clotting zymogen factor C. Zymogen factor C is an LPS-sensitive serine protease (Nakamura et al., 1986). Factor C (123 kDa) is converted to the active form when its light chain (43 kDa) is cleaved to 34 kDa upon addition of LPS. Factor C interacts with at least one other protein (factor B) before activation of the preclotting enzyme and initiation of the clotting cascade. The amebocyte-derived granular protein and peptide components participating in the defense cascade of the horseshoe crab are reviewed by Iwanaga et al. (1994).

Other proteins have been identified that inhibit the LPS-mediated cascade. *Limulus* anti-LPS factor (LALF) is the best characterized anticoagulant LPS-binding protein (Tanaka et al., 1982); it specifically inhibits the LPS-mediated activation of limulus factor C (Nakamura et al., 1986). *Limulus* anti-LPS factor is a 15-kDa amphipathic basic protein consisting of three α-helices packed against a four-stranded β-sheet (Hoess et al., 1993). The proposed LPS binding site in this molecule is the amphipathic loop structure which may be a motif shared by two other LPS-binding proteins, LBP and BPI. Efforts are underway to obtain crystals of LBP and BPI for X-ray analysis to confirm this hypothesis. Hoess et al. (1993) also proposed that the cyclic peptide of polymyxin B is analogous to part of the amphipathic loop of LALF. Nakamura et al. (1988) identified another inhibitor of factor C, an antimicrobial cationic peptide (tachyplesin) from the horseshoe crab *Tachypleus tridentatus*. Tachyplesin

(17 amino acids) inhibits growth of both gram-negative and gram-positive bacteria and forms a complex with LPS. Tachyplesin has a rigid structure imposed by two disulfide linkages. It has three tandem repeats of a tetrapeptide sequence, namely, hydrophobic amino acid-Cys-hydrophobic amino acid-Arg. Its amphipathic nature may be closely associated with its ability to bind and neutralize LPS. *Limulus* endotoxin-binding-protein-protease inhibitor (LEBP-PI) is a 12-kDa LPS-binding protein with protease-inhibitory activity found in the cytoplasm and in secretory granules (Minetti et al., 1991). It inhibits trypsin activity and differs from anti-LPS factor and tachyplesin in its ability to inhibit LBP from activating the clotting cascade, and it may downregulate the clotting process after invading microorganisms are sequestered in the clot. The horseshoe crab coagulation system is a valuable model system providing insights into LPS response mechanisms found in more complicated organisms.

Other proteins that bind LPS have been evaluated based on their ability to alter the activity of LPS in the LAL assay. For example, hemoglobin binds LPS and hemoglobin–LPS complexes show increased LAL activation compared to LPS alone (Kaca et al., 1994). Lactoferrin, secreted from neutrophils, binds LPS and neutralizes the ability of LPS to trigger neutrophil H_2O_2 release but does not alter its LAL activity (Wang et al., 1995). Melittin (David et al., 1992), bacterial outer membrane 39-kDa protein (Weber et al., 1992), lysozyme (Ohno and Morrison, 1989), complement proteins (Galanos et al., 1971), BPI (Marra et al., 1992), polymyxin B (Morrison and Jacobs, 1976), and LBP (Wilde et al., 1994) all are LPS-binding proteins that inhibit LPS activation of LAL.

CONCLUDING REMARKS

Lipopolysaccharide is a potent activator of cells. It is an amphiphilic ligand that interacts with proteins in serum and on the cell surface. The lipid A moiety of the LPS molecule initiates an immune response that can lead to septic shock. Rather than a simple linear flow of signaling information, LPS initiates a network of responses through its binding interactions with proteins in serum and on the surface of cells. Lipopolysaccharide-binding protein acts as a sort of gateway for some of the interactions of LPS with other molecules; it disaggregates LPS and enhances LPS binding to the serum components sCD14 and HDL, and it also may potentiate the bactericidal activity of BPI. Lipopolysaccharide-binding protein facilitates the binding, cellular activation, and internalization of LPS by mCD14. LPS binding and activation of CD11C/CD18 is inhibited by serum. The inhibitory component of serum has not been identified, but LBP and sCD14 are likely candidates. Macrophage scavenger regulators appear to bind and internalize LPS in a serum-independent manner, although this has not been rigorously tested. Lipopolysaccharide seemingly has redundant receptors on the surface of macrophages. However, the presence of multiple receptors for a single ligand is the rule (e.g., adrenergic, dopaminergic, cholinergic, and opiate receptor systems), not the exception (Golenbock et al., 1990). It remains to be elucidated how the LPS-elicited functions of mCD14, scavenger receptors, and CD11/CD18 are integrated among themselves and soluble LPS receptors. Finally,

LPS binding to mCD14 is not sufficient for cellular activation. Formation of multimeric protein complexes for signal transduction systems such as the IL-6 receptor and gp130 (Taga et al., 1989) is frequently observed and a similar model has been proposed for LPS-mCD14 complex signal transduction. The hypothesis of the involvement of a transducer molecule, which signals an event distinct from LPS-mCD14 complex formation, is based on observations such as the following: (1) addition of the anti-CD14 mAb 18E12 inhibits activation with only slight (10%) inhibition of binding (Lee et al., 1992); (2) modified forms of LPS (dLPS, RsDPLA, and lipid IV_A) and taxol alter cellular responsiveness in a mCD14-independent fashion; and (3) LPS hyporesponsive C3H/HeJ mice express CD14 normally. Identification of the proposed LPS transducer should provide insights into the initiation and regulation of early LPS signaling events. Medical advances in the endotoxin field will depend on a multifaceted approach to identify the details of LPS cellular activation and clearance pathways as well as to elucidate how LPS components involved in different aspects of the LPS-initiated immune response are coordinated.

ACKNOWLEDGMENTS

This is publication 9522-IMM from the Scripps Research Institute. This work was supported by an American Heart Association Fellowship (to J. A. G.) and by National Institutes of Health Grants GM37696, AI32021, and HL23584 (to P. S. T.).

REFERENCES

Abrams JM, Lux A, Steller H, Krieger M (1992): Macrophages in Drosophila embryos and L2 cells exhibit scavenger receptor-mediated endocytosis. Proc Nat Acad Sci USA 89:10375–10379.

Arnaout MA (1990): Leukocyte adhesion molecules deficiency: its structural basis, pathophysiology and implications for modulating the inflammatory response. Immunol Rev 114:145–180.

Bazil V, Horejsi V, Baudys M, et al. (1986): Biochemical characterization of a soluble form of the 53-kDa monocyte surface antigen. Eur J Immunol 16:1583–1589.

Bazil V, Strominger JL (1991): Shedding as a mechanism of down-modulation of CD14 on stimulated human monocytes. J Immunol 147:1567–1574.

Beekhuizen H, Blokland I, Corsel-van Tilburg AJ, Koning F, van Furth R (1991): CD14 contributes to the adherence of human monocytes to cytokine-stimulated endothelial cells. J Immunol 147:3761–3767.

Beekhuizen H, Blokland I, van Furth R (1993): Crosslinking of CD14 molecules on monocytes results in a CD11/CD18- and ICAM-1-dependent adherence to cytokine-stimulated human endothelial cells. J Immunol 150:950–959.

Bochsler PN, Maddux JM, Neilsen NR, Slauson DO (1993): Differential binding of bacterial lipopolysaccharide to bovine peripheral-blood leukocytes. Inflammation 17:47–56.

Bone RC (1991): Gram-negative sepsis. Chest 100:802–808.

Calvano SE, Thompson WA, Marra MN, et al. (1994): Changes in polymorphonuclear leukocyte surface and plasma bactericidal/permeability-increasing protein and plasma lipopolysaccharide binding protein during endotoxemia or sepsis. Arch Surg 129:220–226.

Caroff M, Deprun C, Karibian D (1993): ^{252}Cf plasma desorption mass spectrometry applied to the analysis of underivatized rough-type endotoxin preparations. J Biol Chem 268:12321–12324.

Christ WJ, Asano O, Robidoux ALC, et al. (1995): E5531, a pure endotoxin antagonist of high potency. Science 268:80–83.

Cooper JF, Hochstein HD, Seligmann EB Jr (1972): The limulus test for endotoxin (Pyrogen) in radiopharmaceuticals and biologicals. Bull Parenter Drug Assoc 26:153–162.

Cynkin MA, Ashwell G (1960): Estimation of 2-deoxy sugars by means of the malonaldehyde-thiobarbituric acid reaction. Nature 186:155–156.

Dal Nogare A, Yarbrough WC Jr (1990): A comparison of the effects of intact and deacylated lipopolysaccharide on human polymorphonuclear leukocytes. J Immunol 144:1404–1410.

Danner RL, Elin RJ, Hosseini JM, Wesley RA, Reilly JM, Parrillo JE (1991): Endotoxemia in human septic shock. Chest 99:169–175.

Danner RL, Natanson C, Elin RJ, et al. (1990): *Pseudomonas aeruginosa* compared with *Escherichia coli* produces less endotoxemia but more cardiovascular dysfunction and mortality in a canine model of septic shock. Chest 98:1480–1487.

David SA, Mathan VI, Balaram P (1992): Interaction of melittin with endotoxic lipid A. Biochim Biophys Acta 1123:269–274.

Day JR, Albers JJ, Lofton-Day CE, et al. (1994): Complete cDNA encoding human phospholipid transfer protein from human endothelial cells. J Biol Chem 269:9388–9391.

Dear TN, Boehm T, Keverne EB, Rabbitts TH (1991): Novel genes for potential ligand-binding proteins in subregions of the olfactory mucosa. EMBO J 10:2813–2819.

Delude RL, Fenton MJ, Savedra R Jr, et al. (1994): CD14-mediated translocation of nuclear factor-κB induced by lipopolysaccharide does not require tyrosine kinase activity. J Biol Chem 269:22253–22260.

Din ZZ, Mukerjee P, Kastowsky M, Takayama K (1993): Effect of pH on solubility and ionic state of lipopolysaccharide obtained from the deep rough mutant of Escherichia coli. Biochemistry 32:4579–4586.

Ding AH, Porteu F, Sanchez E, Nathan CF (1990): Shared actions of endotoxin and taxol on TNF receptors and TNF release. Science 248:370–372.

Drayna D, Jarnagin AS, McLean J, et al. (1987): Cloning and sequencing of human cholesteryl ester transfer protein cDNA. Nature 327:632–634.

Duchow J, Marchant A, Crusiaux A, et al. (1993): Impaired phagocyte responses to lipopolysaccharide in paroxysmal nocturnal hemoglobinuria. Infect Immun 61:4280–4285.

Dziarski R (1994): Cell-bound albumin is the 70-kDa peptidoglycan-, lipopolysaccharide-, and lipoteichoic acid-binding protein on lymphocytes and macrophages. J Biol Chem 269:20431–20436.

Elomaa O, Kangas M, Sahlberg C, et al. (1995): Cloning of a novel bacteria-binding receptor structurally related to scavenger receptors and expressed in a subset of macrophages. Cell 80:603–609.

Espevik T, Otterlei M, Skjak-Braek G, Ryan L, Wright SD, Sundan A (1993): The involvement of CD14 in stimulation of cytokine production by uronic acid polymers. Eur J Immunol 23:255–261.

Freudenberg MA, Bog-Hansen TC, Back U, Galanos C (1980). Interaction of lipopolysaccharides with plasma high-density lipoprotein in rats. Infect Immun 28:373–380.

Frey EA, Miller DS, Jahr TG, et al. (1992): Soluble CD14 participates in the response of cells to lipopolysaccharide. J Exp Med 176:1665–1671.

Galanos C, Luderitz O (1984): Lipopolysaccharide: properties of an amphipathic molecule. In Rietschel ET (ed): *Chemistry of Endotoxin*, Vol 1. Amsterdam: Elsevier Science Publishers B. V., pp 46–58.

Galanos C, Luderitz O, Rietschel ET, et al. (1985): Synthetic and natural *Escherichia coli* free lipid A express identical endotoxic activities. Eur J Biochem 148:1–5.

Galanos C, Rietschel ET, Luderitz O, Westphal O (1971): Interaction of lipopolysaccharides and lipid A with complement. Eur J Biochem 19:143–152.

Gallay P, Heumann D, Le Roy D, Barras C, Glauser M-P (1994): Mode of action of anti-lipopolysaccharide-binding protein antibodies for prevention of endotoxemic shock in mice. Proc Natl Acad Sci USA 91:7922–7926.

Gallay P, Jongeneel CV, Barras C, et al. (1993): Short time exposure to lipopolysaccharide is sufficient to activate human monocytes. J Immunol 150:5086–5093.

Gazzano-Santoro H, Meszaros K, Birr C, et al. (1994): Competition between rBPI23, a recombinant fragment of bactericidal/permeability-increasing protein, and lipopolysaccharide (LPS)-binding protein for binding to LPS and gram-negative bacteria. Infect Immun 62:1185–1191.

Gegner JA, Ulevitch RJ, Tobias PS (1995): Lipopolysaccharide (LPS) signal transduction and clearance: dual roles for LPS binding protein and membrane CD14. J Biol Chem 270:5320–5325.

Glauser MP, Zanetti G, Baumgartner J-D, Cohen J (1991): Septic shock: pathogenesis. Lancet 338:732–739.

Golenbock DT, Hampton RY, Qureshi N, Takayama K, Raetz CR (1991): Lipid A-like molecules that antagonize the effects of endotoxins on human monocytes. J Biol Chem 266:19490–19498.

Golenbock DT, Hampton RY, Raetz CRH, Wright SD (1990): Human phagocytes have multiple lipid A-binding sites. Infect Immun 58:4069–4075.

Golenbock DT, Liu Y, Millham FH, Freeman MW, Zoeller RA (1993): Surface expression of human CD14 in chinese hamster ovary fibroblasts imparts macrophage-like responsiveness to bacterial endotoxin. J Biol Chem 268:22055–22059.

Gray PW, Flaggs G, Leong SR, et al. (1989): Cloning of the cDNA of a human neutrophil bactericidal protein: structural and functional correlations. J Biol Chem 264:9505–9509.

Grube BJ, Cochane CG, Ye RD, et al. (1994): Lipopolysaccharide binding protein expression in primary human hepatocytes and HepG2 hepatoma cells. J Biol Chem 269:8477–8482.

Grunwald U, Kruger C, Schutt C (1993): Endotoxin-neutralizing capacity of soluble CD14 is a highly conserved specific function. Circ Shock 39:220–225.

Hagen FS, Grant FJ, Kuijer JL, et al. (1991): Expression and characterization of recombinant human acyloxyacyl hydrolase, a leukocyte enzyme that deacylates bacterial lipopolysaccharides. Biochemistry 30:8415–8423.

Hailman E, Lichenstein HS, Wurfel MM, et al. (1994): Lipopolysaccharide (LPS)-binding protein accelerates the binding of LPS to CD14. J Exp Med 179:269–277.

Hall CL, Munford RS (1983): Enzymatic deacylation of the lipid A moiety of *Salmonella typhimurium* lipopolysaccharides by human neutrophils. Proc Natl Acad Sci USA 80:6671–6675.

Hampton RY, Golenbock DT, Penman M, Krieger M, Raetz CRH (1991): Recognition and plasma clearance of endotoxin by scavenger receptors. Nature 352:342–344.

Hampton RY, Golenbock DT, Raetz RH (1988): Lipid A binding sites in membranes of macrophage tumor cells. J Biol Chem 263:14802–14807.

Hampton RY, Raetz CRH (1991): Macrophage catabolism of lipid A is regulated by endotoxin stimulation. J Biol Chem 266:19499–19509.

Han J, Mathison JC, Ulevitch RJ, Tobias PS (1994): Lipopolysaccharide (LPS) binding protein, truncated at Ile-197, binds LPS but does not transfer LPS to CD14. J Biol Chem 269:8172–8175.

Haziot A, Chen C, Ferrero E, Low MG, Silbert R, Goyert SM (1988): The monocyte differentiation antigen, CD14, is anchored to the cell membrane by a phosphatidylinositol linkage. J Immunol 141:547–552.

Haziot A, Ferrero E, Lin X, Stewart C, Goyert SM: (1994). CD14-negative mice: analysis of the response to LPS. J Endotoxin Res 1(suppl):C136 (abstr).

Haziot A, Rong GW, Lin X-Y, Silver J, Goyert SM (1995): Recombinant soluble CD14 prevents mortality in mice treated with endotoxin (lipopolysaccharide). J Immunol 154:6529–6532.

Heumann D, Gallay P, Betz-Corradin S, Barras C, Baumgartner J-D, Glauser MP (1993): Competition between bactericidal/permeability-increasing protein and lipopolysaccharide binding to monocytes. J Infect Dis 167:1351–1357.

Hoess A, Watson S, Siber GR, Liddington R (1993): Crystal structure of an endotoxin-neutralizing protein from the horseshoe crab, Limulus anti-LPS factor, at 1.5 A resolution. EMBO J 12:3351–3356.

Hofer M, Hampton RY, Raetz CRH, Yu H (1991): Aggregation behavior of lipid IV_A in aqueous solutions at physiological pH. 1: simple buffer solutions. Chem Phys Lipids 59:167–181.

Horwitz AH, Williams RE, Nowakowski G (1995): Human lipopolysaccharide-binding protein potentiates bactericidal activity of human bactericidal/permeability-increasing protein. Infect Immun 63:522–527.

Imoto M, Kusumoto S, Shiba T, Rietschel ET, Galanos C, Luderitz O (1985): Chemical structure of Escherichia coli lipid A. Tetrahedron Lett 26:907–908.

Ingalls RR, Golenbock DT (1995): CD11c/CD18, a transmembrane signaling receptor for lipopolysaccharide. J Exp Med 181:1473–1479.

Iwanaga S, Muta T, Shigenaga T, Miura Y, Seki N, Saito T, Kawabata S-I (1994): Role of hemocyte-derived granular components in invertebrate defense. Ann NY Acad Sci 712:102–116.

Johnston KJ, Ward PA, Goralnick S, Osborn MJ (1977): Isolation from human serum of an inactivator of bacterial lipopolysaccharide. Am J Pathol 88:559–574.

Juan TS-C, Hailman E, Kelley MJ, et al. (1995a): Identification of a lipopolysaccharide binding domain in CD14 between amino acids 57 and 64. J Biol Chem 270:5219–5224.

Juan TS-C, Kelly MK, Johnson DA, et al. (1995b): Soluble CD14 truncated at amino acid 152 binds lipopolysaccharide (LPS) and enables cellular response to LPS. J Biol Chem 270:1382–1387.

Kaca W, Roth RI, Levin J (1994): Hemoglobin, a newly recognized lipopolysaccharide (LPS)-binding protein that enhances LPS biological activity. J Biol Chem 269:25078–25084.

Karplus TE, Ulevitch RJ, Wilson CB (1987): A new method for reduction of endotoxin contamination from protein solutions. J Immunol Meth 105:211–220.

Kirikae F, Kirikae T, Qureshi N, Takayama K, Morrison DC, Nakano M (1995): CD14 is not involved in *Rhodobacter sphaeroides* diphosphoryl lipid A inhibition of tumor necrosis factor alpha and nitric oxide induction by taxol in murine macrophages. Infect Immun 63:486–497.

Kirkland TN, Virca GD, Kuus-Reichel T, et al. (1990): Identification of lipopolysaccharide-binding proteins in 70Z/3 cells by photoaffinity cross-linking. J Biol Chem 265:9520–9525.

Kitchens RL, Ulevitch RJ, Munford RS (1992): Lipopolysaccharide (LPS) partial structures inhibit responses to LPS in a human macrophage cell line without inhibiting LPS uptake by a CD14-mediated pathway. J Exp Med 176, 485–494.

Kobe B, Deisenhofer J (1994): The leucine-rich repeat: a versatile binding motif. Trends Biochem Sci 19: 415–421.

Kobe B, Deisenhofer J (1995): A structural basis of the interactions between leucine-rich repeats and protein ligands. Nature 374:183–186.

Kodama T, Freeman M, Rohrer L, Zabrecky J, Matsudaira P, Krieger M (1990): Type 1 macrophage scavenger receptor contains α-helical and collagen-like coiled coils. Nature 343:531–535.

Kovach NL, Yee E, Munford RS, Raetz CRH, Harlan JM (1990): Lipid IV_A inhibits synthesis and release of tumor necrosis factor induced by lipopolysaccharide in human whole blood *ex vivo*. J Exp Med 172: 77–84.

Krieger M, Hertz J (1994): Structure and functions of multiligand lipoprotein receptors: macrophage scavenger receptors and LDL receptor-related protein (LRP). Annu Rev Biochem 63:601–637.

Kruger C, Schutt C, Obertacke U, et al. (1991): Serum CD14 levels in polytraumatized and severely burned patients. Clin Exp Immunol 85:297–301.

Lauener RP, Geha RS, Vercelli D (1990): Engagement of the monocyte surface antigen CD14 induces lymphocyte function-associated antigen-1/intercellular adhesion molecule-1-dependent homotypic adhesion. J Immunol 145:1390–1394.

Lee J-D, Kato K, Tobias PS, Kirkland TN, Ulevitch RJ (1992): Transfection of CD14 into 70Z/3 cells dramatically enhances the sensitivity to complexes of lipopolysaccharide (LPS) and LPS binding protein. J Exp Med 175:1697–1705.

Lee J-D, Kravchenko V, Kirkland TN, et al. (1993): Glycosyl-phosphatidylinositol-anchored or integral membrane forms of CD14 mediate identical cellular responses to endotoxin. Proc Natl Acad Sci USA 90:9930–9934.

Lee J-O, Rieu P, Arnaout MA, Liddington R (1995): Crystal structure of the domain from the α subunit of the integrin CR3 (CD11b/CD18). Cell 80:631–638.

Lei MG, Qureshi N, Morrison DC (1993): Lipopolysaccharide (LPS) binding to 73-kDa and 38-kDa surface proteins on lymphoreticular cells: preferential inhibition of LPS binding to the former by Rhodopseudomonas sphaeroides lipid A. Immunol Lett 36:245–250.

Leturcq DJ, Moriarty AM, Winn RK, Talbott G, Martin TR, Ulevitch RJ (1994): Therapeutic strategies to block LPS interactions with its receptor. ICMED 20:S146.

Liang S-M, Sakmar TP, Liu T-Y (1980): Studies on Limulus amoebocyte lysate. J Biol Chem 255:5586–5590.

Luchi M, Munford RS (1993): Binding, internalization, and deacylation of bacterial lipopolysaccharides by human neutrophils. J Immunol 515, 959–969.

Lynn WA, Raetz CRH, Qureshi N, Golenbock DT (1991): Lipopolysaccharide-induced stimulation of CD11b/CD18 expression on neutrophils. J Immunol 147:3072–3079.

Madonna GS, Peterson JE, Ribi EE, Vogel SN (1986): Early-phase endotoxin tolerance: induction by a detoxified lipid A derivative, monophosphoryl lipid A. Infect Immun 52:6–11.

Maliszewski CR, Ball ED, Graziano RF, Fanger MW (1985): Isolation and characterization of My23, a myeloid cell-derived antigen reactive with the monoclonal antibody AML-2–23. J Immunol 135:1929–1936.

Marra MN, Wilde CG, Collins MS, Snable JL, Thornton MB, Scott RW (1992): The role of bactericidal/permeability-increasing protein as a natural inhibitor of bacterial endotoxin. J Immunol 148:532–537.

Marra MN, Wilde CG, Griffith JE, Snable JL, Scott RW (1990): Bactericidal/permeability-increasing protein has endotoxin-neutralizing activity. J Immunol 144:662–666.

Mathison JC, Ulevitch RJ (1979): The clearance, tissue distribution, and cellular localization of intravenously injected lipopolysaccharide in rabbits. J Immunol 123, 2133–2143.

Mayor S, Rothberg KG, Maxfield FR (1994): Sequestration of GPI-anchored proteins in caveolae triggered by cross-linking. Science 264:1948–1951.

McGinley MD, Narhi LO, Kelly MJ, et al. (1995): CD14: physical properties and identification of an exposed site that is protected by lipopolysaccharide. J Biol Chem 270:5213–5218.

Minetti CASA, Lin Y, Cislo T, Liu T-Y (1991): Purification and characterization of an endotoxin-binding protein with protease inhibitory activity from Limulus amebocytes. J Biol Chem 266:20773–20780.

Morrison DC, Jacobs DM (1976): Binding of polymyxin B to the lipid A portion of bacterial lipopolysaccharide. Immunochemistry 13:813–818.

Munford RS, Hall CL (1985): Uptake and deacylation of bacterial lipopolysaccharides by macrophages from normal and endotoxin-hyporesponsive mice. Infect Immun 48:464–473.

Munford RS, Hall CL (1986): Detoxification of bacterial lipopolysaccharides (endotoxins) by a human neutrophil enzyme. Science 234:203–205.

Munford RS, Hunter JP (1992): Acyloxyacyl hydrolase, a leukocyte enzyme that deacylates bacterial lipopolysaccharides, has phospholipase, lysophospholipase, diacylglycerollipase, and acyltransferase activities in vitro. J Biol Chem 267:10116–10121.

Nakamura T, Furunaka H, Miyata T, et al. (1988): Tachyplesin, a class of antimicrobial peptide from the hemocytes of the horseshoe crab (Tachypleus tridentatus). J Biol Chem 263:16709–16713.

Nakamura T, Morita T, Iwanaga S (1986): Lipopolysaccharide-sensitive serine-protease zymogen (factor C) found in Limulus hemocytes. Eur J Biochem 154:511–521.

Newman SL, Chaturvedi S, Klein BS (1995): The WI-1 antigen of *Blastomyces dermatitidis* yeasts mediate binding to human macrophage CD11b/CD18 (CR3) and CD14. J Immunol 154:753–761.

Nowotny A (1984): Heterogeneity of endotoxins. In Rietschel, ET, (ed). *Chemistry of Endotoxin,* Vol 1. Amsterdam: Elsevier Science Publishers BV, pp 308–338.

Ohno N, Morrison DC (1989): Lipopolysaccharide interactions with lysozyme: binding of lipopolysaccharide to lysozyme and inhibition of lysozyme enzymatic activity. J Biol Chem 264:4434–4441.

Ooi CE, Weiss J, Elsbach P, Fragione B, Mannion B (1987): A 25-kDa NH_2-terminal fragment carries all the antibacterial activities of the human neutrophil 60-kDa bactericidal/permeability-increasing protein. J Biol Chem 262:14891–14894.

Opal SM, Palardy JE, Marra MN, Fisher CJ Jr, McKelligon BM, Scott RW (1994): Relative concentrations of endotoxin-binding proteins in body fluids during infection. Lancet 344:429–431.

Otterlei M, Varum KM, Ryan L, Espevik T (1994): Characterization of binding and TNF-α-inducing ability of chitosans on monocytes: the involvement of CD14. Vaccine 12:825–832.

Parker TS, Levine DM, Chang JCC, Laxer J, Coffin CC, Rubin AL (1995): Reconstituted high-density-lipoprotein neutralizes gram-negative bacterial lipopolysaccharides in human whole blood. Infect Immun 63:253–258.

Pohlman TH, Munford RS, Harlen JM (1987): Deacylated lipopolysaccharide inhibits neutrophil adherence to endothelium induced by lipopolysaccharide in vitro. J Exp Med 165:1393–1402.

Pugin J, Heumann D, Tomasz A, Nishijima M, Glauser MP, Tobias PS, Ulevitch RJ (1994): CD14 is a pattern recognition receptor. Immunity 1:509–516.

Pugin J, Schurer-Maly C-C, Leturcq D, Moriarty A, Ulevitch RJ, Tobias PS (1993): Lipopolysaccharide activation of human endothelial and epithelial cells is mediated by lipopolysaccharide-binding protein and soluble CD14. Proc Natl Acad Sci USA 90:2744–2748.

Qureshi N, Honovich JP, Hara H, Cotter RJ, Takayama K (1988): Location of fatty acids in lipid A obtained from lipopolysaccharide of Rhodopseudomonas sphaeroides ATCC 17023. J Biol Chem 263: 5502–5504.

Raetz CRH (1984): Escherichia coli mutants that allow elucidation of the precursors and biosynthesis of lipid A. In Rietschel, ET (ed). *Chemistry of Endotoxin,* Vol 1. Amsterdam: Elsevier Science Publishers BV, pp 248–268.

Raetz CRH (1990): Biochemistry of endotoxins. Annu Rev Biochem 59:129–170.

Raetz CRH, Purcell S, Meyer MV, Qureshi N, Takayama K (1985): Isolation and characterization of eight lipid A precursors from a 3-deoxy-D-manno-octylosonic acid-deficient mutant of Salmonella typhimurium. J Biol Chem 260:16080–16088.

Ramadori G, Meyer zum Buschenfelde KH, Tobias PS, Mathison JC, Ulevitch RJ (1990): Biosynthesis of lipopolysaccharide-binding protein in rabbit hepatocytes. Pathobiology 58:89–94.

Read MA, Cordle SR, Veach RA, Carlisle CD, Hawiger J (1993): Cell-free pool of CD14 mediates activation of transcription factor NF-κB by lipopolysaccharide in human endothelial cells. Proc Natl Acad Sci USA 90:9887–9891.

Riedo FX, Munford RS, Campbell WB, Reisch JS, Chien KR, Gerard RD (1990): Deacylated lipopolysaccharide inhibits plasminogen activator inhibitor-1, prostacyclin, and prostaglandin E_2 induction by lipopolysaccharide but not by tumor necrosis factor-α. J Immunol 144:3506–3512.

Rietschel ET (1984): *Chemistry of Endotoxin,* Vol 1. Amsterdam: Elsevier Science Publishers B. V.

Rietschel ET, Brade H (1992): Bacterial Endotoxins. Sci Am 267:54–61.

Rietschel ET, Kirikae T, Schade FU, et al. (1994): Bacterial endotoxin: molecular relationship of structure to activity and function. FASEB J 8:217–225.

Rose JR, Christ WJ, Bristol JR, Kawata T, Rossingnol DP (1995): Agonistic and antagonistic activities of bacterially derived Rhodobacter sphaeroides lipid A: comparison with activities of synthetic material of the proposed structure and analogs. Infect Immun 63:833–839.

Roth RI, Tobias PS (1993): Lipopolysaccharide-binding proteins of Limulus amebocyte lysate. Infect Immun 61:1033–1039.

Rothberg KG, Ying Y, Kolhouse JF, Kamen BA, Anderson RGW (1990): The glycophospholipid-linked folate receptor internalizes folate without entering the clathrin-coated pit endocytic pathway. J Cell Biol 110:637–649.

Rudbach JA, Johnson AG (1964): Restoration of endotoxin activity following alteration by plasma. Nature 202:811–812.

Schiff PB (1979): Promotion of microtubule assembly in vitro by taxol. Nature 277:665–667.

Schumann RR, Leong SR, Flaggs GW, et al. (1990): Structure and function of lipopolysaccharide binding protein. Science 249:1429–1431.

Schutt C, Ringel B, Nausch M, et al. (1988): Human monocyte activation induced by anti-CD14 monoclonal antibody. Immunol Lett 19:321–328.

Schutt C, Schilling T, Grunwald U, Schonfeld W, Kruger C (1992): Endotoxin-neutralizing capacity of soluble CD14. Res Immunol 143:71–78.

Shapiro R, Riordan JF, Vallee BL (1995): LRRning the Rite of springs. Struct Biol 2:350–354.

Skelly RR, Munkenbeck P, Morrison DC (1979): Stimulation of T-independent antibody responses by hapten-lipopolysaccharides without repeating polymeric structure. Infect Immun 23:287–293.

Soell M, Lett E, Holveck F, Scholler M, Wachsmann D, Klein J-P (1995): Activation of human monocytes by streptococcal rhamnose glucose polymers is mediated by CD14 antigen, and mannan binding inhibits TNF-α release. J Immunol 154:851–860.

Staab JF, Ginkel DL, Rosenberg GB, Munford RS (1994): A saposin-like domain influences the intracellular localization, stability, and catalytic activity of human acyloxyacyl hydrolase. J Biol Chem 269: 23736–23742.

Stokes DC, Shenep JL, Fishman M, Hildner WK, Bysani GK, Rufus K (1989): Polymyxin B prevents lipopolysaccharide-induced release of tumor necrosis factor-α from alveolar macrophages. J Infect Dis 160:52–57.

Su GL, Freeswick PD, Geller DA, et al. (1994): Molecular cloning, characterization, and tissue distribution of rat lipopolysaccharide binding protein. Evidence for extrahepatic expression. J Immunol 153: 743–752.

Sultzer BM (1968): Genetic control of leukocyte responses to endotoxin. Nature 219:1253–1254.

Taga T, Hibi M, Hirata Y, et al. (1989): Interleukin-6 triggers the association of its receptor with a possible signal transducer, gp130. Cell 58:573–581.

Takayama K, Din ZZ, Mukeree P, Cooke PH, Kirkland TN (1990): Physicochemical properties of the lipopolysaccharide unit that activates B lymphocytes. J Biol Chem 265:14023–14029.

Takayama K, Qureshi N, Beutler B, Kirkland TN (1989): Diphosphoryl lipid A from Rhodopseudomonas sphaeroides ATCC 17023 blocks induction of cachectin in macrophages by lipopolysaccharide. Infect Immun 57:1336–1338.

Tanaka S, Nakamura T, Morita T, Iwanaga S (1982): Limulus anti-LPS factor: an anticoagulant which inhibits the endotoxin-mediated activation of Limulus coagulation system. Biochem Biophy Res Commun 105:717–723.

Theofan G, Horwitz AH, Williams RE, et al. (1994): An amino-terminal fragment of human lipopolysaccharide-binding protein retains lipid A binding but not CD14-stimulatory activity. J Immunol 152: 3623–3629.

Tobias PS, Soldau K, Gegner JA, Mintz D, Ulevitch RJ (1995): Lipopolysaccharide binding protein-mediated complexation of lipopolysaccharide with soluble CD14. J Biol Chem 270:10482–10488.

Tobias PS, Soldau K, Hatlen L, et al. (1992): Lipopolysaccharide binding protein. J Cell Biochem 16C:151

Tobias PS, Soldau K, Kline L, et al. (1993): Cross-linking of lipopolysaccharide (LPS) to CD14 on THP1 cells mediated by LPS-binding protein. J Immunol 150:3011–3021.

Tobias PS, Soldau K, Ulevitch RJ (1986): Isolation of a lipopolysaccharide-binding acute phase reactant from rabbit serum. J Exp Med 164:777–793.

Tobias PS, Soldau K, Ulevitch RJ (1989): Identification of a lipid A binding site in the acute phase reactant lipopolysaccharide binding protein. J Biol Chem 264:10867–10871.

Ulevitch RJ, Johnston AJ (1978): The modification of biophysical and endotoxic properties of bacterial lipopolysaccharides by serum. J Clin Invest 62:1313–1324.

Ulevitch RJ, Johnston AR, Weinstein DB (1979): New function for high density lipoproteins. J Clin Invest 64:1516–1524.

Viriyakosol S, Kirkland TN (1995): A region of human CD14 required for lipopolysaccharide binding. J Biol Chem 270:361–368.

Wang D, Pabst KM, Aida Y, Pabst MJ (1995): Lipopolysaccharide-inactivating activity of neutrophils is due to lactoferrin. J Leukoc Biol 57:865–874.

Wani MC, Taylor HL, Wall ME, Coggon P, McPhail AT (1971): Plant antitumor agents. VI. The isolation and structure of taxol, a novel antileukemic and antitumor agent from *Taxus brevifolia.* J Am Chem Soc 93:2325–2327.

Weber G, Heck D, Bartlett RR, Nixdorff K (1992): Modulation of effects of lipopolysaccharide on macrophages by a major outer membrane protein of *Proteus mirabilis* as measured in a chemilumnescence assay. Infect Immun 60:1069–1075.

Weersink AJL, van Kessel KPM, Torensma R, Van Strijp JAG, Verhoef J (1990): Binding of rough lipopolysaccharides (LPS) to human leukocytes: inhibition by anti-LPS monoclonal antibody. J Immunol 145:318–324.

Weersink AJL, van Kessel KPM, van den Tol ME, et al. (1993): Human granulocytes express a 55-kDa lipopolysaccharide-binding protein on the cell surface that is identical to the bactericidal/permeability-increasing protein. J Immunol 150:253–263.

Weidemann B, Brade H, Rietschel ET, et al. (1994): Soluble peptidoglycan-induced monokine production can be blocked by anti-CD14 monoclonal antibodies and by lipid A partial structures. Infect Immun 62:4709–4715.

Weingarten R, Sklar LA, Mathison JC, et al. (1993): Interactions of lipopolysaccharides with neutrophils in blood via CD14. J Leukoc Biol 53:518–524.

Weiss J, Elsbach P, Shu C, et al. (1992): Human bactericidal/permeability-increasing protein and a recombinant NH2-terminal fragment cause killing of serum-resistant gram-negative bacteria in whole blood and inhibit tumor necrosis factor release induced by the bacteria. J Clin Invest 90:1122–1130.

White ML, Ma JK, Birr CA, Trown PW, Carroll SF (1994): Measurement of bactericidal/permeability-increasing protein in human body fluids by sandwich ELISA. J Immunol Meth 167:227–235.

Wilde CG, Seilhamer JJ, McGrogan M, et al. (1994): Bactericidal/permeability-increasing protein and lipopolysaccharide (LPS)-binding protein. J Biol Chem 269:17411–17416.

Wollenweber H, Morrison DC (1985): Synthesis and biochemical characterization of a photoactivatable, iodinatable, cleavable bacterial lipopolysaccharide derivative. J Biol Chem 260:15068–15074.

Wright SD, Detmers PA, Aida Y, et al. (1990b): CD18-deficient cells respond to lipopolysaccharide in vitro. J Immunol 144:2566–2571.

Wright SD, Jong MT (1986): Adhesion-promoting receptors on human macrophages recognize Escherichia coli by binding to lipopolysaccharide. J Exp Med 164:1876–1888.

Wright SD, Levin SM, Jong MT, Chad Z, Kabbash LG (1989): CR3 (CD11b/CD18) expresses one binding site for Arg-Gly-Asp-containing peptides and a second site for a bacterial lipopolysaccharide. J Exp Med 169:175–183.

Wright SD, Ramos RA, Tobias PS, Ulevitch RJ, Mathison JC (1990a): CD14, a receptor for complexes of lipopolysaccharide (LPS) and LPS binding protein. Science 249:1431–1433.

Wurfel MM, Kunitake ST, Lichenstein H, Kane JP, Wright SD (1994): Lipopolysaccharide (LPS)-binding protein is carried on lipoproteins and acts as a cofactor in the neutralization of LPS. J Exp Med 180: 1025–1035.

Zhang Y, Doerfler M, Lee TC, Guillemin B, Rom WN (1993): Mechanisms of stimulation of interleukin-1β and tumor necrosis factor-α by Mycobacterium tuberculosis components. J Clin Invest 91:2076–2083.

Zuckerman SH, Qureshi N (1992): In vivo inhibition of lipopolysaccharide-induced lethality and tumor necrosis factor synthesis by Rhodobacter sphaeroides diphosphoryl lipid A is dependent on corticosterone induction. Infect Immun 60:2581–2587.

7

Actin Polymerization in Human Neutrophils

Geneva M. Omann
University of Michigan Medical School and Veterans Administration Medical Center, Ann Arbor, Michigan

There are several classes of chemoattractants for neutrophils that induce migration as well as inflammatory functions such as oxidant production and release of proteinases (for reviews see Omann et al., 1987a; Allen et al., 1988; Sha'afi and Molski, 1988; Zigmond, 1989; Durstin and Sha'afi, 1996). These include leukotriene B_4 (LTB_4), platelet activating factor (PAF), C5a, *N*-formylpeptides, and interleukin 8 (IL-8). All these activators induce a rapid, transient increase in polymerized actin. The ability to polymerize and depolymerize actin has been implicated in migration and proposed to regulate various neutrophil functions. In this chapter the characteristics of the response are described in detail for the *N*-formylpeptide receptor, because *N*-formylpeptides are the most studied of the chemoattractants. The significance of the response for neutrophil function is addressed and assays for detection of the response are critiqued. Studies of the signal transduction mechanism regulating the response are reviewed and the current understanding of the mechanisms regulating polymerization and depolymerization is presented. This review assumes the reader has a basic understanding of actin and actin binding protein dynamics in vitro (Carlier, 1991, Pollard, 1990; Pollard and Cooper, 1986; Carlier and Pantaloni,1994; Sohn and Goldschmidt-Clermont, 1994; Hartwig and Kwiatkowski, 1991). Although I have tried to

Phagocyte Function: A Guide for Research and Clinical Evaluation, Edited by J. Paul Robinson and George F. Babcock. ISBN 0-471-12364-1

be thorough in the topics covered in this chapter, it is impossible to cover all aspects of this topic in a single chapter. The reader is referred to other recent excellent reviews (Coates, 1996; Zigmond, 1989, 1993; Durstin and Sha'afi, 1996; Howard and Watts, 1994; Downey et al., 1996).

ACTIVATION OF ACTIN POLYMERIZATION BY *N*-FORMYLPEPTIDES

Characteristics of the Response

When neutrophils in suspension are stimulated with *N*-formylpeptides, there is an approximate twofold increase in polymerized actin, with an initial peak occurring approximately 10 s after activation (at 37°C) and a second prolonged phase that persists for 3–5 min (Fig. 1) (Rao and Varani, 1982; White et al., 1982, 1983; Wallace et al., 1984; Howard and Meyer, 1984; Howard and Oresajo, 1985a; Sklar et al., 1985a). The response is inhibited by *N*-formylpeptide receptor antagonists (White et al., 1983; Wallace et al., 1984; Howard and Oresajo, 1985b), implying specificity for receptor occupancy. The response is inhibited by pertussis toxin treatment, which implies that a guanine nucleotide binding protein (G protein) of the class G_i is involved (Bengtsson et al., 1990; Sha'afi and Molski, 1987; Omann and Porasik-Lowes, 1991). In addition, the response can be induced by nonhydrolyzable GTP analogues in permeabilized cells, further confirming the importance of G proteins in initiating the response (Redmond et al., 1994; Downey et al., 1989; Therrien and Naccache, 1989). The response is inhibited by cytochalasins (Rao and Varani, 1982; White et al., 1983; Wallace et al., 1984), which block polymerization of actin at the plus (or barbed) end of the filament (MacLean-Fletcher and Pollard, 1980). Thus it appears that the increase in polymerization occurs at the plus end of the actin filament.

The F-actin in activated cells depolymerizes at a much faster rate than the F-actin in unstimulated cells, until the basal level of F-actin is reached. If a suspension of cells is first stimulated with a bolus of *N*-formylpeptide, and during the time course of the response occupancy of new receptors is blocked (by either adding a receptor antagonist or depleting the free ligand), then the F-actin rapidly drops to the basal, prestimulus level (Fig. 2) (Sklar et al., 1985b; Omann et al., 1985; Howard and Oresajo, 1985b). In addition, if after the response is initiated cytochalasin B is added, the F-actin levels again drop rapidly to prestimulus levels, with further decreases being very slow (White et al., 1983). That it is the newly polymerized actin that is rapidly depolymerizing under these conditions is suggested by the work of Cassimeris et al. (1990), who showed that the F-actin in *N*-formylpeptide-induced lamellipodia rapidly depolymerizes when agonist is removed or cytochalasin added. Because the rapid depolymerization occurs in the presence of cytochalasins, which block depolymerization from the barbed ends, the rapid depolymerization appears to occur from pointed (minus) ends.

Morphological Correlates

Microscopy studies of neutrophils in suspension show that resting cells are spherical and that within seconds after addition of chemoattractant many shallow ruffles form:

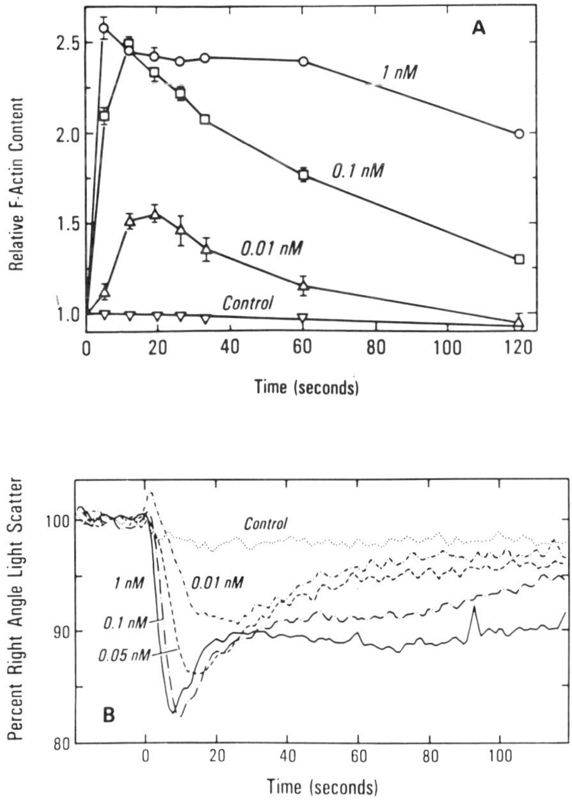

Fig. 1. The *N*-formylpeptide-induced actin polymerization response in human neutrophils in suspension. The *N*-formylpeptide receptor agonist *N*-formyl-norleucyl-leucyl-phenylalanyl-norleucyl-tyrosyl-lysyl-fluorescein was used to stimulate the cells in suspension at 37°C. The relative F-actin content of the cells, detected by the fluorescent phallotoxin assay (A), and right-angle light scatter (B) were measured. Doses of the agonist are listed on the plots. Redrawn from Omann et al., 1987c; copyright 1987, the American Association of Immunologists. Reprinted with permission.

at longer times the cells become polarized with a lamellipodium similar to the leading edge and a uropod similar to the trailing edge of migrating cells on surfaces (Davis et al., 1982: Shields and Haston, 1985; Keller et al., 1991; Watts et al., 1991; Craddock et al., 1977; Hoffstein et al., 1982; Wymann et al., 1987; Coates et al., 1992). The initial increase in F-actin that occurs within seconds of activation is localized under the membrane, whereas at later times the lamellipodium contains polymerized actin and the cell body is relatively deplete of F-actin (Fig. 3) (Oliver et al., 1978; Ryder et al., 1984; Lepidi et al., 1992; Coates et al., 1992; Davis et al., 1982; Keller et al., 1991; Model and Omann, 1996). Kinetic studies relating the distribution of F-actin in cells to the development of polarized morpholgy (Coates et al., 1992) have shown that over the first 45 s of activation of cells with *N*-formyl-methionyl-leucyl-phenylalanine (fMLP) (at 25°C), F-actin increases uniformly beneath the membrane (Coates et al., 1992). An asymmetric distribution of F-actin is seen in round cells at

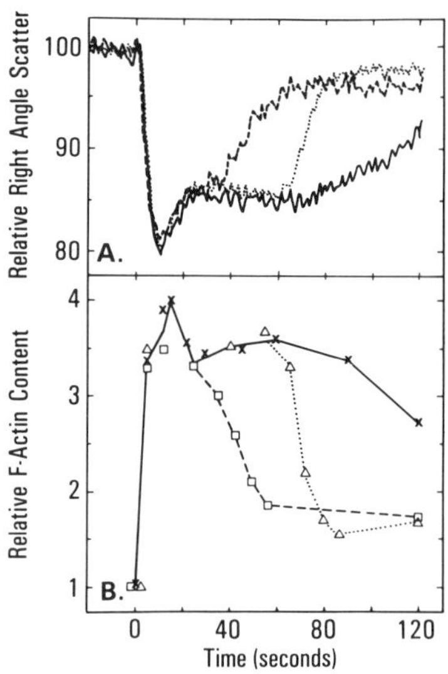

Fig. 2. Pseudosimultaneous right-angle light scatter and actin polymerization in neutrophils: rapid F-actin depolymerization occurs upon termination of ligand-receptor binding. Neutrophils in suspension were stimulated with 1 n*M* *N*-formyl-norleucyl-leucyl-phenylalanyl-norleucyl-tyrosyl-lysyl-fluorescein. At 30 s (squares, dashed line) or 60 s (triangles, dotted line) 2 μ*M* of the receptor antagonist *N-tert*-butoxycarbonyl-Phe-Leu-Phe-Leu-Phe was added to stop binding of the agonist. No *N-tert*-butoxycarbonyl-Phe-Leu-Phe-Leu-Phe was added to the control (solid line, X). Right-angle light scatter was observed (A) while aliquots of sample were removed and rapidly fixed and stained with the fluorescent phallotoxin NBD-phallocidin. These fixed samples were analyzed for NBD-phallocidin staining of F-actin on a flow cytometer (B). From Omann et al., 1985; copyright 1985, Eaton Publishing. Reprinted with permission.

90 s of activation and in polarized cells at later times, suggesting that an asymmetric distribution of F-actin may precede development of polar morphology (Coates et al., 1992). It has also been shown that there is an asymmetric distribution of *N*-formylpeptide receptors on polarized cells, with the bulk of the receptors located not in the lamellipodium where the F-actin is but rather in the cell body (Fig. 4) (Model and Omann, 1996). Thus the mechanism by which F-actin in the lamellipodium is sustained appears to be independent of *N*-formylpeptide receptors or involves a diffusible signal. Mechanisms by which actin polymerization can induce morphological changes in neutrophils are reviewed elsewhere (Cassimeris and Zigmond, 1990; Howard and Watts, 1994).

Ways to Polymerize Actin in Cells

There are two general catagories of mechanisms by which actin could be induced to polymerize in the cells: (1) creation of new free barbed ends which are then available to bind G-actin from the available pool; (2) increasing the ability of G-actin to bind to filaments. Each of these catagories may have several possibilities for producing the

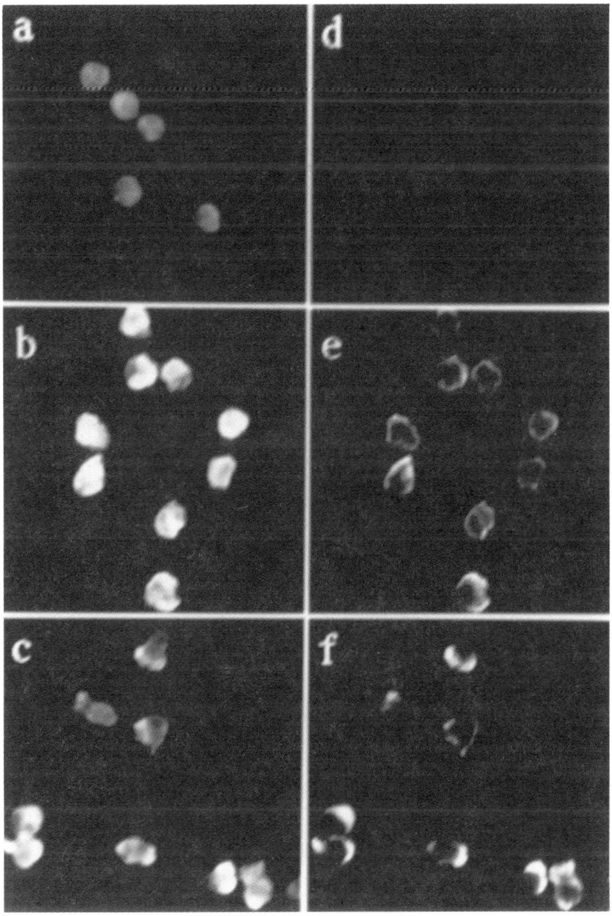

Fig. 3. Localization of F-actin in neutrophils by fluorescence microscopy: total F-actin (*a–c*) and F-actin concentration (*d–f*) in unstimulated neutrophils (*a, d*) and after 0.5 min (*b, e*) and 10 min (*c, f*) of exposure to 1 n*M* *N*-formyl-methionyl-leucyl-phenylalanine. Aliquots of neutrophils, before and after activation, were removed, fixed, and stained with fluorescein isothiocyanate-labeled phalloidin (to label F-actin) and Texas Red (to label cytosolic proteins and indicate cell thickness). Total F-actin was quantified as the fluorescence of FITC-phalloidin. F-Actin concentration was determined by calculating the ratio of the FITC-phalloidin fluorescence to Texas Red fluorescence as described by Coates et al., 1992. From Model and Omann, 1996; copyright 1996, Academic Press, Inc. Reprinted with permission.

effect. For example, creation of new barbed ends could occur by uncapping of existing filments ends and/or severing of filaments to create new, uncapped ends, or by unmasking of nonfilamentous nucleation sites. Increasing the ability of G-actin to bind to filaments could occur by release of sequestered actin monomers (which implies a decrease in the affinity of the actin-sequestering protein complex) or enhancement of the affinity of G-actin for existing uncapped filaments (for example, ATP-actin binds with higher affinity than ADP-actin). Net depolymerization of the F-

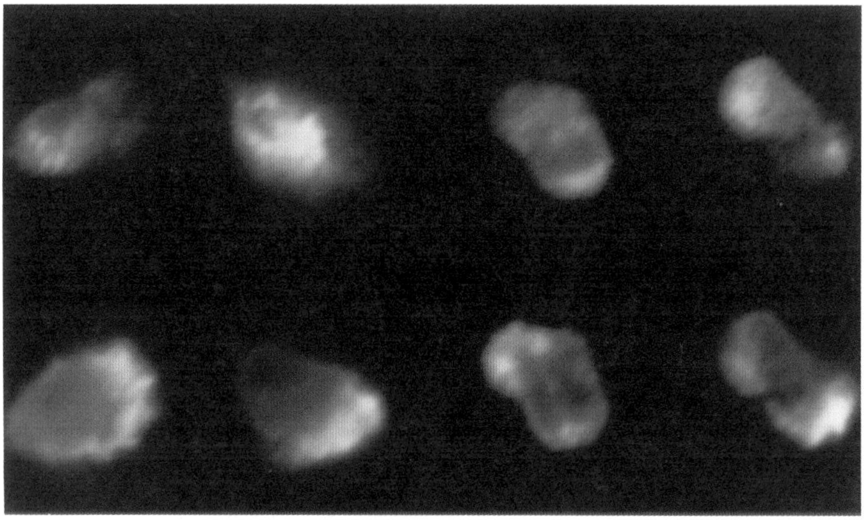

Fig. 4. Sequential staining of *N*-formylpeptide receptors and total F-actin in the same cells. Neutrophils were stained first with *N*-formyl-norleucyl-leucyl-phenylalanyl-norleucyl-tyrosyl-lysyl-fluorescein and imaged (top), then stained with FITC-phalloidin and again imaged (bottom). From Model and Omann, 1996; copyright 1996, Academic Press, Inc. Reprinted with permission.

actin presumably involves a reversal of these effects, although additional mechanisms may regulate depolymerization (Zigmond, 1993). For example, when the *N*-formylpeptide-induced intracellular Ca^{2+} increase is sequestered, the recovery of the actin polymerization response is slowed (Sklar et al., 1985a). This implies that Ca^{2+} may be important, in part, for regulating the depolymerization rate. A hypothetical mechanism could involve gelsolin, which severs F-actin in a Ca^{2+}-dependent manner: the severed filaments are left with capped barbed ends and free pointed ends from which depolymerization can occur. An increase in the number of free pointed ends would increase the depolymerization rate. However, depolymerization in the absence of Ca^{2+} still occurs, although at a lower rate; thus Ca^{2+}-independent depolymerization mechanisms also exist. Excellent reviews by Zigmond and Fechheimer (Zigmond, 1993; Fechheimer and Zigmond, 1993) discuss depolymerization in greater detail.

Studies of the mechanism by which the *N*-formylpeptide-induced increase in F-actin occurs have shown that an increase in the number of actin nucleation sites (measured as the ability of cell lysates to nucleate barbed end polymerization of exogenous actin) correlates with an increase in F-actin (Carson et al., 1986; Cano et al., 1991). Although this work suggested that barbed end nucleating activity is regulated by chemoattractants, it did not clarify for intact cells what fraction of filaments have free barbed ends and whether that fraction is altered by chemoattractants.

To address this question, studies have been performed in streptolysin *O*-permeabilized neutrophils, where exogenous actin binding proteins can be incorporated into the cells (Tardif et al., 1995). An actin polymerization response can be induced in streptolysin *O*-permeabilized cells by chemoattractant and by directly activating G proteins with GTPγS (Redmond et al., 1994; Tardif et al., 1995). The work of Tardif et al. (1995) suggests that there is a GTPγS-induced increase of free barbed ends. This work is based largely on the effect of profilin on actin polymerization in this system. Under certain conditions in vitro, profilin has been shown to promote polymerization of actin (Pantaloni and Carlier, 1993). In the presence of thymosin β_4, G-actin, and F-actin, profilin promotes polymerization if the barbed ends are not capped. In this case, profilin acts as a shuttle, removing actin from the thymosin β_4-bound pool: the profilin–actin complex binds to the barbed end of a filament and the profilin is released as the ATP on the actin is cleaved (because profilin has a low affinity for ADP-actin). If the barbed ends are capped, profilin acts as a monomer-sequestering protein and F-actin depolymerizes. In the Tardif work (Tardif et al., 1995), exogenous profilin in unstimulated cells caused depolymerization of F-actin, suggesting that the barbed ends of the filaments in unstimulated cells are capped. In GTPγS-stimulated cells, exogenous profilin did not cause depolymerization of F-actin, unless the barbed-end capper cytochalasin B was also added. Thus Tardif and co-workers (1995) concluded that activation of actin polymerization in neutrophils involves the generation of free barbed ends. A major barbed-end capping activity in neutrophils has been identified as capping protein-β_2, an analogue of the calcium-independent capZ protein (DiNubile et al., 1995); however, further studies are needed to determine its role in agonist-induced actin polymerization in neutrophils.

Several studies have addressed the question of whether alteration of G-actin-sequestering proteins could also be involved. Southwick and Young (1990) questioned whether profilin, originally identified as a G-actin binding protein (Carlsson et al., 1977), may regulate the availability of monomeric actin. They measured the amount of profilin–actin complexes in resting and stimulated neutrophils. They showed that although there is a decrease in the amount of actin–profilin complexes in stimulated neutrophils, there is far less profilin than monomeric actin in these cells, such that release of all the actin monomer from profilin could account for only less than one-third of the total increase in F-actin. Cassimeris and colleagues (1992) have studied another abundant monomer-sequestering protein, thymosin β_4. Thymosin β_4 sequesters the majority of G-actin in resting neutrophils, but the affinity for actin is not different in lysates from resting and stimulated cells. This result is further supported by studies of streptolysin *O*-permeabilized cells, which show that the affinity of exogenously added thymosin β_4 is not different in resting and GTPγS-stimulated cells (Tardif et al., 1995). These results are consistent with the hypothesis that regulation of the polymerization state of actin occurs by modulation of free barbed ends, although regulation of the pool of G-actin cannot be unequivocally ruled out.

Two Pools of F-Actin

Traditionally, the cytoskeletal elements and cytoplasmic elements of the neutrophil have been studied by fractionating the two based on solubility in the detergent Triton X-100. The detergent appears to solubilize the membrane constituents and release the majority of the cytoplasmic contents, but leaves an insoluble cytoskeletal structure that can be readily sedimented. However, it has recently been shown that there is also a Triton X-100-soluble F-actin pool (Watts and Howard, 1992). The two pools of F-actin differ in their content of actin binding proteins, stability at 4°C, and subcellular distribution (Watts and Howard, 1992). The Triton X-100-insoluble F-actin is localized to submembranous areas, is stable at 4°C, and contains little gelsolin; the Triton X-100-soluble F-actin is distributed diffusely through the cell, is high in gelsolin, and is unstable at 4°C. Upon activation by *N*-formylpeptides, there is an increase in the Triton X-100-insoluble F-actin and a decrease in the Triton X-100-soluble F-actin and G-actin (Watts and Howard, 1993). Thus it appears that both the Triton X-100-soluble F-actin and G-actin are incorporated into the Triton X-100-insoluble pool upon activation.

ASSAYS OF ACTIN POLYMERIZATION IN NEUTROPHILS

Several assays have been utilized to monitor the actin polymerization response, and because proper interpretation of the data depends on an understanding of the assays, they are discussed in detail as follows.

DNAse I Assay

DNAse I enzymatic activity is potently inhibited by G-actin and can be used to quantify the monomeric actin in cell lysates (Blikstad et al., 1978). An fMLP-induced increase in polymerized actin in neutrophils was first inferred by observation of an fMLP-induced decrease in neutrophil G-actin quantified as the ability of cell lysates to inhibit DNAse I (Rao and Varani, 1982; Fechheimer and Zigmond, 1983). The G-actin decreased to about 50% of the prestimulus level within seconds of stimulus addition, the effect was dose-dependent with an ED_{50} of approximately 1 n*M*, and cytochalasin D inhibited the effect (Rao and Varani, 1982). Subsequent studies by Howard and Meyer (1984) and Bengtsson et al. (1986) confirmed this result and showed an inverse correlation with formation of F-actin (by the fluorescent phallotoxin binding assay described below). This assay is rather laborious and has been superseded to a large degree by methods for measuring F-actin directly or indirectly. However, it still remains a useful assay for showing the complementarity of changes in G-and F-actin and for quantifying absolute levels of G-actin and total actin.

Association of Actin with Cytoskeleton

An *N*-formylpeptide-induced increase in F-actin in cells in suspension was first identified as an increase in the actin associated with a Triton X-100-insoluble fraction from the cells. That is, cells were activated, then lysed with Triton X-100 and the suspension centrifuged at low speed. The insoluble pellet containing F-actin was run on an SDS–polyacrylamide gel. The actin band was quantified and showed that chemotactic stimulation induced a rapid, transient 2.5-fold increase in actin in the pellet

(White et al., 1983; Yassin et al., 1985; Sha'afi and Molski, 1987). The ED_{50} for fMLP activation was approximately 10^{-10} *M*. The response was not sensitive to the absence of extracellular calcium, and was inhibited by cytochalasin B, the receptor antagonist boc-Leu-Phe-Leu-Phe, and pertussis toxin. The increase of actin in the Triton X-100-insoluble pellet was mirrored by a decrease of actin in the Triton X-100-soluble fraction. This is also a rather laborious assay, although it is probably the most amenable to absolute quantification of actin in the F-actin and G-actin pools. In addition, this approach is the one that made possible the identification of the Triton X-100-soluble and -insoluble F-actin pools discussed above (Watts and Howard, 1992, 1993).

Fluorescent Phallotoxin Binding Assay

Howard and Meyer (1984) and Wallace et al. (1984) independently developed a fluorescent assay for measuring F-actin in neutrophils. In this assay, cells are fixed and then stained with fluorescent phallotoxins, which bind to filamentous actin but not to G-actin (Wallace et al., 1984). The fluorescence bound to the cells is then quantified by flow cytometry (Howard and Meyer, 1984; Wallace et al., 1984), or by spectrofluorometry after pelleting the stained cells and redissolving them in methanol (Howard and Oresajo, 1985b). When neutrophils are stimulated with fMLP, there is an approximate twofold increase in polymerized actin that occurs with an ED_{50} of approximately 1 n*M*. The response is biphasic with a peak of polymerization at approximately 10 s after addition of fMLP and a sustained second phase that begins to recover after 3–5 min. Fluorescence microscopic examination of cells at the peak of the response shows that the increase in F-actin occurs predominantly in the submembranous cortical actin (Wallace et al., 1984; Howard and Oresajo, 1985a; Coates et al., 1992; Model and Omann, 1996), whereas at later times the F-actin is located predominantly in the lamellipodia. Because this is a relatively easy assay to perform, most of what is known about actin polymerization in neutrophils has been obtained with this assay. A limitation of this assay is that it is not amenable to determination of absolute F-actin levels in cells. Thus most data are reported as changes relative to a control.

It is worth noting that the fluorescent phallotoxin binding assay measures both the Triton X-100-insoluble and -soluble F-actin pools defined by Watts and Howard (1992, 1993), whereas the assay of actin on polyacrylamide gels quantifies only the Triton X-100-insoluble F-actin pool. Presumably the DNAse I assay detects only the G-actin pool.

Right-angle Light Scatter

Yuli and Snyderman (1984a,b) described an fMLP-induced right-angle light scatter (RALS) response. They showed a biphasic response with a rapid decrease in RALS that reaches a minimum after 10 s, and a second, slower decrease that reaches a minimum after 40 s. Scanning electron micrographs of neutrophils at the various stages indicate that little change in overall cell structure occurs early in the response, suggesting that the scattering changes are due to changes in subcellular structure (Yuli and Snyderman, 1984a,b).

Right-angle Light Scatter Response and Actin Polymerization are Tightly Coupled

Sklar and co-workers (1985a) showed that the RALS response and the actin polymerization response are tightly coupled. A pseudosimultaneous protocol for measuring the responses was designed in which RALS was monitored on a suspension of neutrophils in a spectrofluorometer while aliquots were removed and fixed (Sklar et al., 1985a; Omann et al., 1985). The fixed cells were stained with fluorescent phallotoxins and flow cytometric analysis performed. The changes in the side scatter characteristics determined by flow cytometery on the fixed cells were consistant with the changes measured from live cells in the spectrofluorometer (Sklar et al., 1985a). The increase in F-actin as determined by the fluorescent phallotoxin assay was inversely related to the changes in RALS.

Specificity of the Right-angle Light Scatter Assay. Because the intensity of the RALS signal changes by a maximum of 20% upon activation with fMLP, and this correlates with a twofold increase in F-actin, it is clear that the other 80% of the RALS signal is due to something else in the cell that could vary depending upon the treatment conditions of the cells. For example, in the presence of cytochalasin, when the fMLP-induced actin polymerization response is inhibited, degranulation is detectable as a decrease in the RALS (Sklar et al., 1984; Fletcher and Seligman, 1985) presumably because the granules contribute significantly to the light scatter and after fusion with the plasma membrane they disappear. The RALS signal can change as much as 50% owing to degranulation. Thus the conditions under which RALS changes correlate with actin polymerization must be clearly defined. Recent studies have concluded there is no correlation between RALS and F-actin polymerization under very diverse conditions (Keller et al., 1995; Kraus and Niederman, 1990). However, the correlation holds in general for stimulation of neutrophils by chemoattractants for short (< 5 min) times of activation, as has been shown in numerous studies where both responses were measured in a pseudosimultaneous manner (Sklar et al., 1985a,b; Tecoma et al., 1986; Omann et al., 1985, 1987a–c, 1989; Omann and Porasik-Lowes, 1991; Omann and Sklar 1988, 1989). These studies employed differing protocols for stimulus exposure (bolus, ramp, pulse), differing doses of different ligands (*N*-formylpeptide, LTB_4, PAF), and differing levels of calcium. Thus under a wide variety of conditions, RALS is a valid assay that provides high time resolution of the actin polymerization responses in living neutrophils.

Right-angle Light Scatter by Flow Cytometry. Theoretical considerations indicate that RALS results from light reflecting, refracting, and diffracting upon and within a cell (Kerker, 1983). Numerous experimental studies have attempted to relate RALS measured on the flow cytometer with structural features of cells (Brunsting and Mullaney, 1974; Benson et al., 1984; Visser et al., 1980). The main conclusion is that RALS results from scattering from subcellular structures. Few studies have used flow cytometery in real time to quantify RALS changes upon activation by chemoattractants. As noted above, chemoattractant-induced degranulation in cytochalasin-treated cells has been observed by flow cytometry (Fletcher and Seligman, 1985; Sklar et al., 1984). Complications of the scattering signal due to aggregation of neutrophils can easily be avoided by eliminating aggregates from the analysis of data. Although theoretically possible, observation of the RALS response with flow cytometry is limited by the fact that the response occurs very rapidly and few cytometers have an on-line injection system that allows monitoring of the signal within the first 5–10 sec of stimulus addition. However, given this limitation, RALS changes that to a first approximation inversely correlate with F-actin levels have been confirmed by flow cytometery (Sklar et al., 1984; McNeil et al., 1985; Omann et al., 1995).

Significance of the Right-angle Light Scatter Assay. The RALS response has two main advantages over the fluorescent phallotoxin staining assay. It measures the response (1) in real time and (2) in living cells. Thus the time resolution of the RALS assay is considerably higher than that of the fluorescent phallotoxin assay. The major disadvantage of the RALS assay is that it is an indirect measure of actin polymerization. Exactly how the changes in F-actin lead to changes in RALS is not known. Thus it is recommended that any conclusion based on this method be validated with an assay such as the fluorescent phallotoxin assay or the quantification of actin associated with the Triton X-100-insoluble cytoskeleton.

Shape Change by Transmittance

In the study of platelets, changes in light transmission have been extensively used to indicate cell aggregation and changes in cell symmetry (reviewed by Deranleau, 1987). The measurement of light transmission as an assay for neutrophil aggregation has also been employed with some success (Craddock et al., 1977; Hoffstein et al., 1982; Sklar et al., 1985b). The increase in transmitted light at longer times correlates with aggregates observed microscopically (Craddock et al., 1977; Hoffstein et al., 1982). However, the time course of the early kinetics of the response (<1 min) does not correlate well with aggregation. Yuli and Snyderman (1984b) showed that fMLP-induced changes in light transmission mirrored the RALS response (to a first approximation) over short time periods. Under the conditions of this study, changes in the transmitted light did not correlate well with cell aggregation or loss of cell symmetry; thus they concluded that the change of light transmission by activated neutrophils was due to changes in light output from subcellular objects, similar to that seen by RALS (Yuli and Snyderman, 1984a). In contrast, theoretical considerations indicate that the intensity of light scattered from small subcellular organelles is orders of magnitude less than light scattered from an object the size of a cell (Wymann et al., 1987; Latimer et al., 1968). Thus transmittance measurements suggest that a change in the cell size and/or shape occurs. Measurement of light transmission has been employed as an assay of this "shape change" (Ehrengruber et al., 1995a,b, 1996;

Wymann et al., 1987, 1989, 1990). The shape change response observed by transmittance has been shown to correlate with actin polymerization (as assessed by the fluorescent phallotoxin assay) (Wymann et al., 1990).

Summary

Thus it appears that chemoattractants induce some reorganization of the subcellular structure (RALS) and change of cell shape (light transmission assay) that are tightly coupled to the polymerization of actin in the cells. Electron microscopic studies of neutrophils show that resting cells are spherical and that within seconds after adding chemoattractant many shallow ruffles form: at longer times the cells become polarized (Craddock et al., 1977; Hoffstein et al., 1982; Wymann et al., 1987; Watts et al., 1991; Coates et al., 1992). Because membrane ruffling is a morphological change that occurs soon after addition of chemoattractant to cells, it has been postulated that the cellular and subcellular structural changes that are coupled to actin polymerization result from membrane ruffling (Yuli and Snyderman, 1984a; Sklar et al., 1984, 1985a). However, a definitive study has not yet been published.

RELATIONSHIP TO FUNCTION

N-Formylpeptide-induced actin polymerization correlates with migration in terms of the dose-response characteristics of *N*-formylpeptide and LTB_4 (Howard and Meyer, 1984; Harvath, 1990). In addition, it has been known for some time that cytochalasins, which inhibit the actin polymerization response, inhibit migration (Zigmond and Hirsch, 1972). Because actin polymerization and depolymerization play crucial roles in the formation of lamellipodia (reviewed by Zigmond, 1993) it is generally accepted that the actin polymerization response represents a fundamental component of the migratory machinery.

Because cytochalasins enhance the *N*-formylpeptide-induced oxidant response, it has been hypothesized that the actin polymerization response modulates oxidant production. In support of this hypothesis it has been shown that oscillations in *N*-formylpeptide-induced shape change are paralleled by oscillations in oxidant production under certain conditions (Wymann et al., 1989). Contrary to this hypothesis, Bengtsson et al. (1991) have shown that tetracaine, which enhances *N*-formylpeptide-induced actin polymerization, also enhances oxidant production, calling into question the role of the newly polymerized actin on regulating the oxidant response. Jesaitis et al. (1984, 1985, 1986; Klotz et al., 1994; Klotz and Jesaitis, 1994) showed that cytochalasins enhance receptor upregulation and prevent the association of the ligand–receptor complex with the cytoskeleton, a process that leads to desensitization of the receptor. Thus the actin cytoskeleton plays an important role in receptor desensitization. Cytochalasins thus are likely to enhance oxidant production by upregulating the receptors (and components of the NADPH oxidase) to the plasma membrane and inhibiting receptor desensitization rather than by a direct effect on the NADPH oxidase. Thus the direct regulation of oxidant production by F-actin is still an unproven hypothesis.

Because cytochalasins enhance *N*-formyl peptide-induced degranulation, it has been hypothesized that the actin polymerization response modulates degranulation (Zurier et al., 1973). Tetracaine, which enhances *N*-formylpeptide-induced actin polymerization, inhibits degranulation (Bengtsson et al., 1991). This result is consistent with the hypothesis that the actin cytoskeleton forms a barrier to granule fusion with the plasma membrane (Aunis and Bader, 1988; Smolen et al., 1991b). Thus F-actin appears to play an important role in regulating degranulation in neutrophils. How all of these events are coordinated in a neutrophil functioning in vivo is yet to be discovered.

SIGNAL TRANSDUCTION REGULATING ACTIN POLYMERIZATION

Signaling Systems Activated by *N*-Formylpeptides

Activation of neutrophils by several chemoattractants, including *N*-formylpeptides, LTB_4, PAF, C5a, and IL-8, depends on each type of chemoattractant having its own specific cell surface receptor (for reviews see Omann et al., 1987a; Sha'afi and Molski, 1988, Allen et al., 1988; Downey et al., 1996; Durstin and Sha'afi, 1996). These appear to be of the class of seven transmembrane segment receptors. The receptors activate a guanine nucleotide binding protein (G protein) that is of the G_i, pertussis toxin-sensitive, class. There are two subtypes, G_{i2} and G_{i3}, present in the neutrophil.

Signaling systems in the neutrophil are complex and not completely understood. The following summarizes some of what is known. A partial schematic is shown in Figure 5.

Phospholipase C. Substantial evidence exists to support the contention that the G_i proteins couple to phospholipase C, which converts phosphatidylinositol 4,5-bisphosphate (PIP_2) to diacylglycerol (DAG) and inositol 1,4,5-trisphosphate (IP_3) (reviewed in Omann et al., 1987a; Sha'afi and Molski, 1988; Durstin and Sha'afi, 1996). Thus a decrease in PIP_2 is observed after activation of neutrophils with chemoattractants. Inositol 1,4,5-trisphosphate causes the release of Ca^{2+} from intracellular stores, and Ca^{2+} and DAG activate protein kinase C. Protein phosphorylation events appear to be important for stimulation of superoxide production, and Ca^{2+} is important for granule release.

Phosphatidylinositol-3-kinase. Production of phosphatidylinositol 3,4,5-trisphosphate (PIP_3) by phosphatidylinositol-3-kinase (PI-3-kinase) has recently been identified as an important agonist-regulated function in many cells (Stephens et al., 1993a). The best characterized PI-3-kinase is a tyrosine kinase receptor-regulated heterodimer with a 110-kDa catalytic subunit and an 85-kDa regulatory subunit. Phosphatidylinositol 3,4,5-trisphosphate has been shown to increase upon activation of neutrophils with chemoattractants (Traynor-Kaplan et al., 1988,1989; Eberle et al., 1990; Stephens et al., 1991, 1993b; Vlahos and Matter, 1992). The work of Stephens and colleagues (1991), based on kinetic data, strongly supports the hypothesis that phosphatidylinositol 4,5-bisphosphate is the precursor for PIP_3 production. It has been shown that PIP_3 increases are inhibited by pertussis toxin treatment of neu-

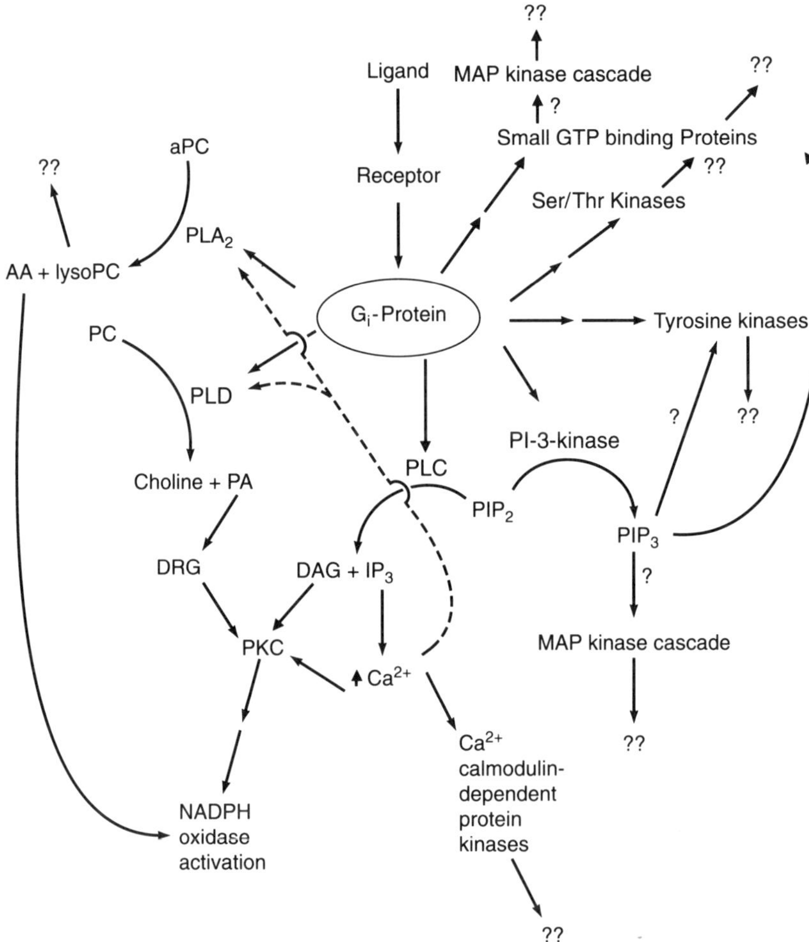

Fig. 5. Chemoattractant-mediated signaling pathways in neutrophils. Ligand–receptor complex activates a pertussis toxin-sensitive G_i protein which then activates numerous events in the cell. Whereas PLC has been most well established as an effector of the G_i protein, evidence suggests the additional pathways shown, although in many cases molecular details are lacking (as indicated by the multiple arrows and question marks). A dashed line represents activation of the enzyme by Ca^{2+}. AA = arachidonic acid; aPC = 1-alkyl-2-arachidonyl-*sn*-glycerophosphocholine; PLA_2 = phospholipase A_2; PC= phosphatidylcholine; PLD = phospholipase D; PLC = phospholipase C; PA = phosphatidic acid; DRG = diradylglycerol; PKC = protein kinase C; DAG = diacylglycerol; IP_3 = inositol 1,4,5-trisphosphate; PIP_2 = phosphatidylinositol 4,5-bisphosphate; PIP_3 = phosphatidylinositol 3,4,5-trisphosphate; $\uparrow Ca^{2+}$ = increased cytosolic Ca^{2+}, MAP = mitogen-activated protein.

trophils and regulated by guanine nucleotides in permeabilized neutrophils (Traynor-Kaplan et al., 1989; Stephens et al., 1993b,c), implying G-protein regulation of this enzyme. Vlahos and Matter (1992) showed that PI-3-kinase activity in neutrophils was stimulated without tyrosine phosphorylation, calling into question the role of tyrosine kinases in the activation cascade. Stephens et al. (1993b) showed that there were chemoattractant-induced tyrosine kinase-associated and tyrosine kinase-independent PI-3-kinase activities that both depended on G-protein activation; of these the tyrosine kinase-independent activity predominated. Subsequently, $G_{\beta\gamma}$ regulated PI-3-kinases from neutrophils and U937 cells (Stephens et al., 1994) and platelets (Thomason et al., 1994) were identified. More recently a 110-kDa, G-protein-regulated PI-3-kinase has been cloned and characterized (Stoyanov et al., 1995). The PI-3-kinase inhibitor wortmannin inhibits both forms of the enzyme (Stephens et al., 1994).

Phospholipase A_2. Phospholipase A_2 (PLA_2) hydrolyzes arachidonic acid from the *sn*-2 position of 1-alkyl-2-arachidonyl-*sn*-glycerophosphocholine (Cockcroft, 1996). Both the arachidonic acid and lysophosphatidyl choline released have been implicated as signaling molecules. The cytosolic PLA_2 is activated by a pertussis toxin-sensitive G protein, requires micromolar intracellular Ca^{2+} for activity, and may involve phosphorylation (Lin et al., 1993; Durstin and Sha'afi, 1996), although the latter has been called into question (Nigam et al., 1995). However, the Ca^{2+} requirement of this enzyme implies that it depends on phospholipase C activation occurring concurrrently.

Phospholipase D. Phospholipase D (PLD) releases the choline moiety from phosphatidylcholine, liberating phosphatidic acid (Cockcroft, 1996; Billah et al., 1989, 1991). The phosphatidic acid released can be converted to 1,2-diradylglycerol, which activates protein kinase C; thus PLD has been shown to be important for sustaining protein kinase C activation. Regulation of PLD appears to be complex, with a pertussis toxin-sensitive G protein, small GTP-binding proteins, and Ca^{2+} being important. However, in spite of its Ca^{2+} requirement, at least one report indicates that PLD is regulated independently of PLC (Mullmann et al., 1993).

Tyrosine Kinases. Several reports have indicated that chemoattractants induce tyrosine phosphorylation in neutrophils (Gomez-Cambronero et al., 1989, 1991; Kraft and Berkow, 1987; Berkow et al., 1989; Berkow and Dodson, 1990). This activity is inhibited by pertussis toxin pretreatment of the cells, implying the involvement of a G protein (Huang et al., 1988; Gomez-Cambronero et al., 1991). Further evidence for the involvement of a G protein is that guanine nucleotides, which activate G proteins, induce tyrosine phosphorylation in electropermeabilized neutrophils (Nasmith et al., 1989). Because the chemoattractant receptors for which the structures are known share no similarity with tyrosine kinase receptors, it seems most likely that nonreceptor tyrosine kinases are involved.

The Search for the Message Regulating Actin Polymerization

The Calcium Ion is Not the Initiating Signal. Because a large increase of intracellular calcium is observed upon activation of neutrophils with chemoattractants, it was initially hypothesized that the increase in Ca^{2+} drove the polymerization of actin by activating the protein gelsolin, which severs actin filaments and thus could promote polymerization by creating more nucleation sites. Using the intracellular Ca^{2+} chelator and fluorescent indicator Quin 2, we showed that sequestration of the cytoplasmic rise in Ca^{2+} did *not* inhibit the actin polymerization response (Sklar et al., 1985a). This observation has been confirmed by several studies using similar and different methods for inhibiting the intracellular Ca^{2+} rise (Howard and Wang, 1987; Al-Mohanna and Hallett, 1990; Meshulam et al., 1986; Sha'afi et al., 1986; Bengtsson et al., 1986) and using a permeabilized cell system (Downey et al., 1990). In addition, studies have shown that migration of neutrophils is insensitive to lowered cytoplasmic calcium levels (Zigmond et al. 1988). Although we showed that sequestration of the intracellular Ca^{2+} rise did not inhibit the actin polymerization response, it did slow the recovery phase (Sklar et al., 1985a). Thus it is possible that Ca^{2+} has a role in regulating actin depolymerization, although it is not the signal that turns on actin polymerization.

Cyclic Adenosine Monophosphate is Not the Initiating Signal. Cyclic AMP is also an important second messenger in many systems. Neutrophils are known to have adenylyl cyclase and stimulatory guanine nucleotide-binding proteins (G_s proteins), and cAMP levels in neutrophils are increased upon activation with chemoattractants (Simchowitz et al., 1980). However, it has been shown that in neutrophils the chemoattractant-induced increase in cAMP results from a Ca^{2+}-dependent mechanism that does not involve a direct G-protein activation of adenylyl cyclase (Verghese et al. 1985). A specific inhibitor of adenylyl cyclase has been shown to inhibit the chemoattractant-induced cAMP rise without inhibiting responses to chemoattractants, including chemotaxis (Simchowitz et al., 1983). In addition, studies have shown that adding exogenous cAMP (e.g., by adding dibutyryl cAMP, a membrane permeable analogue, or by adding cAMP to permeabilized cells) inhibits chemoattractant-induced responses (Takenawa et al., 1986; Downey et al., 1991; Smolen et al., 1991a). It is now well established that there are several nonchemotactic receptors (e.g., prostaglandin E_2 receptors, α-adrenergic receptors) on neutrophils that couple to adenylyl cyclase in a G_s-protein-dependent manner, and activation of these receptors inhibits chemoattractant-induced responses in neutrophils by elevating cAMP (Tecoma et al., 1986; Takenawa et al., 1986; Downey et al., 1991; Harvath et al., 1991). The effects of cAMP appear to be through cAMP-regulated kinases (Mueller et al., 1992). Thus in neutrophils, cAMP plays an inhibitory rather than a stimulatory role with regard to chemoattractant activation.

The Hydrogen Ion is Not the Initiating Signal. Because pH has been shown to regulate the state of actin in other systems, for example, in sea urchin eggs (Begg et al., 1979) and during the acrosomal reaction of echinoderm sperm (Tilney et al., 1978), it was hypothesized to be a signal in neutrophils also. Measurement of the intracellular pH of neutrophils shows an initial chemoattractant-induced drop followed by a rise. Acidification of the cytoplasm by addition of sodium propionate to neutrophils causes a rapid, transient increase in polymerized actin (Yuli and Oplatka, 1987; Molski and Sha'afi, 1987), supporting this idea. However, increased H^+ does not appear to be a sufficient signal (Molski and Sha'afi, 1987). Subsequently, it was shown that chemoattractant-induced acidification of the cytoplasm is inhibited by Ca^{2+} chelators, which do not affect actin polymerization (Naccache et al., 1989). And in correlative studies, Downey and Grinstein (1989) showed a lack of correlation between pH changes and actin polymerization changes. Thus H^+ does not fit the bill for a second messenger for actin polymerization.

Phospholipase C Products and/or Reactants May Not be the Initiating Signals. A great deal of work has shown that PIP_2 binds to several actin-binding proteins, in particular, gelsolin and profilin, and this has been hypothesized to be important for signaling cytoskeletal changes (reviewed by Stossel, 1989). Gelsolin is a protein that binds to actin filaments in the presence of high calcium and severs the filament, leaving the barbed or rapidly growing end of the filament capped. Release of this cap would reveal a nucleation site. Phosphatidylinositol 4,5-bisphosphate binds to gelsolin and inhibits its ability to sever actin filaments (Janmey and Stossel, 1987,1989). The decrease in PIP_2 that occurs during neutrophil activation may result in the release of gelsolin and make it available for severing. However, because high calcium is not required for actin polymerization to occur, this mechanism seems unlikely. In addition, this scheme is not complete without a mechanism for releasing the gelsolin from the barbed ends.

In vitro studies suggest an alternative possible role of gelsolin in creating masked nucleation sites (summarized by Stossel, 1989). Gelsolin binds to G-actin in low (μM) calcium and nucleates pointed-end polymerization such that short actin filaments are formed with the barbed end capped. Because the free end is the slow growing end, these filaments never get very long. However, if the gelsolin molecule dissociates, the barbed end is unmasked and can serve as a nucleation site for (rapid) polymerization. Phosphatidylinositol 4,5-bisphosphate binds to gelsolin in this complex and promotes its dissociation from the gelsolin-actin complex, thereby releasing nucleation sites. In vitro, these short filaments can be driven to depolymerize in buffer containing EGTA, but the last actin monomer bound with the gelsolin remains in a high-affinity complex that does not readily dissociate. These residual gelsolin–actin complexes can be detected experimentally. Howard et al. (1990) and Deaton et al. (1992) have studied gelsolin in neutrophils and have shown that in neutrophils these high-affinity gelsolin–actin complexes exist and their number decreases upon activation with chemoattractants in a manner that correlates with increased actin polymerization. However, PIP_2-induced dissociation of the gelsolin–actin complexes in neutrophils can not explain the increased actin polymerization because PIP_2 levels decrease in activated neutrophils.

Profilin is a protein that has been thought to play a major role in sequestering monomeric actin and preventing it from polymerizing. Phosphatidylinositol 4,5-bisphosphate binds to profilin with high affinity and promotes dissociation of actin–profilin complexes (Lassing and Lindberg 1985, 1988). Thus a PIP_2 increase could induce the release of free actin monomers. However, a recent study has shown that monomeric actin in neutrophils is in large excess over profilin and thus profilin is not likely to be a critical factor in regulation of the state of monomeric actin (Southwick and Young, 1990). In addition, this mechanism is unlikely because it predicts that PIP_2

levels would have to increase to initiate polymerization, rather than decrease as is seen.

As discussed earlier, in the presence of thymosin β_4, G-actin, and F-actin, profilin promotes polymerization if the plus ends are not capped (Pantaloni and Carlier, 1993). Thus it has been hypothesized that the decrease of PIP_2 upon activation could liberate free profilin and make it available to promote polymerization. This does not appear to occur in neutrophils, for Tardif et al. (1995) have shown that simply adding profilin to streptolysin *O*-permeabilized cells does not promote polymerization; rather, depolymerization is seen.

Recent work by Shariff and Luna (1992) has shown that addition of exogenous diacylglycerol (DAG) to membranes from *Dictyostelium discoideum* results in an increase in actin nucleation activity. By analogy, this suggests that phospholipase C activation in the neutrophil, which produces DAG, could be the relevant signal. However, the depletion of phospholipase C substrates or increase in its products seems irrelevant for the actin polymerization response, for Bengtsson et al. (1988) have shown that in the neutrophil phospholipase C activation is not required for actin polymerization to occur. Bengtsson et al. (1988) have shown that phospholipase C can be inhibited by drastically reducing the intracellular calcium concentration. Under these conditions, actin polymerization still occurs. Thus phospholipase C activation does not appear to be required for the actin polymerization response. This further rules out events that are downstream of phospholipase C activation, such as calcium elevation and protein kinase C activation, and is consistent with the data that show inhibition of these signals does not inhibit actin polymerization (Sklar et al., 1985a; Niggli and Keller, 1991).

Phospholipases A_2 and D. Because activation of phospholipase A_2 requires activation of phospholipase C, and inhibition of phospholipase C does not inhibit the actin polymerization response, phospholipase A_2 is an unlikely candidate for the effector regulating actin polymerization. Phospholipase D appears not to mediate the actin polymerization response, for concentrations of wortmannin that inhibit phospholipase D do not inhibit the actin polymerization response (Arcaro and Wymann, 1993; Bonser et al., 1991). In addition, doses of ethanol that inhibit phospholipase D activation do not inhibit the RALS response (R. Rengan and G. M. Omann, unpublished observations).

Phosphatidylinositol-3-kinase Does Not Signal for Actin Polymerization. Like the actin polymerization response, PI-3-kinase activation is inhibited by pertussis toxin and is not inhibited by sequestration of the cytosolic Ca^{2+} rise (Traynor-Kaplan et al., 1989). Because there was a good correlation between actin polymerization and PIP_3 production in terms of the fMLP dose-response characteristics and response kinetics, Eberle et al. (1990) hypothesized that PI-3-kinase activation was required for actin polymerization. However, PI-3-kinase inhibition does not inhibit the *N*-formylpeptide-induced actin polymerization response (Arcaro and Wymann, 1993; Vlahos et al., 1995). Thus PI-3-kinase is not a likely candidate for the G-protein-coupled effector that regulates the actin polymerization response.

Tyrosine Kinases are Not Involved in the Actin Polymerization Response. Studies with the tyrosine kinase inhibitor erbstatin have shown that inhibition of chemoattractant-induced tyrosine phosphorylation correlates with inhibition of chemotaxis (Gaudry et al., 1992). Erbstatin was reported to inhibit the production of superoxide and the intracellular acidification induced by fMLP, LTB_4, and PAF, but did not affect the intracellular calcium rise or the actin polymerization response (Naccache et al., 1990). The tyrosine kinase inhibitor ST638 inhibits fMLP-induced oxidant production (Gomez-Cambronero et al., 1989; Berkow et al., 1989), but not actin polymerization (Gomez-Cambronero et al., 1989). Thus it appears that chemoattractant-induced actin polymerization is not a result of tyrosine kinase activation.

Other Possibilities for Signaling Components

Although a great deal of effort has gone into trying to delineate the signal transduction pathway regulating the actin polymerization response, little is known about what connects the active G protein to the uncapping of free barbed ends.

One caveat that is often cited regarding the above studies is that most of the techniques for measuring potential signaling species measure bulk levels of the signals and would not detect highly localized regions that may be changing. A recent report using chlortetracycline as an indicator of a membrane-bound pool of calcium has suggested that alterations of this Ca^{2+} pool correlate with actin polymerization (Bengtsson, 1990). This problem is likely to be addressed in the near future as methods for looking at localization of signals are being developed. For example, membrane-tethered fluorescent Ca^{2+} indicators (Etter et al., 1994) and total internal fluorescence microscopy techniques (Omann and Axelrod, 1996) have been developed for measuring membrane-proximal Ca^{2+} changes in stimulated cells, and these approaches are readily extendable to other ions.

Other possibilities for what may be involved in the signaling cascade for actin polymerization include small guanosine triphosphate (GTP)-binding proteins. Small GTP-binding proteins have been implicated in many cellular and biochemical functions including membrane ruffling and morphological changes, trafficking of subcellular organelles, and activation of phospholipase D and phosphatidylinositol-3-kinase (Quinn, 1995, 1996). One common feature of all these activities may be the involvement of the small GTP-binding protein in tethering other proteins to the membrane. It has also been shown that small GTP-binding proteins can be activated by G protein-coupled receptors (Knaus et al., 1995). The Rho subfamily of small GTP-binding proteins consists of RhoA, RhoB, RhoC, RhoG, Rac1, Rac2, TC10, and Cdc42Hs and has been implicated in the regulation of cell morphology. It has been shown in Swiss 3T3 cells that Rho, Rac, and Cdc42 regulate the assembly of stress fibers, lamellipodia, and filopodia, respectively, as well as the assembly of multimolecular focal complexes at the plasma membrane (Nobes and Hall, 1995; reviewed by Zigmond, 1996). Rho has been implicated as important for neutrophil migration (Stasia et al., 1991) because pretreatment of neutrophils with *Clostridium botulinum* C3 transferase, which ADP-ribosylates and inhibits rho proteins, inhibits migration. On the contrary, ADP-ribosylation of rho in neutrophils does not inhibit the *N*-

formylpeptide-induced actin polymerization response (Ehrengruber et al., 1995a), and thus small GTP-binding proteins that are not sensitive to *Clostridium botulinum* C3 transferase, including Rac and Cdc42Hs proteins, may be involved. A potential candidate is Cdc42Hs, as suggested by recent studies of patients with Wiskott–Aldrich syndrome (Symons et al., 1996). Patients with this syndrome have defects in hematopoietic cells, and in particular, have defective chemotaxis in neutrophils. The molecular defect in these patients is in a protein that has been identified in neutrophils and shown to be an effector of Cdc42Hs. In transfected cells, the protein localizes to sites of F-actin that are regulated by Cdc42Hs. However, the role of Cdc42Hs in initiating the chemoattractant-induced actin polymerization response in neutrophils still remains to be clarified.

The mitogen-activated protein (MAP) kinase cascade is activated in neutrophils, as are several incompletely characterized serine/threonine kinases (Avdi et al., 1996; reviewed by Downey et al., 1996; Durstin and Sha'afi, 1996). Activation of these kinases is blocked by pertussis toxin and is Ca^{2+}- and phospholipase C-independent. Thus these kinases meet some of the requirements for signals leading to actin polymerization. The MAP kinase cascade may be regulated by small GTP-binding proteins and/or PI-3-kinase (Avdi et al., 1996). However, it remains to be seen if and how these cascades regulate actin polymerization and cell morphology.

A CapZ barbed-end capping (capping protein β2) activity has been purified from neutrophils and shown to be Ca^{2+}-independent and inhibited by PIP_2 or high KCl (Maun et al., 1996). Because phospholipase C activation is not required for chemoattractant-induced actin polymerization, PIP_2 is not likely to be the signal regulating CapZ activity in this response. A role for K^+ in cytoskeletal reorganization has been suggested by studies that show elevated extracellular K^+ causes an increase in cell shape change (Roberts et al., 1984) and enhances chemotaxis and spontaneous movement of neutrophils (Showell and Becker, 1976). Studies of the effect of high K^+ on *N*-formylpeptide-induced oxidant production also showed an enhancement that was correlated with an increase in the number and affinity of the *N*-formylpeptide receptors (Della Bianca et al., 1983; De Togni et al., 1983). Thus additional studies of the potential role of potassium chloride in the actin polymerization response are warranted. Other mechanisms by which this CapZ could be regulated must be explored.

ACTIVATION BY OTHER STIMULI

Other Seven Transmembrane Segment Receptors

Ligands for other seven transmembrane segment receptors also activate an actin polymerization response. These include LTB_4, PAF, C5a, and IL8 (Omann et al., 1987c, 1989; Shalit et al., 1987; Norgauer et al., 1994; Ehrengruber et al., 1995b). These receptors probably all activate the response via the same pathway as the *N*-formylpeptides, because the evidence suggests that they all activate the same pool of G proteins (Wilde et al., 1989; McLeish et al., 1989; Schepers et al., 1992). Whereas *N*-formylpeptide activation of untreated neutrophils results in a two-phased response,

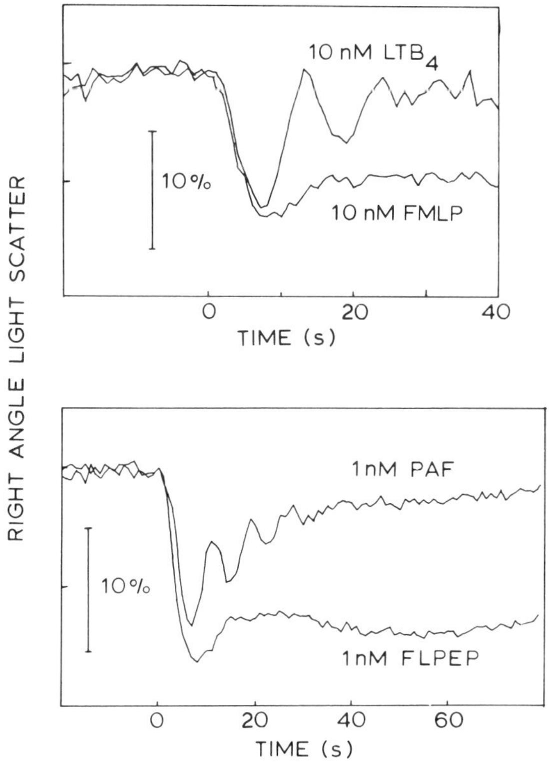

Fig. 6. Kinetics of the right angle light scatter response to the *N*-formylpeptides, *N*-formyl-norleucyl-leucyl-phenylalanyl-norleucyl-tyrosyl-lysyl-fluorescein (FLPEP) and *N*-formyl-methionyl-leucyl-phenylalanine, LTB_4, and PAF. Measurements were performed at 37°C. From Omann et al., 1987; copyright 1987, The American Society for Biochemistry and Molecular Biology, Inc. Reprinted with permission.

activation of actin polymerization by LTB_4 and PAF causes a rapidly oscillating actin polymerization/depolymerization response, detected either as right-angle light scatter or as NBD-phallacidin staining of F-actin, that decays within 1–2 min (Fig. 6) (Omann et al., 1989, 1995). Oscillations can also be observed when the chemoattractant is fMLP, IL-8, or complement component C5a; however, the cells must be pretreated in some way (Erhengruber et al., 1995b, 1996; Wymann et al., 1990). Pretreatment of neutrophils with the inhibitor wortmannin (Wymann et al., 1990) or with low doses of fMLP (Ehrengruber et al., 1995b) have been shown to unmask fMLP-induced oscillations. In contrast, LTB_4 and PAF induce oscillations without pretreatment (Omann et al., 1989). These oscillations are observed in the absence of calcium oscillations (Omann et al., 1989).

The oscillations in actin polymerization/depolymerization and light scattering are thought to reflect alterations in cell shape resulting from the extension and retraction of veil-like lamellipodia (Ehrengruber et al., 1996, Wymann et al., 1990), although the relationship of light scattering to structural changes has not been clearly defined. It has also been shown that random migration of neutrophils on surfaces exhibits oscillatory behavior in shape change and velocity (Hartman et al., 1994) of a similar periodicity to those seen in activated neutrophils in suspension. These results indicate there is a periodic component to the process driving neutrophil movement, and point to the importance of actin polymerization and depolymerization in driving cell movement.

Other Types of Activators

Other types of receptors besides the seven transmembrane segment receptors activate an actin polymerization response. These include integrins (Walzog et al., 1994) and Fc receptors (Brennan et al., 1991). The Fc receptor and integrins signal via tyrosine kinase pathways (Santana et al., 1996; Clark and Brugge, 1995; Walzog et al., 1996). In addition, nonphysiological activators such as phorbol myristate acetate (which activates protein kinase C) and the calcium ionophore A23187 induce an actin polymerization response (Howard and Wang, 1987; Downey et al., 1992; Bengtsson et al., 1986). In all cases, the kinetics of the responses are slower, but the magnitude can be as great as that seen with *N*-formylpeptide. The fact that Ca^{2+} elevation, protein kinase C activation, and tyrosine kinase activation can induce an actin polymerization response, in spite of the fact that these events are not required for the *N*-formyl peptide-induced response, highlights the complexity of signaling regulating the actin cytoskeleton. A common final pathway for all activators is unlikely.

SUMMARY

Chemoattractant-induced actin polymerization in neutrophils is an important event in migration and regulation of degranulation. The response occurs via G-protein-coupled seven transmembrane segment receptors. Evidence indicates that the event that triggers polymerization is the freeing of the barbed end of the actin filaments. However, the G-protein-mediated events that create free barbed ends remain unknown. In addition, the regulation of actin depolymerization that occurs as the response recovers is unknown. Actin polymerization is mediated by non-G-protein-coupled receptors and by nonphysiological activators such as phorbol esters and calcium ionophores. Because these signaling pathways are diverse, a final common pathway for all activators to induce the response is unlikely.

ACKNOWLEDGMENTS

The author thanks Ramesh Rengan for critical review of the manuscript. This work was supported by the Arthritis Foundation, the Office of Research and Development, Medical Research Service, Department of Veterans Affairs, and National Science Foundation grants MCB 945928 and BES 9410403.

REFERENCES

Allen RA, Traynor AE, Omann GM, Jesaitis AJ (1988): The chemotactic peptide receptor: a model for future understanding of chemotactic disorders. Hematal Oncol Clin N Am 2:33–59.

Al-Mohanna FA, Hallett MB (1990): Actin polymerization in neutrophils is triggered without a requirement for a rise in cytoplasmic C^{2+}. Biochem J 266:669–674.

Arcaro A, Wymann MP (1993): Wortmannin is a potent phosphatidylinositol 3-kinase inhibitor: the role of phosphatidylinositol 3,4,5-trisphosphate in neutrophil responses. Biochem J 296:297–301.

Aunis D, Bader M-F (1988): The cytoskeleton as a barrier to exocytosis in secretory cells. J Exp Biol 139: 253–266.

Avdi MJ, Winston BW, Russel M, Young SK, Johnson GL, Worthen GS (1996): Activation of MEKK by formyl-methionyl-leucyl-phenylalanine in human neutrophils. Mapping pathways for mitogen-activated protein kinase activation. J Biol Chem 271:33598–33606.

Begg DA, Rebhun LI (1979): pH regulates the polymerization of actin in the sea urchin egg cortex. J Cell Biol 83:241–248.

Bengtsson T (1990): Correlation between chemotactic peptide-induced changes in chlorotetracycline fluorescence and F-actin content in human neutrophils: a role for membrane-associated calcium in the regulation of actin polymerization? Exp Cell Res 191:57–63.

Bengtsson T, Dahlgren C, Stendahl O, Andersson T (1991): Actin assembly and regulation of neutrophil function: effects of cytochalasin B and tetracaine on chemotactic peptide-induced O_2^- production and degraulation. J Leukoc Biol 49:236–244.

Bengtsson T, Rundquist I, Stendahl O, Wymann MP, Andersson T (1988): Increased breakdown of phophatidylinositol 4,5-bisphosphate is not an initiating factor for actin assembly in human neutrophils. J Biol Chem 263:17385–17389.

Bengtsson T, Sarndahl E, Stendahl O, Andersson T (1990): Involvement of GTP-binding proteins in actin polymerization in human neutrophils. Proc Natl Acad Sci USA 87:2921–2925.

Bengtsson T, Stendahl O, Andersson T (1986): The role of the cytosolic free Ca^{2+} transient for fMet-Leu-Phe induced actin polymerization in human neutrophils. Eur J Cell Biol 42:338–343.

Benson MC, McDougal DC, Coffey DS (1984): The application of perpendicular and forward light scatter to assess nuclear and cellular morphology. Cytometry 5:515–522.

Berkow RL, Dodson RW (1990): Tryrosine-specific protein phosphorylation during activation of human neutrophils. Blood 75:2445–2452.

Berkow RL, Dodson RW, Kraft AS (1989): Human neutrophils contain distinct cytosolic and particulate tyrosine kinase activities: possible role in neutrophil activation. Biochim Biophys Acta 997:292–301.

Billah MM, Anthes JC, Mullmann TJ (1991): Receptor-coupled phospholipase D: regulation and functional significance. Biochem Soc Trans 19:324–329.

Billah MM, Eckel S, Mullmann TJ, Egan RW, Siegel MI (1989): Phosphatidylcholine hydrolysis by phospholipase D determines phosphatidate and diglyceride levels in chemotactic peptide-stimulated human neutrophils. J Biol Chem 264:17069–17077.

Blikstad I, Markey F, Carlsson L, Persson T, Linberg U (1978): Selective assay of monomeric and filamentous actin in cell extracts using inhibition of deoxyribonuclease I. Cell 15:935–943.

Bonser RW, Thompson NT, Randall RW, Tateson JE, Spacey GD, Hodson HF, Garland LG (1991): Demethoxyviridin and wortmannin block phospholipase C and D activation in the human neutrophil. Br J Pharmacol 103:1237–1241.

Brennan PJ, Zigmond SH, Schreiber AD, Smith ER, Southwick FS (1991): Binding of IgG containing immune complexes to human neutrophil FcγRII and FcγRIII induces actin polymerization by a pertussis toxin-insensitive transduction pathway. J Immunol 146:4282–4288.

Brunsting A, Mullaney PF (1974): Differential light scattering from spherical mammalian cells. Biophys J 14:439–453.

Cano ML, Lauffenburger DA, Zigmond SH (1991): Kinetic analysis of F-actin depolymerization in polymorphonuclear leukocytes indicates that chemoattractant stimulation increases actin filament number without altering the filament length distribution. J Cell Biol 115:677–687.

Carlier MF (1991): Actin: protein structure and filament dynamics. J Biol Chem 266:1–4.

Carlier MF, Pantaloni D (1994): Actin assembly in response to extracellular signals. Role of capping proteins, thymosin β_4, and profilin. Semin Cell Biol 5:183–191.

Carlsson L, Nystrom LE, Sundkvist I, Markey F, Lindberg U (1997): Actin polymerizability is influenced by profilin, a low molecular weight protein in nonmuscle cells. J Mol Biol 115:465–483.

Carson M, Weber A, Zigmond SH (1986): An actin nucleating activity in polymorphonuclear leukocytes is modulated by chemotactic peptides. J Cell Biol 103:2707–2714.

Cassimeris L, McNeill H, Zigmond SH (1990): Chemoattractant stimulated polymorphonuclear leukocytes contain two populations of actin filaments which differ in their spatial distributions and relative stabilities. J Cell Biol 110:1067–1075.

Cassimeris L, Safer D, Nachmias VT, Zigmond SH (1992): Thymosin β_4 sequesters the majority of G-actin in resting human polymorphonuclear leukocytes. J Cell Biol 119:1261–1270.

Cassimeris L, Zigmond SH (1990): Chemoattractant stimulation of polymorphonuclear leukocyte locomotion. Semin Cell Biol 1:125–134.

Clark EA, Brugge JS (1995): Integrins and signal transduction pathways: The road taken. Science 268: 233–239.

Coates TD (1996): Behavioral aspects of neutrophil motility. Curr Opin Hematol 3:41–47.

Coates TD, Watts RG, Hartman R, Howard TH (1992): Relationship of F-actin distribution to development of polar shape in human polymorphonuclear neutrophils. J Cell Biol 117:765–774.

Cockcroft S (1996): Phospholipid signaling in leukocytes. Curr Opin Hematol 3:48–54.

Craddock PR, Hammerschmitd D, White JG, Dalmasso AP, Jacob HS (1977): Complement (C5a)-induced granulocyte aggregation in vitro. J Clin Invest 60:260–264.

Davis BH, Walter RJ, Pearson CB, Becker EL, Oliver JM (1982): Membrane activity and topogrophy of f-met-leu-phe-treated polymorphonuclear leukocytes. Am J Pathol 108:206–216.

Deaton JD, Guerrero T, Howard TH (1992): Role of gelsolin interaction with actin in regulation and creation of actin nuclei in chemotactic peptide activated polymorphonuclear neutrophils. Mol Biol Cell 3:1427–1435.

Della Bianca V, Bellavite P, De Togni P, Fumarulo R, Rossi F (1983): Studies on stimulus-response coupling in human neutrophils. 1. Role of monovalent cations in the respiratory and secretory response to N-formylmethionylleucylphenylalanine. Biochim Biophys Acta 755:497–505.

Deranleau DA (1987): Blood platelet shape change ABCs. Trends Biochem Sci 12:439–442.

De Togni P, Della Bianca V, Bellavite P, Grzeskowiak M, Rossi F (1983): Studies on stimulus-response coupling in human neutrophils. II. Relationships between the effects of changes of external ionic composition on the properties of N-formylmethionylleucylphenylalanine receptors and on the respiratory and secretory responses. Biochim Biophys Acta 755:506–513.

DiNubile MJ, Cassimeris L, Joyce M, Zigmond SH (1995): Actin filament barbed-end capping activity in neutrophil lysates: The role of capping protein-β_2. Mol Biol Cell 6:1659–1671.

Downey GP, Chan CK, Grinstein S (1989): Actin assembly in electropermeabilized neutrophils: role of G-proteins. Biochem Biophys Res Commun 164:700–705.

Downey GP, Chan CK, Lea P, Takai A, Grinstein S (1992): Phorbol ester-induced actin assembly in neutrophils: Role of protein kinase C. J Cell Biol 116:695–706.

Downey GP, Chan CK, Trudel S, Grinstein S (1990): Actin assembly in electropermeabilized neutrophils: Role of intracellular calcium. J Cell Biol 110:1975–1982.

Downey GP, Elson EL, Schwab B III, Ersurum SC, Young SK, Worthen GS (1991): Biophysical properties and microfilament assembly in neutrophils: Modulation by cyclic AMP. J Cell Biol 114:1179–1190.

Downey GP, Fukushima T, Fialkow L (1996): Signaling mechanisms in human neutrophils. Curr Opin Hematol 2:76–88.

Downey GP, Grinstein S (1989): Receptor-mediated actin assembly in electropermeabilized neutrophils: role of intracellular pH. Biochem Biophys Res Commun 160:18–24.

Durstin M, Sha'afi RI (1996): Signal transduction in the human neutrophil. In Lad PM, Kaptein JS, Lin C-KE (eds). *Signal Transduction in Leukocytes: G-protein-Related and Other Pathways*. Boca Raton, FL: CRC Press, pp. 253–285.

Eberle M, Traynor-Kaplan AE, Sklar LA, Norgeuer J (1990): Is there a relationship between phophatidylinositol trisphosphate and F-actin polymerization in human neutrophils? J Biol Chem 265:16725–16728.

Ehrengruber MU, Boquet P, Coates TD, Deranleau DA (1995a): ADP-ribosylation of Rho enhances actin polymerization-coupled shape oscillations in human neutrophils. FEBS Lett 372:161–164.

Ehrengruber MU, Coates TD, Deranleau DA (1995b): Shape oscillations: A fundamental response of human neutrophils stimulated by chemotactic peptides? FEBS Lett 359:229–232.

Ehrengruber MU, Deranleau DA, Coates TD (1996): Shape oscillations of human neutrophil leukocytes: Characterization and relationship to cell motility. J Exp Biol 199:741–747.

Etter EF, Kuhn MA, Fay FS (1994): Detection of changes in near-membrane Ca^{2+} concentration using a novel membrane-associated Ca^{2+} indicator. J Biol Chem 269:10141–10149.

Fechheimer M, Zigmond SH (1983): Changes in cytoskeletal proteins of polymorphonuclear leukocytes induced by chemotactinc peptides. Cell Motil 3:349–361.

Fechheimer M, Zigmond SH (1993): Focusing on unpolymerized actin. J Cell Biol 123:1–5.

Fletcher MP, Seligman BE (1985): Monitoring human neutrophil granule secretion by flow cytometry: Secretion and membrane potential changes assessed by light scatter and a fluorescent probe of membrane potential. J Leukoc Biol 37:431–447.

Gaudry M, Caon AC, Gilbert C, Lille S, Naccache PH (1992): Evidence for the involvement of tyrosine kinases in the locomotory responses of human neutrophils. J Leukoc Biol 51:103–108.

Gomez-Cambronero J, Huang C-K, Bonak VA, et al. (1989): Tyrosine phosphorylation in human neutrophils. Biochem Biophys Res Commun 162:1478–1485.

Gomez-Cambronero J, Wang E, Johnson G, Huang C-K, Sha'afi RI (1991): Platelet activating factor induces tyrosine phosphorylation in human neutrophils. J Biol Chem 266:6240–6245.

Hartman RS, Lau K, Chou W, Coates TD (1994): The fundamental motor of the human neutrophil is not random: Evidence for local non-Markov movement in neutrophils. Biophys J 67:2535–2545.

Hartwig JH, Kwiatkowski DJ (1991): Actin-binding proteins. Curr Opin Cell Biol 3:87–97.

Harvath L (1990): Regulation of neutrophil chemotaxis: Correlations with actin polymerization. Cancer Invest 8:651–654.

Harvath L, Robbins JD, Russel AA, Seamon KB (1991): cAMP and human neutrophil chemotaxis. Elevation of cAMP differentially affects chemotactic responsiveness. J Immunol 146:224–232.

Hoffstein ST, Friedman RS, Weissmann G (1982): Degranulation, membrane addition, and shape change during chemotactic factor-induced aggregation of human neutrophils. J Cell Biol 95:234–241.

Howard T, Chaponnier C, Yin H, Stossel T (1990): Gelsolin-actin interaction an actin polymerization in human neutrophils. J Cell Biol 110:1983–1991.

Howard TH, Meyer WH (1984): Chemotactic peptide modulation of actin assembly and locomotion in neutrophils. J Cell Biol 98:1265–1271.

Howard TH, Oresajo CO (1985a): The kinetics of chemotactic peptide-induced change in F-actin content, F-actin distribution, and the shape of neutrophils. J Cell Biol 101:1078–1085.

Howard TH, Oresajo CO (1985b): A method for quantifying F-actin in chemotactic peptide activated neutrophils. Study of the effect of tBOC peptide. Cell Motil 5:545–557.

Howard TH, Wang D (1987): Calcium ionophore, phorbol ester, and chemotactic peptide-induced cytoskeleton reorganization in human neutrophils. J Clin Invest 79:1359–1364.

Howard TH, Watts RG (1994): Actin polymerization and leukocyte function. Curr Opin Hematol 1:61–68.

Huang C-K, Laramee GR, Casnellie JE (1988): Chemotactic factor induced tryrosine phosphorylation of membrane associated proteins in rabbit neutrophils. Biochem Biophys Res Commun 151:794–801.

Janmey PA, Stossel TP (1987): Modulation of gelsolin function by phosphatidylinositol 4,5-bisphosphate. Nature 325:362–364.

Janmey PA, Stossel TP (1989): Gelsolin-polyphophoinositide interations. J Biol Chem 264:4825–4831.

Jesaitis AJ, Naemura JR, Sklar LA, Cochrane CG, Painter RG (1984): Rapid modulation of N-formyl chemotactic peptide receptors on the surface of human granulocytes: function of high-affinity ligand-receptor complexes in transient association with cytoskeleton. J Cell Biol 98:1378–1387.

Jesaitis AJ, Tolley JO, Allen RA (1986): Receptor-cytoskeleton interactions and membrane traffic may regulate chemoattractant-induced superoxide production in human granulocytes. J Biol Chem 261:13662–13669.

Jesaitis AJ, Tolley JO, Painter RG, Sklar LA, Cochrane CG (1985): Membrane-cytoskeleton interactions and the regulation of chemotactic peptide-induced activation of human granulocytes: the effects of dihydrocytochalasin B. J Cell Biochem 27:241–253.

Keller H, Niggli V, Zimmermann A (1991): Diversity in motile responses of human neutrophil granulocytes: functional meaning and cytoskeletal basis. Adv Exp Biol Med 297:23–37.

Keller HU, Fedier A, Rohner R (1995): Relationship between light scattering in flow cytometry and changes in shape, volume, and actin polymerization in human polymorphonuclear leukocytes. J Leukoc Biol 57:519–525.

Kerker M (1983): Elastic and inelastic light scattering in flow cytometry. Cytometry 4:1–10.

Klotz K-N, Jesaitis AJ (1994): Neutrophil chemoattractant receptors and the membrane skeleton. BioEssays 16:193–198.

Klotz K-N, Krotec KL, Gripentrog J, Jesaitis AJ (1994): Regulatory interaction of N-formyl peptide chemoattractant receptors with the membrane skeleton in human neutrophils. J Immunol 152:801–810.

Knaus UG, Morris S, Dong H-J, Chernoff J, Bokoch GM (1995): Regulation of human leukoctye p21-activated kinases through G-protein-coupled receptors. Science 269:221–223.

Kraft AS, Berkow RL (1987): Tyrosine kinase and phosphotyrosine phosphatase activity in human promyelocytic leukemia cells and human polymorphonuclear leukocytes. Blood 70:356–362.

Kraus E, Niederman R (1990): Changes in neutrophil right angle light scatter can occur independent of alterations in cytoskeletal actin. Cytometry 11:272–282.

Lassing I, Lindberg U (1985): Specific interaction between phosphatidylinositol 4,5-bisphosphate and profilactin. Nature 314:472–474.

Lassing I, Lindberg U (1988): Specificity of the interaction between phosphatidylinositol 4,5-bisphosphate and the profilin:actin complex. J Cell Biochem 37:255–267.

Latimer P, Moore DM, Bryant FD (1968): Changes in total light scattering and absorption caused by changes in particle conformation. J Theor Biol 21:348–367.

Lepidi H, Benoliel A-M, Mege JL, Bongrand P, Capo C (1992): Double localization of F-actin in chemoattractant-stimulated polymorphonuclear leukocytes. J Cell Sci 103:145–156.

Lin L-L, Wartmann M, Lin AY, Knopf JL, Seth A, Davis RJ (1993): cPLA2 is phosphorylated and activated by MAP kinase. Cell 72:269–278.

MacLean-Fletcher S, Pollard TD (1980): Mechanism of action of cytochalasin B on actin. Cell 20:329–341.

Maun NA, Speicher DW, DiNubile MJ, Southwick FS (1996): Purification and properties of a Ca^{2+}-independent barbed-end actin filament capping protein, CapZ, from human polymorphonuclear leukocytes. Biochemistry 35:3518–3524.

McLeish KR, Gierschik P, Schepers T, Sidiropoulos D, Jacobs KH (1989): Evidence that activation of a common G-protein by receptors for leukotriene B4 and N-formylmethionyl-leucyl-phenylalanine in HL-60 cells occurs by different mechanisms. Biochem J 260:427–434.

McNeil PL, Kennedy AL, Waggoner AS, Taylor DL, Murphy RF (1985): Light-scattering changes during chemotactic stimulation of human neutrophils: kinetics followed by flow cytometry. Cytometry 6: 7–12.

Meshulam T, Proto P, Diamond RD, Melnick DA (1986): Calcium modulation and chemotactic response:

Divergent stimulation of neutrophil chemotaxis and cytosolic calcium response by the chemotactic peptide receptor. J Immunol 137:1954–1960.

Model MA, Omann GM (1996): Cell polarization as a possible mechanism of response termination. Biochem Biophys Res Commun 224:516–521 (errata 226:943):

Molski TFP, Sha'afi RI (1987): Intracellular acidification, guanine nucleotide binding proteins, and cytoskeletal actin. Cell Motil Cytoskeleton 8:1–6.

Mueller H, Montoya B, Sklar LA (1992): Reversal of inhibitory pathways in neutrophils by protein kinase antagonists: a rational approach to the restoration of depressed cell function? J Leukoc Biol 52:400–406.

Mullmann TJ, Cheewatrakoolpong B, Athes JC, Siegel MI, Egan RW, Billah MM (1993): Phospholipase C and phospholipase D are activated independently of each other in chemotactic peptide-stimulated human neutrophils. J Leukoc Biol 53:630–635.

Naccache PH, Gilbert C, Caon AC, et al. (1990): Selective inhibition of human neutrophil functional responsiveness by erbstatin, an inhibitor of tyrosine protein kinase. Blood 76:2098–2104.

Naccache PH, Therrien S, Caon Ac, Liao N, Gilbert C, McColl SR (1989): Chemoattractant-induced cytoplasmic pH changes and cytoskeletal reorganization in human neutrophils. J Immunol 142:2438–2444.

Nasmith PE, Mills GB, Grinstein S (1989): Guanine nucleotides induce tyrosine phosphorylation and activation of the respiratory burst in neutrophils. Biochem J 257:893–897.

Nigam S, Eskafi S, Garlichs C, Firth S, Zhang H (1995): Role of intracellular calcium in the regulation of phospholipase A_2 in fmet-leu-phe-challenged human polymorph neutrophils. Agents Action Suppl 45:297–301.

Niggli V, Keller H (1991): On the role of protein kinases in regulating neutrophil actin association with the cytoskeleton. J Biol Chem 266:7927–7932.

Nobes CD, Hall A (1995): Rho, Rac, and Cdc42 GTPases regulate the assembly of multimolecular focal complexes associated with actin stress fibers, lamellipodia, and filopodia. Cell 81:53–62.

Norgauer J, Krutman J, Dobos GJ, Traynor-Kaplan AE, Oades ZG, Schraufstatter IU (1994): Actin polymerization, calcium-transients, and phospholipid metabolism in human neutrophils after stimulation with interleukin-8 and N-formyl peptide. Immunology 82:465–472.

Oliver JM, Krawiec JA, Becker EL (1978): The distribution of actin during chemotaxis in rabbit neutrophils. J Reticuloendothel Soc 24:697–704.

Omann GM, Allen RA, Bokoch G, Painter RG, Traynor AE, Sklar LA (1987a): Signal transduction and cytoskeletal activation in the neutrophil. Physiol Rev 67:285–322.

Omann GM, Axelrod D (1996): Membrane proximal calcium transients in stimulated neutrophils detected by total internal reflection fluorescence. Biophy J 71:2885–2891.

Omann GM, Oades ZG, Sklar LA (1985): Simultaneous spectrofluormetric analyses of cell responses. BioTechniques 3:508–512.

Omann GM, Porasik-Lowes MM (1991): Graded G-protein uncoupling by pertussis toxin treatment of human polymorphonuclear leukocytes. J Immunol 146:1303–1308.

Omann GM, Porasik-Lowes MM, Sklar LA (1989): Oscillating actin polymerization/depolymerization responses in human polymorphonuclear leukocytes. J Biol Chem 264:16355–16358.

Omann GM, Rengan R, Hoffman JF, Linderman JJ (1995): Rapid oscillations of actin polymerization/depolymerization in polymorphonuclear leukocytes stimulated by leukotriene B_4 and platelet activating factor. J Immunol 155:5375–5381.

Omann GM, Sklar LA (1988): Response of neutrophils to stimulus infusion: Differential sensitivity of cytoskeletal activation and oxidant production. J Cell Biol 107:951–958.

Omann GM, Sklar LA (1989): Spectrofluorometric analyses of cell responses. Activation of neutrophils by chemoattractants and hexachlorocyclohexanes. In Goldberg MC (ed). *Luminescence Applications in Biological, Chemical, Environmental, and Hydrological Sciences,* Washington, DC: American Chemical Society, pp. 23–44.

Omann GM, Swann WA, Oades ZG, Parkos CA, Jesaitis JA, Sklar LA (1987b): N-formyl peptide receptor dynamics, cytoskeletal activation, and intracellular calcium response in human neutrophil cytoplasts. J Immunol 139:3447–3455.

Omann GM, Traynor AE, Harris A, Sklar LA (1987c): LTB_4-induced activation signals and responses in neutrophils are short-lived compared to formyl peptide. J Immunol 138:2626–2632.

Pantaloni D, Carlier M-F (1993): How profilin promotes actin filament assembly in the presence of thymosin β_4. Cell 75:1007–1014.

Pollard TD (1990): Actin. Curr Opin Cell Biol 2:33–40.

Pollard TD, Cooper JA (1986): Actin and actin-binding proteins, A critical evaluation of mechanisms and functions. Annu Rev Biochem 55:987–1035.

Quinn MT (1995): Low-molecular-weight GTP-binding proteins and leukocyte signal transduction. J Leukoc Biol 58:263–276.

Quinn MT (1996): Role of small GTP-binding proteins in leukocyte signal transduction. In Lad PM, Kaptein JS, Lin C-KE (eds). *Signal Transduction in Leukocytes, G Protein-Related and Other Pathways.* Boca Raton, FL: CRC Press, pp. 75–141.

Rao KMK, Varani J (1982): Actin polymerization induced by chemotactic peptide and concanavalin A in rat neutrophils. J Immunol 129:1605–1607.

Redmond T, Tardif M, Zigmond SH (1994): Induction of actin polymerization in permeabilized neutrophils: Role of ATP. J Biol Chem 269:21657–21663.

Roberts RL, Mounessa NL, Gallin JI (1984): Increasing extracellular potassium causes calcium-dependent shape change and facilitates concanavalin A capping in human neutrophils. J Immunol 132:2000–2006.

Ryder MI, Weinreb RN, Niederman R (1984): The organization of actin filaments in human polymorphonuclear leukocytes. Anat Rec 209:7–20.

Santana C, Noris G, Espinoza B, Ortega E (1996): Protein tyrosine phosphorylation in leukocyte activation through receptors for IgG. J Leukoc Biol 60:433–440.

Schepers TM, Brier ME, McLeish KR (1992): Quantitative and qualitative differences in guanine nucleotide binding protein activation by formyl peptide and leukotriene B_4 receptors. J Biol Chem 267: 159–165.

Sha'afi RI, Molski TFP (1987): Signaling for increased cytoskeletal actin in neutrophils. Biochem Biophys Res Commun 145:934–941.

Sha'afi RI, Molski TFP (1988): Activation of the neutrophil. Prog Allergy 42:1–64.

Sha'afi RI, Shefcyk J, Yassin R, et al. (1986): Is a rise in intracellular concentration of free calcium necessary or sufficient for stimulated cytoskeletal-associated actin? J Cell Biol 102:1459–1463.

Shalit M, Dabiri GA, Southwick FS (1987): Platelet-activating factor both stimulates and "primes" human polymorphonuclear leukocyte actin filament assembly. Blood 70:1921–1927.

Shariff A, Luna EJ (1992): Diacylglycerol-stimulated formation of actin nucleation sites at plasma membranes. Science 256:245–247.

Shields JM, Haston WS (1985): Behaviour of neutrophil leucocytes in uniform concentrations of chemotactic factors: contraction waves, cell polarity, and persistence. J Cell Sci 74:75–93.

Showell HJ, Becker EL (1976): The effects of external K^+ and Na^+ on the chemotaxis of rabbit peritoneal neutrophils. J Immunol 116:99–105.

Simchowitz L, Fischbein LC, Spilberg I, Atkinson JP (1980): Induction of a transient elevation in intracellular levels of adenosine-3′,5′-cyclic monophosphate by chemotactic factors: an early event in human neutrophil activation. J Immunol 124:1482–1491.

Simchowitz L, Spilberg I, Atkinson JP (1983): Evidence that the functional responses of human neutrophils occur independently of transient elevations in cyclic AMP levels. J Cyclic Nucleotide Protein Phosphorylation Res 9:35–47.

Sklar LA, Hyslop PA, Oades ZG, Omann GM, Jesaitis AJ, Cochrane CG(1985b): Signal transduction and ligand receptor dynamics in the human neutrophil. Transient responses and occupancy-response relations at the formyl peptide receptor. J Biol Chem 260:11461–11467.

Sklar LA, Oades ZG, Finney DA (1984): Neutrophil degranulation detected by right angle light scattering: Spectroscopic methods sutiable for simultaneous analysis of degranulation or shape change, elastase release, and cell aggregation. J Immunol 133:1483–1487.

Sklar LA, Omann GM, Painter RG (1985a): Relationship of actin polymerization and depolymerization to light scattering in human neutrophils: Dependence on receptor occupancy and intracellular calcium. J Cell Biol 101:1161–1166.

Smolen JE, Stoehr SJ, Kuczynski B (1991a): Cyclic AMP inhibits secretion from electroporated human neutrophils. J Leukocyte Biol 49:172–179.

Smolen JE, Stoehr SJ, Kuczynski B, Koh EK, Omann GM (1991b): Dual effects of GTPγS on secretion by electroporated human neutrophils. Biochem J 279:657–664.

Sohn RH, Goldschmidt-Clermont PJ (1994): Profilin: At the crossroads of signal transduction and the actin cytoskeleton. BioEssays 16:465–472.

Southwick FS, Young CL (1990): The actin released for profilin-actin complexes is insufficient to account for the increase in F-actin in chemoattractant-stimulated polymorphonuclear leukocytes. J Cell Biol 110:1965–1973.

Stasia M-J, Jouan A, Bourmeyster N, Boquet P, Vignais PV (1991): ADP-ribosylation of a small GTP-binding protein in bovine neutrophils by the C3 exoenzyme of *Clostridium botulinum* and effect on cell motility. Biochem Biophys Res Commun 180:615–622.

Stephens L, Eguinoa A, Corey S, Jackson T, Hawkins PT (1993b): Receptor stimulated accumulation of phosphatidylinositol (3,4,5)-trisphosphate by G-protein mediated pathways in human myeloid derived cells. EMBO J 12:2265–2273.

Stephens L, Jackson T, Hawkins PT (1993c): Synthesis of phosphatidylinositol 3,4,5-trisphosphate in permeablized neutrophils regulated by receptors and G-proteins. J Biol Chem 268:17162–17172.

Stephens L, Smrcka A, Cooke FT, Jackson TR, Sternweis PC, Hawkins PT (1994): A novel phosphoinositide 3 kinase activity in myeloid-derived cells is activated by G protein $\beta\gamma$ subunits. Cell 77:83–93.

Stephens LR, Hughes KT, Irvine RF (1991): Pathway of phosphatidylinositol(3,4,5)-trisphosphate synthesis in activated neutrophils. Nature 351:33–39.

Stephens LR, Jackson TR, Hawkins PT (1993a): Agonist-stimulated synthesis of phosphatidylinositol (3,4,5)-trisphosphate: a new intracellular signalling system? Biochim Biophys Acta 1179:27–75.

Stossel TP (1989): From signal to pseudopod. J Biol Chem 264:18261–18264.

Stoyanov B, Volinia A, Hanck T, et al. (1995): Cloning and characterization of a G protein-associated human phosphainositide-3 kinase. Science 269:690–693.

Symons M, Derry JMJ, Karlak B, et al. (1996): Wiskott-Aldrich syndrome protein, a novel effector for the GTPase CDC42s, is implicated in actin polymerization. Cell 84:723–734.

Takenawa T, Ishitoya J, Nagai Y (1986): Inhibitory effect of prostaglandin E_2, forskolin, and dibutyryl cAMP on arachidonic acid release and inositol phopholipid metabolism in guinea pig neutrophils. J Biol Chem 261:1092–1098.

Tardif M, Huang S, Redmond T, Safer D, Pring M, Zigmond SH (1995): Actin polymerization induced by GTPγS in permeabilized neutrophils is induced and maintained by free barbed ends. J Biol Chem 270: 28075–28083.

Tecoma ES, Motulsky HJ, Traynor AE, Omann GM, Muller H, Sklar LA (1986): Transient catecholamine modulation of neutrophil activation: kinetic and intracellular aspects of isoproterenol action. J Leukoc Biol 40:629–644.

Therrien S, Naccache PH (1989): Guanine nucleotide-induced polymerization of actin in electropermeabilized human neutrophils. J Cell Biol 109:1125–1132.

Thomason PA, James SR, Casey PJ, Downes CP (1994): A G-protein $\beta\gamma$-subunit responsive phosphoinositide 3-kinase activity in human platelet cytosol. J Biol Chem 269:16525–16528.

Tilney LG, Kiehart DP, Sardet C, Tilney M (1978): Polymerization of actin. IV. Role of Ca^{2+} and H^+ in the assembly of actin and in membrane fusion in the acrosomal reaction of echinoderm sperm. J Cell Biol 77:536–550.

Traynor-Kaplan AE, Harris AL, Thompson BL, Taylor P, Sklar LA (1988): An inositol tetrakisphosphate-containing phospholipid in activated neutrophils. Nature 334:353–356.

Traynor-Kaplan AE, Thompson BL, Harris AL, Taylor P, Omann GM, Sklar LA (1989): Transient increase in phosphatidylinositol 3,4-bisphosphate and phosphatidylinositol trisphosphate during activation of human neutrophils. J Biol Chem 264:15668–15673.

Verghese MW, Fox K, MaPhail LC, Snyderman R (1985): Chemoattractant-elicited alterations of cAMP levels in human polymorphonuclear leukocytes require a Ca^{2+}-dependent mechanism which is independent of transmembrane activation of adenlyate cyclase. J Biol Chem 260:6769.

Visser JWM, van den Engh GJ, van Bekkum DW (1980): Light scattering properties of murine hemopoietic cells. Blood Cells 6:391–407.

Vlahos CJ, Matter WF (1992): Signal transduction in neutrophil activation. Phosphatidylinositol 3-kinase is stimulated without tyrosine phosphorylation. FEBS Lett 309:242–248.

Vlahos CJ, Matter WF, Brown RF, et al. (1995): Investigation of neutrophil signal transduction using a specific inhibitor of phosphatidylinositol 3-kinase. J Immunol 154:2413–2422.

Wallace PJ, Wersto RP, Packman CH, Lichtman MA (1984): Chemotactic peptide-induced changes in neutrophil actin conformation. J Cell Biol 99:1060–1065.

Walzog B, Offermanns S, Zakrzewicz A, Gaehtgens P, Ley K (1996): β_2 integrins mediate protein tyrosine phosphorylation in human neutrophils. J Leukoc Biol 59:747–753.

Walzog B, Seifert R, Zakrzewicz A, Gaehtgens P, Ley K (1994): Cross-linking of CD18 in human neutrophils induces an increase of intracellular free Ca^{2+}, exocytosis of azurophilic granules, quantitative up-regulation of CD18, shedding of L-selectin, and actin polymerization. J Leukoc Biol 56:625–635.

Watts RG, Crispens MA, Howard TH (1991): A quantitative study of the role of F-actin in producing neutrophil shape. Cell Motil Cytoskeleton 19:159–168.

Watts RG, Howard TH (1992): Evidence for a gelsolin-rich labile F-actin pool in human polymorphonuclear leukocytes. Cell Motil Cytoskel 21:25–37.

Watts RG, Howard TH (1993): Mechanisms for actin reorganization in chemotactic factor-activated polymorphonuclear leukocytes. Blood 81:2750–2757.

White JR, Naccache PH, Sha'afi RI (1982): The synthetic chemotactic peptide formyl-methionyl-leucyl-phenylalanine causes an increase in actin assocated with the cytoskelet in rabbit neutrophils. Biochem Biophys Res Commun 108:1144–1149.

White JR, Naccache PH, Sha'afi RI (1983): Stimulation by chemotactic factor of actin association with the cytoskeleton in rabbit neutrophils. J Biol Chem 258:14041–14047.

Wilde MW, Carlson KE, Manning DR, Zigmond SH (1989): Chemoattractant-stimulated GTPase activity is decreased on membranes from polymorphonuclear leukocytes incubated in chemoattractant. J Biol Chem 264:190–196.

Wymann MP, Kernen P, Bengtsson T, Andersson T, Baggiolini M, Deranleau DA (1990): Corresponding shape and filamentous actin content. J Biol Chem 265:619–622.

Wymann MP, Kernen P, Deranleau DA, Baggiolini M (1989): Respiratory burst oscillations in human neutrophils and their correlation with fluctuations in apparent cell shape. J Biol Chem 264:15829–15834.

Wymann MP, Kernen P, Deranleau DA, Dewald B, von Tscharner V, Baggioloini M (1987): Oscillatory motion in human neutrophils responding to chemotactic stimuli. Biochem Biophys Res Commun 147: 361–368.

Yassin R, Shefcyk J, White JR, et al. (1985): Effects of chemotactic factors and other agents on the amounts of actin and a 65,000-mol-wt protein associated with the cytoskeleton of rabbit and human neutrophils. J Cell Biol 101:182–188.

Yuli I, Oplatka A (1987): Cytosolic acidification as an early transductory signal of human neutrophil chemotaxis. Science 235:340–342.

Yuli I, Snyderman R (1984a): Rapid changes in light scattering from human polymorphonuclear leukocytes exposed to chemoattractants: Descrete responses correlated with chemotactic and secretory functions. J Clin Invest 73:1408–1417.

Yuli I, Snyderman R (1984b): Light scattering by polymorphonucelar leukocytes stimulated to aggregate under various pharmacologic conditions. Blood 64:649–655.

Zigmond SH (1989): Cell locomotion and chemotaxis. Curr Opin Cell Biol 1:80–86.

Zigmond SH (1993): Recent quantitative studies of actin filament turnover during cell locomotion. Cell Motil Cytoskeleton 25:309–316.

Zigmond SH (1996): Signal transduction and actin filament organization. Curr Opin Cell Biol 8:66–73.

Zigmond SH, Hirsch JG (1972): Effects of cytochalasin B on polymorphonuclear leukocyte locomotion, phagocytosis, and glycolysis. Exp Cell Res 73:383–393.

Zigmond SH, Slonczewski JL, Wilde MW, Carson M (1988): Polymorphonuclear leukocyte locomotion is insensitive to lowered cytoplamic calcium levels. Cell Motil Cytoskeleton 9:184–189.

Zurier RB, Hoffstein S, Weissmann G (1973): Cytochalasin B: effect on lysosomal enzyme release from human leukocytes. Proc Natl Acad Sci USA 70:844–848.

8

Neutrophil Phagocytosis: Influence of Opsonins, Neutrophil Activation, and Microbial Characteristics

Robert Bjerknes
University of Bergen, Haukeland Hospital, Bergen, Norway

Phagocytic leukocytes provide the body with a first-line defense against infection, for they recognize, ingest, and kill invading microorganisms. The principal professional phagocytes are the neutrophils, monocytes, and tissue macrophages; neutrophils are especially involved in acute inflammation and microbial clearance, and monocytes and macrophages in chronic inflammation (see reviews by Klebanoff and Clark, 1978; Gallin et al., 1992). Neutrophils are produced in the bone marrow, where about 55% of the hematopoietic production is dedicated to formation of this cell type (Whintrobe, 1981). After release into the blood, neutrophils circulate about 10 h. In response to inflammatory stimuli, neutrophils adhere to endothelial cells and migrate into affected tissues. Much remains unknown on their fate in the tissues, but neutrophils probably survive there for only 1–2 days (Whintrobe, 1981).

Neutrophils are highly specialized for their primary function, that is, recognition, phagocytosis, and destruction of microorganisms. The interaction between microorganisms and neutrophils induces complex and concerted structural and metabolic alterations of the neutrophils, essential for normal function. The importance of neutrophils for host defense is clearly illustrated by the fact that several syndromes of

Phagocyte Function: A Guide for Research and Clinical Evaluation, Edited by J. Paul Robinson and George F. Babcock. ISBN 0-471-12364-1

increased susceptibility to infection are associated with or caused by defects of neutrophil function (Klebanoff and Clark, 1978; Fletcher et al., 1990; Gallin et al., 1992).

This chapter discusses the main characteristics of neutrophil phagocytosis. The influence of opsonins and their receptors as well as the regulating effects of activating cytokines and growth factors is covered. The chapter also discusses how microbiological characteristics may influence the phagocytic capacity of neutrophils.

NEUTROPHIL PHAGOCYTOSIS

Attachment and Ingestion

The process of phagocytosis may be divided conceptually into recognition and attachment of a particle to the surface of the phagocyte and internalization of the bound particle (Silverstein et al., 1977; Klebanoff and Clark, 1978) (Fig. 1). In contrast to attachment, engulfment is a process that requires energy from glycolysis. Optimal phagocytosis seems also dependent on the presence of Ca^{2+} and Mg^{2+}, although Ca^{2+}-dependent and Ca^{2+}-independent pathways of ingestion have both been described (Lew et al., 1985).

Ingestion results when a microorganism is recognized by specific receptors on the neutrophil surface and requires multiple, successive interactions between the phagocyte and the target. Some microorganisms and inert particles may be ingested by neutrophils in the absence of serum factors, but most microorganisms must be coated with opsonins for optimal attachment and ingestion to occur (Silverstein et al., 1977; Klebanoff and Clark, 1978). Interaction between the opsonic ligands and their receptors on the neutrophil surface induces a circumferential flow of phagocyte membrane which gradually surrounds the microorganism to be ingested (Fig. 1). The elaborating pseudopods meet in a single point, and fusion of opposing membranes at this point pinches off the resulting phagosome into the cytoplasm (Fig. 1). The advancing pseudopods are rich in actin filaments, and blockade of the polymerization of actin by the cytoskeletal poison cytochalasin B completely inhibits phagocytosis (Stossel, 1990). Pseudopods elaborate in response to signals arising from ligated receptors, and polymerized actin appears directly under the area of contact with opsonized particle (Boyles and Bainton, 1981; Stossel, 1990). Little is still known, however, about the specific mechanisms by which phagocytic receptors interact with the cytoskeleton. It has been reported that IgG Fc receptor I (FcγRI) associates with intracellular actin after ligand binding (Ohta et al., 1991), and that cross-linking of FcγRII or FcγRIII induces actin polymerization (Salmon et al., 1991; Brennan et al., 1991).

The signals for phagocytosis do not arise from attachment of a particle alone, for bacteria or red blood cells coated with fibronectin (Bevilacqua et al., 1981) or lipopolysaccharide-binding protein (Wright et al., 1989) may bind to the surface of a phagocyte but not be ingested. However, if the same particles are opsonized with IgG, they are promptly internalized, emphasizing the role of specific receptors initiating ingestion (see also below). The successive interaction between neutrophil receptors and the target to be ingested results in signal-transducing events, confined to the phagocyte membrane and the cytoskeleton around the ligated receptor. These signaling events include activation of serine/threonine and tyrosine kinases, as well as mo-

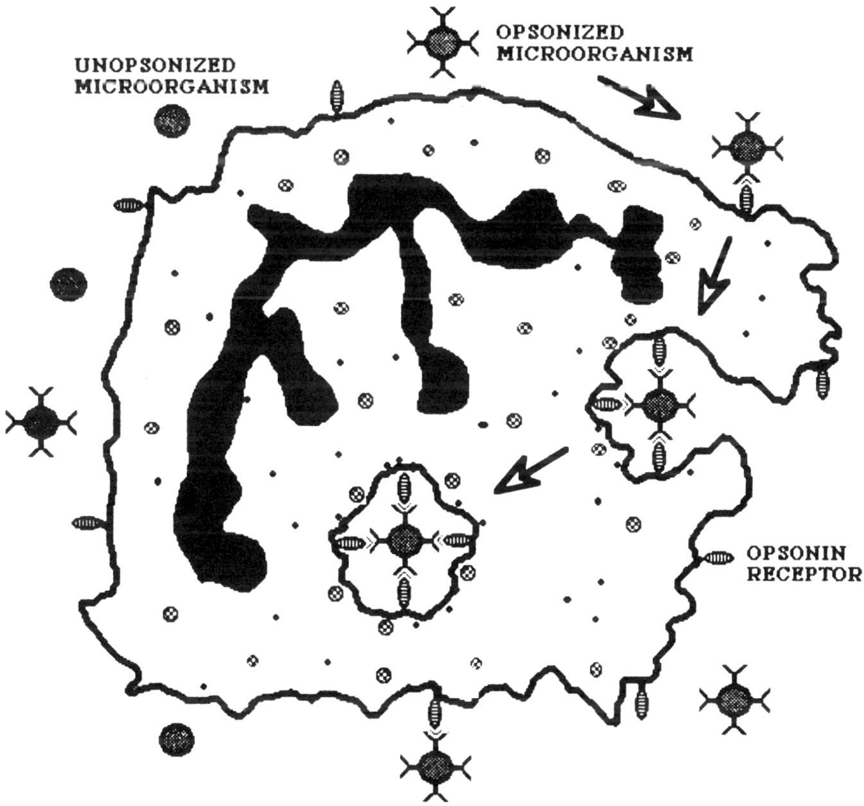

Fig. 1. Neutrophil opsonophagocytosis. Opsonins bound to the microbial surface interact with neutrophil surface receptors and trigger pseudopods to elaborate and surround the microorganism. The resulting phagosome is pinched off into the cytoplasm.

bilization of Ca^{2+} to the cytoplasm and activation of the arachidone metabolism (Andersson et al., 1988; Gresham et al., 1990; Connelly et al., 1991; Lund-Johansen et al., 1991; Dusi et al., 1994; see also reviews by Lew, 1990; and Brown, 1995). The pattern of intracellular signaling during phagocytosis seems to depend on the receptors involved (Murata et al., 1987; Fällman et al., 1993; van de Winkel and Capel, 1993), but much work remains to delineate the details of intracellular signals necessary for phagocytosis to occur.

Postphagocytic Events

The binding of opsonized particles to neutrophils induces intracellular signals that cause cellular activation with a burst of metabolic activity and discharge of granule contents (see reviews by Silverstein et al., 1977; Klebanoff and Clark, 1978; Babior, 1984, and Gallin et al., 1992). During and after phagocytosis, neutrophils produce large amounts of reactive oxygen intermediates (Babior, 1984), as well as leukotrienes and prostaglandins (Hurst and Barrette, 1989; Ford-Hutchinson et al., 1980). The metabolites of arachidonic acid serve to recruit additional phagocytes to a site of inflammation (Samulesson, 1983; Goldman and Goetzl, 1984), whereas the superoxide and hydrogen peroxide produced play an important role for the killing of microorganisms as they are ingested (Klebanoff and Clark, 1978; Babior, 1984; Hurst and Barrette, 1989). During engulfment, cytoplasmic granules also empty their contents of digestive enzymes and microbicidal cationic proteins (e.g., defensins) into the phagosome to destroy the microorganisms ingested (Klebanoff and Clark, 1978; Elsbach and Weiss, 1983; Lehrer and Ganz, 1990). Secretion of these substances does not require complete closure of the phagocytic vacuole, for substrates or microorganisms too large to be ingested readily promote release of reactive oxygen intermediates, enzymes, and arachidonic acid metabolites. Consistently, neutrophils also release a variety of secretory products to the external milieu during phagocytosis, contributing to the formation of inflammatory exudates and to tissue destruction (Weiss, 1989). The postphagocytic processes of neutrophils are covered in more detail elsewhere in this volume.

OPSONINS AND THEIR RECEPTORS ON NEUTROPHILS

Opsonins bridge particles to phagocytes, and the two best characterized and most effective opsonins are IgG and complement factor C3 (Silverstein et al., 1977; Klebanoff and Clark, 1978; Brown, 1995). A variety of microorganisms will not be phagocytosed at all unless they are opsonized, and complete ingestion of a particle requires that opsonins be distributed over the entire surface of the target (Griffin et al., 1975). When phagocytic cells are presented with a hemispherically opsonized particle, pseudopods move to cover only the opsonized part and do not extend onto the unopsonized region (Griffin et al., 1975). Thus the movement of phagocyte plasma membrane along the surface of a ligand-coated particle depends on the availability of phagocyte surface receptors, and is guided by the distribution of opsonic ligands on the surface of the particle. During phagocytosis, some opsonin receptors accumulate at the site of ingestion, and this may enhance the neutrophil opsonophagocytic capacity (Kindzelskii et al., 1994).

In addition to acting as ligands between microorganisms and neutrophils, opsonins may alter the physiochemical characteristics of the microorganism surface and facilitate phagocytosis (van Oss, 1978). The surface of many bacteria is hydrophilic relative to the surface of the neutrophil. With opsonization, the surface of these bacteria becomes more hydrophobic than that of the neutrophil, and they are more readily ingested. Furthermore, opsonization may also promote engulment by reducing the charge repulsion between the microorganism and the neutrophil, both of which are negatively charged (van Oss, 1978).

Neutrophil Receptors for Immunoglobulin G

Immunoglobulin G bridges particles to phagocytes by its distinct binding domains, that is, with the variable antigen-recognizing Fab-region reacting with the particle surface and the Fc-domain recognized by phagocyte surface FcγRs (Metzger and Kinet, 1988). Three classes of FcγRs, termed FcγRI (CD64), FcγRII (CD32), and FcγRIII (CD16), are found on human leukocytes, and all these receptors are members of the immunoglobulin superfamily of proteins (van de Winkel and Anderson, 1991; van de Winkel and Capel, 1993).

FcγRI. This receptor binds IgG1 and IgG3 with high affinity, but it is not normally expressed by resting mature neutrophils (van de Winkel and Anderson, 1991). However, surface expression of FcγRI can be induced in vitro by interferon-γ (IFN-γ) (Perussia et al., 1983) and G-CSF (Gericke et al., 1995), and in vivo by infections (Guyre et al., 1990). The receptor has been cloned, and in contrast to the other members of the family, FcγRI has three (instead of two) extracellular immunoglobulin-like domains (van de Winkel and Capel, 1993).

FcγRII. This surface structure represents low-affinity binding sites on neutrophils that bind IgG subclasses with the following affinity: IgG1 = IgG3 >> IgG2 = IgG4 (van de Winkel and Anderson, 1991). The receptor has a membrane-spanning domain and a cytoplasmic tail. Three different genes code for FcγRII, where FcγRIIA is the major transcript in neutrophils (Brooks et al., 1989; van de Winkel and Capel, 1993). Neutrophil FcγRII exhibits genetically determined structural polymorphism, which may also have consequences for function (Warmerdam et al., 1990; Sanders et al., 1995).

FcγRIII. This receptor binds human IgG1 and IgG3 with intermediate affinity. Two genes encode FcγRIII: *FcγRIIIB*, which encodes a molecule with a glycophosphatidylinositol (GPI) anchor, and *FcγRIIIA*, which encodes a molecule with transmembrane and cytoplasmatic domains (Ravetch and Perussia, 1989; van de Winkel and Capel, 1993). The expression is cell specific, for FcγRIIIB is expressed only by neutrophils and FcγRIIIA is synthesized by natural killer cells and macrophages (Ravetch and Perussia, 1989). As a consequence of the GPI anchor, the FcγRIIIB on neutrophils can be released by digestion with GPI-specific phospholipase C. FcγRIIIB is also shed during activation of neutrophils (Huizinga et al., 1988). The FcγRIII on neutrophils has two allotypic forms, NA1 and NA2, which differ only by four amino acids and two potential glycosylation sites (Ory et al., 1989a; Ory et al., 1989b). Individuals homozygous for NA1 or NA2 have been reported to have different phagocytic capacities independent of their FcγRIIA allotypes (Salmon et al., 1990, 1992), suggesting the FcγRIIIB polymorphism may influence neutrophil function.

Neutrophil FcγR-Mediated Phagocytosis

FcγRII seems to be the most important FcγR for neutrophil phagocytosis of IgG-coated particles and microorganisms (Huizinga et al., 1989; Anderson et al., 1990; Brown, 1995). This capacity has been clearly demonstrated both by blocking of the binding of IgG-coated particles to FcγRII by anti-receptor monoclonal antibodies (mAbs) (Huizinga et al., 1989; Capsoni et al., 1991; Bjerknes, 1992a) (Fig. 2) and by exposing neutrophils to anti-FcγRII opsonized erythrocytes (Anderson et al., 1990).

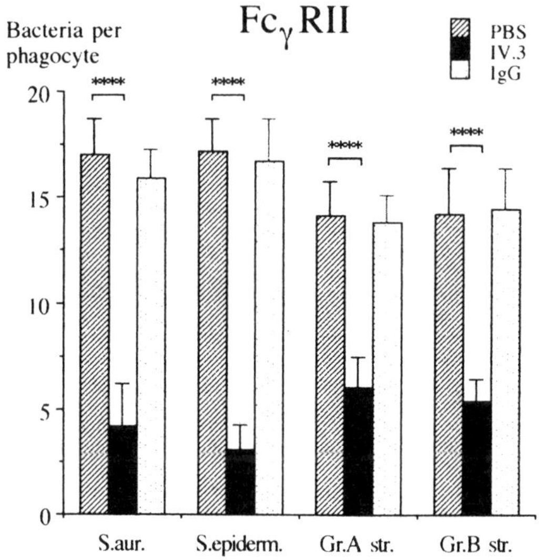

Fig. 2. The effect of FcγRII blockade by an anti-CD32 mAb on the neutrophil ingestion of staphylococci and streptococci. The neutrophils were incubated with saturating amounts of $F(ab)_2$ fragments of the IgG mAb IV.3 (MedaRex) for 15 min, before the cells were exposed to preopsonized FITC-labeled bacteria (clinical isolates) for 15 min at a cell–bacteria ratio of 1:20. After phagocytosis, ingestion was determined by a flow cytometric fluorescence quenching technique (Bjerknes and Bassøe, 1984). The results are given as mean ±95% confidence interval; $n = 6$. ****: $p < .001$.

FcγRII is the sole FcγR class capable of binding human IgG2 complexes (Warmerdam et al., 1990, 1991). Human IgG2 is the predominant IgG subclass containing antibodies to bacterial capsular polysaccharides, and neutrophil ingestion of several encapsulated microorganisms depends on this receptor (Ferrante et al., 1990; Jefferis and Kumararatne, 1990). Recent studies have revealed that FcγRIIA polymorphism is important for the capacity to ingest *Haemophilus influenzae* type b, *Staphylococcus aureus* Wood 46, and encapsulated group B streptococci (Bredius et al., 1993; Sanders et al., 1995). Moreover, from a study of children with recurrent bacterial respiratory tract infections, it seems probable that neutrophil FcγRIIA polymorphism may contribute to increased susceptibility to infections with encapsulated bacteria in patients with low IgG2 anti-carbohydrate antibodies (Sanders et al., 1994).

There have been conflicting results as to whether neutrophil FcγRIII can mediate ingestion (Salmon et al., 1987, 1990; Anderson et al., 1990). Anti-CD16 mAbs have been reported not to affect the neutrophil phagocytosis of IgG-coated particles (Huizinga et al., 1989), and Anderson et al. (1990) repeatedly failed to induce neutrophil phagocytosis of red blood cells coated with an anti-CD16 mAb. However, involvement of neutrophil FcγRIII in phagocytosis is indicated by the reduced uptake of bacteria and concanavalin A-treated erythrocytes in the presence of Fab fragments of the anti-CD16 mAb 3G8 (Salmon et al., 1990), and by a report demonstrating that FcγRIIIB polymorphism may have consequences for neutrophil phagocytic capacity

(Salmon et al., 1990). Moreover, Sanders et al. (1995) recently reported decreased phagocytosis of IgG2-opsonized group B streptococci in the presence of $F(ab)_2$ fragments of the CD16 mAb 3G8. We have also demonstrated reduced phagocytosis of group A streptococci and of staphylococci after blockade of FcγRIII by an IgM anti-CD16 mAb (Fig. 3). Most probably, the role of FcγRIII in phagocytosis is to attach IgG-coated particles to the neutrophil surface, but it seems that FcγRII is the main FcγR that mediates ingestion and triggers the following oxidative burst (Anderson et al., 1990). Other indications of interaction and cooperation between neutrophil FcγRII and FcγRIII are the increased phagocytosis of anti-FcγRII Fab-coated erythrocytes following cross-linking of FcγRIII (Salmon et al., 1991, 1995; Edberg and Kimberley, 1994) and the observation that FcγRII may regulate FcγRIII intracellular signaling (Naziruddin et al., 1992).

The FcγRI mediates antibody-dependent cellular cytotoxicity by IFN-γ stimulated neutrophils (Graziano and Fanger, 1987). In addition, Anderson et al. (1990) have demonstrated that this receptor effectively mediates ingestion of anti-CD64 mAb-coated red blood cells by IFN-γ-exposed neutrophils.

Neutrophil Receptors for Complement Products

Activation of the complement cascade causes proteolytic cleavage of complement factors that liberate split products that are potential ligands for complement receptors on phagocytic leukocytes. The soluble products (e.g., C3a and C5a) induce chemotactic movements to sites of complement activation, whereas surface-bound ligands serve as opsonins (see reviews by Klebanoff and Clark, 1978; Gallin et al., 1992). The main complement-derived opsonins are split products of C3, that is, C3b and C3bi.

Complement Receptor I (CR1). Neutrophil CR1 (CD35) is a glycoprotein consisting of a single membrane-spanning domain, a short cytoplasmatic *C*-terminal domain, and a large extracellular domain composed of at least 30 repeated units arranged in tandem (Ahearn and Fearon, 1989). Four allelic forms of CR1 have been described in humans, and individuals may be either homozygous and express CR1 of a single molecular weight or heterozygous and express CR1 of two distinct molecular weights (Dykman et al., 1985; Ahearn and Fearon, 1989). The expression of any one allele has not so far been linked to functional abnormalities. Complement receptor 1 of unstimulated neutrophils resides in two pools: about 15% is expressed in clusters on the cell surface, and the remainder is in intracellular compartments from which it can be mobilized upon activation (Fearon and Collins, 1983; O'Shea et al., 1985). Complement receptor 1 functions as an opsonic receptor, binding dimeric C3b. In contrast, uncleaved C3, as well as C3dg, is not recognized by the resting state of this receptor (Ahearn and Fearon, 1989). C3bi is recognized by CR1, but the binding is very weak and possibly of little physiological importance (Medof et al., 1982).

Complement Receptor 3 (CR3). The neutrophil glycoprotein CR3 (CD11b/CD18) is a member of the integrin family and consists of a noncovalently linked dimer composed of a 185-kDa α chain (CD11b) and a 95-kDa β chain (CD18)

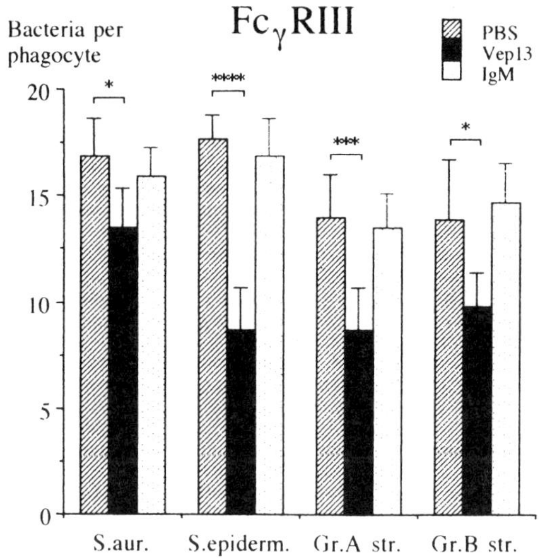

Fig. 3. The effect of FcγRIII blockade by an anti-CD16 mAb on the neutrophil ingestion of staphylococci and streptococci. The neutrophils were incubated with saturating amounts of the IgM mAb Vep-13 (Boehringer) for 15 min, before the cells were exposed to preopsonized FITC-labeled bacteria (clinical isolates) for 15 min at a cell–bacteria ratio of 1:20. After phagocytosis, ingestion was determined by a flow cytometric fluorescence quenching technique (Bjerknes and Bassøe, 1984). The results are given as mean $\pm 95\%$ confidence interval; $n = 6$. *: $p < .05$; ***: $p < .005$; ****: $p < .001$.

(Sanchez-Madrid et al., 1983; Rouslathy, 1991). CR3 shares a β chain with LFA-1 (CD11a) and p150 (CD11c), comprising the leukocyte integrins (also termed the CD11/CD18 complex) (Sanchez-Madrid et al., 1983). CR3, like CR1, resides in two neutrophil pools, either expressed on the surface or located in intracellular compartments from which it can be mobilized upon activation (O'Shea et al., 1985).

Neutrophil CR3 recognizes four types of ligands: (1) *The complement split product C3bi deposited on the surface of microorganisms or other particles* (Klebanoff and Clark, 1978; Wright and Silverstein, 1983; Arnaout et al., 1983; Bjerknes et al., 1984; Cain et al., 1987; Ross and Větvička, 1993; Brown, 1995). In contrast, CR3 does not recognize the precursor C3b. Binding of monomeric or dimeric C3bi to CR3 has not been shown, but multivalent opsonic binding between phagocytes and C3bi-coated particles has been documented in several studies. In contrast to CR1 and FcγRs, CR3 requires divalent cations in order to interact effectively with ligand. (2) *Components of the extracellular matrix.* Binding of CR3 to fibrinogen and fibrin promotes adhesion of neutrophils to extracellular matrix and may aid in the dissolution of clots (Altieri et al., 1988). (3) *Ligands expressed on the surface of other cells, for example, ICAM-1 (CD54) on endothelial cells* (see reviews by Gallin et al., 1992; Harlan and Liu, 1992). Consistently, the CD18-containing integrins are essential for normal movement of neutrophils through an endothelial monolayer into the tissues

(Anderson et al., 1985; Arnaout, 1990). (4) *Surface structures on several microorganisms, mediating attachment to and activation of CR3 in the absence of opsonin.* By the use of mAbs against CR3 and purified CR3, it has been shown that this receptor can bind to several unopsonized microorganisms, such as some staphylococci (Klebanoff et al., 1985), group B streptococci (Smith et al., 1990), *Escherichia coli* (Gbarah et al., 1991), *Bordetella pertussis* (Relman et al., 1990), *Histoplasma capsulatum* (Bullock and Wright, 1987), and *Leishmania* (Wilson and Pearson, 1988), as well as zymosan particles (from yeast cell walls) (Klebanoff et al., 1985; Ross et al., 1985) (see also below).

Neutrophil Complement Receptor-mediated Phagocytosis—Activation of Complement Receptors 1 and 3

The binding activity of CR1 and CR3 is regulated. Complement receptors 1 and 3 on resting neutrophils can attach ligand coated particles, but cannot mediate phagocytosis (Ahern and Fearon, 1989; Ross and Větvička, 1993). However, stimulation of neutrophils with phorbol myristate acetate (PMA) rapidly exposes new binding sites and increases the ability of CR1 and CR3 to mediate phagocytosis (Wright and Meyer, 1986). Consistently, activation of CR3 involves protein kinase C. However, tyrosine kinase-dependent phosphorylation events are also involved (Ross and Větvička, 1993). The increased avidity of neutrophil CR3 is transient, reaching a maximum after 15 min and declining by 60 min. Several stimuli are capable of increasing the avidity of CRs, including endotoxin (Wright et al., 1991), chemoattractants [e.g., C5a (Lo et al., 1989) and interleukin-8 (Detmers et al., 1990)], and cytokines [e.g., TNF-α (Lo et al., 1989)]. In addition, components of extracellular matrix [e.g., laminin (Bohnsack et al., 1985)] can activate CR1 and CR3, suggesting that the acquisition of a capacity for complement-mediated ingestion may develop in vivo as phagocytes migrate to the extravascular space, for example, in response to infection.

Fixed C3bi in zymosan and several strains of bacteria stimulates CR3-dependent phagocytosis (Gordon et al., 1986; Cain et al., 1987; Smith et al., 1990). The opsonin-independent interaction between microorganisms and neutrophil CR3 may also promote the phagocytically active state of CR3, inducing active phagocytosis (Gbarah et al., 1991). It has recently been demonstrated that *E. coli* carrying mannose-specific lectins in the form of type 1 fimbriae bind mannose residues on neutrophil CR3, as well as on the other members of the leukocyte integrin family (CD11a and CD11c), and that those bacteria are effectively ingested by neutrophils (Gbarah et al., 1991). Moreover, endotoxin on *E. coli* may also bind directly to CR3 and stimulate phagocytosis (Wright and Jong, 1986).

Neutrophil phagocytosis is accompanied by changes in surface protein composition (Skubitz and Kinkead, 1987; Lund-Johansen et al., 1991), and the expression of both CRs and FcγRs is altered during the phagocytic process (Figs. 4 and 5). Using clinical isolates of several bacteria, we have demonstrated a rapid initial fall of unoccupied extracellular CRs followed by a significant upregulation of both receptors. After 10 min of phagocytosis of *S. epidermidis,* the neutrophil CR1 and CR3 expression had increased to 132 and 152% of that of unstimulated controls, respectively

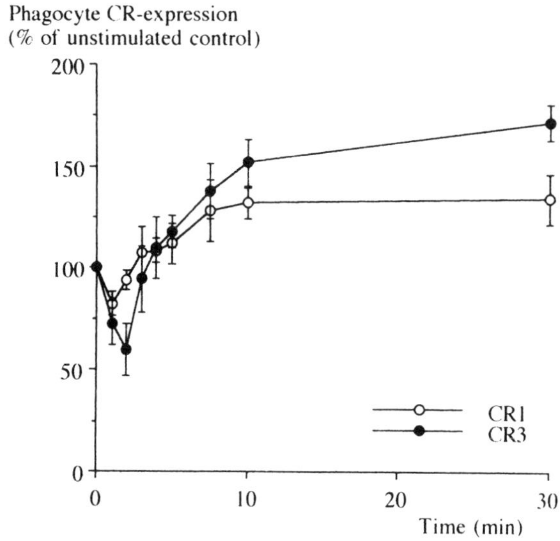

Fig. 4. Kinetics of the changes of neutrophil CR expression during phagocytosis of *S. epidermidis*. The neutrophils were incubated with FITC-labeled preopsonized bacteria at a cell–bacteria ratio of 1:20 for various time intervals, before they were exposed to ice-cold PBS and stained with PE-labeled anti-CR1 (Anti-CR1; Becton-Dickinson) or anti-CR3 (Mol; Coulter). The CR-dependent fluorescence is given as mean fluorescence ±95% confidence interval (n = 3) and expressed as percent of that of unstimulated neutrophils.

(Fig. 4). In contrast, the expression of both FcγRII and FcγRIII was reduced during the phagocytic process (Fig. 5). The initial fall of CR stainability and the permanent reduction of FcγR stainability by receptor-specific mAbs are probably due to a combination of opsonin receptor occupation by ligand and receptor internalization. The upregulation of CRs during the phagocytic process is probably linked to degranulation, and this response seems physiologically advantageous for the host, for increased neutrophil CR availability and function in vivo may increase the phagocytic capacity at sites of infection.

It has been known for over 20 years that when a particle is opsonized with both complement and IgG, the efficiency of IgG effector functions is markedly increased (Ehlenberger and Nussenzweig, 1977). In addition, several studies have demonstrated that mAbs to CR3 inhibit neutrophil FcγR-mediated phagocytosis (Arnaout et al., 1983; Bjerknes et al., 1984; Brown et al., 1988), and that neutrophils from patients with leukocyte adhesion deficiency (LAD) type I (which do not express β_2-integrins) are deficient in both IgG- and CR1-mediated phagocytosis (Gresham et al., 1991) (see also below). However, the LAD defect in IgG-mediated phagocytosis is apparent only when the neutrophils are activated (Brown, 1992), implying that CR3 is involved in the process of maximal ingestion. The cooperation between opsonin receptors is further illustrated by the capacity of CR3 and FcγRIII to form complexes on the surface of neutrophils (Petty and Todd, 1993; Zhou et al., 1993; Stöckl et al., 1995) and the

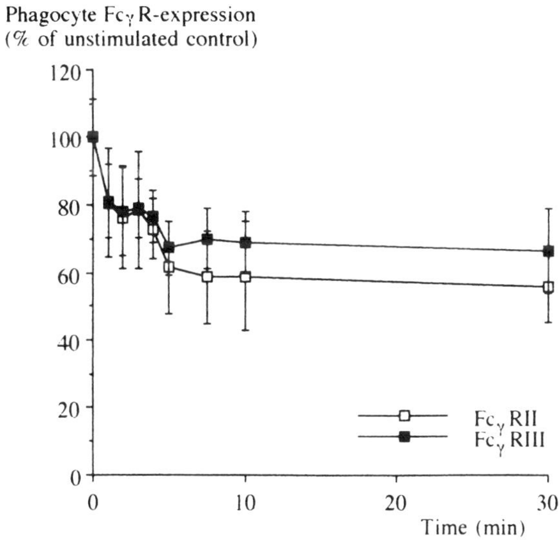

Fig. 5. Kinetics of the changes of neutrophil FcγR expression during phagocytosis of *S. epidermidis*. The neutrophils were incubated with FITC-labeled preopsonized bacteria at a cell–bacteria ratio of 1:20 for various time intervals, before they were exposed to ice-cold PBS and stained with PE-labeled anti-FcγRII (IV.3; MedaRex) or anti-FcγRIII (Vep-13; Boehringer). The FcγR-dependent fluorescence is given as mean fluorescence ±95% confidence interval ($n = 3$) and expressed as percent of that of unstimulated neutrophils.

collaboration of these receptors in the generation of the oxidative burst (Zhou and Brown, 1994).

Other Receptors and Soluble Factors Promoting Neutrophil Phagocytosis

Several proteins other than IgG and complement have been shown to serve an opsonic function or to regulate phagocytosis without being an opsonin. Neutrophils have specific receptors for IgA, and Fanger et al. (1983) have demonstrated that larger percentages of oral neutrophils expressed FcαRs than did blood neutrophils. Oral neutrophils were also able to phagocytose IgA-coated particles (Fanger et al., 1983). Later studies have demonstrated that neutrophil FcαR-mediated phagocytosis is stimulated by GM-CSF (Weisbart et al., 1988), chemoattractants (Hostoffer et al., 1993), and TNF-α (Hostoffer et al., 1994).

Although acting as a true opsonin for the phagocytosis of gram-negative bacteria by macrophages, lipopolysaccharide-binding protein complexed with endotoxin increases neutrophil CR3 binding capacity (Wright et al., 1991). In addition, fibronectin (Yang et al., 1988), laminin (Bohnsack et al., 1985), and C1q (Bobak et al., 1987) enhance phagocytosis. C-reactive protein (CRP) binds to pneumococcal polysaccharides and initiates complement activation and deposition, causing opsonization

(Mortensen et al., 1976). Similarly, there is no receptor for IgM on neutrophils, but IgM can enhance phagocytosis by activating complement through the classical pathway (Reed, 1975; Klebanoff and Clark, 1978).

MODULATION OF NEUTROPHIL FcγR AND COMPLEMENT RECEPTOR EXPRESSION AND PHAGOCYTOSIS BY ACTIVATING CHEMOATTRACTANTS, CYTOKINES, AND HORMONES

The neutrophil phagocytic capacity depends both on the availability of surface receptors that can bind particles and on the state of opsonin receptor and neutrophil activation. A variety of soluble factors act as activators and modulators of neutrophil phagocytosis.

Neutrophil FcγR expression

Resting neutrophils do not express FcγRIs. Interferon-γ induces neutrophil FcγRI expression (Perussia et al., 1983; Buckle and Hogg, 1989), an effect that can be abrogated by dexamethasone in vitro (Petroni et al., 1988). In addition, stimulation of mature neutrophils with G-CSF induces FcγRI expression (Repp et al., 1991; Gericke et al., 1995). Moreover, bacterial pharyngitis has been reported to be associated with significantly elevated expression of neutrophil FcγRI, suggesting that this receptor may play a role in neutrophil function in vivo (Guyre et al., 1990). It is, however, possible that factors other than IFN-γ and G-CSF may be responsible for the FcγRI elevation in patients.

Even though there are differences between donors, IFN-γ, G-CSF, and GM-CSF all have the capacity of increasing the neutrophil expression of FcγRIII (Buckle and Hogg, 1989; Capsoni et al., 1991). In addition, interleukin-8 causes a slight increase of FcγRIII expression in vitro (Detmers et al., 1990). Neither of these factors influences neutrophil FcγRII expression (Buckle and Hogg, 1989; Detmers et al., 1990), but they may prime neutrophils to increase their response to specific FcγRII cross-linking (Roberts et al., 1990). Similarly, TNF-α does not influence the neutrophil expression of FcγRII or FcγRIII (Capsoni et al., 1991). Moreover, growth hormone (Bjerknes, unpublished results) and insulin-like growth factor-I (IGF-I) (Bjerknes and Aarskog, 1995) do not themselves induce FcγRI or alter the expression of FcγRII or FcγRIII on neutrophils. However, the activation-dependent downregulation of FcγRIII by fMLP/cytochalasin B is significantly enhanced by high concentrations of IGF-I (Bjerknes and Aarskog, 1995).

Neutrophil CR expression

GM-CSF, but not G-CSF, increases the synthesis and expression of CR1 and CR3 by neutrophils (Arnaout et al., 1986; Buckle and Hogg, 1989; Neuman et al., 1990). In addition, the chemoattractants fMLP, C5a, interleukin-8, platelet activating factor, and leukotriene B_4 all upregulate the expression of these complement receptors (Arnaout et al., 1986; Shalit et al., 1988; Detmers et al., 1990; Neuman et al., 1990). Tumor necrosis factor-α also upregulates both neutrophil CR1 and CR3 (Berger et al., 1988). IGF-I increases the neutrophil expression of CR1 and CR3 in a concentration- and time-dependent fashion (Bjerknes and Aarskog, 1995). This growth factor also increases the degranulation-dependent upregulation of surface expression of these receptors in response to fMLP, fMLP/cytochalasin B, and PMA (Bjerknes and Aarskog, 1995). Exposure of neutrophils to IFN-γ for 24 h induces a slight increase of surface CR3 expression (Petroni et al., 1988), whereas at shorter incubation periods no such upregulation is seen (Buckle and Hogg, 1989). Interleukin-1 (Berger et al., 1988) and growth hormone (Bjerknes, unpublished results) do not significantly influence neutrophil CR1 and CR3 expression.

Neutrophil Phagocytosis

A variety of cytokines and growth factors modulate phagocytosis. Tumor necrosis factor-α strongly stimulates the neutrophil uptake of serum-opsonized zymosan, fetal calf serum- or IgG/C3b-opsonized latex beads, a variety of IgG-coated bacteria, and IgG-coated sheep erythrocytes (Shalaby et al., 1985; Klebanoff et al., 1986; Ogle et al., 1990; Moxey-Mims et al., 1991; Della Bianca et al., 1993; Ferrante et al., 1993), without altering the expression of FcγRs (Capsoni et al., 1991; Moxey-Mims et al., 1991; Della Bianca et al., 1993). In addition, it has recently been demonstrated that the 55-kDa TNF receptor is the signaling receptor for the potentiation of FcγR-mediated phagocytosis in neutrophils (Della Bianca et al., 1995). Moreover, the phagocytic process modifies the neutrophil's capacity to respond to TNF-α, as the uptake of serum-opsonized *S. epidermidis* and *C. albicans* downregulates the neutrophil expression of TNF-α receptors (Ferrante et al., 1994). This response seems selective for TNF-α, for the neutrophil surface expression of fMLP and CR3 receptors remained unaltered and increased, respectively, during phagocytosis (Ferrante et al., 1994).

Interferon-γ increases the capacity of neutrophils to ingest both IgG-coated and C3-coated particles (Shalaby et al., 1985; Perussia et al., 1987). The augmented neutrophil phagocytosis after IFN-γ may partly be due to increased FcγRI and FcγRIII expression (Buckle and Hogg, 1989). Like fMLP, interleukin-8 has been shown to stimulate FcγR mediated phagocytosis (Detmers et al., 1991).

GM-CSF is also a potent stimulator of neutrophil ingestion of bacteria and fungi (Fleischmann et al., 1986; Kletter et al., 1989; Capsoni et al., 1991). The increased neutrophil phagocytosis after GM-CSF may involve both augmented FcγRIII and CR receptor expression, as well as FcγRII, FcγRIII, and CR function (Fleischmann et al., 1986; Buckle and Hogg, 1989; Kletter et al., 1989; Capsoni et al., 1991). Moreover, interleukin-1α (Ozaki et al., 1987), interleukin-1β (Garner et al., 1996), and interleukin-6 (Mullen et al., 1995) potentiate neutrophil phagocytosis.

We have studied the effect of the growth hormone-dependent peptide IGF-I on neutrophil phagocytosis. This peptide, which is a potent regulator of hematopoiesis, strongly stimulates the neutrophil uptake of *S. aureus* and *C. albicans* (Bjerknes and Aarskog, 1995). Insulin-like growth factor-I increases both the percentage of phagocytosing neutrophils (Fig. 6) and the phagocytic capacity of each leukocyte at phys-

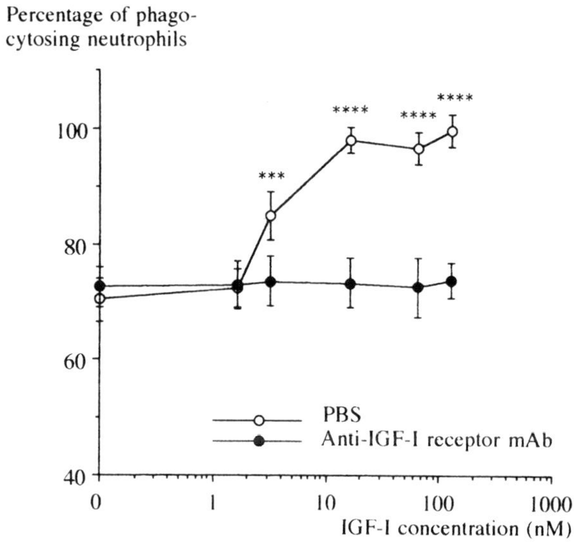

Fig. 6. The influence of IGF-I on the percentage of neutrophils phagocytosing IgG-opsonized *S. aureus*. The neutrophils were incubated with PBS or IGF-I for 3 h at 37°C. In some experiments saturating amounts of an mAb against the IGF-I receptor were added 30 min before the IGF-I. The neutrophils were incubated with FITC-labeled preopsonized bacteria at a cell–bacteria ratio of 1:20 for 7.5 min. The results are given as mean ±95% confidence interval; $n = 6$. ***: $p < .005$; ****: $p < .001$.

iological concentrations (Fig. 7). Insulin-like growth factor-I does not influence the basal neutrophil expression of FcγRs, whereas IGF-I alone induces a slight upregulation of CR1 and CR3 that may contribute to a more efficient phagocytosis of complement-opsonized *C. albicans* (Bjerknes and Aarskog, 1995). However, because the increased uptake of IgG-opsonized *S. aureus* could not have been due to larger availability of these receptors, and the increase of complement-mediated phagocytosis was larger than the upregulation of complement receptors, it is more probable that the priming of neutrophils to increase their phagocytosis by IGF-I was primarily due to augmented receptor function, intracellular signal transduction, or effector systems.

IN VIVO ACTIVATION OF NEUTROPHILS

In vivo, leukocytes are exposed to a complex network of interacting endogenous and exogenous substances that may modify neutrophil function. Consistently, altered neutrophil phagocytic capacity has been reported in patients with infections, mechanical and thermal injuries, inflammatory diseases, and malignancies, as well as after a variety of therapeutic procedures (Alexander and Wixon, 1970; Stossel, 1973; Bjerknes et al., 1989; Simms et al., 1989; Bjerknes, 1992b; Vanholder and Ringoir, 1993; see also review by Klebanoff and Clark, 1978). Altered neutrophil phagocytic capacity demonstrated in patients may indicate a preexisting dysfunction affecting opsonins

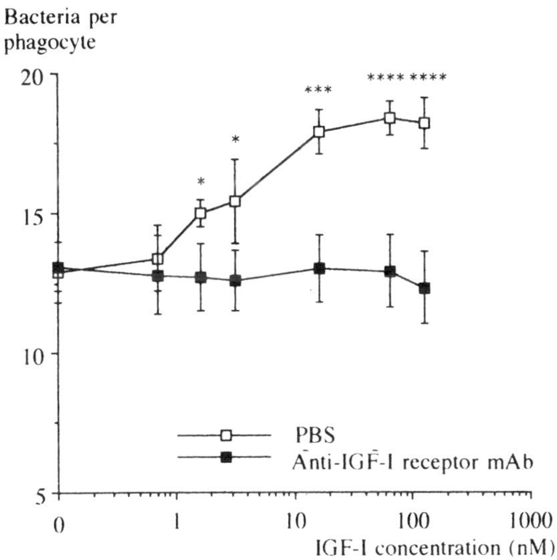

Fig. 7. The influence of IGF-I on the neutrophil ingestion of IgG-opsonized *S. aureus*. The neutrophils were incubated with PBS or IGF-I for 3 h at 37°C. In some experiments saturating amounts of an mAb against the IGF-I receptor were added 30 min before the IGF-I. The neutrophils were incubated with FITC-labeled preopsonized bacteria at a cell–bacteria ratio of 1:20 for 7.5 min. The results are given as mean ±95% confidence interval; $n = 6$. *: $p < .05$; ***: $p < .005$; ****: $p < .001$.

or neutrophils that has contributed to the onset of the illness (e.g., infection in patients with congenital neutrophil defects). However, infection, trauma, or drugs per se also influence neutrophils, and such acquired neutrophil dysfunctions may influence the subsequent course of the illness and the occurrence of superinfections. A variety of the chemoattractants, cytokines, and hormones discussed above (e.g., complement split products, arachidonate metabolites, interleukins, TNF-α, growth factors, and hormones) are present in increased amounts both at the site of disease and in the circulation (see, e.g., Gallin et al., 1992; Bellomo, 1992; Beal and Cerra, 1994; Vindenes et al., 1995, 1996). These molecules play an important regulatory role in the host response to disease, and together with microbial products (e.g., endotoxin) they are important modulators of neutrophil phagocytic capacity during the course of illness. However, much remains to be determined on the specific contribution of each of these factors and the mechanisms involved in neutrophil activation during disease.

INFLUENCE OF MICROBIAL CHARACTERISTICS ON OPSONIC RESPONSE AND NEUTROPHIL OPSONOPHAGOCYTOSIS

Patients with infection develop serum antibodies to microbial antigens that may protect them from new attacks of disease. The developing antibodies may confer opsonic as well as microbicidal function, and their effectiveness in protection may vary significantly between strains of microorganisms. Meningococcal disease is a significant problem in Norway, causing serious infections in children as well as in adults. The ongoing epidemic of meningococcal disease in Norway has been dominated by bacteria of group B and serotype 15 (Frøholm et al., 1985). For years the susceptibility to meningococcal infection has been related to lack of bactericidal antibodies, and less attention has been given to the phagocyte-mediated killing of meningococci. In 1987 we reported increased amounts of opsonins in convalescence sera from patients with group B meningococcal disease (Sjursen et al., 1987). The same year, Ross et al. (1987) reported a series of experiments showing that group B meningococci were resistant to the complement-mediated bactericidal activity of normal human serum, but susceptible to killing by neutrophils. In later studies, we have examined the neutrophil opsonophagocytosis of group B meningococci to assess the opsonic activity in sera from patients with *N. meningitidis* (B:15:P1.16) infections and to determine the nature of these opsonins. A marked increase of the neutrophil opsonophagocytosis of group B meningococci was demonstrated in the presence of convalescence sera as compared to acute sera (Sjursen et al., 1987, 1989; Guttormsen et al., 1992, 1996; Bjerknes et al., 1995). Convalescence sera also induced a significant increase of the neutrophil oxidative burst during phagocytosis (Guttormsen et al., 1992, 1996). By using different bacterial strains we were able to demonstrate that the antibody response following group B meningococcal disease included opsonins reacting with serogroup A, B, C, W-135, and Y meningococci of various serotypes and serosubtypes, and that the serum opsonins recognized surface epitopes other than those determined by the mAbs used for serotyping (Guttormsen et al., 1992). We could also demonstrate that increased opsonic activity, that is, both specific and cross-reacting antibodies, persisted in serum for several years after the acute illness (Guttormsen et al., 1992) and after vaccination (Guttormsen et al., 1993). In a recent study, it was demonstrated that the cross-reaction of antibodies at least in part was due to antibodies reacting with meningococcal surface porins, that is, the outer membrane proteins of class 1 and class 3 (Guttormsen et al., 1996).

Several studies have demonstrated a correlation between the presence of bactericidal antibodies against meningococcal group A and C polysaccharide and protection from disease (Artenstein et al., 1970; Wahdan et al., 1973). Owing to the poor immunogenicity of group B capsular polysaccharide (Mandrell and Zollinger, 1986), other exposed antigens such as outer-membrane proteins (i.e., porins) have been included in vaccines against group B meningococci. We have demonstrated that volunteers immunized with a combined vaccine containing outer-membrane proteins from meningococci serogroup B and serotypes 2b and 15 complexed with polysaccharides from serogroups A, C, Y, and W-135 induced a marked increase in serum opsonic activity (Sjursen et al., 1987). In a subsequent study, we examined whether such a response could be demonstrated in volunteers immunized with an outer-membrane vesicle vaccine prepared from *N. meningitidis* 44/76 (B:15:P1.7,16; with class 3 outer-membrane protein) typical of the Norwegian epidemic (Guttormsen et al., 1993). In this study, we demonstrated long-lasting opsonins capable of increasing neutrophil phagocytosis of class 3 and nontypable strains of different serogroups, serotypes, and serosubtypes. However, no increase in serum opsonins cross-reacting with class 2 (type 2a) meningococci of different serogroups was observed, indicating no increased protective capacity against meningococci expressing class 2, but not class 3, outer-membrane proteins (Guttormsen et al., 1993). Thus we suggest that future meningococcal vesicle vaccines should include both class 2 and class 3 porins in geographic areas where both class 2 and class 3 strains cause disease.

MICROBIAL TACTICS TO OVERCOME NEUTROPHIL PHAGOCYTOSIS

The effective neutrophil phagocytosis and killing of different strains of microorganisms depend on a variety of microbiological characteristics. The tactics used by microorganisms to avoid phagocyte killing include avoidance of recognition, decrement of chemotaxis, inhibition of attachment, ingestion, and intracellular killing, as well as escape from the phagosome and killing of the phagocyte (see reviews by Densen and Mandell, 1980; Spitznagel, 1983; Mandell and Frank, 1992). Because the main topic of this chapter is the process of phagocytosis, we describe those mechanisms involved in attachment and ingestion in more detail.

Microbial Inhibition of Attachment

Structures on the surface of microorganisms, for example, extracellular slime capsules, pili, and fimbriae, convey antiphagocytic properties to the microorganism, preventing it from being attached to the phagocyte (Densen and Mandell, 1980; Spitznagel, 1983; Mandell and Frank, 1992). In these situations, deposition of complement split products or specific anticapsular antibodies produced in response to infection or vaccination is required for phagocytosis to proceed effectively (see also above). However, the opsonic requirements of different microbial species as well as mutants within the same species may vary significantly (Fine, 1975; Verhoef et al., 1977; Guttormsen et al., 1992; Kozel, 1993; Domenico et al., 1994).

Interference with attachment is in most cases related to the presence of a capsule (usually composed of polysaccharides), but may also be due to microbial surface proteins in a few species. The antiphagocytic M protein of group A streptococci inhibits complement activation and binds fibrinogen that can mask C3b-binding sites on the streptococcal surface (Robinson and Kehoe, 1992). The cell wall component protein A of staphylococci can also inhibit neutrophil opsonophagocytosis. Protein A is capable of binding specifically to the Fc region of IgG and activating complement via the classical pathway (Forsgren and Sjöquist, 1966; Peterson et al., 1976). Thus it interferes with the binding of opsonizing IgG to neutrophil FcRs, inhibiting attachment and subsequent engulfment of staphylococci (Peterson et al., 1976).

The group of microorganisms possessing capsules that inhibit attachment to phagocytic cells includes *Streptococcus pneumoniae, Streptococcus pyogenes, Neisseria gonorrhoeae, N. meningitidis, Klebsiella pneumoniae, S. aureus, Haemophilus influenzae* group b, *E. coli, Campylobacter fetus, Bacteroides fragilis, Yersinia pestis,* and *Cryptococcus neoformans* (see review by Densen and Mandell, 1980). The microbial capsules that inhibit attachment to neutrophils seem at least in part to exhibit their effects by interfering with opsonization. Hence Verbrugh et al. (1979) could demonstrate that all encapsulated bacterial strains examined in their study fixed less C3 than did their unencapsulated counterparts. It is also possible that capsules may mask opsonins bound to microorganisms, preventing them from acting as ligands between the organism and the phagocyte. Minor chemical alterations of the surface structures of bacteria may dramatically alter their ability to resist attachment and ingestion by neutrophils (Medearis et al., 1968), although for some pneumococci the resistance of mutants of a given capsular type correlated directly to the amount of capsular material produced (Macleod and Krauss, 1950).

Inhibition of Ingestion after Attachment

Attachment of influenza virus to neutrophils induces inhibition of subsequent ingestion and killing of bacteria (Sawyer, 1969). Moreover, *Mycoplasma* species (Zucker-Franklin et al., 1966; Simberkoff and Elsbach, 1971) and virulent piliated gonococci (Thongthai and Sawyer, 1973; Shafer and Rest, 1989) effectively attach to neutrophils, but they are neither ingested nor killed. The same applies for encapsulated, but not unencapsulated *E. coli* (Horwitz and Silverstein, 1980). The mechanism by which this occurs remains unclear, but it has been suggested that the membrane fluidity of neutrophils exposed to piliated gonococci may be decreased as compared to neutrophils exposed to nonpiliated gonococci (Densen and Mandell, 1980). Because membrane fluidity is essential for ingestion, it is possible that this may partly explain the reduced uptake of virulent gonococci. Consistently, Simberkoff and Elsbach (1971) have reported reduced ingestion of *E. coli* by neutrophils that have *Mycoplasma* already attached to their surface, compared with control neutrophils.

Porins are trimeric proteins that constitute pores that allow diffusion of small solutes through the outer membrane of gram-negative bacteria (Nikaido, 1992). Gonococcal and meningococcal porins are capable of translocating from live bacteria and being directionally inserted into foreign membrane, that is, both lipid bilayers and human cells (Blake, 1985; Lynch et al., 1984). Thus when neutrophils are exposed to purified porins from gonococci (Haines et al., 1988; Bjerknes et al., 1995), meningococci (Bjerknes et al., 1995), or *Yersinia enterocolitica* (Tufano et al., 1994), neutrophil functions are altered. We have demonstrated that exposure of neutrophils to gonococcal outer membrane proteins 1A (P1A) and 1B (P1B) and meningococcal outer membrane proteins of class 1, 2, and 3 induces significant reduction of the neutrophil phagocytic capacity (Bjerknes et al., 1995) (Fig. 8). By the use of a flow cytometric fluorescence quenching technique, it was demonstrated that the meningococcal and gonococcal porins inhibited both the neutrophil attachment and ingestion of meningococci (Bjerknes et al., 1995). The neutrophil expression of FcγRII and FcγRIII, as well as the activation-related upregulation of CR1 and CR3, was reduced by the neisserial porins. It is known from earlier studies that the phagocytosis of meningococci depends on both immunoglobulins and complement (Sjursen et al., 1987; Ross et al., 1987; Guttormsen et al., 1992). The reduced neutrophil expression of opsonin receptors may therefore have contributed to the impaired attachment of bacteria in the presence of porins. In addition, we could demonstrate that the neu-

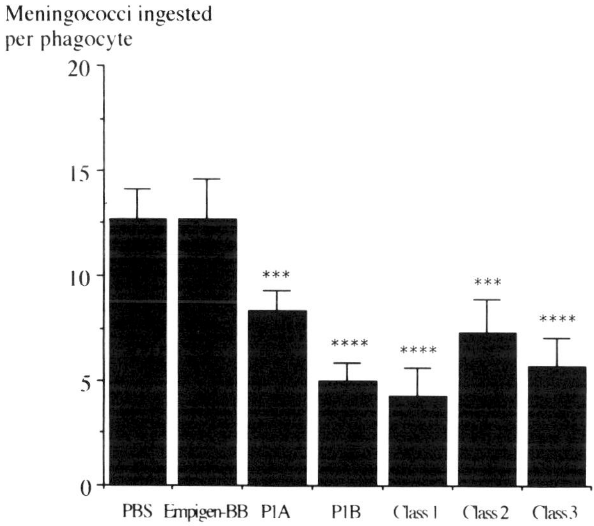

Fig. 8. The effect of neisserial porins on the neutrophil phagocytosis of group B meningococci. The neutrophils were incubated with 500 n*M* gonococcal outer-membrane proteins 1A (P1A) or 1B (P1B), meningococcal outer-membrane proteins class 1 (Class 1), 2 (Class 2), or 3 (Class 3), or 0.001% detergent (Empigen-BB) for 10 min before incubation with preopsonized FITC-labeled meningococci group B (44/76; clinical isolate) at a cell–bacteria ratio of 1:20 for 7.5 min. The number of ingested bacteria per neutrophil was determined by a flow cytometric fluorescence quenching technique (Bjerknes and Bassøe, 1984). The results are given as mean ±95% confidence interval; $n = 6$. ***: $p < .005$; ****: $p < .001$. Dr. L. M. Wetzler, Boston, is thanked for supplying the purified porins.

trophil capacity for actin polymerization was decreased in the presence of neisserial porins (Bjerknes et al., 1995). Reduced actin polymerization may have impaired the neutrophil pseudopod formation during ingestion, further inhibiting the uptake of meningococci. It should be noted that all the neisserial porins examined inhibited neutrophil phagocytosis of meningococci (Fig. 8), and that the inhibition was of similar magnitude irrespective of the content of anti-porin antibodies in the sera used as source of opsonins (Bjerknes et al., 1995). This indicates that the reduced phagocytosis was due to a direct effect on the neutrophils, rather than lack of opsonizing antibodies due to complex formation between free porins and serum anti-porin IgG.

CONGENITAL NEUTROPHIL DEFECTS AFFECTING PHAGOCYTOSIS

Impaired neutrophil phagocytosis may be due to lack of opsonins or neutrophils, as well as cellular defects. Opsonin deficiencies, neutropenias, and acquired deficiencies of neutrophil phagocytosis (e.g., diabetes mellitus, hypophosphatemia, myeloid leukemia, rheumatoid arthritis, systemic lupus erythematosus, and uremia) are not discussed here, and the reader is referred to recent reviews (see reviews by Klebanoff and Clark, 1978; Fletcher et al., 1990; Gallin et al., 1992). The main congenital qualitative neutrophil abnormalities impairing phagocytosis are described briefly because they represent interesting models that to some extent have increased our knowledge of normal phagocyte function. These disease states include LAD type I (lack of CR3), specific granule deficiency, the actin dysfunction syndrome, and the neutrophil dysfunction of newborns.

LAD Type I

This syndrome comprises recurrent or progressive skin, mucous membrane, and subcutaneous infections characterized by decreased pus formation, poor wound healing, delayed separation of the umbilical cord, and persistent granulocytosis (Anderson et al., 1985; Arnaout, 1990; Gallin et al., 1992). These patients lack the β-integrin (CD18), causing defective expression also of the α-subunits (CD11a, CD11b, CD11c), and hence they lack functionally active αβ-molecules. Neutrophils from these patients show impaired adherence to endothelial cells, aggregation, and chemotaxis (Anderson et al., 1985). Lack of CR3 (CD11b) also causes substantial defects in the phagocytosis of C3bi-opsonized particles (Anderson et al., 1985). In addition, the uptake of IgG-coated targets is also reduced in these patients, even though the neutrophil expression of FcγR and the binding of IgG-coated particles attached to the cells are equal to those of normal controls (Gresham et al., 1991). The neutrophil actin assembly is normal in these patients, and reduced ingestive capacity can therefore not be attributed to actin polymerization defects (Southwick et al., 1989). LAD type II, a genetic disorder with lack of sialyl-CD15 on mature neutrophils, is associated with chemotaxis but not with impaired phagocytosis (Etzioni et al., 1992).

Specific Granule Deficiency

These patients have genetic quantitative and functional abnormalities of their specific granula causing increased susceptibility to bacterial and fungal infections of soft tissues not unlike those of LAD type I patients (Gallin, 1985). Neutrophils from these patients exhibit impaired upregulation of CR3 and impaired bacterial killing, and it is possible that this may be partly owing to reduced CR3-dependent phagocytosis (Gallin, 1985).

Neutrophil Actin Dysfunction

Boxer et al. (1974) described a patient with repeated staphylococcal skin infections and visceral infections due to a variety of microorganisms. Neutrophilia was demonstrated, and in vitro the patient's neutrophils showed reduced chemotaxis and phagocytosis of opsonized particles (Boxer et al., 1974). Defective polymerization of actin isolated from neutrophils was demonstrated and suggested to be the underlying cause of the defective neutrophil function. Later studies of the family demonstrated defective actin polymerization (Southwick et al., 1988) and low CR3 expression in two of three family members (Southwick et al., 1989). Reduced phagocytosis in this disorder may thus be due to impaired actin polymerization during pseudopod formation, as well as reduced availability of CR3 receptors.

Neutrophil Dysfunction of the Newborn

Defects in both humoral and cellular aspects of phagocytosis have been described in neonates, although normal results have also been reported (Hill, 1987; Remington and Klein, 1990). Neonates exhibit diminished upregulation and functional activation of CR3 in response to chemoattractants and phorbol esters (Anderson et al., 1987; Bruce et al., 1987). Impaired mobilization of CR3 from intracellular pools, as well as low total content of CR3, is responsible for this defect (Jones et al., 1990; McEvoy et al., 1996). These abnormalities cause impaired ligation of CR3 to ICAM-1 on endothelial cells and reduced transendothelial migration (Anderson et al., 1990). However, the dysregulation of CR3 expression and function may also be responsible for the reduced complement-dependent phagocytosis reported by some authors (Miller, 1979).

Other Neutrophil Defects

Chronic granulomatous disease, glucose-6-phosphatase dehydrogenase deficiency, Chediak–Higashi syndrome, myeloperoxidase deficiency, and Job's syndrome are not clearly associated with impairment of phagocytosis (see reviews by Klebanoff and Clark, 1978; Gallin et al., 1992). In addition, reduced neutrophil phagocytic capacity has not been shown in paroxysmal nocturnal hemoglobinuria, a clonal hematopoietic stem cell disorder in which hematopoietic cells fail to synthesize GPI-anchored proteins and the neutrophils are largely deficient in FcγRIII (Kawakami et al., 1990). Moreover, neutrophil phagocytosis is not compromised in patients with congenital deficiency of FcγRI (Ceuppens et al., 1988).

REFERENCES

Ahearn JM, Fearon DT (1989): Structure and function of the complement receptors CR1 (CD35) and CR2 (CD21). Adv Immunol 46:183–219.

Alexander JW, Wixon D (1970): Neutrophil dysfunction and sepsis in burn injury. Surg Gynecol Obstet 130:431–438.

Altieri DC, Bader R, Mannucci PM, Edgington TS (1988): Oligospecificity of the cellular adhesion receptor MAC-1 encompasses an inducible recognition specificity for fibrinogen. J Cell Biol 107:1893–1900.

Anderson CL, Shen L, Eicher DM, Wewers MD, Gill JK (1990): Phagocytosis mediated by three distinct Fcγ receptor classes on human leukocytes. J Exp Med 171:1333–1345.

Anderson DC, Freeman KLB, Heerdt B, Hughes BJ, Jack RM, Smith CW (1987): Abnormal stimulated adherence of neonatal granulocytes: Impaired induction of surface Mac-1 by chemotactic factors or secretagogues. Blood 70:740–750.

Anderson DC, Rothlein R, Marlin SD, Krater SS, Smith CW (1990): Impaired transendothelial migration by neonatal neutrophils: Abnormalities of Mac-1 (CD11b/CD18)-dependent adherence reactions. Blood 78:2613–2621.

Anderson DC, Schmalsteig FC, Finegold MJ, et al. (1985): The severe and moderate phenotypes of heritable Mac-1, LFA-1 deficiency: Their quantitative definition and relation to leukocyte dysfunction and clinical features. J Infect Dis 152:668–669.

Andersson T, Fällman M, Lew DP, Stendahl O (1988): Does protein kinase C control receptor-mediated phagocytosis in human neutrophils? FEBS Lett 239:371–375.

Arnaout MA (1990): Leukocyte adhesion molecule deficiency: Its structural basis, pathophysiology and implications for modulating the inflammatory response. Immunol Rev 114:145–180.

Arnaout MA, Todd RF III, Dana N, Melamed J, Schlossman SF, Colten HR (1983): Inhibition of phagocytosis of complement C3- or immunoglobulin G-coated particles and of C3bi binding by monoclonal antibodies to a monocyte-granulocyte membrane glycoprotein (Mol). J Clin Invest 72:171–179.

Arnaout MA, Wang EA, Clark SC, Sieff CA (1986): Human granulocyte-macrophage colony-stimulating factor increases cell-to-cell adhesion and surface expression of adhesion-promoting surface glycoproteins on mature granulocytes. J Clin Invest 78:597–601.

Artenstein MS, Gold R, Zimmerly JG, Wyle FA, Schneider H, Harkins C (1970): Prevention of meningococcal disease by group C polysaccharide vaccine. N Engl J Med 282:417–420.

Babior BM (1984): Oxidants from phagocytes: Agents of defense and destruction. Blood 64:959–966.

Bellomo R (1992): The cytokine network in the critically ill. Anaesth Intensive Care 20:288–302.

Beal AL, Cerra FB (1994): Multiple organ failure syndrome in the 1990s: Systemic inflammatory response and organ dysfunction. J Am Med Assoc. 271:226–233.

Berger M, Weizler EM, Wallis RS (1988): Tumor necrosis factor is the major monocyte product that increases complement receptor expression on mature human neutrophils. Blood 71:151–158.

Bevilacqua MP, Amrani D, Mosesson MW, Bianco C (1981): Receptors for cold-insoluble globulin (plasma fibronecin) on human monocytes. J Exp Med 153:42–60.

Bjerknes R (1992a): Analysis of phagocyte-microorganisms interactions. In Lærum OD, Bjerknes R (eds). *Flow Cytometry in Hematology.* London: Academic Press, pp 111–129.

Bjerknes R (1992b): Bacterial and viral infections. In Lærum OD, Bjerknes R (eds). *Flow Cytometry in Hematology.* London: Academic Press, pp 167–181.

Bjerknes R, Aarskog D (1995): Priming of human polymorphonuclear neutrophilic leukocytes by insulin-like growth factor I: Increased phagocytic capacity, complement receptor expression, degranulation and oxidative burst. J Clin Endocrinol Metab 80:1948–1555.

Bjerknes R, Bassøe C-F (1984): Phagocyte C3-mediated attachment and internalization: Flow cytometric studies using a fluorescence quenching technique. Blut 49:315–323.

Bjerknes R, Guttormsen H-K, Solberg CO, Wetzler LM (1995): Neisserial porins inhibit neutrophil actin polymerization, degranulation, opsonin receptor expression, and phagocytosis but prime the neutrophils to increase their oxidative burst. Infect Immun 63:160–167.

Bjerknes R, Laerum OD, Knapp W (1984): Inhibition of phagocytosis by monoclonal antibodies to human myeloid differentiation antigens. Exp Hematol 12:856–862.

Bjerknes R, Vindenes H, Pitkänen J, Ninnemann J, Laerum OD, Åbyholm F (1989): Altered polymorphonuclear neutrophilic granulocyte functions in patients with large burns. J Trauma 29:847–855.

Blake MS (1985): Implications of the active role of gonococcal porins in disease. In Schoolman GK (ed). *The Pathogenic Neisseriae.* Washington DC: The American Society for Microbiology, pp 251–258.

Bobak DA, Gaither TA, Frank MM, Tenner AJ (1987): Modulation of FcR function by complement: Subcomponent C1q enhances the phagocytosis of IgG-opsonized targets by monocytes and culture-derived macrophages. J Immunol 138:1150–1156.

Bohnsack JF, Kleinman HK, Takahashi T, O'Shea JJ, Brown EJ (1985): Connective tissue proteins and phagocyte cell function. Laminin enhances complement and Fc-mediated phagocytosis by cultured human macrophages. J Exp Med 161:912–923.

Boxer LA, Hedley-Whyte ET, Stossel TP (1974): Neutrophil actin dysfunction and abnormal neutrophil behavior. N Engl J Med 291:1093–1099.

Boyles J, Bainton DF (1981): Changes in plasma membrane-associated filaments during endocytosis and exocytosis in polymorphonuclear leukocytes. Cell 24:904–915.

Bredius RG, de Vries CEE, Troelstra A, et al. (1993): Phagocytosis of *Staphylococcus aureus* and *Haemophilus influenzae* type B opsonized with polyclonal human IgG1 and IgG2 antibodies. J Immunol 151:1463–1472.

Brennan PJ, Zigmond S, Schreiber AD, Smith ER, Southwick FS (1991): Binding of IgG containing immune complexes to human neutrophil FcγRII and FcγRIII induces actin polymerization by a pertussis toxin-insensitive transduction pathway. J Immunol 146:4282–4288.

Brooks DG, Qiu WQ, Luster AD, Ravetch JV (1989): Structure and expression of human IgG FcRII (CD32). Functional heterogeneity is encoded by the alternatively spliced products of multiple genes. J Exp Med 170:1369–1385.

Brown EJ (1992): Complement receptors, adhesion, and phagocytosis. Infect Agents Dis 1:63–70.

Brown EJ (1995): Phagocytosis. BioEssays 17:109–117.

Brown EJ, Bohnsack JF, Gresham HD (1988): Mechanism of inhibition of immunoglobulin G-mediated phagocytosis by monoclonal antibodies that recognize the Mac-1 antigen. J Clin Invest 81:365–375.

Bruce MC, Baily JE, Melvik K, Berger M (1987): Impaired surface membrane expression of C3bi, but not C3b, receptors in neonatal neutrophils. Pediatr Res 21:306–311.

Buckle A-M, Hogg N (1989): The effect of IFN-γ and colony-stimulating factors on the expression of neutrophil cell membrane receptors. J Immunol 143:2295–2301.

Bullock WE, Wright SD (1987): Role of the adherence-promoting receptors, CR3, LFA-1, and P150,95, in binding of *Histoplasma capsulatum* by human macrophages. J Exp Med 165:195–210.

Cain JA, Newman SL, Ross GD (1987): Role of complement receptor type three and serum opsonins in the neutrophil response to yeast. Complement 4:75–86.

Capsoni F, Bonara P, Minonzio F, Ongari AM, Colombo G, Rizzardi GP, Zanussi C (1991): The effect of cytokines on human neutrophil Fc receptor-mediated phagocytosis. J Clin Lab Immunol 34:115–124.

Ceuppens JL, Baroja ML, Van Vaeck F, Anderson CL (1988): Defect in the membrane expression of high affinity 72-kD Fcgamma receptors on phagocytic cells in four healthy individuals. J Clin Invest 82: 571–578.

Connelly PA, Farrell CA, Merenda JM, Conkylyn MJ, Showell HJ (1991): Tyrosine phosphorylation is an early signaling event common to Fc receptor crosslinking in human neutrophils and rat basophilic leukemia cells (RBL-2H3). Biochem Biophys Res Commun 177:192–201.

Della Bianca V, Dusi S, Nadalini KA, Donini M, Rossi F (1995): Role of the 55- and 75-kDa TNF receptors in the potentiation of Fc-mediated phagocytosis in human neutrophils. Biochem Biophys Res Commun 214:44–50.

Della Bianca V, Grzeskowiak M, Renzi E, Rossi F (1993): The potentiation by TNF-α and PMA of Fc receptor-mediated phagocytosis in neutrophils is independent of reactive oxygen metabolites produced by NADPH oxidaes and of protein kinase C. Biochem Biophys Res Commun 193:919–926.

Densen P, Mandell GL (1980): Phagocyte strategy vs. microbial tactics. Rev Infect Dis 2:817–838.

Detmers PA, Lo SK, Olsen-Egbert E, Walz A, Baggiolini M, Cohn ZA (1990): NAP1/IL-8 stimulates the binding activity of the leukocyte adhesion receptor CD11b/CD18 on human neutrophils. J Exp Med 171:1155–1162.

Detmers PA, Powell DE, Walz A, Clark-Lewis I, Baggiolini M, Cohn Z (1991): Differential effects of neutrophil activating peptide 1/IL-8 and its homologues on leukocyte adhesion and phagocytosis. J Immunol 147:4211–4217.

Domenico P, Salo RJ, Cross AS, Cuhna BA (1994): Polysaccharide capsule-mediated resistance to opsonophagocytosis in *Klebsiella pneumoniae.* Infect Immun 62:4495–4499.

Dusi S, Donini M, Della Bianca V, Gandini G, Rossi F (1994): In human neutrophils the binding of immunocomplexes induced the tyrosine phosphorylation of FcgammaRII but this phosphorylation is not an essential signal for Fc-mediated phagocytosis. Biochem Biophys Res Commun 201:30–37.

Dykman TR, Hatch JA, Aqua MS, Atkinson JP (1985): Polymorphism of the C3b/C4b receptor (CR1): Characterization of a fourth allele. J Immunol 134:1787–1789.

Edberg J, Kimberley RP (1994): Modulation of Fcγ and complement receptor function by the glycosylphosphatidylinositol-anchored form of FcγRIII. J Immunol 152:5826–5835.

Ehlenberger AG, Nussenzweig V (1977): The role of membrane receptors for C3b and C3d in phagocytosis. J Exp Med 145:357–371.

Elsbach P, Weiss J (1983): A reevaluation of the roles of the O_2-dependent and the O_2-independent microbicidal system of phagocytes. Rev Infect Dis 5:843–853.

Etzioni A, Frydman M, Pollack S, Avidor I, Phillips ML, Paulson JC, Gershoni-Baruch R (1992): Recurrent severe infections caused by a novel leukocyte adhesion deficiency. N Engl J Med 327:1789–1792.

Fällman M, Andersson R, Andersson T (1993): Signaling properties of CR3 (CD11b/CD18) and CR1 (CD35) in relation to phagocytosis of complement-opsonized particles. J Immunol 151:330–338.

Fanger MW, Goldstine SN, Shen L (1983): Cytofluorographic analysis of receptors for IgA on polymorphonuclear cells and monocytes and the correlation of receptor expression with phagocytosis. Mol Immunol 20:1019–1027.

Fearon DT, Collins L (1983): Increased expression of C3b receptors on polymorphonuclear leukocytes induced by chemotactic factors and by purification procedures. J Immunol 130:370–375.

Ferrante A, Beard LJ, Feldman RG (1990): IgG subclass distribution of antibodies to bacterial and viral antigens. Pediatr Infect Dis J 9:S16–S24.

Ferrante A, Martin AJ, Bates EJ, et al. (1993): Killing of *Staphylococcus aureus* by tumor necrosis factor α-activated neutrophils: The role of serum opsonins, integrin receptors, respiratory burst, and degranulation. J Immunol 151:4821–4828.

Ferrante A, Martin AJ, Bates EJ, et al. (1994): Interaction of *Staphylococcus aureus* with human neutrophils and the down-regulation of TNF receptors. J Immunol 152:3998–4004.

Fine DP (1975): Pneumococcal type-associated variability in alternate complement pathway activation. Infect Immun 12:772–778.

Fleischmann J, Golde DW, Weisbart RH, Gasson JC (1986): Granulocyte-macrophage colony-stimulating factor enhances phagocytosis of bacteria by human neutrophils. Blood 68:708–711.

Fletcher J, Haynes AP, Crouch SM (1990): Acquired abnormalities of polymorphonuclear neutrophil function. Blood Rev 4:103–110.

Ford-Hutchinson AW, Bray MA, Doig MV, Shipley ME, Smith MJH (1980): Leukotriene B, a potent chemokinetic and aggregating substance released from polymorphonuclear leukocytes. Nature 286: 264–265.

Forsgren A, Sjöquist J (1966): "Protein A" from *S. aureus*. I. Pseudo-immune reaction with human gamma globulin. J Immunol 97:822–827.

Frøholm LO, Holten E, Poolman JT, Zollinger WD (1985): Typing of Norwegian meningococcal isolates and possible implications for serogroup B vaccination. In Schoolnik GK (ed). *The Pathogenic Neisseriae*. Washington DC: The American Society for Microbiology, pp 541–545.

Gallin JI (1985): Neutrophil specific granule deficiency. Annu Rev Med 36:263–274.

Gallin JI, Goldstein IM, Snyderman R (1992): *Inflammation: Basic Principles and Clinical Correlates*. (2nd ed.) New York: Raven Press.

Garner CV, D'Amico R, Simms HH (1996): Cytokine-mediated human polymorphonuclear neutrophil phagocytosis: Evidence of differential sensitivities to manipulation of intracellular mechanisms. J Surg Res 60:84–90.

Gbarah A, Gahmberg CG, Ofek I, Jacobi U, Sharon N (1991): Identification of leukocyte adhesion molecules CD11 and CD18 as receptors for type 1-fimbriated (mannose specific) *Escherichia coli*. Infect Immun 59:4524–4530.

Gericke GH, Ericson SG, Pan L, Mills LE, Guyre PM, Ely P (1995): Mature polymorphonuclear leukocytes express high-affinity receptors for IgG (FcγRI) after stimulation with granulocyte colony-stimulating factor (G-CSF). J Leukoc Biol 57:455–461.

Goldman DW, Goetzl EJ (1984): Heterogeneity of human polymorphonuclear receptors for leukotriene B_4. Identification of a subset of high affinity receptors that transduce the chemotactic response. J Exp Med 159:1027–1041.

Gordon DL, Johnson GM, Hostetter MK (1986): Ligand-receptor interactions in the phagocytosis of virulent *Streptococcus pneumoniae* by polymorphonuclear leukocytes. J Infect Dis 154:619–626.

Graziano RF, Fanger MW (1987): Fc gamma RI and Fc gamma RII on monocytes and granulocytes are cytotoxic trigger molecules for tumor cells. J Immunol 139:3536–3541.

Gresham HD, Graham IL, Anderson DC, Brown EJ (1991): Leukocyte adhesion-deficient neutrophils fail to amplify phagocytic function in response to stimulation: Evidence for CD11b/CD18-dependent and -independent mechanisms of phagocytosis. J Clin Invest 88:588–597.

Gresham HD, Zheleznyak A, Mormol JS, Brown EJ (1990): Studies on the molecular mechanisms of human neutrophil Fc receptor-mediated phagocytosis: Evidence that a distinct pathway for activation of the respiratory burst results in reactive oxygen metabolite-dependent amplification of ingestion. J Biol Chem 265:7819–7826.

Griffin FM Jr, Griffin JA, Leider JE, Silverstein SC (1975): Studies on the mechanism of phagocytosis. I. Requirements for circumferential attachment of particle-bound ligands to specific receptors on the macrophage plasma membrane. J Exp Med 142:1263–1282.

Guttormsen H-K, Bjerknes R, Halstensen A, Næss A, Høiby EA, Solberg CO (1993): Cross-reacting serum opsonins to meningococci after vaccination. J Infect Dis 167:1314–1319.

Guttormsen H-K, Bjerknes R, Næss A, et al. (1992): Cross-reacting serum opsonins in patients with meningococcal disease. Infect Immun 60:2777–2783.

Guttormsen H-K, Wetzler LM, Solberg CO (1996): Serum opsonins induced during the course of meningococcal disease correlate with anti-outer membrane protein antibodies. Submitted for publication.

Guyre PM, Campbell AS, Kniffin WD, Fanger MW (1990): Monocytes and polymorphonuclear neutrophils of patients with streptococcal pharyngitis express increased numbers of type IIgG Fc receptors. J Clin Invest 86:1892–1896.

Haines KA, Yeh L, Blake MS, Cristello P, Korchack H, Weissmann G (1988): Protein I, a translocatable ion channel from *Neisseriae gonorrhoeae,* selectively inhibits exocytosis from human neutrophils without inhibiting O_2^- generation. J Biol Chem 263:945–951.

Harlan JM, Liu DY (1992): *Adhesion: Its Role in Inflammatory Diseases.* New York: WH Freeman and Company.

Hill HR (1987): Biochemical, structural, and functional abnormalities of polymorphonuclear leukocyte in the neonate. Pediatr Res 22:375–382.

Horwitz MA, Silverstein SC (1980): Influence of the *Escherichia coli* capsule on complement fixation and on phagocytosis and killing by human phagocytes. J Clin Invest 65:82–94.

Hostoffer RW, Krukovets I, Berger M (1993): Increased FcαR expression and IgA-mediated function on neutrophils induced by chemoattractants. J Immunol 150:4532–4540.

Hostoffer RW, Krukovets I, Berger M (1994): Enhancement by tumor necrosis factor-α of a Fcα receptor expression and IgA-mediated superoxide generation and killing of *Pseudomonas aeruginosa* by polymorphonuclear leukocytes. J Infect Dis 170:82–87.

Huizinga TWJ, van der Schoot CE, Jost CE, et al. (1988): PI-linked receptor FcRIII is released on activation of neutropohils. Nature 333:667–669.

Huizinga TWJ, van Kemenade F, Koenderman L, et al. (1989): The 40-kDa receptor (FcRII) on human neutrophils is essential for the IgG-induced respiratory burst and IgG-induced phagocytosis. J Immunol 142:2365–2369.

Hurst JK, Barrette WC Jr (1989): Leukocytic oxygen activation and microbicidal oxidative toxins. Crit Rev Biochem Mol Biol 24:271–328.

Jefferis R, Kumararatne DS (1990): Selective IgG subclass deficiency: Quantification and clinical relevance. Clin Exp Immunol 81:357–367.

Jones DH, Schmalstieg FC, Dempsey K, et al. (1990): Subcellular distribution and mobilization of Mac-1 (CD11b/CD18) in neonatal neutrophils. Blood 75:488–498.

Kawakami Z, Ninomiya H, Tomiyama J, Abe T (1990): Deficiency of glycosyl-phosphatidylinositol anchored proteins on paroxysmal nocturnal hemoglobinuria (PNH) neutrophils and monocytes: Heterogeneous deficiency of decay-accelerating factor (DAF) and CD16 on PNH neutrophils. Br J Haematol 74:508–513.

Kindzelskii AL, Xue W, Todd RF III, Petty HR (1994): Imaging the spatial distribution of membrane receptors during neutrophil phagocytosis. J Struct Biol 113:191–198.

Klebanoff SJ, Beatty PG, Schreiber RD, Ochs HD, Waltersdorph AM (1985): Effect of antibodies directed against complement receptors on phagocytosis by polymorphonuclear leukocytes: Use of iodination as a convenient measure of phagocytosis. J Immunol 134:1153–1159.

Klebanoff SJ, Clark RA (1978): *The Neutrophil: Function and Clinical Disorders.* Amsterdam: North-Holland Publishing Company.

Klebanoff SJ, Vadas MA, Harlan JM, et al. (1986): Stimulation of neutrophils by tumor necrosis factor. J Immunol 136:4220–4225.

Kletter Y, Bleiberg I, Golde DW, Fabian I (1989): Antibody to Mol abrogates the increase in neutrophil phagocytosis and degranulation induced by granulocyte-macropahe colony-stimulating factor. Eur J Haematol 43:389–396.

Kozel TR (1993): Opsonization and phagocytosis of *Cryptococcus neoformans.* Arch Med Res 24:211–218.

Lehrer RI, Ganz T (1990): Antimicrobial polypeptides of human neutrophils. Blood 76:2169–2181.

Lew DP (1990): Receptor and intracellular signaling in human neutrophils. Am Rev Respir Dis 141:S127–S131.

Lew DP, Andersson T, Hed J, Di Virgilio F, Pozzan T, Stendahl O (1985): Ca^{2+}-dependent and Ca^{2+}-independent phagocytosis in human neutrophils. Nature 315:509–511.

Lo SK, Detmers PA, Levin SM, Wright SD (1989): Transient adhesion of neutrophils to endothelium. J Exp Med 169:1779–1793.

Lund-Johansen F, Bjerknes R, Laerum OD (1990): Activation of myeloid cells: altered expression of surface differentiation antigens during phagocytosis. Cytometry 11:(suppl 4) 95.

Lund-Johansen F, Olweus J, Aarli Å, Bjerknes R (1991): The IgG RII and the PI-linked IgG FcRIII trigger cytoplasmic calcium fluxes independently in human granulocytes. Scand J Immunol 33:261–266.

Lynch EC, Blake MS, Gotschlich EC, Mauro A (1984): Studies of porins: Spontaneously transferred from whole cells and reconstituted from purified proteins from *Neisseria gonorrhoeae* and *Neisseria meningitidis.* Biophys J 45:104–107.

Macleod CM, Krauss MR (1950): Relation of virulence of pneumococcal strains for mice to the quantity of capsular polysaccharide formed in vitro. J Exp Med 92:1–9.

Mandell GL, Frank MO (1992): Microbial defenses against killing by phagocytes. Trans Am Clin Climatol Assoc 103:199–209.

Mandrell RE, Zollinger WD (1986): Measurement of antibodies to meningococcal group B polysaccharide: Low avidity binding and equilibrium binding constants. J Immunol 129:2172–2178.

McEvoy LT, Zakem-Cloud H, Tosi MF (1996): Total cell content of CR3 (CD11b/CD18) and LFA-1 (CD11a/CD18) in neonatal neutrophils: Relationship to gestational age. Blood 87:3929–3933.

Medearis DN Jr, Camitta BM, Heath EC (1968): Cell wall composition and the virulence of *Escherichia coli.* J Exp Med 128:399–414.

Medof ME, lida K, Mold C, Nussenzweig V (1982): Unique role of the complement receptor CR1 in the degradation of C3b associated immune complexes. J Exp Med 156:1739–1754.

Metzger H, Kinet JP (1988): How antibodies work: Focus on Fc receptors. FASEB J 2:3–11.

Miller ME (1979): Phagocyte function in the neonate: Selected aspects. Pediatrics 64:709–712.

Mortensen RF, Osmand AP, Lint TF, Gewurz H (1976): Interaction of C-reactive protein with lymphocytes and monocytes: Complement-dependent adherence and phagocytosis. J Immunol 117:774–781.

Moxey-Mims MM, Simms HH, Frank MM, Lin EY, Gaither TA (1991): The effects of IL-1, IL-2, and tumor necrosis factor on polymorphonuclear leukocyte Fcγ receptor-mediated phagocytosis. J Immunol 147:1823–1830.

Mullen PG, Windsor ACJ, Walsch CJ, Fowler III AA, Sugerman HJ (1995): Tumor necrosis factor-α and interleukin-6 selectively regulate neutrophil function in vitro. J Surg Res 58:124–130.

Murata T, Sullivan JA, Sawyer DW, Mandell GL (1987): Influence of type and opsonization of ingested particle on intracellular free calcium distribution and superoxide production by human neutrophils. Infect Immun 55:1784–1791.

Naziruddin B, Duffy BF, Tucker J, Mohanakumar T (1992): Evidence for cross-regulation of FcγRIIIB (CD16) receptor-mediated signaling by FcγRII (CD32) expressed on polymorphonuclear neutrophils. J Immunol 149:3702–3709.

Neuman E, Huleatt JW, Jack RM (1990): Granulocyte-macrophage colony-stimulating factor increases synthesis and expression of CR1 and CR3 by peripheral blood neutrophils. J Immunol 145:3325–3332.

Nikaido H (1992): Porins and specific channels of bacterial outer membranes. Mol Microbiol 6:435–442.

Ogle JD, Noel JG, Sramkoski RM, Ogle CK, Alexander JW (1990): The effects of cytokines, platelet activating factor, and arachidonate metabolites on C3b receptor (CR1, CD35) expression and phagocytosis by neutrophils. Cytokine 2:447–455.

Ohta Y, Stossel TP, Hartwig JH (1991): Ligand-sensitive binding of actin-binding protein to immunoglobulin G Fc receptor I (FcγRI). Cell 67:275–282.

Ory PA, Clark MR, Kwoh EE, Clarkson SB, Goldstein IM (1989a): Sequences of complementary DNAs that encode the NA1 and NA2 forms of Fc receptor III on human neutrophils. J Clin Invest 84:1688–1691.

Ory PA, Goldstein IM, Kwoh EE, Clarkson SB (1989b): Characterization of polymorphic forms of Fc receptor III on human neutrophils. J Clin Invest 83:1676–1681.

O'Shea JJ, Brown EJ, Seligmann BE, Metcalf JA, Frank MM, Gallin JI (1985): Evidence for distinct intracellular pools of receptors for C3b and C3bi in human neutrophils. J Immunol 134:2580–2587.

Ozaki Y, Ohashi T, Kume S (1987): Potentiation of neutrophil function by recombinant DNA-produced interleukin-1-α. J Leukoc Biol 621–627.

Perussia B, Dayton ET, Lazarus R, Fanning V, Trinchieri G (1983): Immune interferon induces the receptor for monomeric IgG1 on human monocytic and myeloid cells. J Exp Med 158:1092–1113.

Perussia B, Kobayashi M, Rossi ME, Anegon I, Trinchieri G (1987): Immune interferon enhances functional properties of human granulocytes: Role of Fc receptors and effect of lymphotoxin, tumor necrosis factor, and granulocyte-macrophage colony-stimulating factor. J Immunol 138:765–774.

Peterson PK, Verhoef J, Sabath LD, Quie PG (1976): Extracellular and bacterial factors influencing staphylococcal phagocytosis and killing by human polymorphonuclear leukocytes. Infect Immun 14:496–501.

Petroni KC, Shen L, Guyre PM (1988): Modulation of human polymorphonuclear leukocyte IgG Fc receptors and Fc receptor-mediated functions by IFN-γ and glucocorticoids. J Immunol 140:3467–3472.

Petty HR, Todd RF III (1993): Receptor-receptor interactions of complement receptor type 3 in neutrophil membranes. J Leukoc Biol 54:492–494.

Ravetch JV, Perussia B (1989): Alternative membrane forms of FcgammaRIII (CD16) on human natural killer cells and neutrophils. Cell type-specific expression of two genes that differ in single nucleotide substitutions. J Exp Med 170:481–497.

Reed WP (1975): Serum factor capable of opsonizing *Shigella* for phagocytosis by polymorphonuclear neutrophils. Immunology 28:1051–1059.

Relman D, Toumanen E, Falkow S, Golenbock DT, Saukkonen K, Wright SD (1990): Recognition of a bacterial adhesin by an integrin: Macrophage CR3 ($\alpha_M\beta_2$, CD11b/CD18) binds filamentous hemagglutinin of Bordetella pertussis. Cell 61:1375–1382.

Remington JS, Klein JO (1990): *Infectious Diseases of the Fetus and the Newborn.* 3rd ed. Philadelphia: WB Saunders Company.

Repp R, Valerius T, Sendler A, et al. (1991): Neutrophils express the high affinity receptor for IgG (FcγRI, CD64) after in vivo application of recombinant human granulocyte colony-stimulating factor. Blood 78:885–889.

Roberts PJ, Devereux S, Pilkington GR, Linch DC (1990): Fc gamma RII-mediated superoxide production of phagocytes is augmented by GM-CSF without a change in Fc gamma RII expression. J Leukoc Biol 48:247–257.

Robinson JH, Kehoe MA (1992): Group A streptococcal M proteins: Virulena factors and protective antigens. Immunol Today 13:362–367.

Ross GD, Cain JA, Lachmann PJ (1985): Membrane complement receptor type three (CR3) has lectin-like properties analogous to bovine conglutinin and functions as a receptor for zymosan and rabbit erythrocytes as well as a receptor for iC3b. J Immunol 134:3307–3315.

Ross GD, Větvička V (1993): CD3 (CD11b,CD18): A phagocyte and NK cell membrane receptor with multiple ligand specificities and functions. Clin Exp Immunol 92:181–184.

Ross SC, Rosenthal PJ, Berberich HM, Densen P (1987): Killing of *Neisseria meningitidis* by human neutrophils: Implications for normal and complement-deficient individuals. J Infect Dis 155:1266–1275.

Rouslathy E (1991): Integrins. J Clin Invest 87:1–5.

Salmon JE, Brogle NL, Edberg JC, Kimberley RP (1991): Fcγ receptor III induces actin polymerization in neutrophils and primes phagocytosis mediated by Fcγ receptor II. J Immunol 146:997–1004.

Salmon JE, Edberg JC, Brogle NL, Kimberley RP (1992): Allelic polymorphisms of human Fcγ receptor IIa and Fcγ receptor IIIb. Independent mechanisms for differences in human phagocyte function. J Clin Invest 89:1274–1281.

Salmon JE, Edberg JC, Kimberley RP, Mensa E, Ryan R (1990): Fcγ receptor III on human neutrophils. Allelic variants have functionally distinct capacities. J Clin Invest 85:1287–1295.

Salmon JE, Kapur S, Kimberley RP (1987): Opsonin-independent ligation of Fcγ receptors. The 3G8-bearing receptors on neutrophils mediate the phagocytosis of concanavalin A-treated erythrocytes and nonopsonized *Escherichia coli*. J Exp Med 166:1798–1813.

Salmon JE, Millard SS, Brogle NL, Kimberley RP (1995): Fcγ receptor IIIb enhances Fcγ receptor IIa function in an oxidant-dependent and allele-sensitive manner. J Clin Invest 95:2877–2885.

Samuelsson B (1983): Leukotrienes: Mediators of immediate hypersensitivity reactions and inflammation. Science 220:568–575.

Sanchez-Madrid F, Nagy JA, Robbins E, Simon P, Springer TA (1983): Characterization of a human leukocyte differentiation antigen family with distinct α subunits and a common β subunit: The lymphocyte-function associated antigen (LFA-1), the C3bi complement receptor (OKM1/Mac1), and the p150,95 molecule. J Exp Med 158:1785–1803.

Sanders LAM, Feldman RG, Voorhorst-Ogink MM, et al. (1995): Human immunoglobulins G (IgG) Fc receptor IIA (CD32) polymorphism and IgG2-mediated bacterial phagocytosis by neutrophils. Infect Immun 63:73–81.

Sanders LAM, van de Winkel JGJ, Rijkers GT, et al. (1994): Fcγ receptor IIa (CD32) heterogeneity in patients with recurrent bacterial respiratory tract infections. J Infect Dis 170:854–861.

Sawyer WD (1969): Interaction of influenza virus with leukocytes and its effect upon phagocytosis. J Infect Dis 119:541–556.

Shafer WM, Rest RF (1989): Interactions of gonococci with phagocytic cells. Annu Rev Microbiol 43:121–145.

Shalaby MR, Aggarwal BB, Rinderknecht E, Svedersky LP, Finkle BS, Palladino MA Jr (1985): Activation of human polymorphonuclear neutrophil functions by interferon-γ and tumor necrosis factors. J Immunol 134:2069–2073.

Shalit M, von Allmen C, Atkins PC, Zweiman B (1988): Platelet activating factor increases expression of complement receptors on human neutrophils. J Leukoc Biol 44:212–217.

Silverstein SC, Steinman RM, Cohn ZA (1977): Endocytosis. Annu Rev Biochem 46:669–722.

Simberkoff MS, Elsbach P (1971): The interaction in vitro between polymorphonuclear leukocytes and mycoplasma. J Exp Med 134:1417–1430.

Simms HH, Frank MM, Quinn TC, Holland S, Gaither TA (1989): Studies on phagocytosis in patients with acute bacterial infections. J Clin Invest 83:252–260.

Sjursen H, Bjerknes R, Halstensen A, Næss A, Frøholm LO, Rosenqvist E, Solberg CO (1987): Serum opsonins to group B meningococci. APMIS (C) 95:283–289.

Sjursen H, Bjerknes R, Halstensen A, Næss A, Sørnes S, Solberg CO (1989): Flow cytometric assay for the measurement of serum opsonins to *Neisseria meningitidis* serogroup B, serotype 15. J Immunol Meth 116:235–243.

Skubitz KM, Kinkead TK (1987): Changes in neutrophil surface proteins composition accompany phagocytosis. Blood 70:60–68.

Smith CL, Baker CJ, Anderson DC, Edwards MS (1990): Role of complement receptors in opsonophagocytosis of group B streptococci by adult and neonatal neutrophils. J Infect Dis 162:489–495.

Southwick FS, Dabiri GA, Stossel TP (1988): Neutrophil actin dysfunction (NAD) is a genetic disorder associated with partial impairment of neutrophil actin assembly in three family members. J Clin Invest 82:1525–1531.

Southwick FS, Howard TH, Holbrook T, Anderson DC, Stossel TP, Arnaout MA (1989): The relationship between CR3 deficiency and neutrophil actin assembly. Blood 73:1973–1979.

Spitznagel JK (1983): Microbial interactions with neutrophils. Rev Infect Dis 5:S806–S822.

Stöckl J, Majdic O, Pickl WF, et al. (1995): Granulocyte activation via a binding site near the C-terminal region of complement receptor 3 α-chain (CD11b) potentially involved in intramembrane complex formation with glycosylphosphatidylinositol-anchored FcγRIIIB (CD16) molecules. J Immunol 154: 5452–5463.

Stossel T (1973): Evaluation of opsonic and leukocyte function with a spectrophotometric test in patients with infection and with phagocytic disorders. Blood 42:121–130.

Stossel T (1990): Actin-membrane interactions in eucaryotic mammalian cells. Curr Top Membr Transport 36:97–107.

Thongthai C, Sawyer WD (1973): Studies on the virulence of *Neisseriae gonorrhoeae*. I. Relation of colonial morphology and resistance to phagocytosis by polymorphonuclear leukocytes. Infect Immun 7:373–379.

Tufano MA, Rossano F, Catalanotti P, et al. (1994): Properties of *Yersinia enterocolitica* porins: Interference with biological functions of phagocytes, nitric oxide production and selective cytokine release. Res Microbiol 145:297–307.

van de Winkel JGJ, Anderson CL (1991): Biology of human immunoglobulin G Fc receptors. J Leukoc Biol 49:511–524.

van de Winkel JGJ, Capel PJA (1993): Human IgG Fc receptor heterogeneity: molecular aspects and clinical implications. Immunol Today 14:215–221.

Vanholder R, Ringoir S (1993): Infectious morbidity and defects of phagocyte function in end-stage renal disease: A review. J Am Soc Nephrol 3:1541–1554.

van Oss CJ (1978): Phagocytosis as a surface phenomenon. Ann Rev Microbiol 32:19–39.

Verbaugh HA, von Dijk WC, van Erne ME, Peters R, Peterson PK, Verhoef J (1979): Quantitation of the thick component of human complement attached to the surface of opsonized bacteria: Opsonin-deficient sera and phagocytosis-resistant strains. Infect Immun 26:808–814.

Verhoef J, Peterson PK, Kim Y, Sabath LD, Quie PG (1977): Opsonic requirements for staphylococcal phagocytosis: Heterogeneity among strains. Immunology 33:191–197.

Vindenes H, Ulvestad E, Bjerknes R (1995): Increased levels of circulating interleukin-8 in patients with large burns: Relation to burn size and sepsis. J Trauma 39:635–640.

Vindenes H, Ulvestad E, Bjerknes R (1997): Concentrations of cytokines in patients with large burns. Their relationship to time after surgery, burn size, inflammatory variables, infection, and outcome. Eur J Surg, in press.

Wahdan MH, Rizk F, El-Akkad AM, et al. (1973): A controlled field trial of a serogroup A meningococcal polysaccharide vaccine. Bull WHO 48:667–673.

Warmerdam PAM, van de Winkel JGJ, Gosselin EJ, Capel PJA (1990): Molecular basis for a polymorphism of human Fcγ receptor II (CD32). J Exp Med 172:19–25.

Warmerdam PAM, van de Winkel JGJ, Vlug A, Westerdaal NAC, Capel PJA (1991): A single amino acid in the second Ig-like domain of the human Fcγ receptor II is critical for human IgG2 binding. J Immunol 147:1338–1343.

Weisbart RH, Kacena A, Schuh A, Golde DW (1988): IgA-mediated phagocytosis by an IgA Fc receptor activation mechanism. Nature 332:647–648.

Weiss SJ (1989): Tissue destruction by neutrophils. N Engl J Med 320:365–376.

Whintrobe MM (1981): *Clinical Hematology.* (8th ed.) Philadelphia: Lea & Fibiger.

Wilson ME, Pearson RD (1988): Roles of CR3 and mannose receptors in the attachment and ingestion of *Leishmania donovani* by human mononuclear phagocytes. Infect Immun 56:363–369.

Wright SD, Jong MT (1986): Adhesion-promoting receptors on human macrophages recognize Escherichia coli by binding to lipopolysaccharide. J Exp Med 164:1876–1888.

Wright SD, Meyer BC (1986): Phorbol esters cause sequential activation and deactivation of complement receptors on polymorphonuclear leukocytes. J Immunol 136:1759–1764.

Wright SD, Ramos RA, Hermanowski-Vosatha A, Rockwell P, Detmers PA (1991): Activation of the adhesion capacity of CR3 by endotoxin: Dependence on lipopolysaccharide binding protein and CD14. J Exp Med 173:1281–1286.

Wright SD, Silverstein SC (1983): Receptors for C3b and C3bi promote phagocytosis but not the release of toxic oxygen from human phagocytes. J Exp Med 158:2016–2023.

Wright SD, Tobias PS, Ulevitch RJ, Ramos RA (1989): Lipopolysaccharide binding protein opsonizes LPS-bearing particles for recognition by a novel receptor on macrophages. J Exp Med 170:1231–1241.

Yang KD, Augustine NH, Gonzales LA, Bohnsack JF, Hill HR (1988): Effects of fibronectin on the interaction of polymorphonuclear leukocytes with unopsonized and antibody-opsonized bacteria. J Infect Dis 158:823–830.

Zhou M, Brown EJ (1994): CR3 (Mac-1, $\alpha_M\beta_2$, CD11b/CD18) and FcγRIII cooperate in generation of a neutrophil respiratory burst: Requirement for FcγRII and tyrosine phosphorylation. J Cell Biol 125: 1407–1416.

Zhou M, Todd RF III, van de Winkel JGJ, Petty HR (1993): Cocapping of the leukoadhesin molecules complement receptor type 3 and lymphocyte function-associated antigen-1 with Fcγ receptor III on human neutrophils. J Immunol 150:3030–3041.

Zucker-Franklin D, Davidson M, Thomas L (1966): The interaction of mycoplasmas with mammalian cells. I. HeLa cells, neutrophils and eosinophils. J Exp Med 124:521–532.

9

Oxygen and Nitrogen Reactive Metabolites and Phagocytic Cells

J. Paul Robinson
Purdue University, West Lafayette, Indiana

The goal of this chapter is to clarify the role of phagocytic cells (mainly neutrophils) as mediators of both protection and destruction. Specifically, the linkage between the oxidative metabolic pathways and protection against bacterial invasion is contrasted with the subsequent tissue damage that ensues when these processes are not properly regulated. Historically, tissue injury, microbial killing, and many pathogenic mechanisms have been ascribed to reactive oxygen intermediates (ROI). However, with the realization of the importance of reactive nitrogen intermediates (RNI), attention has shifted somewhat from ROI. This recent understanding of the role of nitrogen radicals such as nitric oxide and peroxinitrite is compared to the oxygen radical pathways.

Following the discovery of antioxidants such as superoxide dismutase (SOD) by McCord and Fridovich in 1969 (McCord and Fridovich, 1969b), most studies published in neutrophil physiology have been related to the destructive nature of superoxide (O_2^-), hydrogen peroxide (H_2O_2), or hydroxyl radical ($OH\cdot$) with the reasonable understanding that these reactive molecules were also the causative agents in tissue damage involving phagocytic cells. Neutrophils and macrophages have a tremendous capacity for production of these molecules.

Sixty million neutrophils per minute are released into the circulation through the

Phagocyte Function: A Guide for Research and Clinical Evaluation, Edited by J. Paul Robinson and George F. Babcock. ISBN 0-471-12364-1

normal surveillance mechanism of the reticuloendothelial system. Only erythrocytes (RBC) are produced more prolifically in the body, about 1.5 times the rate of the neutrophil (Erslev and Weiss, 1977). Thanks to their longer life-span, however, RBC outnumber neutrophils in the peripheral circulation by a factor of 10^3. Neutrophils are replaced at such a high rate because of their very short half life (several hours) and total life-span (3–4 days). A number of important factors affect the final disposition of these neutrophils, one being a substantial increase—as much as six-to eightfold—in phagocyte production due to stress (Boggs, 1967; Robinson and Mangalik, 1975), and consequently a significant number of immature neutrophils may be present in the circulation. Because some cytokines can delay apoptosis and thereby increase the life-span of neutrophils, vast quantities of these cells can accumulate at inflammatory sites. The mechanism for removal of neutrophils from inflammatory sites requires the neutrophils to progress toward an apoptotic demise rather than becoming necrotic and releasing vast quantities of granule enzymes into surrounding tissuc. The potential for tissue damage is enormous, particularly if the respiratory mechanism of these cells has been activated. A delicate balance between manufacture of reactive oxygen radicals and their removal or detoxification must be maintained. Understanding the nature of this balance mechanism is the key to discerning the difference between protection and destruction in phagocytic cell function.

In this chapter the pathways of oxygen metabolism in phagocytic cells are unraveled, as are those of the parallel nitrogen metabolism, and an attempt is made to show the relationship between protective and destructive mechanisms. Of particular importance is our current understanding of the role of peroxynitrite ($ONOO^-$), a molecule formed by the union of superoxide and nitric oxide.

OXYGEN-RELATED METABOLISM

A key element in the production of ROI by phagocytes is the NADPH oxidase enzyme system, which was originally described in neutrophils (Babior et al., 1973). This enzyme system is known to consist of several components based upon the b-558 cytochrome, a heterodimer consisting of two subunits, gp22-*phox* (α unit) and gp91-*phox* (β unit), as well as a heme moiety and a flavin binding site. It is these α and β units that are missing in neutrophils of most patients with chronic granulomatous disease, in which the neutrophils fail to activate the respiratory burst. Additionally, several other crucial components of the oxidase system must be combined to make the enzyme system active. These include the cytosolic components known as gp47-*phox* (a 47 kDa protein) and gp67-*phox* (a 67 kDa protein), Rac-2 (a GTP-binding protein), and an NADPH-binding protein (Curnutte et al., 1989; Parkos et al., 1987). Phagocytic cell oxidation pathways are thought to be quite different from those traditionally associated with mitochondrial respiration, thus the terminology "phox" for oxidase-related proteins—<u>ph</u>agocyte <u>ox</u>idase, implying that they are somewhat unique to phagocytic cells.

In contrast to mitochondrial respiration, where the entire pathway and all components for respiration are contained within the mitochondria, the phagocyte oxidase

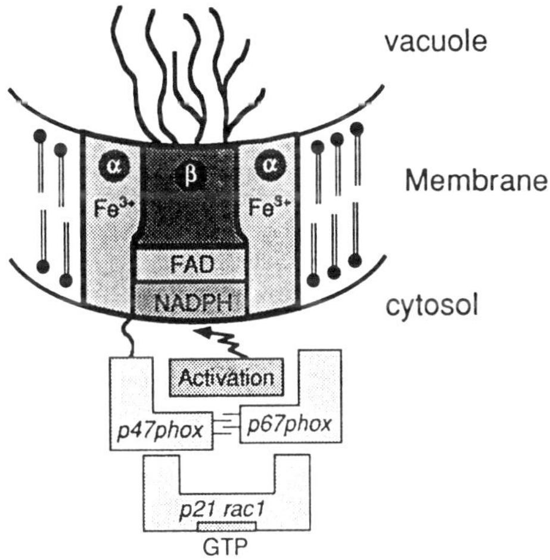

Fig. 1. Schematic representation of NADPH oxidase (after Segal and Nugent). One possible configuration which shows the association of the α and β subunits and the relationship to cytosolic p47-*phox* and p65-*phox* and p21*rac*1. Reprinted from Biological Oxidants: Generation and Injurious Consequences, C. G. Cochrane & M. A. Gimbrone (eds.), 1992.

system must be coupled by transporting cytosolic proteins to the plasma membrane where they are assembled into an operating complex. Figure 1 provides an overview of one possible model of the structure and function of the neutrophil oxidase complex. A series of events takes place, the exact order of which is not known. It is thought that the gp47-*phox* is first phosphorylated via PKC before being integrated into the membrane complex and further phosphorylated (Rotrosen and Leto, 1990). The exact function of these two *phox* proteins is uncertain, as is their location. However, both contain two SH3 domains, suggesting that one or both proteins may be linked to the cytoskeleton rather than to the cytoplasm (Woodman et al., 1991).

Understanding the structure of the oxidase is a key to understanding the respiratory burst in phagocytic cells. As noted above, the cytochrome in phagocytic cells is different from that in other mammalian cells. Firstly, the midpoint redox potential is very low at −245 mV (thus the alternative b_{-245} nomenclature), allowing it to reduce molecular oxygen directly to superoxide. Because the wavelength of the α band of light absorption is at 558 nm, the cytochrome is also referred to as cytochrome b-558 (Wood, 1987; Cross et al., 1981). The greatest concentration of the oxidase system is in the plasma membrane, but in the membranes of specific granules a significant amount is also present and is capable of being transferred to the membrane of the phagocytic vacuole upon fusion, subsequent to activation of the cell.

Is the oxidase system of phagocytes unique to cells such as neutrophils and monocyte/macrophages? Because O_2^- is formed during normal mitochondrial oxidation, clearly all mammalian cells contain the SOD enzyme system necessary to protect the cell from damage and also from external sources of oxidants. Some lines of evidence now suggest that other cell types may contain either the identical phagocyte NADPH oxidase system or components thereof. In most cases the activity is severely reduced

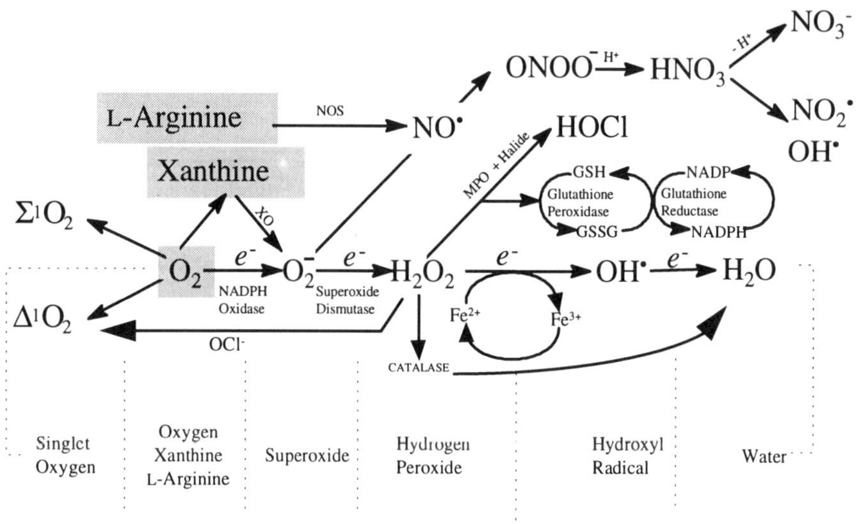

Fig. 2. Oxygen is reduced via the NADPH oxidase flavocytochrome complex to superoxide. Superoxide (O_2^-) is dismutated to hydrogen peroxide (H_2O_2) either by superoxide dismutase or spontaneously. Hydrogen peroxide is further reduced to hydroxyl radical via the iron-catalyzed Haber-Weiss reaction, in which ferrous iron (Fe^{2+}) reduces H_2O_2 to form OH· and OH^-(hydroxide ion). Superoxide may then reduce the Fe^{3+} formed back to Fe^{2+}. H_2O_2 is reduced to water by the catalytic action of catalase, or in the presence of a halide (e.g., Cl^-, Br^-) to the corresponding hypohalous acid by myeloperoxidase catalysis. Oxygen can also be reduced to superoxide in the presence of cells containing the xanthine oxidase system (e.g., endothelial cells). Hydroxyl radical is formed via the Fenton reaction, for which reasonably large amounts of Fe^{2+} must be present (hemoglobin or microorganisms). A more likely pathway in phagocytes is the Haber–Weiss reaction, a "special-case" Fenton reaction in which superoxide anion drives the reaction to produce OH· and OH^-. Removal of H_2O_2 may also be achieved via the glutathione peroxidase–glutathione reductase enzyme system. Of particular interest recently has been the interaction between O_2^- and NO· to form peroxynitrite ($ONOO^-$) and other reactive nitrogen intermediates such as nitrogen dioxide and ultimately nitrite (NO_2^-) or nitrate (NO_3^-).

in comparison with neutrophils. For example, SOD-inhibitable superoxide has been measured in B lymphocytes, although the amount of O_2^- produced was only 5% of that from phorbol myristate acetate (PMA)-stimulated neutrophils (Maly et al., 1989). Similarly, components of the oxidase and measurable O_2^- production have been observed in human fibroblasts (1% of neutrophil production) (Meier et al., 1991), kidney mesangial cells (Radeke et al., 1991; Radeke et al., 1990), endothelial cells (Carter et al., 1994b; Terada et al., 1991; Kontos, 1989; Matsubara and Ziff, 1986), and canine natural killer cells (Carter et al., 1994a). In most of the above cases, the quantities of oxidant produced were extremely small in comparison to that produced by activated neutrophils. Additionally, the role that oxidant-producing mechanisms may play in some of these cells is uncertain.

The overall reaction sequence for oxidative metabolism in phagocytic cells can also involve substrates or reactants produced by associated cells such as macrophages, platelets, and endothelial cells. Figure 2 provides a summary of the most significant reactions within or around phagocytes. In brief, activation of the oxidase is the trigger for the respiratory burst. Although the exact role for each oxidase component is still in doubt, there is clear evidence that the process involves translocation of at least some components from the cytosol to the membrane. Concomitant with this translocation is the activation of several second messenger signaling pathways, including Ca^{2+}, diacylglycerol (DAG), GTP-binding proteins, and tyrosine kinase.

Electronically Excited Species of Oxygen

Singlet Oxygen. Two forms of singlet oxygen are known: Σ^1O_2, with a higher energy but short lifetime, and Δ^1O_2, with a lower energy and a lifetime of 2 μs (Kearns, 1979). For many years the phenomenon of chemiluminescence associated with phagocytes has been ascribed to the production of singlet oxygen, ever since the original observation in phagocytes by Allen (Allen et al., 1972). One of the principal mechanisms for the formation of Δ^1O_2 is thought to be via the interaction of HOCl with O_2^- (Long and Bielski, 1980) or via the myeloperoxidase-catalyzed oxidation of chloride in the presence of excess H_2O_2 (Harrison and Schultz, 1976). Although there are a number of possible mechanisms for reactions to form singlet oxygen, the evidence is still inconclusive that chemiluminescence in neutrophils actually represents singlet oxygen production.

Oxygen Radical Formation

Superoxide. Activated neutrophils are capable of making large quantities of superoxide. For instance, within 1 minute of stimulation with formyl-methionyl-leucyl-phenylalanine (fMLP) in a 1–2 ml l volume, 2×10^6 neutrophils were shown to produce 10 n*M* O_2^-, equivalent to 5–10 m*M* in the absence of dismutation (Schraufstätter and Jackson, 1992). Superoxide either rapidly dismutates to H_2O_2 (via SOD), combines with NO· to form peroxynitrite ($ONOO^-$) (see below), or is protonated ($HO_2\cdot$) in a reaction that is equally probable as SOD-catalyzed dismutation at pH 4.8 (Behar et al., 1970), but less likely at either lower or alkaline pH. Superoxide is formed as a result of respiration, since from 1 to 4% of oxygen used in the mitochondrial electron transport pathway results in the production of superoxide (Cross, 1987) (accounting for the wide distribution of SOD).

Hydrogen Peroxide. Hydrogen peroxide is generated within neutrophils primarily from superoxide dismutase-catalyzed dismutation of O_2^- anions, plus some spontaneous dismutation. The general reactions are as follows:

$$2O_2^- + 2H^+ \xrightarrow{\text{SOD}} H_2O_2 + O_2$$

$$O_2^- + O_2^- \xrightarrow{4H^+} .2H_2O_2$$

In comparison with many other reactive species, H_2O_2 itself is relatively low in reactivity, allowing it to remain in contact with cells and pass through cell membranes

(Frisch et al., 1983). Catalase breaks down H_2O_2 to oxygen and water, thereby removing any potential consequential damage. Interaction with halides such as Cl^- via myeloperoxidase (MPO) and detoxification via the glutathione cycle are the primary mechanisms for removal of H_2O_2 within phagocytes. The glutathione cycle is coupled to hexose monophosphate shunt activity because glucose-6-phosphate dehydrogenase and 6-phosphogluconate dehydrogenase reduce $NADP^+$ to NADPH which must be reoxidized, a task performed by the glutathione cycle (see Fig. 2). Reduced glutathione (GSH), a vital component in phagocytic cells, can be easily measured using a monobromobimane fluorescent probe (Hedley and Chow, 1994). Reduced glutathione is converted to the oxidized form (GSSG) by glutathione peroxidase, and subsequently reduced back to GSH by NADPH. This cycle is an important "detoxification" system for excess H_2O_2 within the cytosolic environment (Voetman and Roos, 1980).

Hypochlorous Acid. The MPO-catalyzed reaction of H_2O_2 with chloride produces a particularly dangerous molecule, hypochlorous acid (HOCl), which acts directly on membrane protein by inactivating sulfhydryl-dependent transporter systems (Schraufstätter et al., 1990). Hypochlorous acid is the predominant species at acidic pH, such as found in activated phagocytes. It also reacts with primary amines to produce monochloramine and taurine monochloramine, *N*-chloramines recognized as very reactive oxidants. The general reaction proceeds as follows:

$$H_2O_2 \xrightarrow{\text{MPO} + 2Cl^-} 2HOCl + 2e^-$$

Neutrophils contain a very significant amount of MPO, estimated to be at least 5% of dry cell weight (Schultz and Kaminker, 1962). This enzyme, found in the azurophilic (primary) granules of neutrophils, has been well characterized biochemically (Johnson and Nauseef, 1991). Other reactions of importance include those with nitrogen-containing compounds to form chloramines. Reactions involving MPO are both numerous and complex and beyond the scope of this chapter.

Hydroxyl Radical. For many years it has been thought that superoxide, while exercising some damaging effects on biological systems, is less important in terms of tissue injury than the more dangerous hydroxyl radical ($OH^{\cdot}$), whose production from H_2O_2 is thought to occur by the iron-catalyzed Haber–Weiss reaction:

$$H_2O_2 \xrightarrow{Fe^{2+}} Fe^{3+} + OH\cdot + OH^-$$

Although this reaction has, to date, provided the most acceptable explanation for tissue damage in oxidative systems, there are some troubling points, such as the fact that the rate constant for this reaction is somewhat lower than that for the competitive reaction of ascorbic acid with iron. It has therefore been proposed that a more realistic molecule for tissue damage is $ONOO^-$ rather than $OH\cdot$ (Beckman et al., 1990) (see below).

Fluorescent Indicators of Intracellular Oxidation

Many assay systems for measuring neutrophil function have been proposed in the literature. Some recently developed methods allow simultaneous measurement of O_2^- and H_2O_2, and NO· as well. These methods are often based upon the use of fluorescent probes. Five specific probes are discussed here: dichlorofluorescin diacetate, dihydrorhodamine 123, hydroethidine, parinaric acid, and monobromobimane. Table 1 summarizes their commonly used excitation and emission wavelengths and target molecules.

Dichlorofluorescin Diacetate. Dichlorofluorescin diacetate (DCFH-DA) has been utilized for H_2O_2 measurement ever since the first application by Keston and Brandt (1965) in a bulk cell assay. The probe was later recognized as a useful one for determination of H_2O_2 in neutrophils (Gubitz et al., 1976; Homan-Muller et al., 1975) and the technique extended to flow cytometry by Bass et al. (1983). The list of subsequent publications, particularly in flow cytometry, is substantial. Dichlorofluorescin diacetate has been used to study H_2O_2 production in human neutrophils (Robinson et al., 1994a; Himmelfarb et al., 1992; Epling et al., 1992; Stelzer and Robinson, 1988a; Wolber et al., 1987; Seeds et al., 1985; Smith and Weidemann, 1993; Vowells et al., 1995), monocytes/macrophages (Holter et al., 1987; Lepoivre et al., 1986), cultured neurons (Saez et al., 1987), renal epithelial cells (Scott et al., 1988), melanocytes (Boissy et al., 1989), chondrocytes (Tiku et al., 1990), rat endothelial cells (Carter et al., 1994b), human umbilical vein endothelial cells (HUVECs) (Niu et al., 1994; Royall and Ischiropoulos, 1993), and bovine aorta endothelial cells (Royall and Ischiropoulos, 1993).

The probe works in the following manner. Dichlorofluorescin diacetate is an esterified molecule that can freely pass through the cell membrane. Once inside the cell, DCFH-DA is deacetylated by cellular esterases to the nonfluorescent dichlorofluorescin (DCFH), which is trapped within the cell by its polar nature. Dichlorofluorescin is converted by intracellular oxidants such as H_2O_2 to the green fluorescent molecule dichlorofluorescein (DCF), a reaction significantly enhanced in the presence of peroxidase. The major difficulties with this probe are its tendency to leak from the cell and its lower sensitivity to H_2O_2 as compared to DHR 123. Many studies have used this probe successfully for intracellular H_2O_2 determination, but careful controls must be made to account for cell leakage.

TABLE 1. Useful Functional Fluorescent Probes

Dye	Excitation (nm)	Emission (nm)	Indicator
DCFH-DA	488	520	H_2O_2
DHR-123	488	520	H_2O_2
HE	488	600	O_2^-
Parinaric acid	325–360	420–450	Lipid peroxidation
Monobromobimane	350–360	420–450	Reduced glutathione

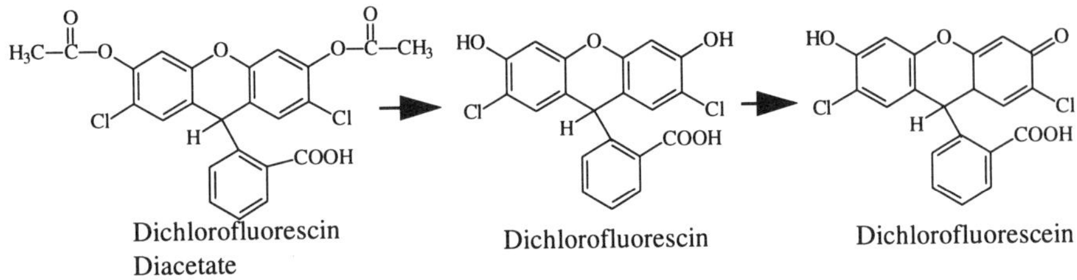

Fig. 3. Dichlorofluorescin diacetate is hydrolyzed by cellular esterases to the nonfluorescent dichlorofluorescin, which is readily oxidized to the fluorescent dichlorofluorescein.

Dihydrorhodamine 123. Dihydrorhodamine (DHR 123) is by far the most-used probe for measurement of intracellular H_2O_2. DHR 123 is oxidized directly to rhodamine 123, which is excitable at 488 nm and emits at 515 nm in the same emission range as DCF and FITC (Rothe et al., 1988). Publications describe its use in human neutrophils (Wenisch et al., 1996; Emmendörffer et al., 1994; Tanigaki et al., 1993; Waddell et al., 1994; Cao et al., 1993; Demaurex et al., 1996; Smith and Weidemann, 1993; Vowells et al., 1995; Rothe et al., 1988; van Pelt et al., 1996; Wenisch et al., 1995), human eosinophils (Elsner et al., 1995), HL60 cells (Kaffenberger and van Beuningen, 1994), rat mast cells (Tsinkalovsky and Laerum, 1994), guinea pig neutrophils (Tanigaki et al., 1993), cultured chondrocytes (Hayem et al., 1994), rat brain cells (LeBel et al., 1992), rat renal proximal tubular cells (van de Water et al., 1995), mesangial cells (Zent et al., 1995), and L929 cells (Goossens et al., 1995). Combinations of DHR123 with surface receptor analysis (Elsner et al., 1994), cell viability using propidium iodide (Clancy et al., 1995), and calcium indicators (Bueb et al., 1995) demonstrate how the probe can be used for simultaneous measurements.

DHR 123 enters the cells as a freely permeable dye that is converted to rhodamine 123 and subsequently localized in the mitochondria. The conversion from the nonfluorescent to the fluorescent molecule depends entirely on oxidation products and does not require enzymatic catalysis. Once oxidized, the probe is identical to rhodamine 123, a common laser dye. An example of the use of rhodamine 123 directly in endothelial cells is shown in Figure 5. One significant advantage of the DHR probe

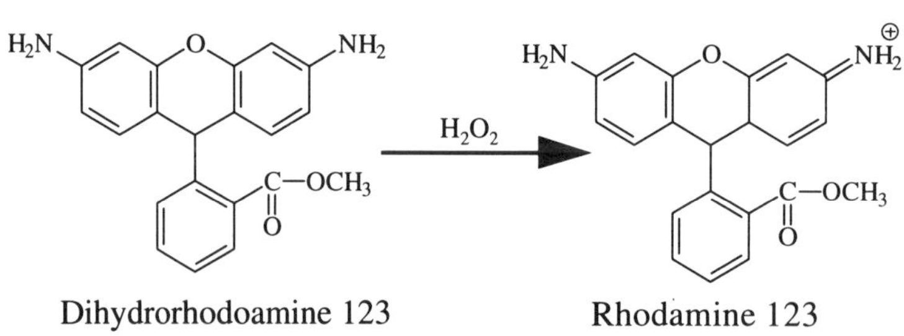

Fig. 4. Dihydrorhodamine enters the cell and is oxidized to rhodamine 123, a fluorescent molecule which emits at 520 nm, the same as FITC. The oxidation is a result of H_2O_2 production.

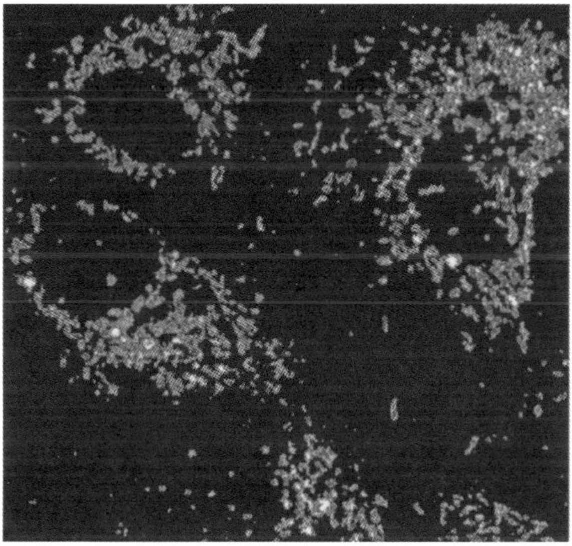

Fig. 5. Mitochondria labeled with rhodamine 123. Attached viable endothelial cells were loaded with 5 μM rhodamine 123 and imaged using a Bio-Rad MRC 1024 confocal microscope.

is that the oxidation product, rhodamine 123, remains essentially within the cell, unlike the oxidation product DCF, which has a strong tendency to leak from cells and requires careful controls to monitor leakage.

A number of publications have made direct comparisons between DCFH-DA and DHR 123 (Smith and Weidemann, 1993; Vowells et al., 1995). In summary, the advantages of using the DHR 123 probe for cellular H_2O_2 production are based upon its increased sensitivity to H_2O_2 (3-to 10-fold) and its general failure to leak from the cells.

Hydroethidine. Hydroethidine (HE) has been proposed by Rothe and Valet (1990) as a probe for measurement of O_2^-. The dye enters cells freely and is dehydrogenated to ethidium bromide (Fig. 6). A major advantage of this probe is its ability to distinguish between O_2^- and H_2O_2. As shown in Figure 7, the specificity of HE for O_2^- is significant. Studies have been performed using neutrophils (Robinson et al., 1994a; Kuypers et al., 1990; Rothe et al., 1991; Rothe and Valet, 1990) and endothelial cells (Carter et al., 1994b), as well as HL60 cells (Robinson et al., 1994b) and macrophages (Kobzik et al., 1990a, b). The probe has been used extensively with NK cell assays (Radcliff et al., 1991; Callewaert et al., 1991; Zanyk et al., 1990; Cavarec et al., 1990) and as a vital dye for identification of proliferation (Saiki et al., 1986; Bucana et al., 1986) and hypoxic cells in tumors (Olive, 1989). Fluorescence emission occurs at around 600 nm.

Parinaric Acid. The parinaric acids are the closest structural analogues of intrinsic membrane lipids among current fluorescent probes (Haugland, 1992). *cis*-Parinaric

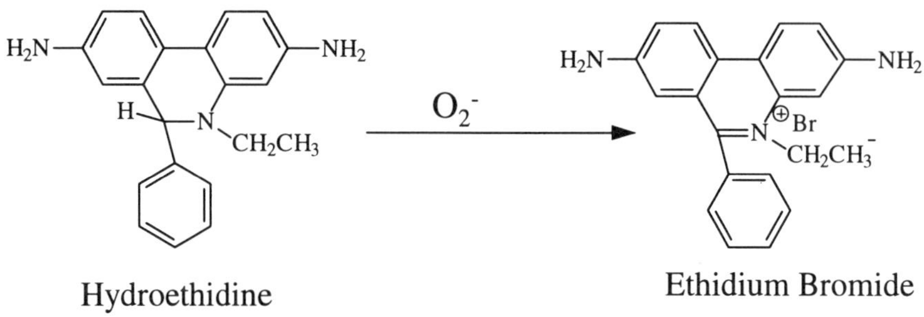

Fig. 6. Hydroethidine enters cells freely and is dehydrogenated by superoxide anion directly to ethidium bromide. Fluorescence emission occurs around 600 nm.

acid (Fig. 8), a naturally fluorescent fatty acid, loses its fluorescence over time when the four conjugated double bonds of the backbone chain become the target of lipid peroxidation reactions (Hedley and Chow, 1992). Thanks to its extensive unsaturation, parinaric acid is quite susceptible to oxidation. It is this property that is utilized in measuring lipid peroxidation. Because a 325-nm ultraviolet excitation source is required, the use of parinaric acid is restricted to spectrofluorometry and flow cytometers with helium–cadmium lasers.

Monobromobimane. Monobromobimane (Fig. 9) and monoclorobimane can be used for the determination of glutathione levels within single cells (Hedley and Chow, 1994). Although monochlorobimane is the most specific GSH probe available be-

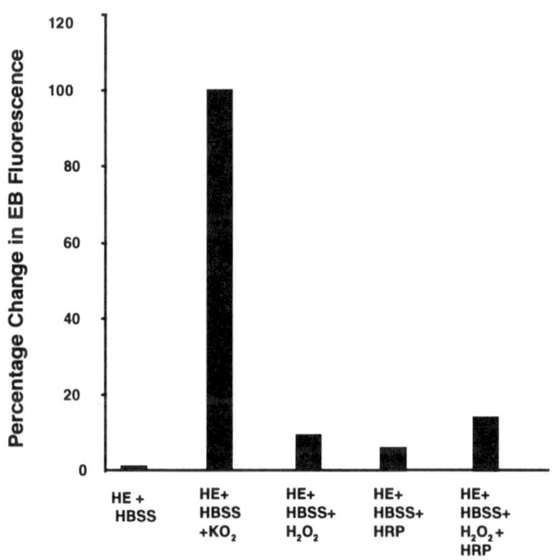

Fig. 7. Cell-free fluorescence generation using hydroethidine. Potassium superoxide oxidizes hydroethidine to ethidium bromide more efficiently than H_2O_2, as shown.

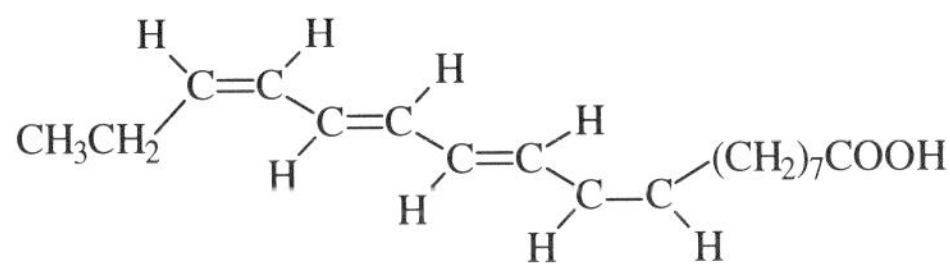

Fig. 8. Parinaric acid is a naturally fluorescent molecule whose fluorescence decreases as the conjugated double bonds are broken via oxidation.This reduction in fluorescence can be monitored kinetically using a He–Cd laser on a flow cytometer with excitation at 325 nm and emission at 420 nm.

cause its binding is catalyzed by glutathione *S*-transferase, human isoenzymes of glutathione *S*-transferase have an unacceptably low affinity for monochlorobimane at attainable physiological concentrations (Ublacker et al., 1991). For human cells (particularly neutrophils or monocytes), the most satisfactory of the probes capable of forming fluorescent adducts with GSH via the sulfhydryl group is monobromobimane. Initial trial of this probe as a tool for flow cytometry showed unacceptably high backgrounds; however, Hedley and Chow reexamined the probe and determined optimal conditions for use with human cells and flow cytometry (Hedley and Chow, 1994).

Differences between Monocytes and Neutrophils

Opinions differ as to whether neutrophils or monocytes are the more reactive. Several factors influence the result, not the least of which is the method used for measuring the particular species. For instance, measurements of external O_2^- are unlikely to provide accurate estimates of O_2^- remaining within the cell. Several early reports indicated that zymosan-stimulated neutrophils produced several times as much reactive species as monocytes based upon oxygen consumption (monocytes consuming around 40% as much as neutrophils) and H_2O_2 production (around 20%) (Roos et al., 1979; Reiss and Roos, 1978). It has been demonstrated that peripheral blood monocytes are far less reactive to soluble activation than neutrophils (Robinson et al., 1988). However, when these monocytes are able to bind to particulates such as bacteria, a significant amount of H_2O_2 is measurable. More recent work has shown that monocytes activated by an identical concentration of PMA not only produce less

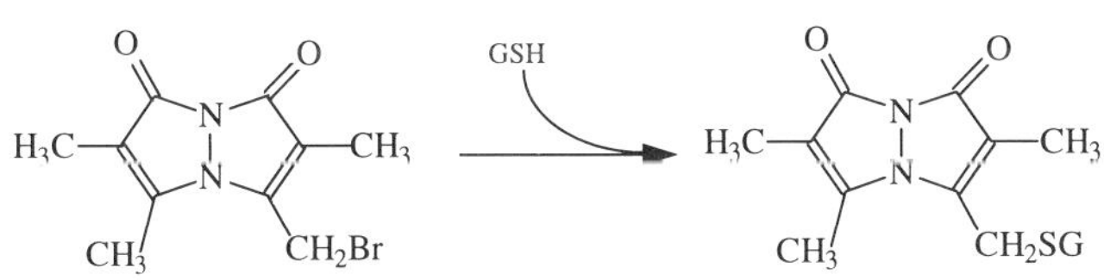

Fig. 9. Monobromobimane combines nonenzymatically with glutathione at low concentrations and is converted from a nonfluorescent molecule to a fluorescent one when so bound. The optimal excitation for flow cytometry is the 350–360 nm UV beam from an argon laser, although the 325-nm line from a helium–cadmium laser will suffice. Emission is at 420–450 nm. A maximum of 10 min incubation is needed for cell staining.

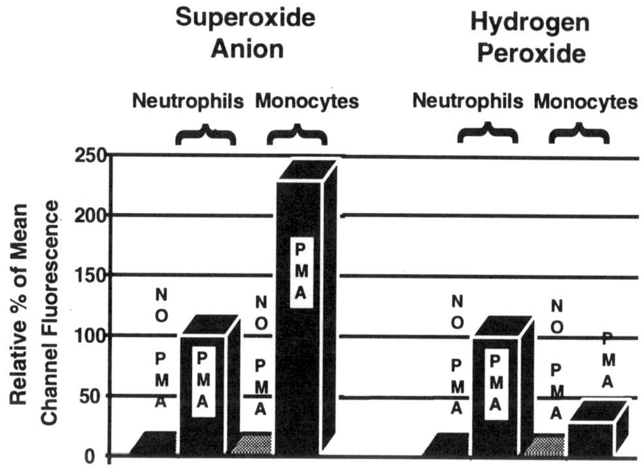

Fig. 10. Comparison between human neutrophils and monocytes of O_2^- and H_2O_2 capacity after stimulation with 10 ng/ml PMA. Data are normalized to set neutrophil production at 100%. Monocytes produce less H_2O_2 but more O_2^- than neutrophils.

H_2O_2 than neutrophils (about a third as much), but also can produce substantially *more* O_2^- as measured intracellularly with the hydroethidine probe (unpublished observation). These data are shown in Figure 10, normalized so that the neutrophil O_2^- or H_2O_2 production is set at 100%.

Oxygen as a Defense Mechanism—Antioxidant Systems

Peroxidases. Peroxidases remove H_2O_2 via peroxidatic mechanisms (glutathione peroxidase is an important example); a number of peroxidases are found in phagocytic cells. In the absence of oxidants, the activity of peroxidases as such is poor. The better known peroxidases within the phagocytic system are myeloperoxidase (which preferentially reacts with Cl^-) in neutrophils and eosinophil peroxidase (which preferentially reacts with Br^-) in eosinophils.

SOD. Superoxide dismutases are a group of metalloproteins/metalloenzymes present in all respiring cells [including mammals, plants, fungi, and bacteria with the exception of some obligate anaerobes (McCord et al., 1971)] that catalyze the dismutation of O_2^- to H_2O_2.

There are three isoenzymes of superoxide dismutase (SOD) in mammals, the first of which was discovered by McCord and Fridovich in 1969 (McCord and Fridovich, 1969a). This CuZn SOD was isolated from cytoplasm, nucleus, and peroxisomes. The second, MnSOD, discovered by Weisiger and Fridovich in 1973 (Weisiger and Fridovich, 1973), is a cytoplasmically synthesized enzyme directed primarily toward the mitochondria. The third isoform, EC-SOD (extracellular), was discovered by Marklund (Marklund, 1982) and is a CuZn SOD with a positively charged binding domain optimized for localization in the extracellular matrix. This isoenzyme has been shown to have particularly high expression in vascular tissue (Oury et al., 1994) and umbilical cord tissue (Sandstrom et al., 1993).

The structure of CuZn SOD in bovine erythrocytes has been determined as a homodimer of 16 kDa with the active site located within a cylinder β structure (Richardson et al., 1975), where it is well protected and is known to retain catalytic activity during isolation procedures (Forman and Fridovich, 1973).

The mechanism of action of SOD is that the copper ion at the active site is reduced by one O_2^- molecule, then reoxidized by another in a continuing cycle (Fridovich, 1981). Thus, copper oscillates between the monovalent and divalent states. Copper-containing SOD is inhibitable by cyanide but the Mn form is not (Haffner and Coleman, 1973). Azide (Misra and Fridovich, 1978) and diethyldithiocarbamate (Heikkila and Cohen, 1977) (which removes the Cu) are also able to inactivate SOD activity in cell preparations, thus rendering the cells susceptible to autodestruction. Recently Naranayan et al. (1998) showed that environmental pollutants such as PCBs reduce the effectiveness of antioxidant systems in human neutrophils, possibly by inactivating SOD activity. If this mechanism is confirmed, it represents the possibility that chronic low levels of antioxidants may be capable of causing more damage than previously understood, because depletion of SOD and/or glutathione may be responsible for a subsequent increase in apoptosis.

Catalase. Catalase is a heme protein present mostly in peroxisomes that rapidly removes H_2O_2 and reduces it to H_2O and $O_2\cdot$ via the general reaction sequence

$$H_2O_2 + ROOH \longrightarrow H_2O + ROH + O_2$$

where R is a short chain alkyl group. The rate of H_2O_2 removal via catalase is 10^8 times faster than the dismutation of H_2O_2 to water and oxygen (Forman and Fisher, 1981), meaning that it is virtually impossible to saturate catalase activity under normal biological conditions. Phagocytic cells contain catalase, which can scavenge not only H_2O_2 produced within the neutrophil but also H_2O_2 added exogenously to cell preparations, because H_2O_2 can freely move across the cell membrane (Voetman and Roos, 1980). Being a large molecule, catalase obviously cannot penetrate the cell membrane and its use as an antioxidant in experimental situations is restricted to extracellular locations.

Glutathione. Glutathione (GSH) is a tripeptide present in most cells that acts as a general antioxidant for the removal of H_2O_2 via the general reaction

$$H_2O_2 + 2GSH \longrightarrow 2H_2O + GSSG$$

Glutathione is generally found in the reduced form GSH as shown in Figure 1 above and its presence in cells affords considerable antioxidant protection. In contrast to catalase, which is located primarily in peroxisomes, glutathione in the cell is ubiquitous. The role of the glutathione cycle has been described in neutrophils (Ohno and

Gallin, 1985) and neutrophil apoptosis (Robinson and Narayanan, 1996), endothelial cells (Andreoli et al., 1986; Tsan et al., 1985), hepatocytes (Keller et al., 1985), platelets (Freedman et al., 1996), and tumor cells (Goossens et al., 1995). Evidence that nitric oxide may play a regulatory role in neutrophils (and perhaps other cells) has recently been demonstrated (Forslund and Sundqvist, 1995a; Nikitovic and Holmgren, 1996; Clancy et al., 1994).

Ascorbic Acid. Although not synthesized in most cells, ascorbic acid can be present in reasonably large quantities in some tissues, where it can act as an antioxidant via the following general reaction:

$$2O_2^- + 2H^+ \xrightarrow{\text{ascorbic acid}} 2H_2O_2 + \text{dehydroascorbate}$$

It should be noted that this reaction is not preferred over the dismutation of O_2^- by SOD; however, in some tissues there may be sufficient ascorbate to afford reasonable antioxidant activity.

Vitamin E. Vitamin E is generally accepted to act as an antioxidant via nonenzymatic reduction of polyunsaturated lipid oxide free radicals in the general reaction

$$\text{vitamin E} + \text{RO}\cdot \longrightarrow \text{vitamin E}\cdot + \text{ROH}$$

The vitamin E· radical must be further reduced by other reaction sequences, possibly via its interaction with ascorbic acid (vitamin C). Intracellular vitamin C can restore the vitamin E radical to reduced vitamin E, thereby limiting lipid peroxidation (Forman and Fisher, 1981).

Polyunsaturated Fats. It is possible that polyunsaturated fats acting as a trap may preferentially reduce certain ROI, preventing further damage to DNA or organelles and subsequently more serious damage to the cell. The reactions may produce conjugated dienes and other molecules that can themselves cause damage under certain conditions. The general reaction sequence for lipid peroxidation would be

$$\text{RH} + O_2^- + \text{OH}\cdot \xrightarrow{+\,H^+} \text{ROOH} + H_2O$$

It has been suggested that O_2^- in particular can cause a chain reaction whereby further lipid peroxidation takes place (Thomas et al., 1978).

Oxygen as a Pathogenic Molecule

Superoxide is produced in significant quantities in many cells; as previously stated, from 1 to 4% of all oxygen consumed in the mitochondrial electron transport chain results in production of O_2^- (Cross, 1987). As discussed above, activated phagocytes can produce a variety of oxidants that may be found both within the phagocytic cells and released into the extracellular milieu. After neutrophils have been recruited to approach, roll, and attach to the vascular endothelium, they transmigrate to the source of inflammation. Once at the site, they usually become further activated, producing additional oxidants. Any neutrophil breakdown releases large quantities of enzymes, with the potential for considerable tissue damage. There are several clear-cut situations in which activated neutrophils cause classic tissue damage, predominantly through the production of reactive oxygen species.

One such case is that of complement-mediated lung injury; invasive neutrophils have been shown to be present in high numbers where severe tissue damage occurs (Till et al., 1982). Similar damage has been observed in a model of immune complex damage to the lung. In this model, a large neutrophil infiltrate subsequent to intratracheal instillation of IgG antibody to bovine serum albumin is thought to be the cause of significant tissue injury (Johnson and Ward, 1974). In terms of free radical generation, it has been shown that alveolar macrophages from patients with asthma have a higher rate of production of superoxide (Jarjour and Calhoun, 1994).

NITROGEN-RELATED METABOLISM

The discovery of nitric oxide is attributed to Joseph Priestley, a clergyman and part-time chemist, who called his find "nitrous air." When exposed to iron, this gas was converted to nitrous oxide (laughing gas, reviewed by Gilbert, 1981). The year of this discovery was 1772, but more than 200 years were to pass before the realization that nitric oxide has a crucial role to play in physiological systems.

The chemistry of nitrogen monoxide (NO) has been elucidated over the past several years. The most studied molecule has been the radical nitric oxide (NO·); however, there are several redox states of NO, including nitrosonium cation (NO^+), nitric oxide (NO·), and nitroxyl anion (NO^-), not dissimilar to the well-known states of oxygen (dioxygen) (O_2), namely, superoxide (O_2^-) and hydrogen peroxide (H_2O_2). NO· itself has the lowest molecular weight of any known mammalian secretory product (Nathan, 1992). The seminal discovery that L-arginine was converted to nitric oxide by macrophages and that this was involved in tumoricidal activity (Hibbs et al., 1987) created an entirely new area of research. Literally tens of thousands of reports have surfaced in the past 10 years associating nitrogen-related metabolites with what was previously an exclusive domain of oxygen metabolism.

Production and Properties of Nitric Oxide

Production. Nitric oxide is produced by oxidation of one of the terminal nitrogens on arginine via nitric oxide synthase, producing *N*-hydroxyarginine, then citrulline, and finally nitric oxide. Each molecule of nitric oxide produced requires one arginine, two oxygens, and 1.5 NADPHs, involving an overall reduction of five electrons.

There are three distinct isoforms of the synthase enzyme, two of which have been termed *constitutive*: one from endothelial cells (ECnos—Type III) and one from neuronal cells (NCnos—Type I) (Bredt et al., 1991). The constitutive form (*c*NOS) is Ca^{2+}-dependent, appears to be mediated by calmodulin (Pollock et al., 1991), requires the presence of co-factors such as tetrahydrobiopterin for activity (Förstermann et al., 1991), and is believed to be located on the cytoplasmic face of the cell membrane (Knowles and Moncada, 1992). The third isoform, calcium-independent and inducible (*i*NOS—Type II), is tightly bound to calmodulin (Nathan and Xie, 1994) and is found in most cells in the body (Stuehr and Griffith, 1992; Nussler and Billiar, 1993).

Nitric oxide (NO·) is formed through a five-electron oxidation of L-arginine via the action of nitric oxide synthase, NADPH, and tetrahydrobiopterin (Stuehr and Griffith, 1992), which proceeds from the hydroxylation of L-arginine to form N^{ω}-hydroxy-L-arginine through subsequent hydrolysis to L-citrulline and hydroxylamine with further oxidation to nitric oxide:

$$\text{L-arginine} + \text{NADPH} \xrightarrow{\text{NOS}} N^{\omega}\text{-OH-L-arginine} \xrightarrow{\text{NOS}} \text{L-citrulline}$$

$$+ NH_2OH + H_2O_2 \xrightarrow{\text{catalase}} NO\cdot + 2H_2O + H^+$$

$$\text{L-Arg} + 3O_2 + 1.5\ \text{NADPH} + H_4\text{-biopterin} \longrightarrow NO\cdot + \text{L-citrulline}$$

$$+ 1.5\ \text{NADP}^+ + H_2\text{-biopterin} + 2H_2O + H_2O_2$$

Hydroxylamine is oxidized by a catalase-like activity to form NO·, utilizing H_2O_2 as a substrate (Stuehr and Griffith, 1992). It is known that NO· reacts rapidly with both oxyhemoglobin and deoxyhemoglobin with a final product of nitrate (Wennmalm et al., 1992):

$$NO\cdot + O_2Hb \longrightarrow NO_3^- + Hb^+$$

As discussed below, the reaction of NO· with superoxide produces nitrates by way of the peroxynitrite free radical intermediate (Knowles and Moncada, 1992):

$$NO\cdot + O_2^- \longrightarrow ONOO^- \longrightarrow NO_3^-$$

or, in the absence of superoxide and hemoglobin at pH 7.4, NO· can utilize other reactive species of oxygen such as singlet oxygen (Knowles and Moncada, 1992):

$$2NO\cdot + {}^1O_2 \longrightarrow N_2O_4 \longrightarrow NO_2^- + NO_3^-$$

Thus it is highly likely that NO· and O_2^- are very closely interactive during the inflammatory response and that the modulation of O_2^- by NO· is most likely an important mechanism for regulation of ROI production.

Physiological Properties. At standard temperature and pressure, pure nitric oxide gas is approximately 40 m*M*, whereas a saturated aqueous solution is 1.9 m*M*; however, physiological levels of NO· range from 10 n*M* (necessary to cause vasodilation) to as high as 1–10 μ*M* around activated macrophages (Beckman, 1995). Nitric oxides have potentially extensive half-lives in tissue, as long as 17 h for a 10 n*M* concentration and around 11 min for a 1 μ*M* concentration of NO· (Beckman and Tsai, 1994).

One of the earlier effects observed for NO· was that it prevented neutrophil aggregation (Kubes et al., 1991; McCall et al., 1988). Thus nitric oxide could regulate the buildup of inflammatory cells and reduce potential tissue damage. The mechanism of leukocyte recruitment is well understood. Leukocyte adhesion to endothelial cells is primarily mediated by the β-integrin glycoprotein complex on neutrophils (Tonnesen, 1989). NO· plays a role in P-selectin—dependent leukocyte rolling, for N^{ω}-nitro-L-arginine methyl ester pretreatment of rats with L-NAME caused increased leukocyte rolling (Davenpeck et al., 1994). It is known that once an inflammatory state is signaled (via a chemoattractant, for instance), neutrophils slow down and gently adhere to the wall of the blood vessel via L-selectin. Because the adherence is of low affinity, the effect is a rolling along the endothelial layer. The L-selectin (on neutrophils) binding via E(ELAM-1)- and P(CD62)-selectins (on endothelial cells) is essentially intermittent until the neutrophil approaches sufficiently close to the source of the inflammatory mediator. At this point, the neutrophil slows and then adheres to the endothelial surface via adhesion glycoproteins (CD11/CD18) on neutrophils and their ligands (ICAM-1) on endothelial cells. Now the neutrophil migrates through the microvessel wall, a process that takes several minutes. Thus the regulatory role that NO· appears to play in expression of P-selectin is physiologically very significant, with the implication that NO· acts as a homeostatic regulator of neutrophil-endothelial interactions (Beckman, 1995).

Other physiological effects of nitric oxide on neutrophils include direct inhibition of superoxide production, possibly by direct action on the NADPH oxidase (Clancy et al., 1992), providing additional evidence of protection against oxidative injury. In a similar manner, nitric oxide has been shown to act against mast cells, demonstrating a diminution of mast cell reactivity. It has been shown that exvenous addition of NO· via the NO· producer nitroprusside decreased the amount of histamine released by mast cells (Salvemini et al., 1991) as well as the amount of platelet activating factor produced by them (Hogaboam et al., 1993).

Peroxynitrite

Peroxynitrite ($ONOO^-$) is formed from a radical–radical reaction between O_2^- and NO· at a diffusion-limited rate ($k = 6.7 \times 10^9 M^{-1}s^{-1}$) (Huie and Padmaja, 1993). It has been known for many years that superoxide and nitric oxide could combine to form peroxynitrite, as well as the conditions under which this will occur (Blough and Zafiriou, 1985). Making accurate measurements of $ONOO^-$ in biological systems, however, proved more difficult. Beckman et al. (1990) have determined many of the properties of peroxynitrite. The pK_a is 7.49 at 37°C, and once protonated it rapidly decomposes with a half-life of 1.9 s. Macrophages are capable of producing at least 1mmol/min of peroxynitrite; and the effects on bacterial survival are not altered by mannitol, ethanol, or benzoate, but are enhanced by dimethyl sulfoxide or deferoxamine (Zhu et al., 1992).

When $ONOO^-$ is protonated, the resultant peroxynitrous acid (ONOOH) decomposes to several toxic species with reactivity similar to hydroxyl radical and nitrogen dioxide (Volk et al., 1995; Radi et al., 1991; Beckman et al., 1990; Ischiropoulos et al., 1992a,b). Peroxynitrite has also been shown to have high bactericidal capacity in macrophages (Zhu et al., 1992; Ischiropoulos et al., 1992a,b) and to replace the active site of CuZn superoxide dismutase, removing its dismutation capacity for superoxide anion by blocking the O_2^- binding site (Beckman et al., 1992; Ischiropoulos et al., 1992a,b). DNA strand breakage (Inoue and Kawanishi, 1995; Szabó et al., 1996) demonstrated in macrophages (Szabó et al., 1996), smooth muscle (McCauley and Hartmann, 1984; Begley et al., 1985), and thymocytes (Salgo et al., 1995) has been attributed to $ONOO^-$. Furthermore, $ONOO^-$ can directly oxidize sulfhydryl groups such as cysteine and BSA at physiological pH (7.4) at rates 1000 times faster than H_2O_2 (Radi et al., 1991a).

The situation is further complicated by the fact that transition metals can catalyze $ONOO^-$ to form OH^- (hydroxyl anion) plus nitronium ion (NO_2^+), which can subsequently nitrate protein tyrosine residues to produce 3-nitrotyrosine in a reaction catalyzed by SOD (Beckman et al., 1992), whose normal role is to dismutate O_2^-.

It has been suggested that the toxicity of either the O_2^- or NO· radical is significantly enhanced when the two combine to form $ONOO^-$ (Beckman et al., 1990). Peroxynitrite has been shown to exhibit strong activity with a number of biological molecules, and a substantial amount of evidence demonstrates that $ONOO^-$ is considerably more reactive than NO·. The strong oxidizing capability has been shown to act on peroxidation of lipids in the absence of iron (Rubbo et al., 1994; Radi et al., 1991a), cause formation of malondialdehyde and conjugated dienes during lipid peroxidation (Radi et al., 1991b), and also act directly on carbohydrates (Beckman et al., 1990).

Peroxynitrite inhibits pulmonary α1-proteinase inhibitor and therefore oxidizes critical methionine residues in the active site of the enzyme (Moreno and Pryor, 1992). Peroxynitrite is also responsible for the oxidation of arachidonic acid and formation of F2-isoprostanes via oxidation of low-density lipoproteins (Moore et al., 1995) and causes inhibition of mitochondrial respiratory chain enzymes, including cytosolic aconitase (Radi et al., 1994; Hausladen and Fridovich, 1994).

Deferoxamine and glutathione are scavengers of peroxynitrite, whereas H_2O_2 has been shown to prolong its half-life (Alvarez et al., 1995). It is likely that most of the damaging effects against mitochondria result from the activity of peroxynitrite, not OH· or other radicals, for antioxidants against O_2^- or OH· provide little protection.

Under certain conditions the NO·/O_2^- interaction may behave as a significant antioxidant mechanism for the detoxification of O_2^-. What is clear is that the most important criteria for the role of the NO· and O_2^- combination may be the relative fluxes of each molecule and the nature of the tissue in which they are formed.

Fluorescent Indicators of Excited Nitrogen Species

Some confusion exists as to what particular fluorescent probes actually measure. A number of reports in the literature claim to measure one or another oxidant, when in fact there is no clear evidence as to what molecule is actually being measured. This problem is witnessed by the vast quantity of studies described in the early 1990s purporting to measure nitric oxide, but in reality measuring an accumulation of nitrites. This has led many investigators to the conclusion that NO· was indeed produced and was the key biological species. Many publications have assumed that the species contributing the biological effect was NO· when it could well have been, and most probably was, a combination of excited molecules. Regardless, the conclusion has been drawn on many occasions that NO· was the key to the biological mystery being unraveled. In a similar manner, NO· and $ONOO^-$ measurements using fluorescent probes may have resulted in the same problem of crossreactivity already observed with probes for O_2^- and H_2O_2.

Dihydrorhodamine 123. DHR 123 has recently become a popular probe for measuring peroxynitrite. Haddad et al. (1994) have used DHR 123 to measure peroxynitrite production, claiming that the simultaneous generation of O_2^- and NO· resulted in the oxidation of DHR 123 to rhodamine 123. They claim further that DHR 123 is not oxidized by NO·, O_2^-, or H_2O_2 alone (Kim et al., 1996; Haddad et al., 1994). Clearly there are contradictory publications on the nature and use of certain fluorescent probes, and the choice of such probes depends entirely on the biological system in question. Many studies have used DHR 123 as a direct measure of peroxynitrite (Szabó et al., 1995); for cases involving noncellular systems where highly defined reactions can be studied, the efficacy of the probe is acceptable. In one such controlled system the authors concluded that in the presence of oxygen, nitric oxide induces a relatively slow oxidation of dihydrorhodamine owing to the formation of nitrogen dioxide and consequently that dihydrorhodamine was a sensitive and efficient trap for peroxynitrite (Miles et al., 1996; Kooy et al., 1994).

When using these probes for more complex biological systems such as cell suspensions or for in vivo use, care must be exercised in drawing and interpreting conclusions from the data. Experiments such as injecting DHR 123 directly into rats and measuring subsequent plasma levels of rhodamine 123 spectrofluorometrically are not straightforward cases of cause and effect. In that instance, the conclusions that the resulting measurement was a direct measure of peroxynitrite (Szabó et al., 1995) must be interpreted with great caution, and in the case of peroxynitrite measurements, are only relative to NOS-inhibitable changes in fluorescence as discussed by Szabó (Szabó et al., 1995).

Dichlorofluorescin Diacetate. Evidence exists that DCFH-DA under some circumstances may actually measure NO· in addition to H_2O_2. Such evidence includes the observation for human neutrophils that a calmodulin inhibitor (W-13) inhibits O_2^- production as measured by cytochrome c and nitroblue tetrazolium assays but enhances the formation of fluorescent DCF (Rao et al., 1992). Furthermore, this reaction was inhibitable by NMMA (N^{ω}-monomethyl-L-arginine), an L-arginine analogue that inhibits production of NO·. Additionally, it was demonstrated that pure NO· gas could directly oxidize DCFH to the fluorescent DCF (Rao et al., 1992).

It has also been observed that DCFH-DA–loaded neutrophils incubated with a

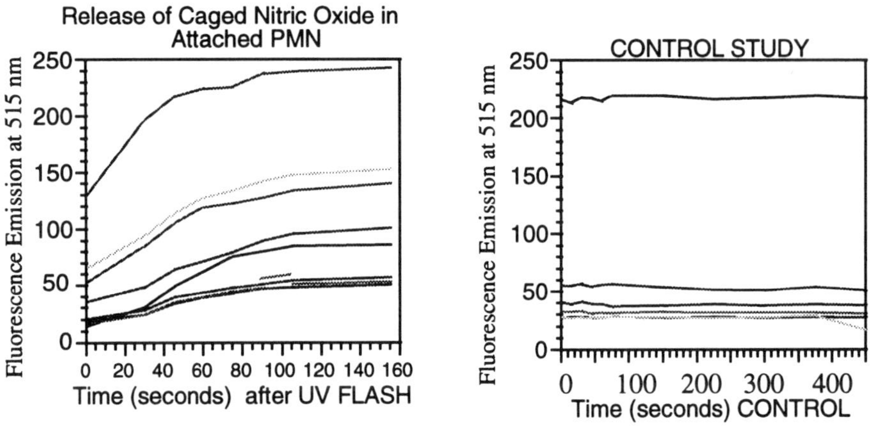

Fig. 11. The change in fluorescence over time after a controlled UV flash (360 nm excitation) on a small population of neutrophils previously loaded with DCFH-DA. Neutrophils attached to coverslips coated with polylysine were loaded with DCFH-DA (20 μM) and incubated at 37°C on the heated stage of a Bio-Rad MRC 1024 confocal microscope for 20 min to hydrolyze the probe. Cells were then incubated with potassium nitrosylpentachlororuthenate (PNPCR, a "caged nitric oxide" compound). Control cells not flashed with UV light but given equal exposure to 488-nm excitation demonstrated no fluorescence shift. Neither did cells incubated with DCFH-DA alone and given the same UV flash.

caged nitric oxide showed an increase in DCF fluorescence after the cage was released by ultraviolet light. In these (unpublished) studies it appears that the DCFH-DA probe was measuring the released NO·, for the cells were otherwise not stimulated to produce reactive oxygen species (Fig. 11). Ultraviolet light alone did not appear to change the fluorescent nature of the DCFH probe.

Nitric Oxide Production in Neutrophils

Although there is no doubt that neutrophils can produce staggering quantities of oxygen radicals, their ability to produce nitric oxide has been disputed. Before the biological properties of nitric oxide had been elucidated, reports of the impact of rat neutrophils on relaxation of smooth muscle (Rimele et al., 1988), elevated cGMP levels (Lee et al., 1988), and platelet antiagreggation factor (McCall et al., 1989) would suggest involvement of NO· based upon present-day knowledge. Subsequently, the "factor" present in these neutrophils was shown to be nitric oxide (Schmidt et al., 1989; Wright et al., 1989). The controversy, however, has continued.

The question as to whether the presence of nitrites indicates intracellular NO· production or results from alternative metabolic pathways has been addressed by a number of groups. Klebanoff and Nathan (1993) measured nitrite production in human neutrophils in the presence of azide and catalase; nitrite production was unaffected by addition of either SOD or monomethylarginine. When stimulated neutrophils were replaced with the H_2O_2-generating system glucose–glucose oxidase, nitrite was also produced, leading these authors to conclude that nitrite production did not reflect nitric oxide synthase activity, but rather the catalase-catalyzed conversion of azide in the presence of H_2O_2 generated by stimulated neutrophils. Another study by Padgett and Pruett (1995) found that rat, mouse, and human neutrophils were able to produce very small amounts of nitrite, far less than would be required for antimicrobial activity. They conclude that the small amount of NO· produced by these cells may be related to intercellular signaling rather than playing a role as a defense mechanism.

These data are contrasted with a significant literature confirming the presence of both *i*NOS and subsequent production of significant amounts of NO·. Identification of *i*NOS expression has been confirmed in human neutrophils by measuring reverse transcription polymerase chain reaction (RT-PCR) products (Kolls et al., 1994; Cook et al., 1994; Evans et al., 1996). In one neutrophil sudy, PCR (polymerase chain reaction) products to both the endothelial constitutive (756 bp) and neuronal constitutive (629 bp) NOS isoforms were identified (Chen and Mehta, 1996). Confirmation of *i*NOS or NO· production has been by Southern blot (Chen and Mehta, 1996), spectrophotometric measurement of nitric oxide-dependent methemoglobin formation from oxyhemoglobin (Larfars and Gyllenhammar, 1995), and antibodies against *i*NOS in rat neutrophils (Clark et al., 1996). Detailed immunohistochemistry with a specific anti-nitrotyrosine antibody showed intense staining in both macrophages and neutrophils in mouse lung (Akaike et al., 1996) and rat macrophages and neutrophils (Goldman et al., 1996), and reports of the conversion of ^{3}H-arginine to ^{3}H-citrulline in human monocytes and neutrophils (Laffi et al., 1995) have been published.

Secondary evidence exists for the effect of NOS inhibitors on neutrophil function. For example, NOS inhibition attenuated chemotaxis of both unstimulated and primed neutrophils, suggesting a role for NO· synthesis in neutrophil emigration (Wildhirt et al., 1995) as well as in monocyte chemotaxis (Belenky et al., 1993), and exogenous NO· has been shown to induce chemotactic locomotion in human neutrophils (Beauvais et al., 1995). Other evidence of NO· production is based upon nitrite production alone (Ahmed and Weidemann, 1996; Dias-Da-Motta et al., 1996; Carreras et al., 1994a; Biswas et al., 1993) and chemiluminescence (Forslund and Sundqvist, 1995a; Catz et al., 1995; Carreras et al., 1994b).

Wright et al. (1989) showed that human neutrophils were able to generate nitric oxide at a rate of 1.8 nmol/2 $\times$ 10^6 cells/30 min. Direct evidence of NO· through detection in the gas phase of the specific chemiluminescence resulting from the reaction of nitric oxide with ozone showed categorically that neutrophils produce NO·.

It is becoming increasingly clear that NO·, present either exogenously or intracellularly, exerts serious influence on the physiology of neutrophil function. Production of O_2^- is reduced by the direct effects of extracellular NO· on the neutrophil membrane-bound oxidase (Forslund and Sundqvist, 1995b). Augmentation of bacterial phagocytosis by human neutrophils in the presence of L-arginine (but not D-arginine or glycine) has been demonstrated (Moffat et al., 1996). NO· has been shown to reduce neutrophil adherence (Clancy et al., 1995; Egdell et al., 1994; McCall et al., 1988) and cause depletion in intracellular glutathione (Clancy et al., 1994), but not to affect neutrophil-mediated killing by fMLP-activated rat neutrophils (Wagner et al., 1996). Anesthetics such as lidocaine stimulate NO· production in human neutrophils (Mamiya et al., 1995).

Both O_2^- and H_2O_2 production decrease in neutrophils pre-incubated with L-argi-

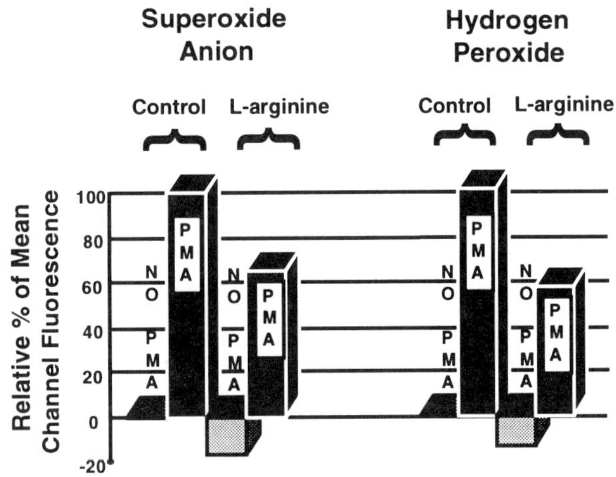

Fig. 12. Relative percent of O_2^- and H_2O_2 in human neutrophils with and without added L-arginine. 2 ng/ml PMA was used for cell stimulation. Both basal and stimulated O_2^- and H_2O_2 levels decreased after preincubation of PMN with 1 m*M* L-arginine.

nine and stimulated with low concentrations of PMA (2 ng/ml) (Fig. 12). In one set of experiments, reductions were 35 and 43% for O_2^- and H_2O_2, respectively. In unstimulated cells, L-arginine also caused a reduction in the basal O_2^-- and H_2O_2-induced fluorescence by approximately 15% each. These data provide evidence that L-arginine can modulate both O_2^- and H_2O_2 production in (human) neutrophils.

The effect of nitric oxide modulators on rat neutrophil O_2^- activity is demonstrated using the fluorescent indicator hydroethidine in Figure 13, which shows the rela-

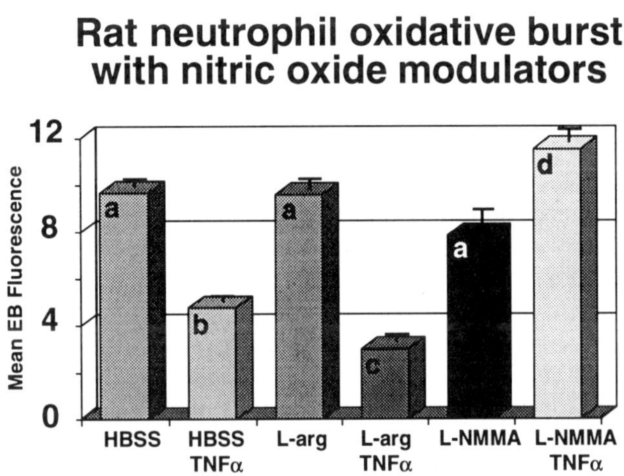

Fig. 13. O_2^- in rat neutrophils measured by flow cytometry after 15 min incubation with 4 ng/ml PMA. NOS inhibitor L-NMMA significantly reduced intracellular O_2^- of neutrophils.

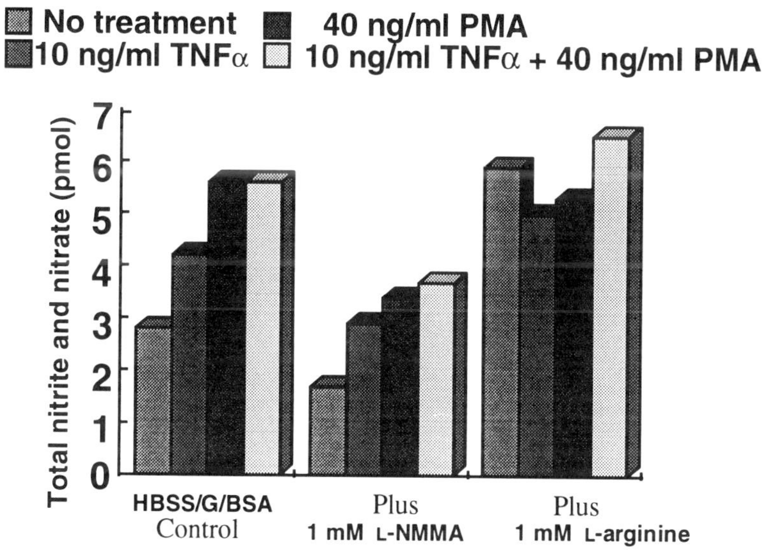

Fig. 14. Total nitrite and nitrate production in NO·-modulated rat peripheral blood neutrophil. Measurements after the treatments shown indicate significant NO· production.

tive change in ethidium bromide (EB) fluorescence (reflecting O_2^- production) in rat neutrophils incubated with L-NMMA (N^{ω}-monomethyl-L-arginine), an inhibitor of *i*NOS. A low concentration of phorbol myristate acetate (4 ng/ml, approximate V_{max} for the rat oxidative burst) was sufficient to elicit measurable changes in O_2^- production, which was significantly decreased by stimulating NO· production with 1 m*M* L-arginine and TNF-α (10 ng/ml). Upon addition of L-NMMA to inhibit NO· production, O_2^- was significantly increased. Confirmation of these data came from measuring total nitrites and nitrates in the same neutrophils. Figure 14 shows the differences in three groups: control, addition of L-NMMA, and L-arginine. A significant difference was observed between addition of L-NMMA to inhibit *i*NOS and all other treatments. This suggests the production of NO· in significant quantities by neutrophils and supports the hypothesis that the interactions between ROI and RNI play an important role in the activity of both systems.

Nitric Oxide, Peroxynitrite, and Disease

Nitric oxide is a free radical that has recently received considerable attention as an important messenger molecule in the body. In reviewing the physiological roles of the molecule, attention should first be directed to its function as a vasodilator through its production in vascular endothelial cells and action on underlying smooth muscle (Palmer et al., 1987). Confirmation that the effect was mediated by NO· was obtained by utilizing an inhibitor of the synthase enzyme for NO·, namely N^{ω}-monomethyl-L-arginine (L-NMMA) (Moncada et al., 1991).

It should be noted that NO·, being a gas, can freely diffuse throughout the cell, and indeed from one cell to another, unless bound by protein. The molecule can survive for about 10–30 s in typical physiological environments, during which time it has been hypothesized to travel between 200 and 600 μm (based upon diffusion coefficient of oxygen in tissue sections) (Knowles and Moncada, 1992).

The recognition that nitric oxide regulation can affect some disease processes has brought about an interest in the therapeutic use of high L-arginine diets (Becker et al., 1993; Saito et al., 1987; Gianotti et al., 1993) or L-arginine infusion in traumatized patients. Studies in L-arginine–treated rats have demonstrated significant alterations in the translocation of gut bacteria. In one study (Becker et al., 1993), the amount of NO· was substantially increased in L-arginine–fed rats after burn. In another (Gianotti et al., 1993), the authors concluded that one of the more significant effects of L-arginine administration was influence on immune function, most likely the macrophage. Current dogma suggests that increases in intracellular and extracellular nitric oxide may be beneficial to the appropriate microenvironment since the nitric oxide can act as an O_2^- scavenger, essentially being converted to nitrite and nitrate and forestalling the conversion of O_2^- via SOD to H_2O_2 and OH·.

A similar situation might exist in ischemia/reperfusion injury, in which the role of neutrophils has been demonstrated, particularly in remote injury such as seen in the lung (Mulligan et al., 1992). In addition, it has recently been demonstrated that endotoxin can cause a two-to fivefold increase in arginine transport by pulmonary artery endothelial cells (Lind et al., 1993), suggesting that vast amounts of NO· can be made, at least in the lung, which consists of approximately 50% endothelial cells (Crapo et al., 1978), creating significant quantities of potentially dangerous molecules. Administration of a recombinant human TNF soluble receptor type 1 to lipopolysaccharide-treated rats significantly reduced the damaging effects of endotoxin in lung tissue, probably by preventing the upregulation of IL-6 by TNF (Ulich et al., 1993).

Thus we have a curious situation. On the one hand, superoxide and its dismutation products, hydrogen peroxide and subsequent hydroxyl radical, have been considered to be the dangerous species of oxygen-mediated tissue damage. On the other hand, it has been hypothesized that SOD, by removal of superoxide anions, creates a protective barrier by directly preventing the reaction of O_2^- with NO· to produce peroxynitrite (Beckman et al., 1990) in a reaction known to be very rapid with a rate constant of at least $3.7 \times 10^7\ \mathrm{M}^{-1}\mathrm{s}^{-1}$ (Zhu et al., 1992).

Nitrogen as a Defense Mechanism

The concept that NO· can act as a defense mechanism is based upon the knowledge that NO· can remove superoxide from a system by a diffusion-limited reaction that forms peroxynitrite ($ONOO^-$):

$$O_2^- + NO\cdot \longrightarrow ONOO^-$$

Furthermore, this reaction occurs faster than the dismutation of O_2^- by SOD and therefore is highly likely to take place in biological systems. It is thought that NO· can react with transition metals, especially Fe^{2+} (which may also be the most available metal), thus competing with the Haber–Weiss reaction between H_2O_2 and Fe^{2+}. If this is true, NO· may be acting as an antioxidant.

Nitric oxide has been identified as playing a protective role by a number of studies, such as that of Wink et al. (1993), who showed that in the presence of NO·, the cytotoxicity of H_2O_2 and O_2^- against Chinese hamster lung fibroblasts was abrogated in a dose-responsive manner. Nitric oxide has been shown to promote ADP ribosylation of actin, thus inhibiting cytoskeletal assembly in neutrophils. The result is the regulation of neutrophil adhesion in margination, adhesion, and diapedesis (Clancy et al., 1995). Suppression of NO· in arteriolar and venular endothelium resulted in an increase in adhesion of leukocytes, but also a significant increase in oxidative stress. Furthermore, this L-NAME–induced enhancement of endothelial adhesiveness was mediated by intracellular oxidative stress rather than by direct action of NO· suppression (Hausladen and Fridovich, 1994).

Prolonged NOS inhibition in HUVECs caused an oxidative- and platelet activating factor (PAF)-associated rise in adhesion of neutrophils on the surface of endothelial cells (Niu et al., 1994), activated mast cells in the mucosa (Kanwar et al., 1994), and mast cells in rat mesenteric postcapillary venules (Kubes et al., 1993). Inhibition of nitric oxide synthesis increased leukocyte and endothelial interaction in rat mesenteric venules as measured by an increase in P-selectin expression (Davenpeck et al., 1994). Likewise Gauthier et al. (1994) demonstrated a reduction in adhesion and decreased P-selectin expression following infusion of exogenous nitric oxide. Volk et al. (1995) have demonstrated that extracellularly produced H_2O_2, but not O_2^-, enhanced the toxicity of NO· against endothelial cells. Decreased basal release of NO· after myocardial ischemia/reperfusion preceded enhanced neutrophil adherence to the coronary endothelium, leading to neutrophil-induced myocardial injury (Ma et al., 1993).

Nitric oxide overproduction was shown to be a major protective mechanism in the T cell-dependent shock induced by staphylococcal enterotoxin B in mice, and NOS inhibition might have detrimental consequences in T cell-mediated inflammatory disorders by enhancing both production and toxicity of inflammatory cytokines (Florquin et al., 1994). These data indicated that NO· production was able to exert a direct effect on the production of some cytokines (TNF-α and IFN-γ). Additionally, there is a significant amount of evidence that the presence of NO· regulates the adhesion of neutrophils, and in the absence of NO·-producing systems, a potentially damaging accumulation of inflammatory cells is likely.

Nitrogen as a Pathogenic Molecule

Direct tissue damage by NO· has also been demonstrated, but usually at higher than physiological concentrations. Administration of high concentrations of NO· (via nitroprusside, 10–40 μg/kg/min for 15 min) was shown to cause rat mucosal damage, which was attributed to excessive formation of peroxynitrite and subsequent formation of superoxide and hydroxyl radicals (Lamarque and Whittle, 1995).

The primary evidence for a role of nitric oxide in pathogenesis must come from the effect of peroxynitrite, a potent and reactive oxidant (Beckman and Tsai, 1994).

Peroxynitrite has been shown to damage endothelial cells directly (Palmer et al., 1992; Kooy and Royall, 1994; Kooy et al., 1994), and to impair vascular permeability (Villa et al., 1994) and play a role in the pathogenesis of atherogenesis (White et al., 1994).

CONCLUSIONS

It is clear that many of the observations previously considered to be related solely to reactive oxygen species are more likely associated with both ROI and RNI and the interactions between the two. There is a strong body of evidence that peroxynitrite is a mediator of serious consequences, and that NO· plays a powerful role in regulating many of the inflammatory responses, particularly those involving interaction between neutrophils and endothelial cells.

The benefits of using NOS inhibitors or NO· donors are yet to be made clear. If peroxynitrite is a particularly dangerous molecule, then prevention of its formation by NOS inhibitors is logical. However, the consequences of blocking NO· formation are themselves quite serious, leading to increased neutrophil adhesion and simultaneous removal of a significant pathway for elimination of superoxide.

Neutrophils and macrophages are particularly reactive cells. Neutrophils know but the one role, which they play to perfection: they are designed with powerful protective mechanisms for antioxidant defenses. They produce reactive oxygen and nitrogen species as a means of destroying invading microorganisms, while attempting at the same time to stay alive, at least until they have completed their disinfection task. It is during this period—between the attachment phase and destruction of microbes, and their subsequent removal from inflammatory sites—that significant tissue damage can occur. Alternatively, when vast numbers of these cells respond, as in ischemia/reperfusion injury, regulation of their reactive nature is necesasary to prevent rapid and destructive tissue injury. How exactly this regulation can be achieved without further compromising the host is a question still seeking answers.

REFERENCES

Ahmed N, Weidemann MJ (1996): Interaction of reactive nitrogen and oxygen intermediates in HL60 and dimethylsulphoxide-differentiated HL60 cells. Leuk Res 20:271–279.

Akaike T, Noguchi Y, Ijiri S, Setoguchi K, Suga M, Zheng YM, Dietzschold B, Maeda H (1996): Pathogenesis of influenza virus-induced pneumonia: involvement of both nitric oxide and oxygen radicals. Proc Natl Acad Sci USA 93:2448–2453.

Allen RC, Stjernholm RL, Steele RH (1972): Evidence for the generation of an electronic state(s) in human polymorphonuclear leukocytes and its participation in bactericidal activity. Biochem Biophys Res Commun 47:679–684.

Alvarez B, Denicola A, Radi R (1995): Reaction between peroxynitrite and hydrogen peroxide: formation of oxygen and slowing of peroxynitrite decomposition. Chem Res Toxicol 8:859–864.

Andreoli SP, Mallett CP, Bergstein JM (1986): Role of glutathione in protecting endothelial cells against hydrogen peroxide oxidant injury. J Lab Clin Med 108:190–198.

Babior BM, Kipnes RS, Curnutte JT (1973): Biological defense mechanisms. The production by leukocytes of superoxide, a potential bactericidal agent. J Clin Invest 52:741–744.

Bass DA, Parce JW, DeChatelet LR, Szejda P, Seeds MC, Thomas M (1983): Flow cytometric studies of oxidative product formation by neutrophils: a graded response to membrane stimulation. J Immunol 130:1910–1917.

Beauvais F, Michel L, Dubertret L (1995): Exogenous nitric oxide elicits chemotaxis of neutrophils in vitro. J Cell Physiol 165:610–614.

Becker WK, Shippee RL, McManus AT, Mason AD Jr., Pruitt BA Jr (1993): Kinetics of nitrogen oxide production following experimental thermal injury in rats. J Trauma 34:855–862.

Beckman JS (1995): Biochemistry of nitric oxide and peroxynitrite. In Kubes P (ed). *Nitric Oxide: A Modulator of Cell–Cell Interactions in the Microcirculation.* Austin, TX: R. G. Landes Company, pp 1–18.

Beckman JS, Beckman TW, Chen J, Marshall PA, Freeman BA (1990): Apparent hydroxyl radical production by peroxynitrite: Implications for endothelial injury from nitric oxide and superoxide. Proc Natl Acad Sci USA 87:1620–1624.

Beckman JS, Ischiropoulos H, Zhu L, et al. (1992): Kinetics of superoxide dismutase and iron catalyzed nitration of phenolics by peroxynitrite. Arch Biochem Biophys 298:438–445.

Beckman JS, Tsai JHM (1994): Reaction rates and diffusion in the toxicity of peroxynitrite. Biochemist 16:8–10.

Begley CG, Lopez AF, Vadas MA, Metcalf D (1985): The clonal proliferation *in vitro* of enriched populations of human promyelocytes and myelocytes. Blood 65:951–958.

Behar D, Czapski G, Rabani J, Dorfman LM, Schwartz HA (1970): The acid dissociation constant and decay kinetics of the perhydroxyl radical. J Phys Chem 74:3209.

Belenky SN, Robbins RA, Rubinstein I (1993): Nitric oxide synthase inhibitors attenuate human monocyte chemotaxis in vitro. J Leukoc Biol 53:498–503.

Biswas SK, Bhelwa AP, Upadhyay AU, George A, Nath N (1993): Status of nitric oxide free radicals in diabetic neutrophils: effect of diabetic serum factor on the generation of these species in normal neutrophils and their relation to lysosomal degranulation. Indian J Biochem Biophys 30:293–296.

Blough NV, Zafiriou OC (1985): Reaction of superoxide with nitric oxide to form peroxonitrite in alkaline aqueous solution. Inorg Chem 24:3502–3504.

Boggs DR (1967): The kinetics of neutrophilic leukocytes in health and disease. Semin Hematol 4:359–386.

Boissy RE, Trinkle LS, Nordlund JJ (1989): Separation of pigmented and albino melanocytes and the concomitant evaluation of endogenous peroxide content using flow cytometry. Cytometry 10:779–787.

Bredt DS, Glatt CE, Hwang PM, Fotuhi M, Dawson TM, Snyder SH (1991): Nitric oxide synthase protein and mRNA are discretely localized in neuronal populations of the mammalian CNS together with the NADPH diaphorase. Neuron 7:615–624.

Bucana C, Saiki I, Nayar R (1986): Uptake and accumulation of the vital dye hydroethidine in neoplastic cells. J Histochem Cytochem 34:1109–1115.

Bueb J-L, Gallois A, Schneider J-C, Parini J-P, Tschirhart E (1995): A double-labelling fluorescent assay for concomitant measurements of $[Ca^{2+}]_i$ and $O_2\cdot$ production in human macrophages. Biochim Biophys Acta 1244:79–84.

Callewaert DM, Radcliff G, Waite R, LeFevre J, Poulik MD (1991): Characterization of effector-target conjugates for cloned human natural killer and human lymphokine activated killer cells by flow cytometry. Cytometry 12:666–676.

Cao D, Boxer LA, Petty HR (1993): Deposition of reactive oxygen metabolites onto and within living tumor cells during neutrophil-mediated antibody-dependent cellular cytotoxicity. J Cell Physiol 156:428–436.

Carreras MC, Catz SD, Pargament GA, Del Bosco CG, Poderoso JJ (1994a): Decreased production of nitric oxide by human neutrophils during septic multiple organ dysfunction syndrome. Comparison with endotoxin and cytokine effects on normal cells. Inflammation 18:151–161.

Carreras MC, Pargament GA, Catz SD, Poderoso JJ, Boveris A (1994b): Kinetics of nitric oxide and hydrogen peroxide production and formation of peroxynitrite during the respiratory burst of human neutrophils. FEBS Lett 341:65–68.

Carter WO, Knapp DW, Snyder P, et al. (1994a): Oxidative burst-like activity in canine natural killer cells. Cytometry 7(Suppl):28 (Abstract).

Carter WO, Narayanan PK, Robinson JP (1994b): Intracellular hydrogen peroxide and superoxide anion detection in endothelial cells. J Leukoc Biol 55:253–258.

Catz SD, Carreras MC, Poderoso JJ (1995): Nitric oxide synthase inhibitors decrease human polymorphonuclear leukocyte luminol-dependent chemiluminescence. Free Radic Biol Med 19:741–748.

Cavarec L, Quillet MA, Fradelizi D, Conjeaud H (1990): An improved double fluorescence flow cytometry method for the quantification of killer cell/target cell conjugate formation. J Immunol Meth 130:251–261.

Chen LY, Mehta JL (1996): Variable effects of L-arginine analogs on L-arginine-nitric oxide pathway in human neutrophils and platelets may relate to different nitric oxide synthase isoforms. J Pharmacol Exp Ther 276:253–257.

Clancy R, Leszczynska J, Amin A, Levartovsky D, Abramson SB (1995): Nitric oxide stimulates ADP ribosylation of actin in association with the inhibition of actin polymerization in human neutrophils. J Leukoc Biol 58:196–202.

Clancy RM, Leszczynska-Piziak J, Abramson SB (1992): Nitric oxide, an endothelial cell relaxation factor, inhibits neutrophil superoxide anion production via a direct action on the NADPH oxidase. J Clin Invest 90:1116–1121.

Clancy RM, Levartovsky D, Leszczynska-Piziak J, Yegudin J, Abramson SB (1994): Nitric oxide reacts with intracellular glutathione and activates the hexose monophosphate shunt in human neutrophils: evidence for S-nitrosoglutathione as a bioactive intermediary. Proc Natl Acad Sci USA 91:3680–3684.

Clark RS, Kochanek PM, Schwarz MA, et al. (1996): Inducible nitric oxide synthase expression in cerebrovascular smooth muscle and neutrophils after traumatic brain injury in immature rats. Pediatr Res 39:784–790.

Cook HT, Ebrahim H, Jansen AS, Foster GR, Largen P, Cattell V (1994): Expression of the gene for inducible nitric oxide synthase in experimental glomerulonephritis in the rat. Clin Exp Immunol 97:315–320.

Crapo JD, Peters-Golden M, Marsh-Salin J, Shelburne, JS (1978): Pathologic changes in the lungs of oxygen-adapted rats: a morphometric analysis. Lab Invest 39:640–653.

Cross AR, Jones OTG, Harper AM, Segal AW (1981): Oxidation-reduction properties of the cytochrome *b* found in the plasma-membrane fraction of human neutrophils. Biochem J 194:599–606.

Cross CE (1987): Oxygen radicals and human disease. Ann Intern Med 107:526–545.

Curnutte JT, Scott PJ, Mayo LA (1989): Cytosolic components of the respiratory burst oxidase: resolution of four components, two of which are missing in complementing types of chronic granulomatous disease. Proc Natl Acad Sci USA 86:825–829.

Davenpeck KL, Gauthier TW, Lefer AM (1994): Inhibition of endothelial-derived nitric oxide promotes P-selectin expression and actions in the rat microcirculation. Gastroenterology 107:1050–1058.

Demaurex N, Downey GP, Waddell TK, Grinstein S (1996): Intracellular pH regulation during spreading of human neutrophils. J Cell Biol 133:1391–1402.

Dias-Da-Motta P, Arruda VR, Muscara MN, et al. (1996): The release of nitric oxide and superoxide anion by neutrophils and mononuclear cells from patients with sickle cell anaemia. Br J Haematol 93:333–340.

Egdell RM, Siminiak T, Sheridan DJ (1994): Modulation of neutrophil activity by nitric oxide during acute myocardial ischaemia and reperfusion [review]. Basic Res Cardiol 89:499–509.

Elsner J, Dichmann S, Kapp A (1995): Activation of the respiratory burst in human eosinophils by chemotaxins requires intracellular calcium fluxes. J Invest Dermatol 105:231–236.

Elsner J, Oppermann M, Czech W, Kapp A (1994): C3a activates the respiratory burst in human polymorphonuclear neutrophilic leukocytes via pertussis toxin-sensitive G-proteins. Blood 83:3324–3331.

Emmendörffer A, Nakamura M, Rothe G, Spiekermann K, Lohmann-Matthes M-L, Roesler J (1994): Evaluation of flow cytometric methods for diagnosis of chronic granulomatous disease variants under routine laboratory conditions. Cytometry 18:147–155.

Epling CL, Stites DP, McHugh TM, Chong HO, Blackwood LL, Wara DW (1992): Neutrophil function screening in patients with chronic granulomatous disease by a flow cytometric method. Cytometry 13:615–620.

Erslev AJ, Weiss L (1977): Structure and function of the bone marrow. In Williams WJ, Beutler E, Erslev AJ, Rundles RW (eds). *Hematology*. 2nd ed. New York: McGraw-Hill Book Co., pp 57–64.

Evans TJ, Buttery LD, Carpenter A, Springall DR, Polak JM, Cohen J (1996): Cytokine-treated human neutrophils contain inducible nitric oxide synthase that produces nitration of ingested bacteria. Proc Natl Acad Sci USA 93:9553–9558.

Florquin S, Amraoui Z, Dubois C, Decuyper J, Goldman M (1994): The protective role of endogenously synthesized nitric oxide in staphyloccal enterotoxin B-induced shock in mice. J Exp Med 180:1153–1158.

Forman HJ, Fisher AB (1981): Antioxidant defenses. In Gilbert DL (ed). *Oxygen and Living Processes. An Interdisciplinary Approach.* New York: Springer Verlag, pp 235–249.

Forman HJ, Fridovich I (1973): On the stability of bovine superoxide dismutase. The effects of metals. J Biol Chem 248:2645–2649.

Forslund T, Sundqvist T (1995a): Nitric oxide regulates the chemiluminescence from stimulated human neutrophils. APMIS 103:813–817.

Forslund T, Sundqvist T (1995b): Nitric oxide reduces hydrogen peroxide production from human polymorphonuclear neutrophils. Eur J Clin Invest 25:9–14.

Förstermann U, Schmidt HH, Pollock JS, Sheng H (1991): Isoforms of nitric oxide synthase. Characterization and purification from different cell types. Biochem Pharmacol 42:1849–1857.

Freedman JE, Loscalzo J, Benoit SE, Valeri CR, Barnard MR, Michelson AD (1996): Decreased platelet inhibition by nitric oxide in two brothers with a history of arterial thrombosis. J Clin Invest 97:979–987.

Fridovich I (1981): Superoxide radical and superoxide dismutases. In Gilbert DL (ed). *Oxygen and Living Processes. An Interdisciplinary Approach.* New York: Springer Verlag, pp 250–272.

Frisch H, Forman A, Borg DC (1983): H2O2-diffusion through lysosomes. Israel J Chem 23:442–445.

Gauthier TW, Davenpeck KL, Lefer AM (1994): Nitric oxide attenuates leukocyte-endothelial interaction via P-selectin in splanchnic ischemia-reperfusion. Am J Physiol 267:G562–G568.

Gianotti L, Alexander JW, Pyles T, Fukushima R (1993): Arginine-supplemented diets improve survival in gut-derived sepsis and peritonitis by modulating bacterial clearance. The role of nitric oxide. Ann Surg 217:644–653; discus.

Gilbert DL (1981): Perspective on the history of oxygen and life. In Gilbert DL (ed). *Oxygen and Living Processes: An Interdisciplinary Approach.* New York: Springer Verlag, pp 1–43.

Goldman D, Cho Y, Zhao M, Casadevall A, Lee SC (1996): Expression of inducible nitric oxide synthase in rat pulmonary Cryptococcus neoformans granulomas. Am J Pathol 148:1275–1282.

Goossens V, Grooten J, De Vos K, Fiers W (1995): Direct evidence for tumor necrosis factor-induced mitochondrial reactive oxygen intermediates and their involvement in cytotoxicity. Proc Natl Acad Sci USA 92:8115–8119.

Gubitz G, Frei RW, Bethke H (1976): Fluorescence densitometric method for the determination of gluconic and lactobionic acids ("sugar acids") in pharmaceutical preparations. J Chromatogr 117:337–343.

Haddad IY, Crow JP, Hu P, Ye Y, Beckman J, Matalon S (1994): Concurrent generation of nitric oxide and superoxide damages surfactant protein A. Am J Physiol 267:L242–L249.

Haffner PH, Coleman JE (1973): Cu(II)-carbon bonding in cyanide complexes of copper enzymes. 13C splitting of the Cu(II) electron spin resonance. J Biol Chem 248:6626–6629.

Harrison JE, Schultz J (1976): Studies on the chlorinating activity of myeloperoxidase. J Biol Chem 251:1371–1374.

Haugland RP (1992): *Handbook of Fluorescent Probes and Research Chemicals.* 5th ed. Eugene, OR: Molecular Probes, Inc.

Hausladen A, Fridovich I (1994): Superoxide and peroxynitrite inactivate aconitases, but nitric oxide does not. J Biol Chem 269:29405–29408.

Hayem G, Petit PX, Levacher M, Gaudin C, Kahn M-F, Pocidalo J-J (1994): Cytofluorometric analysis of chondrotoxicity of fluoroquinolone antimicrobial agents. Antimicrob Agents Chemother 38:243–247.

Hedley DW, Chow S (1992): Flow cytometric measurement of lipid peroxidation in vital cells using parinaric acid. Cytometry 13:686–692.

Hedley DW, Chow S (1994): Evaluation of methods for measuring cellular glutathione content using flow cytometry. Cytometry 15:349–358.

Heikkila RE, Cohen G (1977): The inactivation of copper-zinc superoxide dismutase by diethyldithiocarbamate. In Michelson AM, McCord JM, Fridovich I (eds). *Superoxide and Superoxide Dismutases.* New York: Academic Press, pp 367–373.

Hibbs JB, Taintor RR, Vavrin Z (1987): Macrophage cytotoxicity: role for L-arginine deiminase and imino nitrogen oxidation to nitrite. Science 235:473–476.

Himmelfarb J, Hakim RM, Holbrook DG, Leeber DA, Ault KA (1992): Detection of granulocyte reactive oxygen species formation in whole blood using flow cytometry. Cytometry 13:83–89.

Hogaboam CM, Bissonnette E, Befus AD, et al. (1993): Modulation of rat mast cell reactivity by IL-1 beta. Divergent effects on nitric oxide and platelet-activating factor release. J Immunol 151:3767–3774.

Holter W, Goldman CK, Casabo L, Nelson DL, Greene WC, Waldmann TA (1987): Expression of functional IL 2 receptors by lipopolysaccharide and interferon-gamma stimulated human monocytes. J Immunol 138:2917–2922.

Homan-Muller JW, Weening RS, Roos D (1975): Production of hydrogen peroxide by phagocytizing human granulocytes. J Lab Clin Med 85:198–206.

Huie RE, Padmaja S (1993): The reaction of NO with superoxide. Free Radical Res Commun 18:195–199.

Inoue S, Kawanishi S (1995): Oxidative DNA damage induced by simultaneous generation of nitric oxide and superoxide. FEBS Lett 371:86–88.

Ischiropoulos H, Zhu L, Beckman JS (1992a): Peroxynitrite formation from macrophage-derived nitric oxide. Arch Biochem Biophys 298:446–451.

Ischiropoulos H, Zhu L, Chen J, et al. (1992b): Peroxynitrite-mediated tyrosine nitration catalyzed by superoxide dismutase. Arch Biochem Biophys 298:431–437.

Jarjour NN, Calhoun WJ (1994): Enhanced production of oxygen radicals in asthma. J Lab Clin Med 123:131–136.

Johnson KJ, Ward PA (1974): Acute immunologic pulmonary alveolitis. J Clin Invest 54:349–357.

Johnson KR, Nauseef WM (1991): Molecular Biology of MPO. In Everse J, Everse KE, Grisham MB (eds). *Peroxidases in Chemistry and Biology.* Boca Raton, FL: CRC Press, pp 63–81.

Kaffenberger W, van Beuningen D (1994): Differences in radiosensitivity of the respiratory burst generated in HL-60 cells via different signal transduction pathways. Int J Radiat Biol 66:767–774.

Kanwar S, Wallace JL, Befus D, Kubes P (1994): Nitric oxide synthesis inhibition increases epithelial permeability via mast cells. Am J Physiol 266:G222–G229.

Kearns DR (1979): Solvent and solvent isotope effects on the lifetime of singlet oxygen. In Wasserman HH, Murray RW (eds). *Singlet Oxygen.* New York: Academic Press, pp 115–137.

Keller GA, Barke R, Harty JT, Humphrey E, Simmons RL (1985): Decreased hepatic glutathione levels in septic shock. Predisposition of hepatocytes to oxidative stress: an experimental approach. Arch Surg 120:941–945.

Keston AS, Brandt R (1965): The fluorometric analysis of ultramicro quantities of hydrogen peroxide. Anal Biochem 11:1–5.

Kim Y-M, Hong S-J, Billiar TR, Simmons RL (1996): Counterprotective effect of erythrocytes in experimental bacterial peritonitis is due to scavenging of nitric oxide and reactive oxygen intermediates. Infect Immun 64:3074–3080.

Klebanoff SJ, Nathan CF (1993): Nitrite production by stimulated human polymorphonuclear leukocytes supplemented with azide and catalase. Biochem Biophys Res Commun 197:192–196.

Knowles RG, Moncada S (1992): Nitric oxide as a signal in blood vessels. Trends Biochem Sci 17:399–402.

Kobzik L, Godleski JJ, Brain JD (1990a): Oxidative metabolism in the alveolar macrophage: analysis by flow cytometry. J Leukoc Biol 47:295–303.

Kobzik L, Godleski JJ, Brain JD (1990b): Selective down-regulation of alveolar macrophage oxidative response to opsonin-independent phagocytosis. J Immunol 144:4312–4319.

Kolls J, Xie J, LeBlanc R, Malinski T, Nelson S, Summer W, Greenberg SS (1994): Rapid induction of messenger RNA for nitric oxide synthase II in rat neutrophils in vivo by endotoxin and its suppression by prednisolone. Proc Soc Exp Biol Med 205:220–229.

Kontos HA (1989): Oxygen radicals in CNS damage. Chem Biol Interact 72:229–255.

Kooy NW, Royall JA (1994): Agonist-induced peroxynitrite production from endothelial cells. Arch Biochem Biophys 310:352–359.

Kooy NW, Royall JA, Ischiropoulos H, Beckman JS (1994): Peroxynitrite-mediated oxidation of dihydrorhodamine 123. Free Radical Biol Med 16:149–156.

Kubes P, Kanwar S, Niu XF, Gaboury JP (1993): Nitric oxide synthesis inhibition induces leukocyte adhesion via superoxide and mast cells. FASEB J 7:1293–1299.

Kubes P, Suzuki M, Granger DN (1991): Nitric oxide: An endogenous modulator of leukocyte adhesion. Proc Natl Acad Sci USA 88:4651–4655.

Kuypers TW, Koenderman L, Weening RS, Verhoeven AJ, Roos D (1990): Continuous cell activation is necessary for stable interaction of complement receptor type 3 with its counter-structure in the aggregation response of human neutrophils. Eur J Immunol 20:501–508.

Laffi G, Foschi M, Masini E, et al. (1995): Increased production of nitric oxide by neutrophils and monocytes from cirrhotic patients with ascites and hyperdynamic circulation. Hepatology 22:1666–1673.

Lamarque D, Whittle BJ (1995): Role of oxygen-derived metabolites in the rat gastric mucosal injury induced by nitric oxide donors. Eur J Pharmacol 277:187–194.

Larfars G, Gyllenhammar H (1995): Measurement of methemoglobin formation from oxyhemoglobin. A real-time, continuous assay of nitric oxide release by human polymorphonuclear leukocytes. J Immunol Meth 184:53–62.

LeBel CP, Ali SF, Bondy SC (1992): Deferoxamine inhibits methyl mercury-induced increases in reactive oxygen species formation in rat brain. Toxicol Appl Pharmacol 112:161–165.

Lee DKH, Faunce D, Henry D, Sturm R, Rimele T (1988): Rat polymorphonuclear leukocytes (PMN) increase cGMP levels in rat aorta. FASEB J 2:A518

Lepoivre M, Roche AC, Tenu JP, Petit JF, Nolibe D, Monsigny M (1986): Identification of two macrophage populations by flow cytometry monitoring of oxidative burst and phagocytic functions. Biol Cell 57:143–146.

Lind DS, Copeland EM, Souba WW (1993): Endotoxin stimulates arginine transport in pulmonary artery endothelial cells. Surgery 114:199–205.

Long CA, Bielski BHJ (1980): Rate of reaction of superoxide radical with chloride-containing species. J Phys Chem 84:555–557.

Ma X-L, Weyrich AS, Lefer DJ, Lefer AM (1993): Diminished basal nitric oxide release after myocardial ischemia and reperfusion promotes neutrophil adherence to coronary endothelium. Circ Res 72:403–412.

Maly F.-E., Nakamura M, Gauchat J-F, et al. (1989): Superoxide-dependent nitroblue tetrazolium reduction and expression of cytochrome b_{-245} components by human tonsillar B lymphocytes and B cell lines. J Immunol 142:1260–1267.

Mamiya K, Tomoda MK, Edashige K, Ueda W, Manabe M (1995): Local anesthetics enhance nitric oxide production by human peripheral neutrophils. Physiol Chem Phys Med NMR 27:111–119.

Marklund SL (1982): Human copper-containing superoxide dismutase of high molecular weight. Proc Natl Acad Sci USA 79:7634–7638.

Matsubara T, Ziff M (1986): Increased superoxide anion released from human endothelial cells in response to cytokines. J Immunol 137:3295–3298.

McCall T, Whittle BJR, Boughton-Smith NK, et al. (1988): Inhibition of FMLP-induced aggregation of rabbit neutrophils by nitric oxide. Br J Pharmacol 95:517P (Abstract).

McCall TB, Boughton-Smith NK, Palmer RMJ, Whittle BJR, Moncada S (1989): Synthesis of nitric oxide from L-arginine by neutrophils. Release and interaction with superoxide anion. Biochem J 261:293–296.

McCauley I, Hartmann PE (1984): Changes in piglet leucocytes, B lymphocytes and plasma cortisol from birth to three weeks after weaning. Res Vet Sci 37:234–241.

McCord JM, Fridovich I (1969a): Superoxide dismutase: An enzymatic function for erythrocuprein (hemocupriene). J Biol Chem 244:6049–6055.

McCord JM, Fridovich I (1969b): The utility of superoxide dismutase in studying free radical reactions. I. Radicals generated by the interaction of sulfite, dimethyl sulfoxide, and oxygen. J Biol Chem 244:6056–6063.

McCord JM, Keele BBJ, Fridovich I (1971): An enzyme-based theory of obligate anaerobiosis: the physiological function of superoxide dismutase. Proc Natl Acad Sci USA 68:1024–1027.

Meier B, Cross AR, Hancock JT, Kaup FJ, Jones OTG (1991): Identification of a superoxide-generating NADPH oxidase system in human fibroblasts. Biochem J 275:241–245.

Miles AM, Bohle DS, Glassbrenner PA, Hansert B, Wink DA, Grisham MB (1996): Modulation of superoxide-dependent oxidation and hydroxylation reactions by nitric oxide. J Biol Chem 271:40–47.

Misra HP, Fridovich I (1978): Inhibition of superoxide dismutase by azide. Arch Biochem Biophys 189:317–322.

Moffat FL, Jr., Han T, Li ZM, et al. (1996): Supplemental L-arginine HCl augments bacterial phagocytosis in human polymorphonuclear leukocytes. J Cell Physiol 168:26–33.

Moncada S, Palmer RMJ, Higgs EA (1991): Nitric oxide: physiology, pathophysiology, and pharmacology. Pharmacol Rev 43:109–142.

Moore KP, Darley-Usmar V, Morrow J, Roberts LJ II (1995): Formation of F_2-isoprostanes during oxidation of human low-density lipoprotein and plasma by peroxynitrite. Circ Res 77:335–341.

Moreno JJ, Pryor WA (1992): Inactivation of α-1 proteinase inhibitor by peroxynitrite. Chem Res Toxicol 5:425–431.

Mulligan MS, Varani J, Warren JS, et al. (1992): Roles of β_2 integrins of rat neutrophils in complement and oxygen radical-mediated acute inflammatory injury. J Immunol 148:1847–1857.

Narayan PK, Carter WO, Ganey PE, Roth RA, Voytik-Harbin SL, Robinson JP (1998): Impairment of human neutrophil oxidative burst by polychlorinated biphenyls: inhibition of superoxide dismutase activity. J Leukoc Biol 63:216–224.

Nathan CF (1992): Nitric oxide as a secretory product of mammalian cells. FASEB J 6:3051–3064.

Nathan CF, Xie Q-w (1994): Regulation of biosynthesis of nitric oxide. J Biol Chem 269:13725–13728.

Nikitovic D, Holmgren A (1996): S-nitrosoglutathione is cleaved by the thioredoxin system with liberation of glutathione and redox regulating nitric oxide. J Biol Chem 271:19180–19185.

Niu XF, Smith CW, Kubes P (1994): Intracellular oxidative stress induced by nitric oxide synthesis inhibition increases endothelial cell adhesion to neutrophils. Circ Res 74:1133–1140.

Nussler AK, Billiar TR (1993): Inflammation, immunoregulation, and inducible nitric oxide synthase. J Leukoc Biol 54:171–178.

Ohno Y, Gallin JI (1985): Diffusion of extracellular hydrogen peroxide into intracellular compartments of human neutrophils. Studies utilizing the inactivation of myeloperoxidase by hydrogen peroxide and azide. J Biol Chem 260:8438–8446.

Olive PL (1989): Hydroethidine: a fluorescent redox probe for locating hypoxic cells in spheroids and murine tumours. Br J Cancer 60:332–338.

Oury TD, Chang L, Marklund SL, Day BJ, Crapo JD (1994): Immunocytochemical localization of extracellular superoxide dismutase in human lung. Lab Invest 70:889–898.

Padgett EL, Pruett SB (1995): Rat, mouse and human neutrophils stimulated by a variety of activating agents produce much less nitrite than rodent macrophages. Immunology 84:135–141.

Palmer RMJ, Bridge L, Foxwell NA, Moncada S (1992): The role of nitric oxide in endothelial cell damage and its inhibition by glucocorticoids. Br J Pharmacol 105:11–12.

Palmer RMJ, Ferrige AG, Moncada S (1987): Nitric oxide release accounts for the biological activity of endothelium-derived relaxing factor. Nature 327:524–526.

Parkos CA, Allen RA, Cochrane CG, Jesaitis AJ (1987): Purified cytochrome b from human granulocyte plasma membrane is comprised of two polypetides with relative molecular weights of 91,000 and 22,000. J Clin Invest 80:732–742.

Pollock JS, Förstermann U, Mitchell JA, et al. (1991): Purification and characterization of particulate endothelium-derived relaxing factor synthase from cultured and native bovine aortic endothelial cells. Proc Natl Acad Sci USA 88:10480

Radcliff G, Waite R, LeFevre J, Poulik MD, Callewaert DM (1991): Quantification of effector/target conjugation involving natural killer (NK) or lymphokine activated killer (LAK) cells by two-color flow cytometry. J Immunol Meth 139:281–292.

Radeke HH, Cross AR, Hancock JT, et al. (1991): Functional expression of NADPH oxidase components (α and β subunits of cytochrome b_{558} and 45-kDa flavoprotein) by intrinsic human glomerular mesangial cells. J Biol Chem 266:21026–21029.

Radeke HH, Meier B, Topley N, Floge J, Habermehl GG, Resch K (1990): Interleukin 1α and tumor necrosis factor-α induce oxygen radical production in mesangial cells. Kidney Int 37:767–775.

Radi R, Beckman JS, Bush KM, Freeman BA (1991a): Peroxynitrite oxidation of sulfhydryls: The cytotoxic potention of superoxide and nitric oxide. J Biol Chem 266:4244–4250.

Radi R, Beckman JS, Bush KM, Freeman BA (1991b): Peroxynitrite-induced membrane lipid peroxidation: the cytotoxic potential of superoxide and nitric oxide. Arch Biochem Biophys 288:481–487.

Radi R, Rodriguez M, Castro L, Telleri R (1994): Inhibition of mitochondrial electron transport by peroxynitrite. Arch Biochem Biophys 308:89–95.

Rao KMK, Padmanabhan J, Kilby DL, Cohen HJ, Currie MS, Weinberg JB (1992): Flow cytometric analysis of nitric oxide production in human neutrophils using dichlorofluorescein diacetate in the presence of a calmodulin inhibitor. J Leukoc Biol 51:496–500.

Reiss M, Roos D (1978): Differences in oxygen metabolism of phagocytosing monocytes and neutrophils. J Clin Invest 61:480–488.

Richardson JS, Thomas KA, Rubin BH, Richardson DC (1975): Crystal structure of bovine Cu, Zn superoxide dismutase at 3Å reduction: chain tracing and metal ligands. Proc Natl Acad Sci USA 72:1349–1353.

Rimele T, Sturm RJ, Adams LM, Henry DE, Heaslip RJ, Weichman BM, Grimes D (1988): Interaction of neutrophils with vascular smooth muscle: identification of a neutrophil-derived relaxing factor. Exp Ther 245:102–111.

Robinson JP, Bruner LH, Bassøe C-F, Hudson JL, Ward PA, Phan SH (1988): Measurement of intracellular fluorescence of human monocytes relative to oxidative metabolism. J Leukoc Biol 43:304–310.

Robinson JP, Carter WO, Narayanan PK (1994a): Oxidative product formation analysis by flow cytometry. In Darzynkiewicz Z, Robinson JP, Crissman HA (eds). *Methods in Cell Biology: Flow Cytometry.* 2nd ed. New York: Academic Press, pp 437–447.

Robinson JP, Narayanan PK, Carter WO (1994b): Functional measurements using HL-60 cells. In Darzynkiewicz Z, Robinson JP, Crissman HA (eds). *Methods In Cell Biology. Flow Cytometry.* 2nd ed. New York: Academic Press, pp 423–436.

Robinson JP, Narayanan PK (1996): Defects in intracellular oxidative metabolism of neutrophils undergoing apoptosis. J Leukoc Biol (suppl):38. (abstr).

Robinson WA, Mangalik A (1975): The kinetics and regulation of granulopoiesis. Semin Hematol 12:7–25.

Roos D, Reiss M, Balm AJ, Palache AM, Cambier PH, Van Der Stijl-Neijenhuis JS (1979): A metabolic comparison between human blood monocytes and neutrophils. Adv Exp Med Biol 121:29–36.

Rothe G, Emmendörffer A, Oser A, Roesler J, Valet G (1991): Flow cytometric measurement of the respiratory burst activity of phagocytes using dihydrorhodamine 123. J Immunol Meth 138:133–135.

Rothe G, Oser A, Valet GK (1988): Dihydrorhodamine 123: a new flow cytometric indicator for respiratory burst activity in neutrophil granulocytes. Naturwissenschaften 75:354–355.

Rothe G, Valet G (1990): Flow cytometric analysis of respiratory burst activity in phagocytes with hydroethidine and 2′,7′-dichlorofluorescin. J Leukoc Biol 47:440–448.

Rotrosen D, Leto TL (1990): Phosphorylation of neutrophil 47kDa cytosolic oxidase factor. Translocation to membrane is associated with distinct phosphorylation events. J Biol Chem 265:19910–19915.

Royall JA, Ischiropoulos H (1993): Evaluation of 2′,7′-dichlorofluorescin and dihydrorhodamine 123 as fluorescent probes for intracellular H2O2 in cultured endothelial cells. Arch Biochem Biophys 302:348–355.

Rubbo H, Radi R, Trujillo M, et al. (1994): Nitric oxide regulation of superoxide and peroxynitrite-dependent lipid peroxidation. Formation of novel nitrogen-containing oxidized lipid derivatives. J Biol Chem 269:26066–26075.

Saez JC, Kessler JA, Bennett MV, Spray DC (1987): Superoxide dismutase protects cultured neurons against death by starvation. Proc Natl Acad Sci USA 84:3056–3059.

Saiki I, Bucana CD, Tsao JY, Fidler IJ (1986): Quantitative fluorescent microassay for identification of antiproliferative compounds. J Natl Cancer Inst 77:1235–1240.

Saito H, Trocki O, Wang S, Gonce SJ, Joffe SN, Alexander JW (1987): Metabolic and immune effects of dietary arginine supplementation after burn. Arch Surg 122:784–789.

Salgo MG, Bermúdez E, Squadrito GL, Pryor WA (1995): DNA damage and oxidation of thiols peroxynitrite causes in rat thymocytes. Arch Biochem Biophys 322:500–505.

Salvemini D, Masini E, Pistelli A, Mannaioni PF, Vane J (1991): Nitric oxide: A regulatory mediator of mast cell reactivity. J Cardiovasc Pharmacol 17:S258–S264.

Sandstrom J, Karlsson K, Edlund T, Marklund SL (1993): Heparin-affinity patterns and composition of extracellular-superoxide dismutase in human plasma and tissues. Biochem J 294:853–857.

Schmidt HHHW, Seifert R, Böhme E (1989): Formation and release of nitric oxide from human neutrophils and HL-60 cells induced by a chemotactic peptide, platelet activating factor and leukotriene B_4. FEBS Lett 244:357–360.

Schraufstätter IU, Browne K, Harris A, et al. (1990): Mechanism of hypochlorite injury of target cells. J Clin Invest 85:554–562.

Schraufstätter IU, Jackson JH (1992): Early injury of cells by external oxidants and the consequences of DNA damage. In Cochrane CG, Gimbrone MA (eds). *Biological Oxidants: Generation and Injurious Consequences.* New York: Academic Press, pp 21–41.

Schultz J, Kaminker K (1962): Myeloperoxidase of the leukocyte of normal human blood: content and localization. Arch Biochem Biophys 96:465–467.

Scott JA, Homcy CJ, Khaw BA, Rabito CA (1988): Quantitation of intracellular oxidation in a renal epithelial cell line. Free Radic Biol Med 4:79–83.

Seeds MC, Parce JW, Szejda P, Bass DA (1985): Independent stimulation of membrane potential changes and the oxidative metabolic burst in polymorphonuclear leukocytes. Blood 65:233–240.

Smith JA, Weidemann MJ (1993): Further characterization of the neutrophil oxidative burst by flow cytometry. J Immunol Meth 162:261–268.

Stelzer GT, Robinson JP (1988): Flow cytometric evaluation of leukocyte function. Diag Clin Immunol 5:223–231.

Stuehr DJ, Griffith OW (1992): Mammalian nitric oxide synthases. Adv Enzymol Mol Related Areas Biol 65:287–346.

Szabó C, Day BJ, Salzman AL (1996): Evaluation of the relative contribution of nitric oxide and peroxynitrite to the suppression of mitochondrial respiration in immunostimulated macrophages using a manganese mesoporphyrin superoxide dismutase mimetic and peroxynitrite scavenger. FEBS Lett 381:82–86.

Szabó C, Salzman AL, Ischiropoulos H (1995): Peroxynitrite-mediated oxidation of dihydrorhodamine 123 occurs in early stages of endotoxic and hemorrhagic shock and ischemia-reperfusion injury. FEBS Lett 372:229–232.

Tanigaki T, Suzuki Y, Heimer D, Sussman HH, Ross WG, Raffin TA (1993): Attenuation of acute lung injury and oxygen radical production by the 21-aminosteroid, U-78518F. J Appl Physiol 74:2155–2160.

Terada LS, Willingham IR, Rosandich ME, Leff JA, Kindt GW, Repine JE (1991): Generation of superoxide anion by brain endothelial cell xanthine oxidase. J Cell Physiol 148:191–196.

Thomas MJ, Mehl KS, Pryor WA (1978): The role of the superoxide anion in the xanthine oxidase-induced autooxidation of linoleic acid. Biochem Biophys Res Commun 83:927

Tiku ML, Liesch JB, Robertson FM (1990): Production of hydrogen peroxide by rabbit articular chondrocytes. Enhancement by cytokines. J Immunol 145:690–696.

Till GO, Johnson KJ, Kunkel R, Ward PA (1982): Intravascular activation of complement and acute lung injury. Dependency on neutrophils and toxic oxygen metabolites. J Clin Invest 69:1126–1135.

Tonnesen MG (1989): Neutrophil-endothelial cell interactions: mechanisms of neutrophil adherence to vascular endothelium. J Invest Dermatol 93:53S–58S.

Tsan MF, Danis EH, Del Vecchio PJ, Rosano CL (1985): Enhancement of intracellular glutathione protects endothelial cells against oxidant damage. Biochem Biophys Res Commun 127:270–276.

Tsinkalovsky OR, Laerum OD (1994): Flow cytometric measurement of the production of reactive oxygen intermediate in activated rat mast cells. APMIS 102:474–480.

Ublacker GA, Johnson JA, Siegel FL, Mulcahy RT (1991): Influence of glutathione S-transferases on cellular glutathione determination by flow cytometry using monochlorobimane. Cancer Res 51:1783–1788.

Ulich TR, Yin S, Remick DG, Russell D, Eisenberg SP, Kohno T (1993): Intratracheal administration of endotoxin and cytokines. IV. The soluble tumor necrosis factor receptor type I inhibits acute inflammation. Am J Pathol 142:1335–1338.

van de Water B, Zoeteweij JP, de Bont HJ, Nagelkerke JF (1995): Inhibition of succinate:ubiquinone reductase and decrease of ubiquinol in nephrotoxic cysteine S-conjugate-induced oxidative cell injury. Mol Pharmacol 48:928–937.

van Pelt LJ, van Zwieten R, Weening RS, Roos D, Verhoeven AJ, Bolscher BG (1996): Limitations on the use of dihydrorhodamine 123 for flow cytometric analysis of the neutrophil respiratory burst. J Immunol Methods 191:187–196.

Villa LM, Salas E, Darley-Usmar VM, Radomski MW, Moncada S (1994): Peroxynitrite induces both vasodilation and impaired vascular relaxation in the isolated perfused rat heart. Proc Natl Acad Sci USA 91:12383–12387.

Voetman AA, Roos D (1980): Endogenous catalase protects human blood phagocytes against oxidative damage by extracellularly generated hydrogen peroxide. Blood 56:846–852.

Volk T, Ioannidis I, Hensel M, deGroot H, Kox WJ (1995): Endothelial damage induced by nitric oxide: synergism with reactive oxygen species. Biochem Biophys Res Commun 213:196–203.

Vowells SJ, Sekhsaria S, Malech HL, Shalit M, Fleisher TA (1995): Flow cytometric analysis of the granulocyte respiratory burst: a comparison study of fluorescent probes. J Immunol Meth 178:89–97.

Waddell TK, Fialkow L, Chan CK, Kishimoto TK, Downey GP (1994): Potentiation of the oxidative burst of human neutrophils. A signaling role for L-selectin. J Biol Chem 269:18485–18491.

Wagner JG, Ganey PE, Roth RA (1996): Nitric oxide is not involved in hepatocyte killing by neutrophils activated by N-formyl-methionyl-leucylphenylalanine or phorbol myristate acetate in vitro. Hepatology 23:803–810.

Weisiger RA, Fridovich I (1973): Superoxide dismutase: Organelle specificity. J Biol Chem 248:3582–3592.

Wenisch C, Narzt E, Sessler DI, et al. (1996): Mild intraoperative hypothermia reduces production of reactive oxygen intermediates by polymorphonuclear leukocytes. Anesth Analg 82:810–816.

Wenisch C, Parschalk B, Hasenhündl M, Wiesinger E, Graninger W (1995): Effect of cefodizime and ceftriaxone on phagocytic function in patients with severe infections. Antimicrob Agents Chemother 39:672–676.

Wennmalm A, Benthin G, Petersson A-S (1992): Dependence of the metabolism of nitric oxide (NO) in healthy human whole blood on the oxygenation of its red cell. Br J Pharmacol 106:507–508.

White CR, Brock TA, Chang LY, et al. (1994): Superoxide and peroxynitrite in atherosclerosis. Proc Natl Acad Sci USA 91:1044–1048.

Wildhirt SM, Dudek RR, Suzuki H, Pinto V, Narayan KS, Bing RJ (1995): Immunohistochemistry in the identification of nitric oxide synthase isoenzymes in myocardial infarction. Cardiovasc Res 29:526–531.

Wink DA, Hanbauer I, Krishna MC, deGraaf W, Gamson J, Mitchell JB (1993): Nitric oxide protects against cellular damage and cytotoxicity from reactive oxygen species. Proc Natl Acad Sci USA 90:9813–9817.

Wolber RA, Duque RE, Robinson JP, Oberman HA (1987): Oxidative product formation in irradiated neutrophils. A flow cytometric analysis. Transfusion 27:167–170.

Wood PM (1987): The two redox potentials for oxygen reduction to superoxide. Trends Biochem Sci 12:250–251.

Woodman RC, Ruedi JM, Jesaitis AJ, et al. (1991): Respiratory burst oxidase and three of four oxidase-related polypeptides are associated with the cytoskeleton of human neutrophils. J Clin Invest 87:1345–1351.

Wright CD, Mülsch A, Busse R, Osswald H (1989): Generation of nitric oxide by human neutrophils. Biochem Biophys Res Commun 160:813–819.

Zanyk MJ, Banerjee D, McFarlane DL (1990): A flow cytometric assay for target binding by NKH1A+ cells using a single laser system. Anal Cell Pathol 2:241–251.

Zent R, Ailenberg M, Waddell TK, Downey GP, Silverman M (1995): Puromycin aminonucleoside inhibits mesangial cell-induced contraction of collagen gels by stimulating production of reactive oxygen species. Kidney Int 47:811–817.

Zhu L, Gunn C, Beckman JS (1992): Bactericidal activity of peroxynitrite. Arch Biochem Biophys 298:452–457.

10

Neutrophil–Endothelial Cell Interactions

Wayne O. Carter
Pfizer Medical Technology Group, Groton, Connecticut

J. Paul Robinson
Purdue University, West Lafayette, Indiana

This chapter discusses endothelial cell research with application to neutrophil–endothelial interactions; it focuses on research that has utilized various cytometric methodologies in the investigations. Implicit in the intent are the reader's basic knowledge of endothelial cell physiology and pathology and interest in utilizing new or alternative cytometric techniques or a creative combination of techniques, particularly those enabling single cell measurement.

Endothelial research originated with light microscopists who recognized the existence of a thin layer of cells that line blood vessels. Yet it was not until the development of in vitro endothelial culturing techniques by Pomerat, Maruyama, and others that the complex role played by endothelial cells in physiology and pathology came to be appreciated (Jaffe et al., 1973; Pomerat and Slick, 1963; Maruyama, 1963). Light microscopy remained the mainstay of endothelial research until the development of electron microscopy greatly enhanced our visual understanding of the endothelial substructure. The molecular understanding of endothelial cells was expanded by the development of monoclonal antibodies recognizing specific epitopes. Fluorochromes conjugated to monoclonal antibodies provided simultaneous identification of one or more epitopes on endothelial cells. More recently other fluorochromes have been used

Phagocyte Function: A Guide for Research and Clinical Evaluation, Edited by J. Paul Robinson and George F. Babcock. ISBN 0-471-12364-1

to investigate facets of intracellular physiology such as reactive oxygen species (ROS), pH, glutathione content, and calcium flux. With physiology comes the realization that endothelial cells are actively changing in response to their environment and to stimuli. Accordingly, with flow cytometry and confocal microscopy we now have the technology to assess cellular physiology and pathology in terms of active, real-time processes, with three-dimensional capabilities afforded by laser scanning confocal microscopy and computer reconstruction. An additional cytometric tool, termed intravital microscopy, is the rediscovered technique of videomicroscopy in combination with in vivo techniques. Intravital microscopy provides the capability for real-time visualization of previously hypothesized mechanisms. Finally, the integration of fluorescence and confocal microscopy with intravital/videomicroscopy has opened new horizons for the study of cell–cell interactions.

MARGINATION TO EMIGRATION

The circulating neutrophil undergoes a complex series of interactions with endothelial cells before emerging as a tissue neutrophil. The initial neutrophil–endothelial interactions were classically referred to as margination. Recently this process has been more aptly described as "tethering" by several groups who have described the visual tethering of the neutrophil to endothelial cells for an instant, essentially providing a deceleration mechanism that initiates leukocyte rolling along the endothelium. Molecular tools and the characterization of several adhesion molecule antibodies along with intravital microscopy have created a truly visual appreciation of this process, shown schematically in Figure 1. As the initial molecular tethers are created, the slowing neutrophil begins to roll along the endothelial surface and continues to decelerate as more tethers are created between endothelial and leukocyte selectins and their respective ligands (Tables 1a and 1b). Tethers are created between the neutrophil L-selectin (also referred to as LECAM-1, LAM-1, or CD62L) and carbohydrate ligands on endothelial cells (Imai et al., 1990; von Andrian et al., 1991; Ley et al., 1991). Another "tethering" receptor is P-selectin (also referred to as GMP-140 and CD62P), which is expressed on endothelial cells and like L-selectin, initiates deceleration and rolling of neutrophils along the endothelial surface. P-Selectin is stored in Weibel–Palade bodies in endothelial cells (and in secretory granules in platelets) and is rapidly translocated within minutes of cell activation to the plasma membrane via exocytosis (McEver et al., 1989; Lorant et al., 1991). The ligand for P-selectin is also a carbohydrate ligand (fucosylated and sialylated), with evidence that it is the same sialyl-Lewis x (sLe^x, CD15s) ligand to which E-selectin binds, and is also associated with L-selectin on neutrophils (Polley et al., 1991; Picker et al., 1991; Foxall et al., 1992). Using P-selectin–deficient mice, Mayadas and co-workers (1993) demonstrated virtually total absence of leukocyte rolling in mesenteric venules, increased circulating neutrophil numbers, and decreased recruitment of neutrophils to the peritoneal cavity during inflammation. The third selectin molecule identified in neutrophil–endothelial interactions is E-selectin (also referred to as ELAM-1, LECAM-2, or CD62E,) which is also expressed on endothelial cells. The ligand for E-selectin is the carbohydrate ligand sLe^x located on neutrophils (Walz et al., 1990; Phillips et al.,

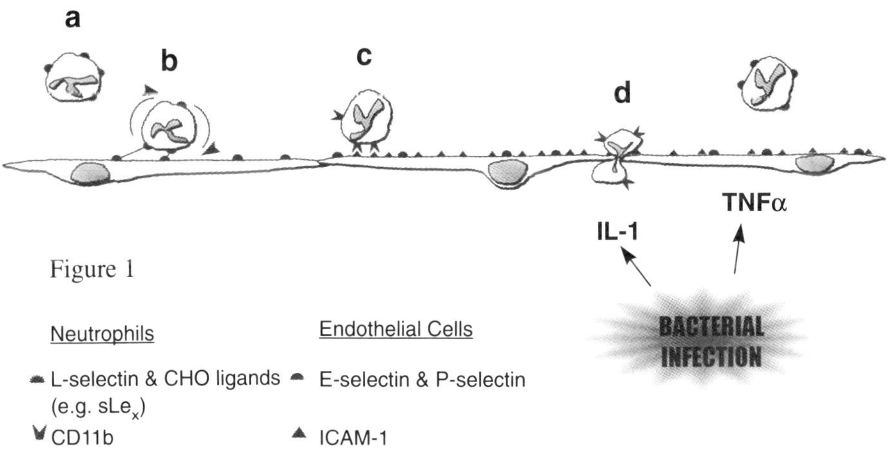

Fig. 1. Diagram showing the circulating neutrophil (a) and the initiation of rolling (b) as molecular tethers are formed between selectin and CHO ligands on neutrophils and endothelial cells. If an adequate number of tethers are formed, the neutrophil completely decelerates. With chemotactic stimulation of the neutrophil, L-selectin is rapidly shed whereas other receptors such as E-selectin, CD11b, and ICAM-1 are upregulated by cytokines and other inflammatory mediators (c). Firm neutrophil/endothelial cell adhesion is mediated by CD11b and ICAM-1 and is followed by emigration of the neutrophil through the endothelium (d).

1990). Other reports have suggested that L-selectin, CD66, and β_2 integrins may also be ligands for E-selectin (Picker et al., 1991; Kuijpers et al., 1992; Kotovuori et al., 1993, Patel et al., 1995). The interaction of sLe^x with soluble E-selectin has been very elegantly demonstrated by Jacob et al. (1995) using fluorescence polarization. Their results confirmed weak binding between the carbohydrate and protein entities, consistent with other carbohydrate–protein interactions; however, they also acknowledged additional binding interactions, such as clustering, that may be factors. It is also important to recognize that this deceleration and rolling process is reversible. Without a critical mass of tethering attachments between the neutrophil and endothelial cell, the neutrophil can escape back into the circulating pool. Of utmost importance in developing this critical mass of attachments is the upregulation and expression of P- and E-selectin on endothelial cells and the development of a firm adhesion via β_2 integrins and ICAM-1 molecules.

Once the neutrophil has come to a complete stop, chemotactic stimulation (e.g., GM-CSF, TNF-α, fMLP, LTB4, and IL-8) or ROS induces a rapid loss of L-selectin from the neutrophil surface (Smith et al., 1991; Griffin et al., 1990; Rao et al., 1994; Huber et al., 1991; Fraticelli et al., 1996) (Fig. 1). This loss is temporally associated with an upregulation of CD11b, which develops a firm adhesion with its endothelial ligand ICAM-1 (CD54) (Diamond et al., 1990; Griffin et al., 1990; Simon et al., 1995; Fraticelli et al., 1996). However, CD11b is only a subunit of the heterodimeric integral membrane protein CD11b/CD18, which consists of an alpha and a beta subunit. On neutrophils, the two heterodimers of most importance for adhesion to endothelial cells are CD11a/CD18 (also referred to as LFA-1 or $a_L\beta_2$) and CD11b/CD18 (also referred to as MAC-1 or $a_M\beta_2$). These β_2 integrins, which have unique alpha and

TABLE 1A. Endothelial and Neutrophil Adhesion Molecules and their Respective Ligands

Endothelial Adhesion Molecules	Neutrophil Counter Ligand
P-Selectin (CD62P)	s-Le^x (CD15s)
E-Selectin (CD62E)	s-Le^x, CD66, L-selectin, β2 integrins

identical beta subunits, both bind to ICAM-1 on endothelial cells. Additionally, CD11a/CD18 binds to ICAM-1 and ICAM-2 (CD102) on endothelial cells (Staunton et al., 1989). CD11c/CD18 (p150,95 or $a_X\beta_2$), the third β_2 integrin on neutrophils, is generally considered to make a smaller contribution to adhesion to endothelial cells (Carlos and Harlan, 1994). A definitive endothelial ligand has not been well defined; however, many ligands that may act as a bridge between endothelial cells and neutrophils have been identified, including iC3b and fibrinogen (Stacker and Springer, 1991; Loike et al., 1991; Bilsland et al., 1994). CD11b/CD18 has also been shown to bind to iC3b, fibrinogen, and factor X, which may also bridge to endothelial cells (Beller et al., 1982; Wright et al.,1988; Altieri et al.,1988). Once this firm adhesion is established between the neutrophil integrins and the endothelial counter ligands, migration begins through the endothelium and the transition from the circulating pool to the tissue is complete.

Cytokine Modulation of Adhesion

The adhesion described above is very complex, yet it is only a fraction of the entire process. The elements missing are the sensory stimuli (the chemoattractants and cytokines) to which the neutrophil and endothelial cells respond by altering binding avidity or expression or shedding of adhesion molecules.

The binding avidity of neutrophil L-selectin can be upregulated by G-CSF, GM-CSF, and TNF-α (Spertini et al., 1991). As mentioned above, however, neutrophils shed L-selectin receptors following this upregulation. E-Selectin on endothelial cells is an inducible selectin whose expression can be upregulated by IL-1, TNF-α, and lipopolysaccharide (Bevilacqua et al., 1987; Gimbrone, Jr., et al., 1989). Following intradermal lipopolysaccharide injection, E-selectin expression is evident after 30 min, peaks at 8 h, and is sustained for up to 72 h; as expected, neutrophil tissue lo-

TABLE 1B. Endothelial and Neutrophil Adhesion Molecules and their Respective Ligands

Neutrophil Adhesion Molecules	Endothelial Counter Ligand
L-Selectin (CD62L)	s-Le^x (CD15s)
CD11a/CD18	ICAM-1 (CD54), ICAM-2 (CD102)
CD11b/CD18	ICAM-1 (CD54), [iC3b, fibrinogen, factor X]
CD11c/CD18	?, [iC3b, fibrinogen]

calization parallels the E-selectin expression (Silber et al., 1994). Kuijpers et al. (1994) investigated the effect of TNF-α stimulation on E-selectin expression in human umbilical vein endothelial cells (HUVEC) monolayers. Using confocal microscopy, they determined that monoclonal antibodies (mAbs) to E-selectin were internalized in a rapid, energy-dependent fashion, in contrast to ICAM-1 and VCAM-1, which remained surface-bound. After internalization, E-selectin was identified in intracellular tubular compartments and was co-localized with cathepsin B. A fluorescently labeled sLe^x-oligosaccharide was similarly internalized. The internalization process was not inhibited by protein kinase C, cAMP-dependent protein kinase A, tyrosine kinase, or cytochalasin B. However, colchicine and vinblastine inhibited formation of the intracellular tubular structures and increased surface expression, thus indicating a tubulin-driven process (Kuijpers et al., 1994). In *Rickettsia rickettsii*-infected endothelial cells, E-selectin expression has also been demonstrated to aid in recruitment of neutrophils (Sporn et al., 1993). Tat protein, an HIV gene product, has been shown to increase E-selectin expression on central nervous system-derived endothelial cells (Hofman et al., 1994).

Subramaniam and co-workers (1993) compared P-selectin and E-selectin and confirmed that E-selectin was not stored but synthesized after induction by cytokines, and after a brief cell surface exposure was internalized and degraded by lysosomes. In contrast, P-selectin was stored, recycled to storage granules, and reused. The release of P-selectin from Weibel–Palade bodies is stimulated by thrombin, histamine, phorbol esters, and trypsin, but not by IL-1 (Collins et al., 1993; Geng et al., 1990). Asako et al. (1994) infused histamine into rat mesentery and demonstrated increased leukocyte rolling and decreased velocity and increased albumin leakage. The effects were inhibited by anti–P-selectin mAb, soluble sLe^x oligosaccharide, and H_1 but not H_2 receptor antagonists. Recent evidence also indicates that inhibition of nitric oxide (NO·) promotes P-selectin expression and leukocyte rolling whereas NO· donors can reduce P-selectin expression on endothelial cells (Davenpeck, et al., 1994, Gauthier et al., 1994) (Fig. 2).

Some β_2 integrin/ICAM-1 adhesions are also modulated by various inflammatory cytokines. CD11a/CD18 is not upregulated by inflammatory mediators. However, CD11b/CD18 molecules are stored in secondary and tertiary granules of neutrophils and the integrins are mobilized and quantitatively upregulated by TNF-α, LTB_4, and phorbol esters (Miller et al., 1987; Jones et al., 1988; Freyer et al., 1988; Huber et al., 1991). After stimulation of endothelial cells with TNF-α, IL-1α, IL-1β, γ-IFN, or LPS, a dose-dependent increase in ICAM-1 expression is detectable within 2 h and peaks 24 h after activation (Fabry et al., 1992; de Fougerolles et al., 1991). Although unstimulated HUVEC have a 10-fold greater expression of ICAM-2 than ICAM-1, ICAM-2 expression is unaffected by inflammatory cytokines (de Fougerolles et al., 1991). Fabry et al. (1992) demonstrated expression of ICAM-1 and ICAM-2 in 30–40 and 15–20%, respectively, of nonactivated murine brain microvascular endothelial cells. Additionally, they showed that transforming growth factor beta (TGF-β) did not increase ICAM-1 expression in murine brain microvascular endothelial cells and that ICAM-2 was not upregulated by pro-inflammatory cytokines. In contrast, Suzuki et al. (1994) have recently reported that TGF-β1 increased the expression of ICAM-1

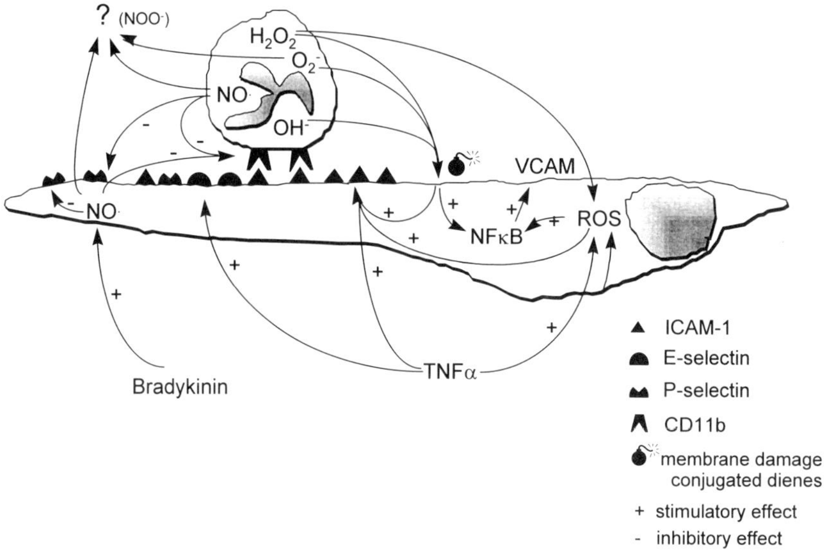

Fig. 2. Diagram of some of the known and unknown interactions between neutrophils and endothelial cells. Nitric oxide (NO·) and reactive oxygen species (ROS) are produced by both neutrophils and endothelial cells.

on HUVEC and induced the potential for neutrophil-mediated lung injury as indicated by increases in ^{125}I-labeled albumin and neutrophils in lung tissue and bronchoalveolar lavage fluid. Margiotta et al. (1992) reported increased ICAM-1 expression in human saphenous vein endothelial cells exposed to Dacron, an observation with some obvious implications for vascular prostheses and other implants.

Another key cytokine in neutrophil and endothelial cell interactions is IL-8. Interleukin-8 has been reported to inhibit neutrophil adhesion to endothelial cells and yet induce neutrophil migration into tissues (Gimbrone et al., 1989; Huber et al., 1991; Kunkel et al., 1991). As part of these investigations, IL-8 has been shown to induce L-selectin shedding and β_2 integrin upregulation (Huber et al., 1991). Solomkin et al. (1994) investigated these findings in vivo and reported that 3 h after humans were infused with endotoxin, IL-8 concentrations were elevated but TNF-α concentrations were not. Neutrophils isolated from these subjects had decreased L-selectin expression, which would reduce their ability to adhere to activated endothelial cells. Thus an inflammatory lesion with IL-8 as the predominant cytokine may impair neutrophil delivery to that area. In addition to lipopolysaccharide and TNF-α, IL-1 induces production of IL-8 by endothelial cells that can be inhibited by IL-1ra (Kaplanski et al., 1994). Westlin et al. (1992) used confocal microscopy to demonstrate that IL-8 induced rapid F-actin polymerization in neutrophils and that the rate of depolymerization was inversely related to the IL-8 concentration; chemotactic factors C5a and fMLP showed similar effects. The authors concluded that these agonists may help regulate actin polymerization in neutrophils, which is crucial for neutrophil adhesion to endothelium.

Interleukin-2 has recently been proposed in many antineoplastic therapies; however, various toxicities including a "vascular leak" phenomenon have proved problematic. Recognizing that IL-2 has limited direct effects on neutrophils but may have mediating ones through TNF-α, Edwards et al. investigated this effect (1992). Using fluorescein isothiocyanate-labeled albumin and intravital microscopy, they verified macromolecular leakage and increased leukocyte–endothelial adhesion that was inhibited by a polyclonal antibody against TNF-α.

Many other interleukins have been investigated for their effect on endothelial cells, but interleukins that affect non-neutrophilic granulocytes (basophils and eosinophils) are of specific interest. Interleukin-4 and IL-13 induce the expression of VCAM-1 on endothelial cells but not an increase in ELAM-1 or ICAM-1 expression (Schleimer et al., 1992; Bochner et al., 1995). The ligand for VCAM-1 on eosinophils and basophils is VLA-4, which is lacking on neutrophils; this may partly explain the enrichment of basophils and eosinophils versus neutrophils in allergic diseases (Schleimer et al., 1992; Bochner et al., 1991; Walsh et al., 1991; Bochner et al., 1995). Interleukin-3 has been shown to increase surface expression of CD11b on basophils and adhesion to HUVEC but has no effect on neutrophil adhesiveness (Bochner et al., 1990). Interleukin-5 induces upregulation of eosinophil CR3 but not neutrophil CR3 and thus IL-5 enhances eosinophil (not neutrophil) adhesion to endothelial cells via β_2 integrins (Walsh et al., 1990). In contrast, IL-1 can induce increased adhesion of eosinophils, basophils, and neutrophils by inducing expression of VCAM-1, ICAM-1, and E-selectin (Bochner et al., 1991; Walsh et al., 1991).

In neutrophil–endothelial cell experiments, with current techniques it is very difficult if not impossible to control for all variables, but crucial to realize their effect. Silvestro et al. recently reported (1994) that heparin and its desulfated derivatives inhibit neutrophil adhesion to endothelial cells in a dose-dependent manner. The effects observed appeared to be related to inhibitory properties on the neutrophils and not the endothelial cells. Others have reported the direct or indirect inhibitory effect of heparin on the complement system, which may obviously also affect neutrophil adhesion or neutrophil activation (Cheung et al., 1994; Moen et al.,1995). Temperature has also been shown to have a significant effect on CD11b and CD11c expression in neutrophils (Le Deist et al., 1994).

Cytoskeletal Factors Regulating Adhesion, Emigration, and Transport

The intention behind the term "integrin" receptor was to describe an integration of the extracellular matrix with the intracellular cytoskeleton via these receptors (Tamkun et al., 1986). However, relatively little research has focused on these interactions, especially compared to the information known about the β_2 integrin subunit, ICAM, and the cytoskeleton. Several reports have indicated different responses by neutrophils in solution compared to adherent neutrophils. Nathan et al. (1989) demonstrated that adherent neutrophils released H_2O_2 in response to TNF-α, whereas neutrophils in suspension failed to respond accordingly. It was further shown that the production of ROS upon TNF-α stimulation was mediated through β_2 integrin receptors via tyrosine phosphorylation of paxillin (Fuortes et al., 1994). Richter et al. (1990)

investigated TNF-α-induced degranulation in adherent neutrophils and demonstrated that degranulation depended on CD11b/CD18-triggered oscillatory cycles of unbound cytosolic calcium and that this could be inhibited by anti-CD18 mAb. β_1 and β_3 cytoplasmic domains also contain a tyrosine kinase phosphorylation site, suggesting a functional or regulatory interaction of these integrins (Tamkun et al., 1986). Asako et al. (1992) assessed the role of the cytoskeleton on adhesion and emigration by utilizing intravital microscopy of rat mesentery and phalloidin, a microfilament stabilizer. Their findings demonstrated reduced emigration but normal adherence with low-dose phalloidin and reduced leukocyte adherence and emigration with high-dose phalloidin. Another group evaluated macromolecular transport in bovine endothelial monolayers with a fluid-phase endocytic tracer, fluorescein isothiocyanate-labeled dextran-70 (Liu et al., 1993). Using pharmacological modulation, they determined that translocation occurred by chains of vesicles that required microtubules and that microfilaments hinder the vesicle transport. Clearly this research area is ripe with opportunities to utilize the capabilities of intracellular and extracellular fluorescent markers, especially with the visual discrimination possible with confocal microscopy and physiological impact of time-resolved studies.

THE EFFECTS OF NUCLEAR FACTOR-κB, REACTIVE OXYGEN SPECIES, AND NITRIC OXIDE ON ENDOTHELIAL CELL PHYSIOLOGY AND PATHOLOGY

Nuclear Factor-κB

Many publications have linked nuclear factor-κB (NF-κB) activation to reactive oxygen species (Staal et al., 1990; Schreck et al., 1991). Weber et al. (1994) investigated the role of antioxidants in inhibiting VCAM-1 and ICAM-1 expression in TNF-α-stimulated endothelial cells. The antioxidant pyrrolidine dithiocarbamate (PDTC) inhibited NF-κB and blocked VCAM-1 but not ICAM-1 expression in TNF-α-stimulated HUVEC, suggesting that only VCAM-1 was controlled by NF-κB. Moynagh et al. (1994) reported similar findings on human glial cells, yet a binding site for NF-κB has been identified in the 5′-regulatory region of the human gene for ICAM-1 (Voraberger et al., 1991). Aspirin has been reported to inhibit NF-κB by preventing the degradation of the NF-κB inhibitor I kappa B (Kopp and Ghosh, 1994). Accordingly, aspirin has been shown to induce the inhibition of VCAM-1 and E-selectin but not ICAM-1 by TNF-α-stimulated HUVEC (Weber et al., 1995). Salicylates have since been shown to reduce transendothelial migration of neutrophils (Pierce et al., 1996).

ICAM-1 and L-Selectin

The ability of H_2O_2 to increase adhesion to HUVEC through ICAM-1 has been investigated. A dose-dependent response was observed with a two- to threefold increase in ICAM-1 expression noted between 0.5 and 1 h after H_2O_2 challenge (Lo et al.,

1993). Suzuki *et al.* (1991) showed an increase in leukocyte adhesion induced by neutrophil oxidants H_2O_2 and monochloramine (NH_2Cl) that was inhibited by an anti-CD18 mAb. Sellak et al. (1994) evaluated adherence of unstimulated neutrophils on HUVEC exposed to nonlytic concentrations of ROS. Their data suggested increased adhesion involving ICAM-1 but independent of ICAM-1 upregulation and included a non-ICAM-1 mechanism that recognized L-selectin on neutrophils. However, it has been demonstrated that ROS induce a rapid loss of L-selectin from the neutrophil surface (Fraticelli et al., 1996).

When investigating ICAM expression on endothelial cells and the effects of neutrophils, it is important to remember that many cells other than endothelial cells express ICAM-1. For example, Ikeda et al. (1994) investigated the expression of ICAM-1 in keratinocytes and showed that ROS increased epidermal ICAM-1 expression and that some antioxidants prevented the expression. Furthermore, it has been demonstrated using intracellular cardiac myocyte ROS with 2′,7′-dichlorofluorescein and myocyte ICAM-1 expression that binding of CD11b/CD18 and ICAM-1 was necessary for activation of the neutrophil oxidative burst and thus ROS-mediated injury (Entman et al., 1992).

In neutrophils, a link between L-selectin binding and ROS generation was shown when Waddell et al. (1994) determined that cross-linking L-selectin with DREG and secondary antibody failed to generate ROS but enhanced the response of fMLP and TNF-α to produce ROS. Additionally an intracellular rise in Ca^{2+} was measured upon L-selectin cross-linking. Similar signaling roles for other adhesion molecules in neutrophils and endothelial cells are likely.

Reactive Oxygen Species Generation

A combination of two fluorochromes, 2′,7′-dichlorofluorescin diacetate (DCFH-DA) and hydroethidine (HE), has been used to measure rat pulmonary artery endothelial ROS generation after stimulation with H_2O_2 (Carter et al., 1994). It was determined that superoxide anion (O_2^-) was generated partially by the xanthine oxidase pathway after stimulation with H_2O_2. Additional studies with confocal microscopy have suggested that O_2^- generation occurs primarily in mitochondria. Using DCFH-DA and laser scanning confocal microscopy, Egawa et al. (1994) demonstrated an increased intracellular oxidation state in immortalized human endothelial cells (ECV304) after stimulation with TNF-α. This effect was blocked by free-radical scavengers *N*-acetylcysteine (a precursor of glutathione) and pyrrolidine dithiocarbamate. These findings appear to accord with results suggesting that TNF-α induced free-radical production that was necessary for NF-κB induction (Menon et al., 1993). Suematsu et al. (1993) used intravital fluorescence microscopy and DCFH-DA–loaded rat mesentery to evaluate oxidative changes in microvessels during neutrophil activation with platelet activating factor (PAF). By alternating light sources, they observed both neutrophil adhesion and significant increases in DCF fluorescence in endothelial cells as early as 10 min after PAF superfusion. An iron chelator and hydroxyl radical scavenger both significantly attenuated the PAF-induced oxidative changes, suggesting a possible role for hydroxyl radical in endothelial oxidative stress. Al-Mehdi et al. (1994)

also used DCFH-DA to determine if oxidizing species were generated in lung ischemia-reperfusion (I/R) experiments in which rat lungs were isolated, oxygen ventilated, and perfused with artificial media. Dichlorofluorescin diacetate was added to the perfusate and tissue fluorescence following the ischemia/reperfusion injury was measured with a confocal laser scanning microscope. The authors showed that oxidizing species were generated after the I/R injury with predominant localization in endothelial, type II, Clara, and ciliated cells and macrophages.

The ROS generated by neutrophils, macrophages, and other cells obviously have potential for direct damage to cell membranes. Block (1991) investigated the direct effects of H_2O_2 on membrane fluidity and function of porcine pulmonary artery endothelial cells. A 30-min exposure to 50 μM H_2O_2 increased conjugated dienes and significantly altered membrane fluidity 0.5 and 6 h post-exposure, but conditions returned to normal within 24 h (Block, 1991) (Fig. 2). Lipid membrane peroxidation and functional alterations were inhibited by pretreatment with α-tocopherol or dimethylthiourea.

Bradykinin is a vasoactive peptide and inflammatory mediator that also stimulates nitric oxide (NO·) generation (Fig. 2). Bradykinin has also been shown to stimulate O_2^- generation via a mechanism that is partially inhibited by cyclooxygenase inhibitors (Holland et al., 1990). Shimizu et al. (1994) used DCFH-DA to assess ROS generation by endothelial cells after stimulation by bradykinin. The oxidation was inhibited by hydroxyl radical (OH^-) scavengers and cyclooxygenase inhibitors but unaffected by inhibitors of NO·. Furthermore, using Fluo-3 to measure intracellular calcium, they identified an intracellular calcium peak following bradykinin stimulation. These findings suggested that the ROS generated after bradykinin stimulation was OH^- and not associated with NO· generation but instead may be related to a cyclooxygenase pathway.

In addition to neutrophils and other phagocytic cells, cancer cells (specifically Walker 256 cells) have been shown to generate ROS and thus may directly damage endothelial cells (Soares et al., 1994). Masuda et al. (1990) investigated the mechanism of endothelial injury in inherited-cataract rats. Using bromodeoxyuridine pulse labeling they showed a reduced cell cycle traverse rate for inherited-cataract rats and that neutrophil-generated ROS may be responsible for endothelial cell injury.

Nitric Oxide

Nitric oxide (NO·) has been shown to exhibit a multitude of physiological and pathological effects. The interactions between phagocytic cells (neutrophils and macrophages) and endothelial cells become very complex when one attempts to understand the effects of ROS, including NO·, not least because neutrophils, macrophages, and endothelial cells produce both NO· and superoxide anion (O_2^-), and these ROS can react with each other to form peroxynitrite ($ONOO^-$). An in-depth discussion of peroxynitrite is beyond the scope of this chapter (see Chapter 9), yet potentially very important considering the oxidizing capability of this molecule and the potential for cellular damage; for a general review we refer the reader to Szabó (1996). The complexity of these interactions is also difficult to segregate when one realizes the relative nonspecificity of many fluorochromes to measure individual oxidizing agents. The authors have shown that rat pulmonary artery endothelial cells and human neutrophils loaded with DCFH-DA (2′,7′-dichlorofluorescin diacetate, a fluorochrome commonly used in phagocytic cell experiments to measure H_2O_2) demonstrated a significant increase in DCF fluorescence after the addition of 100 n*M* NO· or caged nitric oxide (potassium nitrosylpentachlororuthenate) (Robinson et al., 1994). The dose-dependent response observed suggested that NO· was directly oxidizing DCFH to DCF; however, another possible explanation was that peroxynitrite or hydroxyl radical produced by the combination of nitric oxide and O_2^- oxidized DCFH to DCF. In other studies 3-morpholinosydnonimine *N*-ethylcarbamide (SIN-1) was used to generate NO· and O_2^- simultaneously, with the conclusion that hydroxyl radical was formed (Hogg et al., 1992).

Intravital microscopy has been used to assess the effect of two NO· inhibitors (N^{ω}-monomethyl-L-arginine (L-NMMA) and N^{ω}-nitro-L-arginine methyl ester (L-NAME)) on leukocyte adhesion in feline mesenteric vessels (Kubes et al., 1991). The study demonstrated a 15-fold increase in leukocyte adhesion induced by the NO· inhibition that was blocked by an anti-CD18 mAb or L-arginine. Gaboury et al. (1993) evaluated the effect of NO· on leukocyte adhesion in a superoxide-generating system of hypoxanthine and xanthine oxidase (HX-XO) infusion, PAF infusion, or LTB_4 infusion into mesenteric vessels. They demonstrated that both superoxide dismutase and SIN-1 inhibited leukocyte adhesion in the HX-XO infusion and PAF infusion but not the LTB_4 infusion and concluded that the anti-adhesive properties of NO· were likely related to the ability of NO· to inactivate O_2^-. Niu et al. (1994) recently investigated neutrophil adhesion in HUVEC by blocking NO· synthesis with L-NAME. At 1 h, these researchers observed no effect, but at 4 h after L-NAME administration, a dose-dependent increase in adhesion that was prevented with L-arginine and NO· donors. The adhesion was inhibited by mAb against CD18 and ICAM-1, and by PAF receptor antagonist. In addition, intracellular oxygen radical scavengers inhibited the adhesion but extracellular oxygen radical scavengers had no effect. An increase in intracellular endothelial ROS generation in L-NAME–treated cells was confirmed with an increase in DCFH oxidation to DCF in NO·-blocked cells. Recently Biffl et al. (1996) demonstrated that NO· reduces endothelial expression of constitutive ICAM-1, but does not affect lipopolysaccharide-induced upregulation of ICAM-1 on endothelial cells. As mentioned above, inhibition of NO· has been shown to promote P-selectin expression and leukocyte rolling whereas nitric oxide (NO·) donors reduce P-selectin expression on endothelial cells (Davenpeck et al., 1994, Gauthier et al., 1994).

ENDOTHELIAL RESEARCH AND CONFOCAL MICROSCOPY

Technical Aspects of Live Endothelial Confocal Microscopy Research

Although confocal microscopy has been available for several years, performing experiments on live endothelial cells under physiological conditions creates special

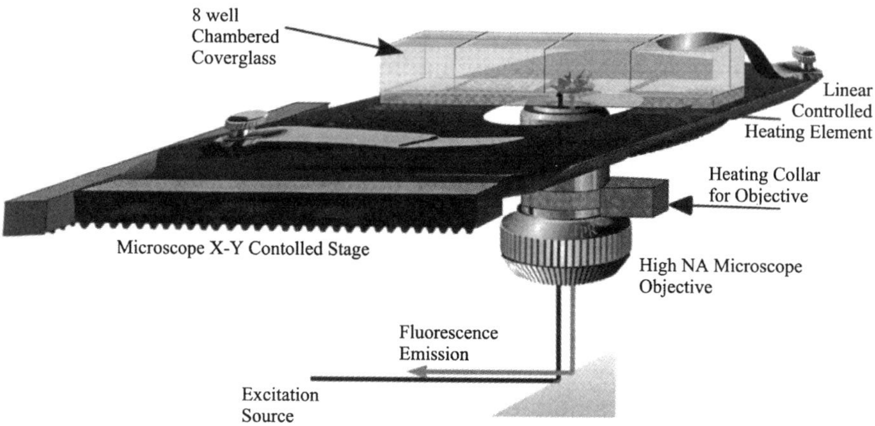

Fig. 3. Inverted microscope setup for endothelial cell studies. Reproduced with permission from Purdue Cytometry CDROM Volume 3, 1997.

problems. For proper optical assessment, these experiments require the use of an inverted scope with a heated stage that must not only be maintained within 0.1°C but also prevent electrical on–off cycle oscillations that result in specimen-focusing aberrations and subsequent poor image quality. Additionally, the cells must be grown in culture chambers made with optically pure coverglass (Lab-Tek® culture chambers are one commercial option) to allow close positioning of high-NA (numerical aperture) fluorescence microscope objectives. If an oil or water interface is used between the objective and coverglass, a heated objective is also recommended to prevent local heat loss and subsequent altered physiological conditions for the cells of most interest (those directly over the objective). Heated chambers that enclose the entire stage and objectives are an alternative but can be cumbersome and physically interfering to the operator. Figure 3 shows a successful setup using a BioRad MRC 1024 confocal microscope, a Nikon 300 inverted microscope, and endothelial cells cultured in a Lab-Tek® eight-well chambered coverglass.

Confocal Studies Assessing Endothelial Damage and Vascular Leakage

The development and evolution of confocal microscopy has provided many new techniques for scientists seeking understanding of endothelial cell physiology and pathology. Many fluorochromes are available to assess intracellular changes in a single cell and extracellular effects. For example, macromolecular leakage of intact endothelium has been studied utilizing a mesenteric window in rats injected with fluorescein isothiocyanate (FITC)-labeled albumin (Thurston et al., 1995). After being harvested and fixed, tissues were stained with rhodamine phalloidin to label filamentous actin and assess endothelial cytoskeleton. Thurston showed that focal leaks occurred in small areas of cell–cell contact and involved disruption of the endothelial peripheral actin rim. "Mid-leaks" involved one endothelial cell with extended disruption of the peripheral actin rim, and "extended leaks" involved two or more adjacent endothelial cells (Thurston et al., 1995). Phalloidin labeled with FITC has been used to examine cytoskeletal organization of F-actin microfilaments during cell migration and translocation of corneal endothelial repair (Ichijima et al., 1993). After corneal injury, tissues were harvested, fixed, and labeled with FITC-phalloidin. Results indicated two separate, injury-dependent mechanisms of corneal repair: cell spreading and cell migration. Nagashima et al. (1994) used confocal microscopy to study the role of leukotriene C_4 on brain edema. They isolated cerebral capillary endothelial cells from rats and cultured the cells on an optically clear collagen membrane mounted in a plastic frame. Their study found an increase in permeability and no intracellular calcium flux with leukotriene C_4.

Andries and Brutsaert (1994) used confocal microscopy to investigate the selective permeability of ventricular endocardial endothelium versus myocardial capillary endothelium. Intercellular clefts were shorter and deeper in endocardial endothelium than in capillary endothelium, suggesting that endocardial endothelium may be less permeable. However, Lucifer Yellow-labeled dextran 10,000 penetrated endocardial endothelium faster than capillary endothelium, but dextran 40,000 did not penetrate either endothelium. The authors concluded the structural differences observed may function to limit diffusion in the endocardial endothelial owing to the high hydrostatic pressure in the heart (Andries and Brutsaert, 1994).

Three-color analysis is also possible with a single-laser confocal microscope. Uchihara et al. (1995) used phycoerythrin-labeled HLA-DR, FITC-labeled von Willebrand factor (vWF), and cyanin 5-labeled amyloid beta protein to assess histopathological changes in brain tissue of Alzheimer's disease. Using an argon laser, these investigators sequentially imaged the three antigens and performed a computer reconstruction of the individual images for simultaneous evaluation of three distinct structures: microglia, endothelium, and A beta deposits.

Endothelial Responses to Shear Stress

Endothelial adhesion to substratum in real time has been assessed using tandem scanning confocal microscopy (Davies et al., 1993). By subtraction analysis of consecutive images the researchers recorded continuous remodeling of focal adhesion sites of endothelial cells during periods of less than 1 min yet with less than 10% variation in cell adhesion for the entire cell. Cytoskeletal disruption with cytochalasin or exposure of the endothelial cells to proteolytic enzymes resulted in a rapid reduction in cell adhesion. Later experiments by this same group assessed quantitative adhesion of endothelial cells exposed to shear stress (Davies et al., 1994). They demonstrated focal adhesion-site remodeling starting within 7–9 h in the direction of laminar flow (10 dyne/cm^2). Additionally they observed redistribution of intracellular stress fibers and coalescence of small focal adhesions into larger focal adhesions, resulting in fewer total adhesion sites per cell.

Walpola et al. (1993) evaluated the effect of low and high shear stress in vivo by partially ligating the left carotid artery in rabbits. There was an adaptive reduction in left carotid (low shear) artery diameter, desquamation of endothelial cells, and fewer and shorter stress fibers than in normal cells. Although the right carotid artery (high shear) diameter was unchanged, their findings indicated an increase in stress fibers and a reduction in peripheral actin staining. Another group demonstrated an inducible cell membrane complex that developed strong adhesion to elastin fibers (Perdomo et al., 1994).

Surface Receptor Studies and Confocal Microscopy

Syn-capping of human T-lymphocyte adhesion molecules during interaction with endothelial cells has been identified through confocal microscopy (Rosenman et al., 1993). Of particular interest was the finding that CD2 co-capped with CD44 and CD11a/CD18, suggesting functional cooperation of these molecules. Furthermore, CD2, CD44, and L-selectin redistributed to pseudopodia when exposed to HUVEC, but CD11a/CD18 remained globally distributed on the cell surface. Similar studies with neutrophils and endothelial cells exposed to ROS and activated with cytokines could provide key information to help understand physiological and pathological neutrophil–endothelial interactions.

Both confocal microscopy and immunogold electron microscopy were used to localize tumor necrosis factor (TNF) receptors on HUVEC in time-lapse experiments to demonstrate physiological changes (Bradley et al., 1995). In this study the authors identified a 75-kDa receptor on the cell surface, endocytosis, and transport to lysosomes through the coated pits and coated vesicles. Additionally they showed that TNF may accelerate clustering and internalization of the 75-kDa receptor. The 55-kDa receptor, expressed at much lower concentrations, was mostly unaffected by TNF.

The alpha 6 beta 4 integrin complex is expressed by epithelial cells and localized in hemidesmosomes (Stepp et al., 1990). Alpha 6 beta 4 was recently reported to be expressed in a subset of endothelial cells in mice (Kennel et al., 1992). However, Cremona et al. (1994) demonstrated by confocal microscopy that alpha 6 beta 4 was not expressed by endothelial cells, because they did not express von Willebrand factor. The higher resolution of confocal microscopy compared to conventional immunofluorescence permitted determination of the mAb to alpha 6 beta 4-labeled smooth muscle cells of small vessels.

ENDOTHELIAL RESEARCH IN SPECIFIC CLINICAL DISEASES

Trauma, Sepsis, and Adult Respiratory Distress Syndrome

Most clinicians and researchers agree that neutrophil–endothelial interaction is also a key contributor to the pathophysiology of adult respiratory distress syndrome (ARDS), which can be associated with many disease syndromes, including sepsis and trauma. Vogel et al. (1993) investigated the role of trauma and plasma products in the induction of neutrophil-independent endothelial cell damage. They determined that both fibrin degradation products (FDP-D and FDP-E) significantly decreased neutrophil adhesion to human endothelial cells in vitro and that endothelial cell membrane integrity was disturbed by serum from trauma patients. Another study evaluated microvascular function and rheological changes in hyperdynamic sepsis and showed that increased leukocyte adherence to microvascular endothelial cells contributed to rheological changes and compromised capillary cross sectional area (Astiz et al., 1995). Fabian et al. (1994) investigated the effect of blocking adhesion through the CD18 adhesion complex after traumatic shock and endotoxemia. An anti-CD18 mAb decreased lung injury and improved hemodynamic parameters in endotoxic pigs. Using intravital microscopy, Harris et al. (1994) assessed leukocyte–endothelial adhesion in feline mesenteric venules exposed to lipopolysaccharide. An anti-CD11/CD18 mAb and superoxide dismutase (SOD) significantly attenuated leukocyte adhesion to endothelial cells, whereas an anti-ICAM-1 mAb and platelet activating factor receptor antagonist (PAFra) had no effect. Additionally, anti-CD11/CD18, anti-ICAM-1, SOD, and PAFra did not affect leukocyte emigration or leukocyte rolling. Kurose and co-workers induced intestinal mucosal erosions and increased tissue neutrophils and protein leakage with intravenous lactoferrin infusion; these alterations were reduced by anti-P-selectin mAb but not by anti-E-selectin (Kurose et al., 1994b).

Several disease models for pulmonary injury exist. One utilizes the intrapulmonary deposition of IgG immune complexes to elicit damage and has been used to investigate the effect of blocking several adhesion molecules by intravenous or intratracheal administration of mAb (Mulligan et al., 1995). The data suggested therapeutic compartmentalization. Intravenous mAbs against CD11a, L-selectin, and ICAM-1 effectively reduced pulmonary neutrophils and lung injury. In contrast, with intratracheal administration (alveolar compartment), only CD11b and ICAM-1 were protective and only CD11b significantly reduced TNF-α concentrations in bronchoalveolar lavage fluid. In a different model of cobra venom factor-induced acute lung injury, Mulligan and co-workers reported in 1993 that sLe^x significantly reduced lung injury and tissue neutrophil accumulation.

Reactive oxygen species are believed to be a major contributor to the pathogenesis of pulmonary oxygen toxicity. In addition, ICAM-1 expression is increased on inflamed alveolar epithelium, whereas an anti-ICAM-1 mAb in pulmonary oxygen toxicity can reduce pulmonary neutrophil infiltration and pulmonary damage (Wegner et al., 1992).

Ischemia/Reperfusion Injuries

Another clinical syndrome that appears to involve endothelial cells directly is ischemia/reperfusion injuries. The primary pathophysiological process is still the subject of much debate; however, it is generally agreed that neutrophil–endothelial interactions and ROS generation contribute to the pathology. Although some scientists still debate the existence of the ischemia/reperfusion injury, it is important to recognize that experimental findings depend somewhat on the choice of animal species and organ.

Granger and co-workers have used intravital microscopy to observe neutrophil adhesion to endothelial cells following ischemia/reperfusion (I/R) injuries and assess the molecular components of adhesion, the role of ROS, and potential therapies. They demonstrated decreased I/R-induced capillary permeability, leukocyte adhesion, and extravasation with adenosine and similar responses with dimethyl sulfoxide (Grisham et al., 1989; Sekizuka et al., 1989). Kurose et al. (1994a) used intravital fluorescence microscopy to evaluate the ability of several mAbs to block the reperfusion-induced leukocyte–endothelial interactions in rat mesentery. They verified that the magnitude of leukocyte adhesion was related to the duration of ischemia, as was albumin leakage. Monoclonal antibodies against CD11b and CD18 consistently reduced leukocyte adhesion, whereas anti-ICAM-1 was effective only after 10 and 30 min of ischemia and anti-L-selectin only after 10 min of ischemia. P- and E-selectin did not reduce adhesion. Post-ischemic venules contained platelet–leukocyte aggregates that were blocked by mAbs against P-selectin, CD11b, CD18, and ICAM-1. A feline model of small intestine transplantation has been used to assess I/R injuries and the therapeutic potential of mAbs against several adhesion molecules (Slocum and Granger, 1993). Catalase, superoxide dismutase, mAbs against ICAM-1, and P-selectin had no protective effect, although anti-CD18 mAb prevented much of the microvascular dysfunction. In another study, HUVEC exposed to anoxia followed by reoxygenation (A/R) did not increase ICAM-1 or ELAM-1 expression; however, neutrophil adherence was blocked by mAbs against CD11a, CD11b, CD18, and ICAM-1 but not ELAM-1 (Yoshida et al., 1992). Furthermore, A/R HUVEC-conditioned medium increased CD11b and CD18 expression on neutrophils whereas catalase and PAF antagonist diminished the increased adhesion. These results suggested that hydrogen peroxide and PAF mediated increased neutrophil adherence via CD11a/CD18 and CD11b/CD18 with ICAM-1 on endothelial cells. Farhood et al. (1995) used a rat model of hepatic I/R to assess ICAM-1 expression and the therapeutic efficacy of an anti-ICAM-1 mAb. ICAM-1 mRNA was evident in the ischemic liver lobes during ischemia and in the ischemic and nonischemic lobes during reperfusion. Therapeutically, anti-ICAM-1 mAb reduced hepatocyte injury as indicated by a 51% reduction in plasma alanine aminotransferase concentrations and reduced hepatic necrosis; however, no significant reduction in hepatic neutrophil infiltration occurred during reperfusion.

The endothelial changes following myocardial I/R injuries have been studied in several animal models, and recently extensive research has been published on the feline model. Ma et al. (1992) investigated the cardioprotective effect of anti-ICAM-1 in a feline myocardial I/R injury. Their results indicated reduced myocardial necrosis, reduced tissue neutrophil infiltration, and increased endothelial-dependent relaxation. Another feline myocardial I/R study demonstrated cardioprotective effects of P-selectin (Weyrich et al., 1993). These results suggested that cardioprotection was directly related to reduced neutrophil adhesion to endothelial cells and demonstrated significant upregulation of P-selectin in endothelial cells that lined coronary vessels. Weyrich et al. (1995) also demonstrated sequential expression of P-selectin, ICAM-1, and E-selectin, respectively, after feline myocardial I/R injury and suggested that P-selectin and ICAM-1 expression coordinated with neutrophil trafficking. Cardioprotection has also been demonstrated with anti-L-selectin mAb, sLe^x oligosaccharide, and L-arginine (Ma et al., 1993; Buerke et al., 1994; Weyrich et al., 1992). Using the Langendorff model of an isolated perfused rat heart, Kamikubo (1993) demonstrated that in the early phase of myocardial ischemia, myocardial dysfunction and cytokine gene expression can occur without the infiltration of inflammatory cells. Kamikubo also measured an increased in the expression of IL-6, IL 8, γ-IFN, TNF-α, and IL-1a, yet found no detectable increase in the expression of IL-1β, MCP-1, and IL-1ra measured by semiquantitative reverse transcription/polymerase chain reaction (RT-PCR).

CONCLUSION

Endothelial and neutrophil research continues to evolve with the cytometric tools and molecular techniques available today. Integrating endothelial and neutrophil (and other cell) research at the cytoskeletal and molecular level with techniques discussed above will allow us more fully to understand these cellular interactions and signalling processes. Creative combinations of these techniques and future developments and refinements will initiate new cytometric evaluations of cellular interactions. These capabilities will certainly yield novel understandings of endothelial pathophysiology and endothelial–neutrophil interactions.

REFERENCES

Al-Mehdi A, Shuman H, Fisher AB (1994): Fluorescence microtopography of oxidative stress in lung ischemia-reperfusion. Lab Invest 70:579–587.

Altieri DC, Morrissey JH, Edgington TS (1988): Adhesive receptor Mac-1 coordinates the activation of factor X on stimulated cells of monocytic and myeloid differentiation: an alternative initiation of the coagulation protease cascade. Proc Natl Acad Sci USA 85:7462–7466.

Andries LJ, Brutsaert DL (1994): Endocardial endothelium in the rat: junctional organization and permeability. Cell Tissue Res 277:391–400.

Asako H, Kurose I, Wolf R, et al. (1994): Role of H1 receptors and P-selectin in histamine-induced leukocyte rolling and adhesion in postcapillary venules. J Clin Invest 93:1508–1515.

Asako H, Wolf RE, Granger DN, Korthuis RJ (1992): Phalloidin prevents leukocyte emigration induced by proinflammatory stimuli in rat mesentery. Am J Physiol 263:H1637–H1642.

Astiz ME, DeGent GE, Lin RY, Rackow EC (1995): Microvascular function and rheologic changes in hyperdynamic sepsis. Crit Care Med 23:265–271.

Beller DI, Springer TA, Schreiber RD (1982): Anti-Mac-1 selectively inhibits the mouse and human type three complement receptor. J Exp Med 156:1000–1009.

Bevilacqua MP, Pober JS, Mendrick DL, Cotran RS, Gimbrone MA, Jr. (1987): Identification of an inducible endothelial-leukocyte adhesion molecule. Proc Natl Acad Sci USA 84:9238–9242.

Biffl WL, Moore EE, Moore FA, Barnett C (1996): Nitric oxide reduces endothelial expression of intercellular adhesion molecule (ICAM)-1. J Surg Res 63:328–332.

Bilsland CA, Diamond MS, Springer TA (1994): The leukocyte integrin p150,95 (CD11c/CD18) as a receptor for iC3b. Activation by a heterologous beta subunit and localization of a ligand recognition site to the I domain. J Immunol 152:4582–4589.

Block ER (1991): Hydrogen peroxide alters the physical state and function of the plasma membrane of pulmonary artery endothelial cells. J Cell Physiol 146:362–369.

Bochner BS, Klunk DA, Sterbinsky SA, Coffman RL, Schleimer RP (1995): IL-13 selectively induces vascular cell adhesion molecule-1 expression in human endothelial cells. J Immunol 154:799–803.

Bochner BS, Luscinskas FW, Gimbrone MA, Jr., Newman W, Sterbinsky SA (1991): Adhesion of human basophils, eosinophils, and neutrophils to interleukin 1-activated human vascular endothelial cells: contributions of endothelial cell adhesion molecules. J Exp Med 173:1553–1557.

Bochner BS, McKelvey AA, Sterbinsky SA, et al. (1990): IL-3 augments adhesiveness for endothelium and CD11b expression in human basophils but not neutrophils. J Immunol 145:1832–1837.

Bradley JR, Thiru S, Pober JS (1995): Disparate localization of 55-kd and 75-kd tumor necrosis factor receptors in human endothelial cells. Am J Pathol 146:27–32.

Buerke M, Weyrich AS, Zheng Z, Gaeta FC, Forrest MJ, Lefer AM (1994): Sialyl Lewisx-containing oligosaccharide attenuates myocardial reperfusion injury in cats. J Clin Invest 93:1140–1148.

Carlos TM, Harlan JM (1994): Leukocyte-endothelial adhesion molecules. Blood 84:2068–2101.

Carter WO, Narayanan PK, Robinson JP (1994): Intracellular hydrogen peroxide and superoxide anion detection in endothelial cells. J Leukoc Biol 55:253–258.

Cheung AK, Faezi-Jenkin B, Leypoldt JK (1994): Effect of thrombosis on complement activation and neutrophil degranulation during in vitro hemodialysis. J Am Soc Nephrol 5:110–115.

Collins PW, Macey MG, Cahill MR, Newland AC (1993): von Willebrand factor release and P-selectin expression is stimulated by thrombin and trypsin but not IL-1 in cultured human endothelial cells. Thromb Haemost 70:346–350.

Cremona O, Savoia P, Marchisio PC, Gabbiani G, Chaponnier C (1994): The alpha 6 and beta 4 integrin subunits are expressed by smooth muscle cells of human small vessels: a new localization in mesenchymal cells. J Histochem Cytochem 42:1221–1228.

Davenpeck KL, Gauthier TW, Lefer AM (1994): Inhibition of endothelial-derived nitric oxide promotes P-selectin expression and actions in the rat microcirculation. Gastroenterology 107:1050–1058.

Davies PF, Robotewskyj A, Griem ML (1993): Endothelial cell adhesion in real time. Measurements in vitro by tandem scanning confocal image analysis. J Clin Invest 91:2640–2652.

Davies PF, Robotewskyj A, Griem ML (1994): Quantitative studies of endothelial cell adhesion. Directional remodeling of focal adhesion sites in response to flow forces. J Clin Invest 93:2031–2038.

de Fougerolles AR, Stacker SA, Schwarting R, Springer TA (1991): Characterization of ICAM-2 and evidence for a third counter-receptor for LFA-1. J Exp Med 174:253–267.

Diamond MS, Staunton DE, de Fougerolles AR, et al. (1990): ICAM-1 (CD54): a counter-receptor for Mac-1 (CD11b/CD18). J Cell Biol 111:3129–3139.

Edwards MJ, Abney DL, Heniford BT, Miller FN (1992): Passive immunization against tumor necrosis factor inhibits interleukin-2-induced microvascular alterations and reduces toxicity. Surgery 112:480–486.

Egawa K, Yoshiwara M, Nose K (1994): Effect of radical scavengers on TNF alpha-mediated activation of the uPA in cultured cells. Experientia 50:958–962.

Entman ML, Youker K, Shoji T, et al. (1992): Neutrophil induced oxidative injury of cardiac myocytes. A compartmented system requiring CD11b/CD18-ICAM-1 adherence. J Clin Invest 90:1335–1345.

Fabian TC, Croce MA, Stewart RM, Dockter ME, Proctor KG (1994): Neutrophil CD18 expression and blockade after traumatic shock and endotoxin challenge. Ann Surg 220:552–561.

Fabry Z, Waldschmidt MM, Hendrickson D, et al. (1992): Adhesion molecules on murine brain microvascular endothelial cells: expression and regulation of ICAM-1 and Lgp 55. J Neuroimmunol 36:1–11.

Farhood A, McGuire GM, Manning AM, Miyasaka M, Smith CW, Jaeschke H (1995): Intercellular adhesion molecule 1 (ICAM-1) expression and its role in neutrophil-induced ischemia-reperfusion injury in rat liver. J Leukocyte Biol 57:368–374.

Foxall C, Watson SR, Dowbenko D, et al. (1992): The three members of the selectin receptor family recognize a common carbohydrate epitope, the sialyl Lewis(x) oligosaccharide. J Cell Biol 117:895–902.

Fraticelli A, Serrano CV, Bochner BS, Capogrossi MC, Zweier JL (1996): Hydrogen peroxide and superoxide modulate leukocyte adhesion molecule expression and leukocyte endothelial adhesion. Biochim Biophys Acta 1310:251–259.

Freyer DR, Morganroth ML, Rogers CE, Arnaout MA, Todd RF 3d (1988): Modulation of surface CD11/CD18 glycoproteins (Mo1, LFA-1, p150, 95) by human mononuclear phagocytes. Clin Immunol Immunopathol 46:272–283.

Fuortes M, Jin WW, Nathan C (1994): Beta 2 integrin-dependent tyrosine phosphorylation of paxillin in human neutrophils treated with tumor necrosis factor. J Cell Biol 127:1477–1483.

Gaboury J, Woodman RC, Granger DN, Reinhardt P, Kubes P (1993): Nitric oxide prevents leukocyte adherence: role of superoxide. Am J Physiol 265:H862–H867.

Gauthier TW, Davenpeck KL, Lefer AW (1994): Nitric oxide attenuates leukocyte-endothelial interaction via P-selectin in splanchnic ischemia-reperfusion. Am J Physiol 267:G562–G568.

Geng JG, Bevilacqua MP, Moore KL, McIntyre TM, Prescott SM, Kim (1990): Rapid neutrophil adhesion to activated endothelium mediated by GMP-140. Nature 343:757–760.

Gimbrone MA, Jr., Obin MS, Brock AF, et al. (1989): Endothelial interleukin-8: a novel inhibitor of leukocyte-endothelial interactions. Science 246:1601–1603.

Griffin JD, Spertini O, Ernst TJ, et al. (1990): Granulocyte-macrophage colony-stimulating factor and other cytokines regulate surface expression of the leukocyte adhesion molecule-1 on human neutrophils, monocytes, and their precursors. J Immunol 145:576–584.

Grisham MB, Hernandez LA, Granger DN (1989): Adenosine inhibits ischemia-reperfusion-induced leukocyte adherence and extravasation. Am J Physiol 257:H1334–H1339.

Harris NR, Russell JM, Granger DN (1994): Mediators of endotoxin-induced leukocyte adhesion in mesenteric postcapillary venules. Circ Shock 43:155–160.

Hofman FM, Dohadwala MM, Wright AD, Hinton DR, Walker SM (1994): Exogenous tat protein activates central nervous system-derived endothelial cells. J Neuroimmunol 54:19–28.

Hogg N, Darley Usmar VM, Wilson MT, Moncada S (1992): Production of hydroxyl radicals from the simultaneous generation of superoxide and nitric oxide. Biochem J 281:419–424.

Holland JA, Pritchard KA, Pappolla MA, Wolin MS, Rogers NJ, Stemerman MB (1990): Bradykinin induces superoxide anion release from human endothelial cells. J Cell Physiol 143:21–25.

Huber AR, Kunkel SL, Todd RF 3d, Weiss SJ (1991): Regulation of transendothelial neutrophil migration by endogenous interleukin-8. Science 254:99–102 [published errata appear in Science 1991 254(5032):631 and 1991 254(5037):1435].

Ichijima H, Petroll WM, Barry PA, et al. (1993): Actin filament organization during endothelial wound healing in the rabbit cornea: comparison between transcorneal freeze and mechanical scrape injuries. Invest Ophthalmol Vis Sci 34:2803–2812.

Ikeda M, Schroeder KK, Mosher LB, Woods CW, Akeson AL (1994): Suppressive effect of antioxidants on intercellular adhesion molecule-1 (ICAM-1) expression in human epidermal keratinocytes. J Invest Dermatol 103:791–796.

Imai Y, True DD, Singer MS, Rosen SD (1990): Direct demonstration of the lectin activity of gp90MEL, a lymphocyte homing receptor. J Cell Biol 111:1225–1232.

Jacob GS, Kirmaier C, Abbas SZ, Howard SC, Steininger CN, Welply (1995): Binding of sialyl Lewis x to E-selectin as measured by fluorescence polarization. Biochemistry 34:1210–1217.

Jaffe EA, Nachman RL, Becker CG, Minick CR (1973): Culture of human endothelial cells derived from umbilical veins. Identification by morphologic and immunologic criteria. J Clin Invest 52:2745–2756.

Jones DH, Anderson DC, Burr BL, et al. (1988): Quantitation of intracellular Mac-1 (CD11b/CD18) pools in human neutrophils. J Leukoc Biol 44:535–544.

Kamikubo Y (1993): [Cardiac dysfunction and endogenous cytokines in global ischemia and reperfusion injury]. Hokkaido Igaku Zasshi 68:813–826.

Kaplanski G, Farnarier C, Kaplanski S, et al. (1994): Interleukin-1 induces interleukin-8 secretion from endothelial cells by a juxtacrine mechanism. Blood 84:4242–4248.

Kennel SJ, Godfrey V, Ch'ang LY, Lankford TK, Foote LJ, Makkinje A (1992): The beta 4 subunit of the integrin family is displayed on a restricted subset of endothelium in mice. J Cell Sci 101:145–150.

Kopp E, Ghosh S (1994): Inhibition of NF-kappa B by sodium salicylate and aspirin. Science 265:956–959.

Kotovuori P, Tontti E, Pigott R, et al. (1993): The vascular E-selectin binds to the leukocyte integrins CD11/CD18. Glycobiology 3:131–136.

Kubes P, Suzuki M, Granger DN (1991): Nitric oxide: an endogenous modulator of leukocyte adhesion. Proc Natl Acad Sci USA 88:4651–4655.

Kuijpers TW, Hoogerwerf M, van der Laan LJW, et al. (1992): CD66 nonspecific cross-reacting antigens are involved in neutrophil adherence to cytokine-activated endothelial cells. J Cell Biol 118:457–466.

Kuijpers TW, Raleigh M, Kavanagh T, et al. (1994): Cytokine-activated endothelial cells internalize E-selectin into a lysosomal compartment of vesiculotubular shape. A tubulin-driven process. J Immunol 152:5060–5069.

Kunkel SL, Standiford T, Kasahara K, Strieter RM (1991): Interleukin-8 (IL-8): the major neutrophil chemotactic factor in the lung. Exp Lung Res 17:17–23.

Kurose I, Anderson DC, Miyasaka M, et al. (1994a): Molecular determinants of reperfusion-induced leukocyte adhesion and vascular protein leakage. Circ Res 74:336–343.

Kurose I, Yamada T, Wolf R, Granger DN (1994b): P-selectin-dependent leukocyte recruitment and intestinal mucosal injury induced by lactoferrin. J Leukoc Biol 55:771–777.

Le Deist F, Menasche P, Kucharski K, Bel A, Piwnica A, Bloch G (1994): Hypothermia during cardiopulmonary bypass delays but does not prevent neutrophil-endothelial cell adhesion. Circulation 90:1-48.

Ley K, Gaehtgens P, Fennie C, Singer MS, Lasky LA, Rosen SD (1991): Lectin-like cell adhesion molecule 1 mediates leukocyte rolling in mesenteric venules in vivo. Blood 77:2553–2555.

Liu SM, Magnusson KE, Sundqvist T (1993): Microtubules are involved in transport of macromolecules by vesicles in cultured bovine aortic endothelial cells. J Cell Physiol 156:311–316.

Lo SK, Janakidevi K, Lai L, Malik AB (1993): Hydrogen peroxide-induced increase in endothelial adhesiveness is dependent on ICAM-1 activation. Am J Physiol 264:L406–412.

Loike JD, Sodeik B, Cao L, et al. (1991): CD11c/CD18 on neutrophils recognizes a domain at the N terminus of the A alpha chain of fibrinogen. Proc Natl Acad Sci USA 88:1044–1048.

Lorant DE, Patel KD, McIntyre TM, McEver RP, Prescott SM, Zimmerman GA (1991): Coexpression of GMP-140 and PAF by endothelium stimulated by histamine or thrombin: a juxtacrine system for adhesion and activation of neutrophils. J Cell Biol 115:223–234.

Ma XL, Lefer DJ, Lefer AM, Rothlein R (1992): Coronary endothelial and cardiac protective effects of a monoclonal antibody to intercellular adhesion molecule-1 in myocardial ischemia and reperfusion. Circulation 86:937–946.

Ma XL, Weyrich AS, Lefer DJ, et al. (1993): Monoclonal antibody to L-selectin attenuates neutrophil accumulation and protects ischemic reperfused cat myocardium. Circulation 88:649–658.

Margiotta MS, Robertson FS, Greco RS (1992): The adherence of endothelial cells to Dacron induces the expression of the intercellular adhesion molecule (ICAM-1). Ann Surg 216:600–604.

Maruyama Y (1963): The human endothelial cell tissue culture. Z Zellforsch Mikrosk Anat 60:69–79.

Masuda M, Ihara N, Komiyama Y, Murakami T, Murata K (1990): Flow cytometric study of injuries in cultured endothelial cells by neutrophils of the inherited cataract rats. Atherosclerosis 84:245–254.

Mayadas TN, Johnson RC, Rayburn H, Hynes RO, Wagner DD (1993): Leukocyte rolling and extravasation are severely compromised in P selectin-deficient mice. Cell 74:541–554.

McEver RP, Beckstead JH, Moore KL, Marshall Carlson L, Bainton DF (1989): GMP-140, a platelet alpha-granule membrane protein, is also synthesized by vascular endothelial cells and is localized in Weibel-Palade bodies. J Clin Invest 84:92–99.

Menon SD, Qin S, Guy GR, Tan YH (1993): Differential induction of nuclear NF-kappa B by protein phosphatase inhibitors in primary and transformed human cells. Requirement for both oxidation and phosphorylation in nuclear translocation. J Biol Chem 268:26805–26812.

Miller LJ, Bainton DF, Borregaard N, Springer TA (1987): Stimulated mobilization of monocyte Mac-1 and p150,95 adhesion proteins from an intracellular vesicular compartment to the cell surface. J Clin Invest 80:535–544.

Moen O, Fosse E, Brockmeier V, et al. (1995): Disparity in blood activation by two different heparin-coated cardiopulmonary bypass systems. Ann Thorac Surg 60:1317–1323.

Moynagh PN, Williams DC, O'Neill LA (1994): Activation of NF-kappa B and induction of vascular cell adhesion molecule-1 and intracellular adhesion molecule-1 expression in human glial cells by IL-1. Modulation by antioxidants. J Immunol 153:2681–2690.

Mulligan MS, Paulson JC, De Frees S, Zheng ZL, Lowe JB, Ward PA (1993): Protective effects of oligosaccharides in P-selectin-dependent lung injury. Nature 364:149–151.

Mulligan MS, Vaporciyan AA, Warner RL, et al. (1995): Compartmentalized roles for leukocytic adhesion molecules in lung inflammatory injury. J Immunol 154:1350–1363.

Nagashima T, Shigin W, Mizoguchi A, Arakawa M, Yamaguchi M, Tamaki N (1994): The effect of leukotriene C4 on the permeability of brain capillary endothelial cell monolayer. Acta Neurochir Suppl Wien 60:55–57.

Nathan C, Srimal S, Farber C, et al. (1989): Cytokine-induced respiratory burst of human neutrophils: dependence on extracellular matrix proteins and CD11/CD18 integrins. J Cell Biol 109:1341–1349.

Niu XF, Smith CW, Kubes P (1994): Intracellular oxidative stress induced by nitric oxide synthesis inhibition increases endothelial cell adhesion to neutrophils. Circ Res 74:1133–1140.

Patel KD, Moore KL, Nollert MU, McEver RP (1995): Neutrophils use both shared and distinct mechanisms to adhere to selectins under static and flow conditions. J Clin Invest 96:1887–1896.

Perdomo JJ, Gounon P, Schaeverbeke M, et al. (1994): Interaction between cells and elastin fibers: an ultrastructural and immunocytochemical study. J Cell Physiol 158:451–458.

Phillips ML, Nudelman E, Gaeta FC, et al. (1990): ELAM-1 mediates cell adhesion by recognition of a carbohydrate ligand, sialyl-Lex. Science 250:1130–1132.

Picker LJ, Warnock RA, Burns AR, Doerschuk CM, Berg EL, Butcher EC (1991): The neutrophil selectin LECAM-1 presents carbohydrate ligands to the vascular selectins ELAM-1 and GMP-140. Cell 66:921–933 [published erratum appears in Cell 1991 67(6):1267].

Pierce JW, Read MA, Ding H, Luscinskas FW, Collins T (1996): Salicylates inhibit I kappa B-alpha phosphorylation, endothelial-leukocyte adhesion molecule expression, and neutrophil transmigration. J Immunol 156:3961–3969.

Polley MJ, Phillips ML, Wayner E, et al. (1991): CD62 and endothelial cell-leukocyte adhesion molecule 1 (ELAM-1) recognize the same carbohydrate ligand, sialyl-Lewis x. Proc Natl Acad Sci USA 88:6224–6228.

Pomerat CM, Slick WC (1963): Isolation and growth of endothelial cells in tissue culture. Nature 198:859–861.

Rao KM, Currie MS, Cohen HJ, Peters WP (1994): Alterations in L-selectin expression and elastase activity in neutrophils from patients receiving granulocyte colony-stimulating factor alone or in conjunction with high-dose chemotherapy with autologous bone marrow transplantation. Lymphokine Cytokine Res 13:383–390.

Richter J, Ng Sikorski J, Olsson I, Andersson T (1990): Tumor necrosis factor-induced degranulation in adherent human neutrophils is dependent on CD11b/CD18-integrin-triggered oscillations of cytosolic free Ca^{2+}. Proc Natl Acad Sci USA 87:9472–9476.

Robinson JP, Cordy W, Narayanan PK, Carter WO (1994): Fate of intracellular DCFH in human neutrophils after release of caged nitric oxide using UV confocal microscopy. Cytometry 7:76.

Rosenman SJ, Ganji AA, Tedder TF, Gallatin WM (1993): Syn-capping of human T lymphocyte adhesion/activation molecules and their redistribution during interaction with endothelial cells. J Leukoc Biol 53:1–10.

Schleimer RP, Sterbinsky SA, Kaiser J, et al. (1992): IL-4 induces adherence of human eosinophils and basophils but not neutrophils to endothelium. Association with expression of VCAM-1. J Immunol 148:1086–1092.

Schreck R, Rieber P, Baeuerle PA (1991): Reactive oxygen intermediates as apparently widely used messengers in the activation of the NF-kappa B transcription factor and HIV-1. EMBO J 10:2247–2258.

Sekizuka E, Benoit JN, Grisham MB, Granger DN (1989): Dimethylsulfoxide prevents chemoattractant-induced leukocyte adherence. Am J Physiol 256:H594–H597.

Sellak H, Franzini E, Hakim J, Pasquier C (1994): Reactive oxygen species rapidly increase endothelial ICAM-1 ability to bind neutrophils without detectable upregulation. Blood 83:2669–2677.

Shimizu S, Ishii M, Yamamoto T, Kawanishi T, Momose K, Kuroiwa Y (1994): Bradykinin induces generation of reactive oxygen species in bovine aortic endothelial cells. Res Commun Chem Pathol Pharmacol 84:301–314.

Silber A, Newman W, Reimann KA, Hendricks E, Walsh D, Ringler DJ (1994): Kinetic expression of endothelial adhesion molecules and relationship to leukocyte recruitment in two cutaneous models of inflammation. Lab Invest 70:163–175.

Silvestro L, Viano I, Macario M, et al. (1994): Effects of heparin and its desulfated derivatives on leukocyte-endothelial adhesion. Semin Thromb Hemost 20:254–258.

Simon SI, Burns AR, Taylor AD, et al. (1995): L-selectin (CD62L) cross-linking signals neutrophil adhesive functions via Mac-1 (CD11b/CD18) beta 2-integrin. J Immunol 155:1502–1514.

Slocum MM, Granger DN (1993): Early mucosal and microvascular changes in feline intestinal transplants. Gastroenterology 105:1761–1768.

Smith CW, Kishimoto TK, Abbassi O, Hughes B, Rothlein R, McIntire (1991): Chemotactic factors regulate lectin adhesion molecule 1 (LECAM-1)-dependent neutrophil adhesion to cytokine-stimulated endothelial cells in vitro. J Clin Invest 87:609–618 [published erratum appears in J Clin Invest 1991 87(5):1873].

Soares FA, Shaughnessy SG, MacLarkey WR, Orr FW (1994): Quantification and morphologic demonstration of reactive oxygen species produced by Walker 256 tumor cells in vitro and during metastasis in vivo. Lab Invest 71:480–489.

Solomkin JS, Bass RC, Bjornson HS, Tindal CJ, Babcock GF (1994): Alterations of neutrophil responses to tumor necrosis factor alpha and interleukin-8 following human endotoxemia. Infect Immun 62:943–947.

Spertini O, Kansas GS, Munro JM, Griffin JD, Tedder TF (1991): Regulation of leukocyte migration by activation of the leukocyte adhesion molecule-1 (LAM-1) selectin. Nature 349:691–694.

Sporn LA, Lawrence SO, Silverman DJ, Marder VJ (1993): E-selectin-dependent neutrophil adhesion to Rickettsia rickettsii-infected endothelial cells. Blood 81:2406–2412.

Staal FJ, Roederer M, Herzenberg LA (1990): Intracellular thiols regulate activation of nuclear factor kappa B and transcription of human immunodeficiency virus. Proc Natl Acad Sci USA 87:9943–9947.

Stacker SA, Springer TA: Leukocyte integrin P150,95 (CD11c/CD18) functions as an adhesion molecule binding to a counter-receptor on stimulated endothelium. J Immunol 146:648–655.

Stacker, SA, Springer TA (1991): Leukocyte integrin P150,95 (CD11c/CD18) functions as an adhesion molecule binding to a counter-receptor on stimulated endothelium. J Immunol 146:648–655.

Staunton DE, Dustin ML, Springer TA (1988): Functional cloning of ICAM-2, a cell adhesion ligand for LFA-1 homologous to ICAM-1. Nature 339:61–64.

Stepp MA, Spurr Michaud S, Tisdale A, Elwell J, Gipson IK (1990): Alpha 6 beta 4 integrin heterodimer is a component of hemidesmosomes. Proc Natl Acad Sci USA 87:8970–8974.

Subramaniam M, Koedam JA, Wagner DD (1993): Divergent fates of P- and E-selectins after their expression on the plasma membrane. Mol Biol Cell 4:791–801.

Suematsu M, Schmid Schonbein GW, Chavez Chavez RH, et al. (1993): In vivo visualization of oxidative changes in microvessels during neutrophil activation. Am J Physiol 264:H881–H891.

Suzuki M, Asako H, Kubes P, Jennings S, Grisham MB, Granger DN (1991): Neutrophil-derived oxidants promote leukocyte adherence in postcapillary venules. Microvasc Res 42:125–138.

Suzuki Y, Tanigaki T, Heimer D, et al. (1994): TGF-beta 1 causes increased endothelial ICAM-1 expression and lung injury. J Appl Physiol 77:1281–1287.

Szabo C (1996): The pathophysiological role of peroxynitrite in shock, inflammation, and ischemia-reperfusion injury. Shock 6:79–88.

Tamkun JW, DeSimone DW, Fonda D, et al. (1986): Structure of integrin, a glycoprotein involved in the transmembrane linkage between fibronectin and actin. Cell 46:271–282.

Thurston G, Baldwin AL, Wilson LM (1995): Changes in endothelial actin cytoskeleton at leakage sites in the rat mesenteric microvasculature. Am J Physiol 268:H316–H329.

Uchihara T, Kondo H, Akiyama H, Ikeda K (1995): Single-laser three-color immunolabeling of a histological section by laser scanning microscopy: application to senile plaque-related structures in post-mortem human brain tissue. J Histochem Cytochem 43:103–106.

Vogel P, v.d.Beek J, Marohl K, Fischer EG, Kirkpatrick CJ (1993): In vitro studies on PMN-independent endothelial cell damage in trauma: decrease of PMN-endothelial cell adherence by fibrinogen degradation products and disturbance of endothelial cell membrane integrity by trauma serum. Eur Surg Res 25:83–90.

von Andrian UH, Chambers JD, McEvoy LM, Bargatze RF, Arfors KE, Butcher EC (1991): Two-step model of leukocyte-endothelial cell interaction in inflammation: distinct roles for LECAM-1 and the leukocyte beta 2 integrins in vivo. Proc Natl Acad Sci USA 88:7538–7542.

Voraberger G, Schafer R, Stratowa C (1991): Cloning of the human gene for intercellular adhesion molecule 1 and analysis of its 5′-regulatory region. Induction by cytokines and phorbol ester. J Immunol 147:2777–2786.

Waddell TK, Fialkow L, Chan CK, Kishimoto TK, Downey GP (1994): Potentiation of the oxidative burst of human neutrophils. A signaling role for L-selectin. J Biol Chem 269:18485–18491.

Walpola PL, Gotlieb AI, Langille BL (1993): Monocyte adhesion and changes in endothelial cell number, morphology, and F-actin distribution elicited by low shear stress in vivo. Am J Pathol 142:1392–1400.

Walsh GM, Hartnell A, Wardlaw AJ, Kurihara K, Sanderson CJ, Kay AB (1990): IL-5 enhances the in vitro adhesion of human eosinophils, but not neutrophils, in a leucocyte integrin (CD11/18)-dependent manner. Immunology 71:258–265.

Walsh GM, Mermod JJ, Hartnell A, Kay AB, Wardlaw AJ (1991): Human eosinophil, but not neutrophil, adherence to IL-1-stimulated human umbilical vascular endothelial cells is alpha 4 beta 1 (very late antigen-4) dependent. J Immunol 146:3419–3423.

Walz G, Aruffo A, Kolanus W, Bevilacqua M, Seed B (1990): Recognition by ELAM-1 of the sialyl-Lex determinant on myeloid and tumor cells. Science 250:1132–1135.

Weber C, Erl W, Pietsch A, Strobel M, Ziegler Heitbrock HW, Weber (1994): Antioxidants inhibit monocyte adhesion by suppressing nuclear factor-kappa B mobilization and induction of vascular cell adhesion molecule-1 in endothelial cells stimulated to generate radicals. Arterioscler Thromb 14:1665–1673.

Weber C, Erl W, Pietsch A, Weber PC (1995): Aspirin inhibits nuclear factor-kappa B mobilization and monocyte adhesion in stimulated human endothelial cells. Circulation 91:1914–1917.

Wegner CD, Wolyniec WW, LaPlante AM, et al. (1992): Intercellular adhesion molecule-1 contributes to pulmonary oxygen toxicity in mice: role of leukocytes revised. Lung 170:267–279.

Westlin WF, Kiely JM, Gimbrone MA Jr (1992): Interleukin-8 induces changes in human neutrophil actin conformation and distribution: relationship to inhibition of adhesion to cytokine-activated endothelium. J Leukoc Biol 52:43{endash}51.

Weyrich AS, Buerke M, Albertine KH, Lefer AM (1995): Time course of coronary vascular endothelial adhesion molecule expression during reperfusion of the ischemic feline myocardium. J Leukoc Biol 57:45–55.

Weyrich AS, Ma XL, Lefer AM (1992): The role of L-arginine in ameliorating reperfusion injury after myocardial ischemia in the cat. Circulation 86:279–288.

Weyrich AS, Ma XL, Lefer DJ, Albertine KH, Lefer AM (1993): In vivo neutralization of P-selectin protects feline heart and endothelium in myocardial ischemia and reperfusion injury [see comments]. J Clin Invest 91:2620–2629.

Wright SD, Weitz JI, Huang AJ, Levin SM, Silverstein SC, Loike JD (1988): Complement receptor type three (CD11b/CD18) of human polymorphonuclear leukocytes recognizes fibrinogen. Proc Natl Acad Sci USA 85:7734–7738.

Yoshida N, Granger DN, Anderson DC, Rothlein R, Lane C, Kvietys PR (1992): Anoxia/reoxygenation-induced neutrophil adherence to cultured endothelial cells. Am J Physiol 262:H1891–H1898.

11

The HL-60 Cell: A Cell Culture Model for Studying Phagocyte Cell Functions

Padma Kumar Narayanan
SmithKline Beecham, King of Prussia, Pennsylvania

J. Paul Robinson
Purdue University, West Lafayette, Indiana

The phagocyte respiratory burst is one of the immune system's most important mechanisms for intercepting invading microbial pathogens. The molecular machinery that is essentially dormant in resting cells becomes activated when phagocytes encounter a bacterial invasion, resulting in the generation of reactive oxygen species (ROS)—superoxide anion (O_2^-), hydrogen peroxide (H_2O_2), and/or nitric oxide (NO·). The biochemical events leading to the respiratory burst depend on the successful assembly of the NADPH oxidase complex in the plasma membrane of activated neutrophils from components located at different sites in the cell. Morel et al. (1991) describe two redox components thought to be present in the active oxidase complex, a flavoprotein and a low-potential cytochrome b. The resting NADPH oxidase contains redox proteins localized in the plasma membrane, including an NADPH flavodehydrogenase and cytochrome b_{558}. This cytochrome is a heterodimer consisting of a small 22-kDa subunit (α chain) which carries the heme and a large 91-kDa subunit (β chain) which is highly glycosylated (Morel et al., 1991; Kleinberg et al., 1989). The transition from the resting state to the active state results from the translocation of cytosolic factors p46-*phox* and p67-*phox* to redox components in the membrane.

Phagocyte Function: A Guide for Research and Clinical Evaluation, Edited by J. Paul Robinson and George F. Babcock. ISBN 0-471-12364-1

Chronic granulomatous disease (CGD) patients have defective expression of either the cytochrome b subsets or the cytosolic factors (Curnutte et al. 1989). Efficient oxidase activation can be recovered by mixing the cytosols of neutrophils from CGD patients deficient in p47-*phox* and p67-*phox*, respectively. The majority of patients with the autosomal form of CGD show a deficiency in p47-*phox*. At present, statistical data indicate that p47-*phox* deficiency represents 30–35% of all CGD cases, compared to less than 5% for the p67-*phox* deficiency. Another cytosolic protein believed to play an important role in the NADPH oxidase-mediated oxidative burst is p40-*phox*. Even though the precise functions of p40-*phox* are relatively unknown, it is presumed to play a major role in the stabilization of other proteins that migrate from the cytosol to the plasma membrane complex. Rap1A and Rac2 are two low-molecular-weight guanosine triphosphate-binding proteins of the Ras superfamily that assist in regulating the function of one or more constituent proteins of the NADPH oxidase complex (DeLeo and Quinn, 1996).

Early precursor myeloid leukemic cells can be induced to differentiate into monocyte- or granulocyte-like cells and thereby acquire functional responsiveness. Phagocytic and respiratory burst activity is acquired early in the metamyelocyte stage, whereas chemotaxis appears only at the banded or segmented stage of development (Laskin et al., 1991). Thus leukemic cells may serve as a model for studying the acquisition of NADPH oxidase activity during myeloid differentiation. Correlation between expression and activity of these factors may help in understanding the role of ROS generation.

THE HL-60 CELL CULTURE MODEL

Myeloid leukemic cell lines, in particular HL-60, have been used as models for the study of neutrophil and monocyte differentiation (Collins et al., 1977; Collins et al., 1978; Forsbeck et al., 1985; Ross, 1985a,b; Pilz et al., 1987; Todd et al., 1987; Dufer et al., 1989; Chou et al., 1990; Cowen et al., 1991; Goldstein et al., 1991; Zhou et al., 1991; Becker and Tyobeka, 1992). Because these cells are stable and amenable to continous culture, significant progress has been made in understanding neutrophil differentiation in terms of respiratory burst and phagocytic capacity.

In an attempt to develop long-term suspension cultures of human myeloid leukemic cells, Collins et al. (1977) isolated myeloid leukemic cells from the peripheral blood of an adult female with acute promyelocytic leukemia. They seeded these cells, known as HL-60 cells, in RPMI-1640 medium supplemented with 15% heat-inactivated fetal calf serum. The unique feature of this cell line is that it not only proliferated but also differentiated along the myeloid series without an exogenous source of conditioned medium (Dalton et al., 1988). Previous attempts to raise cells that differentiated along the myeloid series required appropriate conditioned media or a feeder layer of peripheral blood leukocytes. Since then the HL-60 cell line has proved a productive model for investigating the process of myeloid differentiation and has also been a convenient source of various mRNAs and proteins such as tumor necrosis factor (TNF) and myeloperoxidase (MPO).

REACTIVE OXYGEN INTERMEDIATES

Blair et al. (1985) initiated the original studies by flow cytometry to assess the extent of differentiation in terms of acquisition of respiratory burst capacity, as indicated by the reduction of nitroblue tetrazolium (NBT) during the stimulus-induced respiratory burst by maturing HL-60 promyelocytes. Microscopic determination of NBT-positive cells had been the method for measuring the respiratory burst activity of HL-60 cells acquiring mature granulocyte status until the advent of flow cytometry, which made possible the measurement of changes in the scatter properties of maturing HL-60 cells. Figure 1 sketches the pathways followed by HL-60 cells induced to differentiate by various agents. Simultaneous measurement of forward and 90° light scatter (90°LS) is routinely used in flow cytometry to distinguish monocytes and granulocytes from lymphocytes (Shapiro, 1995). HL-60 cells grown in the presence of 1.5% dimethyl sulfoxide (DMSO) differentiate into mature granulocytes able to reduce NBT following stimulation with phorbol myristate acetate (PMA). Nitroblue tetrazolium positivity in the population with high 90°LS, that is, granulocytes, increased with prolonged exposure to DMSO. In another study on HL-60 maturation induced by anticancer drug therapy (Ross, 1986), an automated flow cytochemistry system (Hemalog-D) was employed to measure mean cell volume, myeloperoxidse (MPO), and nonspecific esterase (NSE) content. Granulocytic maturation induced by

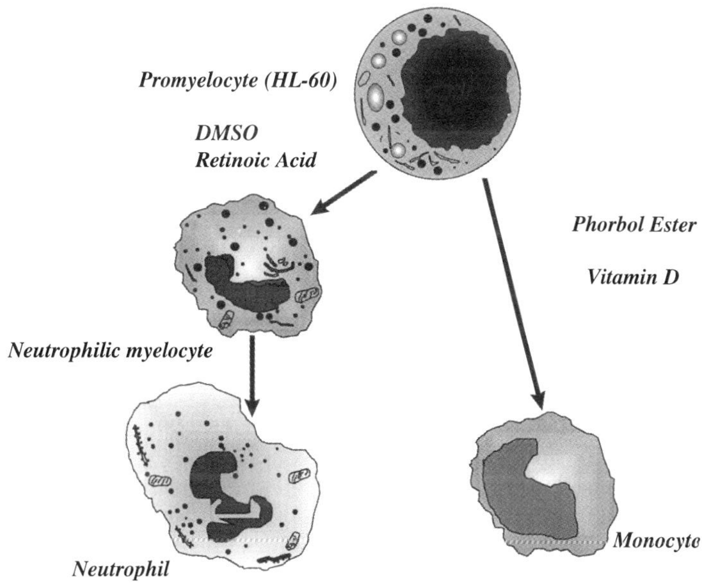

Fig. 1. Cartoon illustrating morphological changes induced by DMSO (1.25%) in an HL-60 culture. Undifferentiated HL-60 cells are predominantly promyelocytes with large round nuclei and prominent azurophilic granules. Dimethyl sulfoxide in appropriate concentration induces the undifferentiated HL-60 cells to mature into myelocytes, then metamyleocytes, and finally neutrophils. The cells undergo a reduction in size, reduction in nuclear/cytoplasmic ratio, pyknosis of nuclei, increase in phagocytic capacity, and decrease in prominence of basophilic granules.

vitamin A or DMSO resulted in a 50% decrease in cell volume, whereas mean cellular concentration of MPO remained constant throughout from promyelocyte to neutrophil-like forms. For monocytic maturation induced by low-dose ARA-c, the mean NSE increased substantially, whereas cell volume remained constant. The first studies to quantify oxygen metabolism of the HL-60 cell line were performed by Thompson et al. (1988), who compared granulocyte-like and macrophage-like differentiation of HL-60 cells. Dimethyl sulfoxide-induced differentiation produced cells with morphological characteristics akin to human neutrophils but with a reduced capacity to consume oxygen (O_2) and produce superoxide (O_2^-). Even though vitamin D-treated cells had a morphology similar to that of monocytes, O_2 consumption and O_2^- production were less than those for DMSO-treated cells. Exposure to mitochondrial inhibitors potassium cyanide (KCN) and antimycin A did not alter the response of monocyte-like differentiated HL-60 cells to stimuli such as PMA, opsonized zymosan, or a combination of opsonized zymosan and cytochalasin B. In contrast to these results, human monocyte O_2 consumption is known to decrease in the presence of these mitochondrial inhibitors (Thompson et al., 1988). Hadjokas et al. (1992) reported that repetitive passaging of HL-60 cells was accompanied by an increased rate of replication which resulted in significant impairment in respiratory burst, calcium mobilization, and β-glucuronidase release on exposure to agonists such as *N*-formylmethionyl-leucyl-phenylalanine (fMLP) and PMA.

Newburger and Tauber (1982) showed that changes in ethylene production by neutrophilic HL-60 cells differentiated with *N,N*-dimethylformamide (DMF) indicated hydroxyl radical formation. On the other hand, a spin-trapping system utilized by Thompson et al. (1988) found no evidence that either DMSO- or vitamin D-treated HL-60 cells generated hydroxyl radical. Speier and Newberger (1986) also reported that HL-60 cells differentiated with DMF possess considerably higher concentrations of glutathione than human neutrophils do. Luminol-dependent luminescence in monocyte-like and granulocyte-like HL-60 cells, which likely reflects production of O_2^-, hypochlorous acid, and other ROS, was markedly less than that observed in neutrophils.

In a time-course study by Levy and co-workers (1990) expression of membrane and cytosolic NADPH oxidase components was monitored as HL-60 cells were induced into neutrophilic and monocytic pathways. The authors observed that the rate of development of O_2^- generating capacity varied with the inducing agent employed. Over a 6-day differentiation period the ability of these agents to induce PMA-stimulated O_2^- production ranked as interferon-γ (IFN-γ > vitamin D > retinoic acid (RA) > DMSO. In this regard, Thompson et al. (1988) claimed that the response to PMA of HL-60 cells treated with vitamin D was less than the response of those treated with DMSO. Moreover, O_2^- production in monocyte-like cells was increased compared to neutrophil-like HL-60 cells, indicating that the model system differed from the normal counterpart. The reason for this effect might have been the suboptimal concentrations of differentiating agents used, which resulted in partial differentiation. Kitagawa et al. (1984) and Newburger et al. (1979) reported a higher rate of O_2^- production in DMSO-differentiated HL-60 cells than did Levy et al. (1990). One restriction of these and other studies is that only extracellular generation of ROS was measured.

The literature describes in great detail many assay systems utilized for measurement of intracellular ROS production in neutrophils. Often based upon the use of fluorescent probes and flow cytometry, these methods allow simultaneous measurement of O_2^- and H_2O_2, and NO· as well. Dichlorofluorescin diacetate (DCFH-DA), dihydrorhodamine 123 (DHR 123), and hydroethidine (HE) are the principal probes used for detecting ROS production in stimulated neutrophils (Rothe and Valet, 1990). Hydroethidine and DCFH-DA are the probes specific for the detection of O_2^- and H_2O_2, respectively (Rothe and Valet, 1990; see also a more detailed discussion in Chapter 9 of this volume). A significant advantage of these methods is the ability to measure individual cells, identifying physical characteristics (size, granularity) as well as fluorescence. In an attempt to characterize the acquisition of ROS production in general, we monitored H_2O_2 production in HL-60 cells induced to differentiate along the neutrophilic pathway on exposure to 1.25% DMSO. Phorbol 12-myristate 13-acetate (PMA, 8 and 50 ng/ml) was used to stimulate HL-60 cells at different stages of maturation. As shown in Figure 2, a significant increase in intracellular production of H_2O_2 (DCF fluorescence) in response to 50 ng/ml PMA is seen in HL-60 cells at 24 hr of differentiation, with a maximum response at 96 h. A dose-dependent increase in H_2O_2 production is also seen at all time points. This amply demonstrates the progressive acquisition of full functionality of the NADPH oxidase complex from an immature promyelocytic level to a mature neutrophilic stage. The nature of the possible mechanism responsible for this increase in H_2O_2 production in differentiated, stimulated HL-60 cells was most likely related to activation of the NADPH oxidase complex, which leads to increased free-radical generation independent of mitochondrial respiration. This conclusion was supported by studies employing diphenyleneiodonium (DPI), an inhibitor of the plasma membrane NADPH oxidase (Hancock and Jones, 1987). Fully differentiated HL-60 cells, assessed previously by morphological

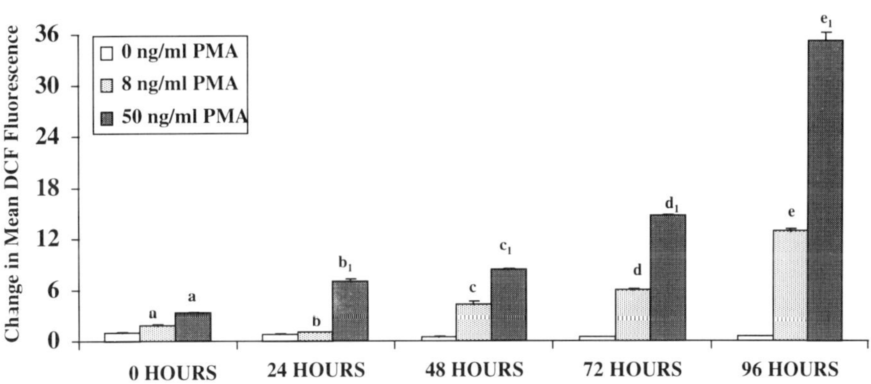

Fig. 2. Intracellular H_2O_2 production following stimulation of HL-60 cells by 8 and 50 ng/ml of PMA on different days of differentiation. Hydrogen peroxide was measured using DCFH-DA with a Coulter *EPICS* ELITE flow cytometer (Coulter Cytometry, Hialeah, Florida). Intracellular H_2O_2 production was increased on exposure to PMA ($p < .05$). Different alphabets indicate significant difference from control and HL-60 cells exposed to 8 ng/ml PMA.

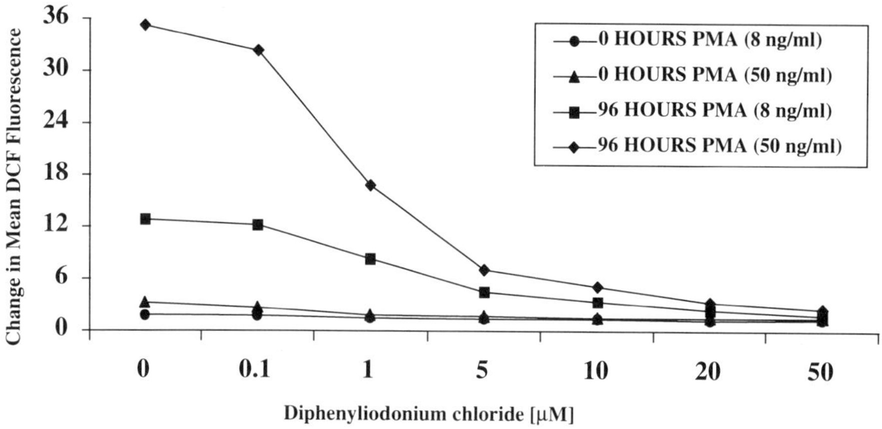

Fig. 3. Intracellular H_2O_2 production in 0-h and 96-h differentiated HL-60 cells on exposure to PMA (8 and 50 ng/ml) in the presence or absence of diphenyliodonium (DPI, 0.1–50 μ*M*). DPI, a specific inhibitor of cytochrome b_{558}, significantly inhibits the production of intracellular H_2O_2, indicating that the pathway primarily responsible for the elaboration of this ROS is the NADPH oxidase complex.

characteristics, stimulated by PMA (8 and 50 ng/ml) showed a dose-dependent decrease in H_2O_2 production on exposure to DPI (0.1–50 μ*M*) (Figure 3). It is also interesting to note that HL-60 cells in different passages have different levels of basal intracellular O_2^- and H_2O_2 production (Figure 4), possibly from impairment acquired by HL-60 cells when carried over long periods of time. This substantiates the findings of Hadjokas et al. (1992) that repetitive passaging of HL-60 cells was accompanied by an increased rate of replication resulting in significant impairment in metabolism.

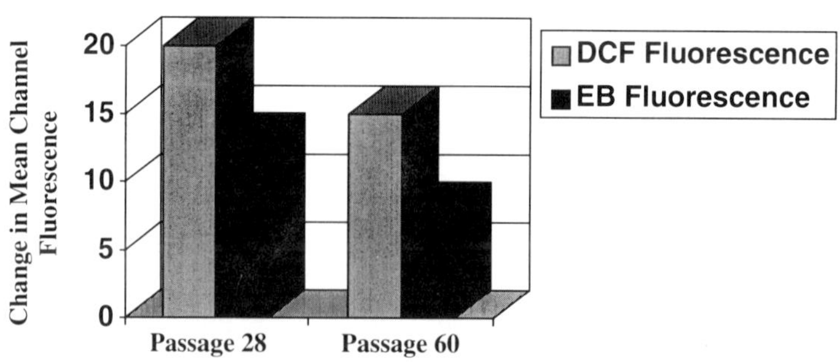

Fig. 4. Basal intracellular O_2^- and H_2O_2 production in HL-60 cells over a 45-min interval from different passages in culture. Basal ROS production in passage-60 cells was lower than basal ROS production in passage-28 cells. This study lends support to the findings of Hadjokas et al. (1992) that correlated impairment in cell function with increased number of passages.

MOLECULAR DISSECTION OF HL-60 DIFFERENTIATION

Immunoblot analysis of HL-60 membranes and cytosol was used to assess the amount of specific phagocyte oxidase factors (91- and 22-kDa subunits of membrane cytochrome b_{558} (gp91 and p22), and 47- and 67-kDa cytosol oxidase factors (p47-*phox* and p67-*phox*) by Levy et al. in 1990. A common feature of all induction regimens was that the cytosolic factor p47-*phox* was expressed earlier and increased more rapidly than p67-*phox*. In addition, p47-*phox* appeared after 16 h and peaked at 3–4 days of differentiation, whereas p67-*phox* did not appear for 2–6 days, depending on the inducing agent. The early expression of p47-*phox* was not sufficient to provide the necessary support for O_2^- production on stimulation. Only after the appearance of p67-*phox* did differentiated HL-60 cells exhibit complete cytosol oxidase activity. An elegant experiment conducted by combining cytosol from 3-day RA-treated HL-60 cells (containing p47-*phox*, but devoid of p67-*phox* immunoreactivity) with CGD demonstrated this (Levy et al. 1990). Examination of cytosolic factors in IFN-γ- or vitamin D-differentiated HL-60 cells at 6 days demonstrated that the amount of p47-*phox* was similar to that present in an equal number of neutrophils, whereas the amount of p67-*phox* was much lower. Despite this, the rate of O_2^- production of these differentiated HL-60 cells stimulated by PMA approaches that of intact neutrophils (Levy et al., 1990).

Despite the small quantity of both subunits of cytochrome b_{558} detected in uninduced HL-60 cell membranes (Kleinberg et al., 1989; Newburger et al., 1984), a substantial amount of O_2^- can be produced in combination with normal neutrophil cytosol. This phenomenon was also observed when uninduced HL-60 cell membranes together with differentiated HL-60 cytosols were capable of supporting O_2^- production in a cell-free assay (Seifert et al., 1989). This study contradicts others in which spectral assays failed to detect cytochrome b_{558} in uninduced HL-60 cells (Roberts et al., 1982). Levy and co-workers argue that this may be due to the interference of mitochondrial cytochromes and the inherently lower sensitivity of detection of the assay used compared to immunoblot studies.

The α- and β-subunits of cytochrome b_{558} increase in parallel during differentiation (Levy et al., 1990), consistent with their tight association in the native cytochrome b_{558} molecule. Compared with RA, IFN-γ was a more potent inducer of membrane factors resulting in the augmentation of O_2^- production on exposure to PMA. In a study that combined examination of accumulation of transcripts of the gene encoding gp91-phox and the cytosolic factors p47-*phox* and p67-*phox* in IFN-γ- and/or TNF-induced HL-60 cells, a direct correlation was found between accumulation of transcripts, appearance of cell surface differentiation antigens CD11b and CD14, and respiratory burst activity (Gupta et al., 1992).

There is evidence that a low-molecular-weight guanosine triphosphate-binding protein of unknown character may be involved in regulating NADPH oxidase activation (Bokoch and Prossnitz, 1992). During post-translational processing, the low-molecular-weight protein incorporates an isoprenyl group, which might be required for activation of the oxidase. Although the functional significance is unclear, a guanosine triphosphate-binding protein known as Rap1A has been shown to bind to the cytochrome b component of the NADPH oxidase system (Quinn et al., 1989). When differentiated HL-60 cells or Epstein–Barr virus (EBV)-transformed B cells were transfected with Rap1A N17 and 63 E mutants, a normal respiratory burst was abrogated, suggesting that Rap1A may function as a curb on the components necessary to maintain the respiratory burst (DeLeo and Quinn, 1996).

DIFFERENTIATION-INDUCING FACTORS (DIF)

Leftwich and Hall (1990) reported characterization of a cytokine distinct from CSF, IFN-α or -γ, TNF-α or -β, or IL-6 that induced differentiation and cytolytic activity in HL-60 cells and that may play a role in the differentiation of normal and malignant cells of the myelomonocytic lineage. Previously, several groups had illustrated the presence of differentiation-inducing factors (DIFs) that induced differentiation of HL-60 cells. A 45-kD factor was purified by Olsson et al. (1984) from conditioned medium of the human cell line HUT-102. The DIF activity of this factor was blocked by anti-TNF-β and appeared to bind to the same receptors as TNF-α and TNF-β. Leung and Chiao (1985) described another differentiation-inducing factor found in conditioned medium from peripheral blood mononuclear cells. Takeda et al. (1986) later showed that this factor was identical to TNF-α in terms of its *N*-terminal amino acid sequence and it was further demonstrated by Gupta et al. (1992) that the differentiation-inducing effect of leukocyte-conditioned medium is due mostly to the synergistic effect of IFN-γ combined with lymphotoxin or TNF. This cytokine was subsequently identified as P48, a 48-kDa differentiation-inducing cytokine isolated from the culture medium of the human leukemic cell line Reh (Kestler et al., 1995).

PHAGOCYTOSIS

HL-60 cells represent a useful in vitro model system for dissection of the molecular events leading to the changes at the cell surface that accompany phagocytosis. Undifferentiated HL-60 cells completely lack the functions of both chemotaxis and degranulation (Cotter and Henson, 1983). For example, undifferentiated HL-60 cells that were unable to bind the anti-neutrophil monoclonal antibody NCD-1 showed an increased capacity to do so upon differentiation into granulocytes by DMSO (Cotter and Henson, 1983). Studies on complement receptors (CR) have shown that CR1 appear relatively late during myeloid differentiation and can be detected in a fraction of metamyelocytes and in almost all band cells and granulocytes (Perussia et al., 1981; Ross et al., 1978). Receptors for the Fc fragment (FcR) of IgG molecules appear earlier than CR1 during terminal differentiation of myeloid cells (Theofilopoulos and Dixon, 1979). Expression of FcR has been demonstrated on promyelocytes, myelocytes, and HL-60 cells (Perussia et al., 1981; Crabtree et al., 1979). In an attempt to delineate the suppressive effects of glucocorticoids on the phagocytosis and destruction of autologous tissue, Crabtree et al. reported in 1979 that glucocorticoids inhibited the expression of FcR on HL-60 cells. The earliest studies investigating the appearance of specific antigens on terminally differentiated cells came from the laboratory of Perussia, who demonstrated that the monoclonal antibodies B9.8.1 and B13.4.1 bound specifically to metamyelocytes and myelocytes (Perussia et al., 1981).

LEUKOCYTE INTEGRINS

Leukocyte (β_2) integrins, especially CD11b/CD18 (Mac-1) and CD11a/CD18 (LFA-1), have been shown in vitro and in vivo to be important in inflammation (Hynes, 1992; Gresham et al., 1991). Adherence of either stimulated or unstimulated human neutrophils to human umbilical vein endothelial cells is diminished significantly by antibodies to CD11b or CD18 (Mulligan et al., 1992). Differentiation of HL-60 cells along the granulocytic pathway with DMSO results in enhanced CD11b expression, consistent with the high levels expressed on human granulocytes (Back et al., 1992; Hickstein et al., 1987; Hickstein et al., 1989). Differentiation of HL-60 cells along the monocyte/macrophage pathway with PMA results in enhanced CD11a expression, consistent with the high levels of CD11a expressed on human macrophages (Back et al., 1992).

The capacity for phagocytosis is a primary function of mature monocytes and granulocytes and therefore is commonly used as a marker for myeloid cell differentiation. Phagocytic ability can be determined by quantitative flow cytometric measurement of fluorescent target uptake. Blair et al. (1986) induced differentiation with DMSO and showed that attainment of phagocytic capability was coincident with attainment of respiratory burst. Simultaneous monitoring of ingestion and determination of DNA content using propidium iodide showed that following DMSO treatment, accumulation of cells in the G_0/G_1 or (G_0-G_1) phase of the cell cycle treatment was correlated with acquisition of the capacity to phagocytize. This is evidence that the ability to phagocytize fluorescent microspheres is associated with end-stage differentiation.

Chemotactic ability has also been used as a yardstick to evaluate HL-60 maturation along the monocytic and neutrophilic pathway with respect to different inducers. Despite the early development of phagocytosis and respiratory burst during the metamyelocyte stage, chemotaxis is a later stage development. Differentiation of HL-60 cells into neutrophils with dibutyryl cyclic adenosine 3′,5′-monophosphate (dbcAMP) induced expression of fMLP receptors but no chemotactic activity to fMLP (Sirak et al., 1990). This lack of chemotaxis was attributed to the incomplete differentiation of the HL-60 cells by dbcAMP. Incompletely developed signal transduction pathways might also contribute to the delayed appearance of functional responses associated with certain cell surface receptors.

CONCLUSIONS

HL-60 cells can be utilized to clarify the intermolecular connections among the different protein components of the NADPH oxidase complex occurring during neutrophilic differentiation, thereby leading to a better understanding and/or control of their function in both microbicidal and inflammatory responses (DeLeo and Quinn, 1996). Flow cytometry is a particularly suitable technology for monitoring maturation of neutrophils, especially when functional assays can be used to distinguish between populations in different stages of maturation.

REFERENCES

Back AL, Gollahon KA, Hickstein DD (1992): Regulation of expression of the leukocyte integrin CD11a (LFA-1) molecule during differentiation of HL-60 cells along the monocyte/macrophage pathway. J Immunol 148:710–714.

Becker RW, Tyobeka EM (1992): Effects of lithium on dimethyl sulfoxide induced differentiation of HL-60 promyelocytic leukemia cells. Leuk Res 16:823–828.

Blair OC, Carbone R, Sartorelli AC (1985): Differentiation of HL-60 promyelocytic leukemia cells monitored by flow cytometric measurement of nitro blue tetrazolium (NBT) reduction. Cytometry 6:54–61.

Blair OC, Carbone R, Sartorelli AC (1986): Differentiation of HL-60 promyelocytic leukemia cells: simultaneous determination of phagocytic activity and cell cycle distribution by flow cytometry. Cytometry 7:171–177.

Bokoch GM, Prossnitz V (1992): Isoprenoid metabolism is required for stimulation of the respiratory burst oxidase of HL-60 cells. J Clin Invest 89:402–408.

Chou S, Kaneko M, Nakaya K, Nakamura Y (1990): Induction of differentiation of human and mouse myeloid leukemia cells by camptothecin. Biochem Biophys Res Commun 166:160–167.

Collins SJ, Gallo RC, Gallagher RE (1977): Continuous growth and differentiation of human myeloid leukaemic cells in suspension culture. Nature 270:347–349.

Collins SJ, Ruscetti FW, Gallagher RE, Gallo RC (1978): Terminal differentiation of human promyelocytic leukemia cells induced by dimethyl sulfoxide and other polar compounds. Proc Natl Acad Sci USA 75(5):2458–2462.

Cotter TG, Henson PM (1983): A cell surface differentiation antigen involved in human neutrophil chemotaxis and degranulation. Clin Exp Immunol 53:249–253.

Cowen DS, Berger M, Nuttle L, Dubyak GR (1991): Chronic treatment with P_2-purinergic receptor agonists induces phenotypic modulation of the HL-60 and U937 human myelogenous leukemia cell lines. J Leukoc Biol 50: 109–122.

Crabtree GR, Munck A, Smith KA (1979): Glucocorticoids inhibit expression of Fc receptors on the human granulocytic cell line HL-60. Nature 279:338–339.

Curnutte JT, Scott PJ, Mayo LA (1989). Cytosolic components of the respiratory burst oxidase: resolution of four components, two of which are missing in complementing types of chronic granulomatous disease. Proc Natl Acad Sci USA 86:825–829.

Dalton WT, Ahearn MJ, McCredie KB, Freireich EJ, Stass SA, Trujillo JM (1988): HL-60 cell line was derived from a patient with FAB-M2 and not FAB-M3. Blood 71:242–247.

DeLeo FR, Quinn MT (1996): Assembly of the phagocyte NADPH oxidase: molecular interaction of oxidase proteins. J Leukoc Biol 60:677–691.

Dufer J, Biakou D, Joly P, Benoist H, Carpentier Y, Desplaces A (1989): Quantitative morphological aspects of granulocytic differentiation induced in HL-60 cells by dimethylsulfoxide and retinoic acid. Leuk Res 13:621–627.

Forsbeck K, Nilsson K, Hansson A, Skoglund G, Ingelman-Sundberg M (1985): Phorbol ester-induced alteration of differentiation and proliferation in human hematopoietic tumor cell lines: relationship to the presence and subcellular distribution of protein kinase c. Cancer Res 45:6194–6199.

Goldstein BM, Leary JF, Farley BA, Marquez VE, Levy PC, Rowley PT (1991): Induction of HL60 cell differentiation by tiazofurin and its analogues: characterization and efficacy. Blood 78:593–598.

Gresham HD, Graham IL, Anderson DC, Brown EJ (1991): Leukocyte adhesion-deficient neutrophils fail to amplify phagocytic function in response to stimulation. J Clin Invest 88:588–597.

Gupta JW, Kubin M, Hartman L, Cassatella M, Trinchieri G (1992): Induction of expression of genes encoding components of the respiratory burst oxidase during differentiation of human myeloid cell lines induced by tumor necrosis factor and interferon-γ. Cancer Res 52:2530–2537.

Hadjokas N, Bayer C, Nielson CP (1992): Impaired stimulus-response coupling in association with increased growth rate of HL60 cells. J Leukoc Biol 52:157–160.

Hancock JT, Jones TG (1987): The inhibition by diphenyleneiodonium and its analogues of superoxide generation by macrophages. Biochem J 242:103–107.

Hickstein DD, Back AL, Collins SJ (1989): Regulation of expression of the CD11b and CD18 subunits of the neutrophil adherence receptor during myeloid differentiation. J Biol Chem 264.21812.

Hickstein DD, Smith AJ, Beatty PG, et al. (1987): Expression of leukocyte-adhesion-related glycoproteins during differentiation of HL-60 promyelocytic leukemia cells. J Immunol 138:513.

Hynes RO (1992): Integrins: versatility, modulation, and signaling in cell adhesion. Cell 69:11–25.

Kestler DP, Agarwal S, Hall RE (1995): Up-regulation of cytokine messenger-rna in human monocytes and myeloid cell-lines by the differentiation activation factor P48. Immunology 86:463–468

Kitagawa S, Ohta M, Nojiri K, et al. (1984): Functional maturation of membrane potential changes and superoxide-producing capacity during differentiation of human granulocytes. J Clin Invest 73:1062.

Kleinberg ME, Rotrosen D, Malech HL (1989): Asparagine-linked glycosylation of cytochrome b_{558} large subunit varies in different human phagocytic cells. J Immunol 143:4152–4157.

Laskin DL, Sirak AA, Robertson FM, Laskin JD (1991): Distinct patterns of chemotactic peptide-induced calcium mobilization in differentiated myeloid leukemia cells and peripheral blood neutrophils. J Leukoc Biol 49:369–379.

Leftwich JA, Hall RE (1990): Initial characterization of a cytokine which induces differentiation and cytolytic activity in HL-60 promyelocytic leukemia cells: Evidence that the cytokine is distinct from other differentiation-active cytokines. J Leukoc Biol 47:87–96.

Leung K, Chiao JW (1985): Human leukemia cell maturation induced by a T-cell lymphokine isolated from medium conditioned by normal lymphocytes. Proc Natl Acad USA 82:1209–1213.

Levy R, Rotrosen D, Nagauker O, Leto TL, Malech HL (1990): Induction of the respiratory burst in HL-60 cells. J Immunol 145:2595–2601.

Morel F, Doussiere J, Vignais P (1991): The superoxide-generating oxidase of phagocytic cells. Eur J Biochem 201:523–546.

Mulligan MS, Varani J, Warren JS, et al. (1992): Roles of β_2 integrins of rat neutrophils in complement and oxygen radical-mediated acute inflammatory injury. J Immunol 148:1847–1857.

Newburger PE, Chavaniec ME, Greenberger JS, Cohen HJ (1979): Functional changes in human leukemic cell line HL-60. A model of differentiation. J Cell Biol 82:315.

Newburger PE, Speier C, Borregaard N, Walsh CE, Whitin JC, Simons ER (1984): Development of the superoxide generating system during differentiation of the HL-60 human promyleocytic leukemia cell line. J Biol Chem 259:3771.

Newburger PE, Tauber AI (1982): Heterogeneous pathways of oxidizing radical production in human neutrophils and the HL-60 cell line. Pediatr Res 16:856.

Olsson IL, Sarngadharan MG, Breitman TR, Gallo, RC (1984): Isolation and characterization of a lymphocyte-derived differentiation inducing factor for the myeloid leukemic cell line HL-60. Blood 63:510–517.

Perussia B, Lebman D, Ip SH, Rovera G, Trinchieri G (1981): Terminal differentiation surface antigens of myelomonocytic cells are expressed in human promyelocytic leukemia cells (HL60) treated with chemical inducers. Blood 58:836–843.

Pilz RB, Van den Berghe G, Boss GR (1987): Induction of HL-60 differentiation by starvation for a single essential amino acid but not by protein synthesis inhibitors. J Clin Invest 79:1006–1009.

Quinn MT, Parkos CA, Walker L, Orkin SH, Dinauer MC, Jesaitis AJ (1989): Association of Ras-related protein with cytochrome b of human neutrophils. Nature 342:198–200.

Roberts PJ, Cross AR, Jones OTG, Segal AW (1982): Development of cytochrome b and an active oxidase system in association with maturation of human promyelocytic (HL-60) cell line. J Cell Biol 95:720.

Ross DW (1985a): Differences in cell cycle kinetics during induced granulocytic *versus* monocytic maturation of HL-60 leukemia cells. Cancer Res 45:1308–1313.

Ross DW (1985b): Leukemic cell maturation. Arch Pathol Lab Med 109:309–313.

Ross DW (1986): Measurement of myeloid maturation by flow cytochemistry in HL-60 leukemia: esterase is inducible, myeloperoxidase is not. Cytometry 7:263–267.

Ross GD, Jarowski CI, Rabellino EM, Winchester RJ (1978): The sequential appearance of Ia-like antigens and two different complement receptors during the maturation of human neutrophils. J Exp Med 147:730.

Rothe G, Valet G (1990): Flow cytometric analysis of respiratory burst activity in phagocytes with hydroethidine and 2′,7′-dichlorofluorescin. J Leukoc Biol 47:440–448.

Seifert R, Jungblut P, Schultz G (1989): Differential expresion of cytosolic activation factors for NADPH oxidase in HL-60 leukemic cells. Biochem Biophys Res Commun 161:1109.

Shapiro HM (1995): *Practical Flow Cytometry.* 3rd ed. New York: Wiley-Liss, Inc.

Sirak AA, Laskin JD, Robertson FM, Laskin DL (1990): Failure of f-Met-Leu-Phe to induce chemotaxis in differentiated promyelocytic (HL60) leukemia cells. J Leukoc Biol 48:333–342.

Speier C, Newburger PE (1986): Changes in superoxide dismutase, catalase, and the glutathione cycle during induced myeloid differentiation. Arch Biochem Biophys 251:551–557.

Takeda K, Iwamoto S, Sugimoto H, et al. (1986): Identity of differentiation inducing factor and tumor necrosis factor. Nature 323:338–340.

Theofilopoulos AN, Dixon FJ (1979): The biology and detection of immune complexes. Adv Immunol 28:89.

Thompson BY, Sivam G, Britigan BE, Rosen GM, Cohen MS (1988): Oxygen metabolism of the HL-60 cell line: comparison of the effects of monocytoid and neutrophilic differentiation. J Leukoc Biol 43:140–147.

Todd RF, Bury M, Liu D (1987): Expression of an activation antigen, Mo3e, associated with the cellular response to migration inhibitory factor by HL-60 promyelocytes undergoing monocyte-macrophage differentiation. J Leukoc Biol 41:492–499.

Zhou J-Y., Norman AW, Akashi M, et al. (1991): Development of a novel 1,25$(OH)_2$-vitamin D_3 analog with potent ability to induce HL-60 cell differentiation without modulating calcium metabolism. Blood 78: 75–82.

12

Risk Assessment for Intensive Care Patients by Automated Classification of Flow Cytometric Data

Günter K. Valet
Max-Planck-Institut für Biochemie, Martinsried, Germany

Gabriel Roth
Institut für Anästhesiologie, Klinikum Grosshadern der Universität, Munich, Germany

Wolfgang Kellermann
Institut für Anästhesiologie, Klinikum Grosshadern der Universität, Munich, Germany

SUMMARY

The early recognition of endangered sepsis patients in intensive care medicine is of great clinical importance. The risk assessment by clinical chemistry or by other clinical parameter does not, however provide this information at an early stage. It was therefore investigated whether metabolic parameters of blood leukocytes such as oxidative burst activity of monocytes and granulocytes as well as leukocyte proteinase activities in conjunction with data pattern classification are more informative.

Phagocyte Function: A Guide for Research and Clinical Evaluation, Edited by J. Paul Robinson and George F. Babcock. ISBN 0-471-12364-1

The oxidative burst activity in vital leukocyte preparations following 1g sedimentation of fresh heparinized blood was determined with dihydrorhodamine 123 (DHR 123). Furthermore the serine and cathepsin B,H,L proteinase activities were measured using fluorogenic $(Z\text{-AlaAla})_2$-rhodamine 110 (R 110) and $(Z\text{-ArgArg})_2$R 110 substrates.

The differences between septic and nonseptic patients were significant and septic patients could be correctly diagnosed in 100% of the cases by automated evaluation of the DHR 123, $(Z\text{-AlaAla})_2$ R 110, and $(Z\text{-ArgArg})_2$ R 110 assays with the self-learning data pattern classifier CLASSIF1.

The clinically more important risk assessment of septic patients at admission was equally possible by the DHR 123 and $(Z\text{-AlaAla})_2$ R 110 assays with a slight advantage for DHR 123. The findings emphasize the potential for early recognition of endangered patients. Nonsurviving patients died between 5 and 14 days after admission to the intensive care unit. The systematic optimization of the classification indicates that the TNF-α- and fMLP-stimulated DHR 123 assays were most informative whereas the serine proteinase, DHR 123 ex vivo, TNF-α + fMLP, and PMA assays contained redundant information that did not further improve the overall classification results.

INTRODUCTION

The management of preseptic or septic patients in intensive care units constitutes an important challenge for the treating physician. Although the determination of sepsis scores as well as biophysical and biochemical measurements of constituents or functions of humoral body compartments such as blood, blood plasma, or urine provides valuable diagnostic information, they are of comparatively low value for a patient's risk assessment, that is, for prognostic purposes.

As a conceptual alternative, the investigation of biochemically specific features of blood cell functions by flow cytometry is of great interest for the individual patient's risk assessment. Disease processes derive from biochemical deviations in cellular system or organs. Once recognized at the cellular level, immediate therapeutic intervention may considerably lower the degree of intracorporeal destruction. This effect increases the chances for fast and complete patient recovery.

Granulocytes are particularly important in sepsis and shock events because of their capacity for phagocytosis and destruction of microorganisms as well as for unbalancing the regulation of the microvasculatory system by the overshooting liberation of enzymes or mediators and by the exposure of new antigens. Some of these antigens may provide increased interaction with the endothelial cell layer.

Early multiparametric determinations of granulocyte parameters by flow cytometry permitted the prediction of the danger of sepsis or noninfectious shock in intensive care patients 2–3 days in advance (Rothe et al., 1993c), that is, at a time when humoral or other patient parameters are entirely unable to indicate the imminent danger.

The measurement of phagocytosis of live *E. coli* by simultaneous acridine orange/propidium iodide (AO/PI) staining (Rothe and Valet, 1988) as well as of the intracellular pH values and esterase activities by 1,4-diacetoxy-2,3-dicyanobenzene/propidium iodide (ADB/PI) staining (Valet et al., 1981) needs no centrifugation steps; that is, such assays are suitable for automated flow cytometric instrumentation.

The handling of live bacteria, however, requires special care and a separate 37°C incubation step prior to fluorescent cell staining. Furthermore, the UV-excitation of the ADB dye is not available in most of the current analytical flow cytometers, and the learning process with statistical classifiers (Rothe et al., 1990) takes substantial time. In addition, statistical classifiers have a tendency to be unstable when comparatively few informative parameters are buried in a majority of "noise," that is, noninformative parameters. The resulting classifiers are not standardized; that is, they are not easily portable.

To overcome these problems, the goals of subsequent work were fourfold:

1. Development of the 488-nm exciting and sensitive oxidative burst indicator dye dihydrorhodamine 123 (DHR 123) (Rothe and Valet, 1993); DHR 123 has become a widespread fluorogenic dye in research on neutrophils (Werusch et al., 1996a,b; Löhrke et al., 1996; Woodman et al., 1995), in general cell biology (Sobreira et al., 1966; Wang et al., 1996; Wersto et al., 1996), and in immunology (Williams and Henkart, 1996), but also in investigations on the formation and action of intracellular peroxinitrite (Zinarelli et al., 1997; Kooy and Lewis, 1996; Briviba et al., 1996).
2. Development of rhodamine 110 (R 110) proteinase substrates for the specific determination of intracellular proteolytic activities in viable cells (Rothe and Valet, 1993a,b; Ganesh et al., 1995; Klingal et al., 1994).
3. Differential exposure of blood leukocytes to humoral stimulators such as bacterial peptide *N*-formyl-L-methionyl-L-leucyl-L-phenylalanine (fMLP), tumor necrosis factor-alpha (TNF-α) and phorbol myristate acetate (PMA) as a positive control instead of stimulation by bacteria.
4. Use of triple matrix classification (Valet et al., 1993) with its capacity to provide instrument- and laboratory-independent, that is, standardized classifiers for routine classification processes in hospitals.

Using these new tools, substantial improvements can be obtained. This is exemplified in the measurement and prospective classification of a series of patients admitted to the intensive care unit either without or with infection of various degree including sepsis. The question of the present work was to what extent the flow cytometrically determined DHR 123 burst and R 110 protease activity measurements were useful for infection diagnosis as well as for predictions on ultimate survivors and nonsurvivors in sepsis.

MATERIALS AND METHODS

Patients

Patients were admitted noninfected, infected, or in the state of sepsis to the anesthesia intensive care unit following general abdominal, orthopedic, polytraumatic or gynecologic surgery. Sepsis was defined as recommended by the ACCP + SCCM consensus conference (American College of Chest Physicians, 1992). The CLASSIF1 classified patients were unselected; that is, the data of all patients in this series were used, with the exception of the data of one survivor and one nonsurvivor patient. Both patients were selected prior to the learning process and constituted a small test set, unknown to previously learned classifiers.

All patients were treated according to current principles of intensive care medicine such as maintenance of sufficient tissue oxygenation, cardiac output, electrolyte balance, and nutrition status, as well as antibiotic therapy, search and removal of septic foci, and so on.

Blood Samples and Cell Staining

Buffy coat leukocytes were obtained following 40-min sedimentation of 2-ml heparinized blood samples (10 IU/ml) at room temperature after overlayering them at a 1:1 volume ratio onto 1.077 g/ml lymphocyte separation medium (Sigma, Deisenhofen, Germany) in polycarbonate test tubes.

Leukocyte samples were diluted 1/50 in 5 m*M* Hepes (Sigma) buffered saline (HBS) pH 7.35 and incubated for 5 min at 37°C in the presence of 1 μ*M* DHR 123 (Molecular Probes, Eugene, Oregon). They were further incubated for 20 min either unstimulated or in separate assays in the presence of physiological stimulators such as 10 ng/ml TNF-α (Sigma, recombinant in yeast 2×10^7 U/mg protein), 100 n*M* fMLP (Sigma), or a combination of both as well as with 100 n*M* PMA (Sigma) as a positive control (Rothe and Valet, 1993c).

Leukocyte samples were further incubated without stimulation in 1/50 HBS dilution for 30 min at 37°C in the presence of 4 μ*M* of either $(\text{Z-AlaAla})_2$-R 110 serine proteinase or $(\text{Z-ArgArg})_2$-R 110 cathepsin B,H,L substrate. The substrates were produced by synthesis in our laboratory (Rothe et al., 1992).

Dead cells in all assays were counterstained for 5 min at room temperature with 60 μ*M* propidium iodide (PI, Serva, Heidelberg, Germany).

Flow Cytometry

The stained leukocytes were analyzed in a FACScan flow cytometer (Becton Dickinson, Heidelberg) with linear amplification for sideward light scatter (SSC) and four-decade logarithmic amplification for forward light scatter (FSC) as well as in the FITC (F1) and PI (F3) fluorescence channels. 10,000 cells were collected with Consort30 software as FCS1.0 list mode files. The resulting data files were transferred by the Fastlink program (Becton Dickinson) to a personal computer for further analysis. Day-to-day precision of the flow cytometer was assured by repeated measurements of the same lot of monosized 4.5-μm YG FITC-like calibration particles (Polysciences, Eppelheim, Germany).

List Mode Data Analysis

List mode data were analyzed by the CLASSIF1 program (Partec, Münster, Germany) in a standardized way.

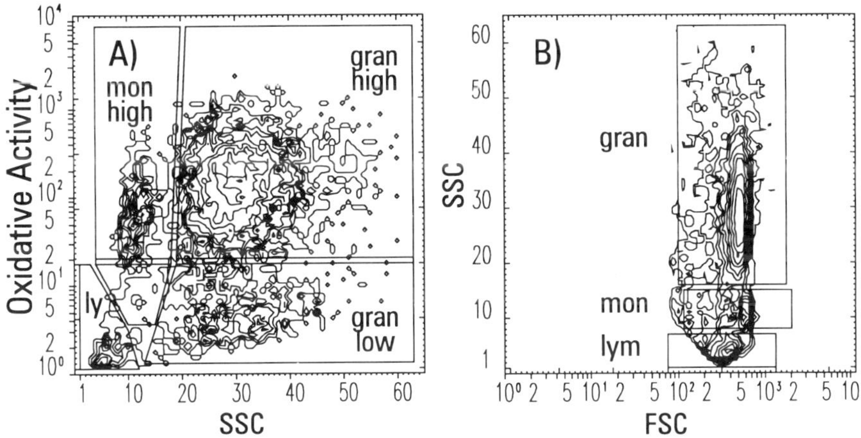

Fig. 1. Standardized evaluation of the DHR 123 oxidative burst measurement of a PMA-stimulated leukocyte sample from a healthy person. The evaluation windows were fixed (A), that is, identical for all list mode files of the learning and test sets to assure the unbiased evaluation. The gates in the light scatter histogram (B) were autogated according to following procedure: the lymphocyte evaluation window is used as leader gate. The evaluation program detects the maximum of the lymphocyte cluster and sets thresholds in predetermined distances to the peak in the *x* and *y* directions (up/down/left/right). The monocyte gate is localized immediately on top of the lymphocyte gate with initially 1.2 times the *y*-span of the lymphocyte gate. The maximum of the monocyte peak is determined and thresholds are set in predetermined distances of the peak. If no monocyte peak is detected, the initial default gate is used. The granulocyte gate is finally placed immediately above the monocyte gate. It spans until the highest *y*-axis channel and has predetermined outreaches to the left and right of the maximum of the granulocyte cluster on the *x*-axis.

Vital and dead cells were discriminated in the F1/F3 histograms (not shown), and the DHR 123 (Fig. 1) and R 110 (Fig. 2) stain of vital lymphocytes, monocytes, and granulocytes were evaluated within the indicated windows. Suitable but fixed evaluation windows for the SSC/FITC histograms and self adjusting gates for the separation of lymphocytes, monocytes, and granulocytes in the FSC/SSC histograms were used for all list mode files to assure their unattended and automated evaluation. Means from linear or linearized logarithmic analogue-to-digital-converter (ADC) channels as well as absolute cell counts or percent cell frequency values were automatically introduced into data bases (Tables 1, 2).

Triple Matrix Data Base Classification

Data-base data were classified according to the triple matrix concept (Valet et al., 1993) which works in short as follows: Symmetrical upper and lower percentile values (e.g., 15 and 85%) are calculated for the reference value distribution of each database column. All individual data-base column values are then reexpressed as: +, 0, or − according to their position above, within, or below the respective reference percentile thresholds.

A confusion matrix for each classification (disease) state is then set up by the computer program with the different clinical diagnostic or prognostic states on the ordi-

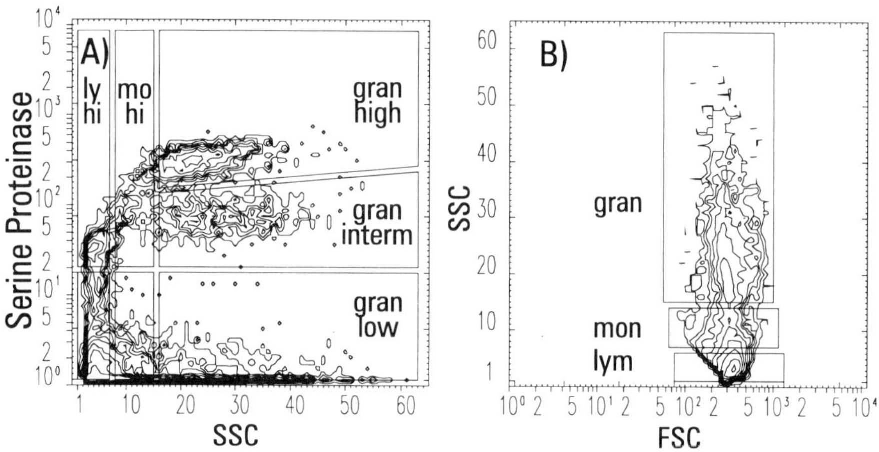

Fig. 2. Standardized evaluation of the $(Z\text{-AlaAla})_2R$ 110 serine proteinase activity measurement of the leukocyte sample of a septic patient. The evaluation of the FSC/R 110 fluorescence and of the FSC/SSC histograms is similar to that in Figure 1. An additional fixed window for the intermediate activity granulocytes is introduced, however, in the FSC/R 110 histogram. Infection and sepsis lead frequently to the formation of a distinct population of vital granulocytes with low serine protease activity (ca 30%). They are CD63 positive; that is, they have secreted a substantial amount of their proteases. The secretion process is paralleled by a significant reduction of SSC.

nate and the corresponding CLASSIF1 classification states on the abscissa. For a correct classification, all diagonal values of the confusion matrix are 100%, with 0% values in the remaining boxes of the confusion matrix.

To achieve as good results as possible, reference classification masks for the various disease states are determined from the learning set in the following way: the most frequent triple matrix character of all samples of a given disease state is introduced for each data-base column into the reference classification mask of this disease. Following the establishment of a reference classification mask for all disease states, all samples of the learning set are individually reclassified according to their highest positional coincidence with the different reference classification masks. Once all samples are classified, the sum of the diagonal values of the confusion matrix is computed to evaluate the quality of the classification. This is followed by the sequential but temporary removal of individual or combinations of data-base columns and by the recalculation of the diagonal sum after each temporary omission step.

Improvement of the diagonal sum means that temporarily eliminated single or combined data-base columns carry no significant classification information, but deterioration of the classification result identifies informative single or combined columns. After all single data-base columns as well as all column pairs and triplets are checked for their information content in all permutations, data-base columns that have not improved the classification result either alone or in conjunction with others are permanently removed, thus leaving the pattern of informative data-base columns in the final reference classification masks of each disease state (e.g., tables 5 and 6 "parameter change" and fig. 5 upper box).

TABLE 1. Columns of DHR 123 Data Base[a]

1 = TOTAL CELLS

Vital Lymphocytes

2 = % LYMPHO
3 = FSC LYMPHO
4 = SSC LYMPHO
5 = SSC/FSC LYMPHO
6 = DHR LYMPHO
7 = DHR/FSC LYMPHO

Vital Low DHR Monocytes

8 = % LOW DHR MONO
9 = FSC LOW DHR MONO
10 = SSC LOW DHR MONO
11 = SSC/FSC LOW DHR MONO
12 = DHR LOW DHR MONO
13 = DHR/FSC LOW DHR MONO

Vital High DHR Monocytes

14 = % HIGH DHR MONO
15 = FSC HIGH DHR MONO
16 = SSC HIGH DHR MONO
17 = SSC/FSC HIGH DHR MONO
18 = DHR HIGH DHR MONO
19 = DHR/FSC HIGH DHR MONO

All Vital Monocytes

20 = % MONO OF VITAL CELLS
21 = FSC ALL DHR MONO
22 = SSC ALL DHR MONO
23 = SSC/FSC ALL DHR MONO
24 = DHR ALL DHR MONO
25 = DHR/FSC ALL DHR MONO

Vital Low DHR Granulocytes

26 = % LOW DHR ACT.VIT.GRANULO
27 = FSC LOW DHR GRANULO
28 = SSC LOW DHR GRANULO
29 = SSC/FSC LOW DHR GRANULO
30 = DHR LOW DHR GRANULO
31 = DHR/FSC LOW DHR GRANULO

Vital High DHR Granulocytes

32 = % HIGH DHR ACT.VITAL GRAN
33 = FSC HIGH DHR GRANULO
34 = SSC HIGH DHR GRANULO
35 = SSC/FSC HIGH DHR GRANULO
36 = DHR HIGH DHR GRANULO
37 = DHR/FSC HIGH DHR GRANULO

All Vital Granulocytes

38 = % GRAN OF ALL VITAL CELLS
39 = FSC ALL DHR GRANULO
40 = SSC ALL DHR GRANULO
41 = SSC/FSC ALL DHR GRANULO
42 = DHR ALL DHR GRANULO
43 = DHR/FSC ALL DHR GRANULO

Dead Cells

44 = TOTAL DEAD CELLS
45 = %TOTAL DEAD CELLS

[a]All values except % and total cell counts represent arithmetic means of linear or linearized logarithmic analogue-to-digital converter (ADC) channel values for all cells in the respective evaluation windows (see Fig. 1B). Columns 12, 18, 24, 30, 36, and 42 represent total cellular oxidative activity whereas columns 13, 19, 25, 31, 37, and 43 are normalized on FSC; that is, they indicate equivalents of intracellular oxidative activity concentration.

Normalization of the learning and test-set data bases as well as of the classification percentiles onto the data-column means of the respective reference samples provides standardized, that is, instrument- and laboratory-independent data bases and classifiers. The measurement of a set of valid reference samples at a new classification site permits the application of a standardized classifier without relearning on abnormal samples, provided that no difference between the normalized new and original reference group has been detected by the CLASSIF1 algorithm.

TABLE 2. Columns of (Z-AlaAla)R 110 Serine Proteinase Data Base[a]

1 = TOTAL CELLS
2 = % VITAL CELLS
3 = % DEAD CELLS
4 = % VITAL LYMPHOCYTES
5 = % VITAL MONOCYTES
6 = % VITAL GRANULOCYTES

Low Serine Proteinase Lymphocytes

7 = % LOW SER LYM/VITAL CELLS
8 = % LOW SER LYM/VIT.LYMPHO
9 = FSC LOW SER LYMPHO
10 = SSC LOW SER LYMPHO
11 = SSC/FSC LOW SER LYMPHO
12 = SER.PROT LOW SER LYMPHO
13 = SER/FSC LOW SER LYMPHO

High Serine Proteinase Lymphocytes

14 = % HIGH SER LYM/VITAL CELLS
15 = % HIGH SER LYM/VIT.LYMPHO
16 = FSC HIGH SER LYMPHO
17 = SSC HIGH SER LYMPHO
18 = SSC/FSC HIGH SER LYMPHO
19 = SER.PROT HIGH SER LYMPHO
20 = SER/FSC HIGH SER LYMPHO

Low Serine Proteinase Monocytes

21 = % LOW SER MON/VITAL CELLS
22 = % LOW SER MON/VIT.MONO
23 = FSC LOW SER MONO
24 = SSC LOW SER MONO
25 = SSC/FSC LOW SER MONO
26 = SER PROT LOW SER MONO
27 = SER/FSC LOW SER MONO

High Serine Proteinase Monocytes

28 = % HIGH SER MON/VITAL CELLS
29 = % HIGH SER MON/VIT.MONO
30 = FSC HIGH SER MONO
31 = SSC HIGH SER MONO
32 = SSC/FSC HIGH SER MONO
33 = SER HIGH SER MONO
34 = SER/FSC HIGH SER MONO

Low Serine Proteinase Granulocytes

35 = % LOW SER GRAN/VITAL CELLS
36 = % LOW SER GRAN/VIT GRANULO
37 = FSC LOW SER GRANULO
38 = SSC LOW SER GRANULO
39 = SSC/FSC LOW SER GRANULO
40 = SER LOW SER GRANULO
41 = SER/FSC LOW SER GRANULO

High Serine Proteinase Granulocytes

42 = % HIGH SER GRAN/VITAL CELLS
43 = % HIGH SER GRAN/VIT GRANULO
44 = FSC HIGH SER GRANULO
45 = SSC HIGH SER GRANULO
46 = SSC/FSC HIGH SER GRANULO
47 = SER HIGH SER GRANULO
48 = SER/FSC HIGH SER GRANULO

[a]All values except % and total cell counts represent arithmetic means of linear or linearized logarithmic analogue-to-digital converter (ADC) channel values for all cells in the respective evaluation windows (see Fig. 2b). Columns 12, 19, 26, 33, 40, and 47 represent total cellular enzyme activity whereas columns 13, 20, 27, 34, 41, and 48 are normalized on FSC; that is, they indicate equivalents of intracellular enzyme activity concentration.

RESULTS

The standardized thresholding during the automated list mode evaluation is quite robust (Figs. 1–3), provided the flow cytometer is daily calibrated to the same sensitivity by monosized fluorescent calibration particles. Although granulocytes and lymphocytes show clear changes in DHR 123 assays following stimulation (Fig. 1A), lymphocytes remain visually close to background. The serine proteinase assay (Fig. 2) displays significant activities for lymphocytes, with usually a high- and low-ac-

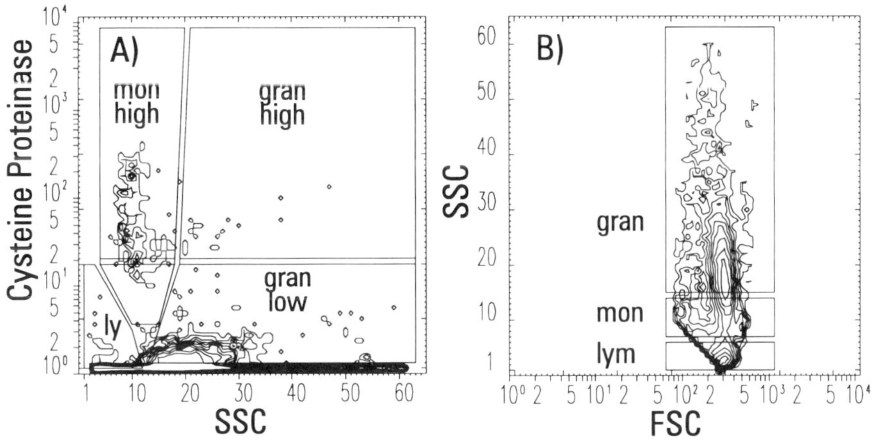

Fig. 3. Standardized evaluation of the (Z-ArgArg)$_2$R 110 cathepsin B,H,L activity measurement of a leukocyte sample from a septic patient. The evaluation is identical to the DHR 123 evaluation of Figure 1.

tivity lymphocyte population (Fig. 2A). Monocytes and granulocytes form a single cluster each in healthy persons. An additional cell cluster of vital intermediate-activity granulocytes at about one-third of the normal activity is seen at variable frequency in infected or septic patients. The low-activity cells are CD63 positive (not shown), indicating that the intermediate-activity granulocytes have secreted intracellular proteases.

Forty-five data-base columns (Table 1) are extracted from each of the four DHR 123 assays, that is, a total of $4 \times 45 = 180$ data-base columns per patient and time point, and 48 data columns (Table 2) + 8 data-base columns for the intermediate-activity cells (not shown) are derived from the serine proteinase as well as 45 data columns for the cathepsin B,H,L (Fig. 3) measurements.

The results from the three assays for the sepsis diagnosis at admission showed that all three assays identify the septic patients in 100% of the cases (Table 3). Because the patient's diagnosis is clinically known, the flow cytometric assay is confirmatory, that is, of little practical interest for the clinician.

The interesting question, however, is whether the assays permit the early identification of nonsurviving patients. Such risk assessment is of great importance for therapy improvement and clearly not available by other clinical or clinical chemistry parameters. If all DHR 123 assay information is used as well as all serine proteinase parameters separately, both assays are capable of predicting the fatal and nonfatal outcome quite correctly on admission, that is, on day 0 (Table 4), although the nonsurviving patients die only between 5 and 14 days later.

This conclusion is reached with a pattern of 11 parameters out of the initial 180 parameters (Table 5). It is unexpected that four DHR parameters are combined with seven scatter, scatter ratio, and percent cell frequency parameters. Furthermore, only three parameters concern granulocytes and three derive from monocytes and five

TABLE 3. Sepsis Diagnosis

Clinical Diagnosis	Patients (*n*)	CLASSIF1 Classification (%)[a]	
		No Infection	Sepsis
	Oxidative Activity		
No infection	15	100.0	0.0
Sepsis	17	0.0	100.0
	Serine Proteinases		
No infection	6	100.0	0.0
Sepsis	23	0.0	100.0
	Cathepsin B, H, L		
No infection	6	100.0	0.0
Sepsis	21	0.0	100.0

[a]25 and 75% percentile thresholds.

from lymphocytes. A similar result is obtained from the 17 parameters of the serine proteinase assay (Table 6) where five serine proteinase activity parameters contrast with 12 scatter and percent cell frequency parameters being distributed into four granulocyte, six monocyte, and seven lymphocyte parameters.

The normalization of all data-base values onto the respective reference values means in each column provides a standardized classifier formula, for example, for the DHR 123 assay (Table 7). The majority of the means of the survivor and nonsurvivor groups are statistically significantly different. The display of the individual values for each of the 11 selected classification parameters for survivors and nonsurvivors does

TABLE 4. Prediction for Sepsis Survival on Admission

Clinical Diagnosis	Patients (*n*)	CLASSIF1 classification (%)	
		Survivors	Nonsurvivors
	A. Oxidative Activity[a]		
Survivor	12	100.0	0.0
Nonsurvivor	5	20.0	80.0
	B. Serine Proteinases[b]		
Survivor	13	100.0	0.0
Non survivor	5	20.0	80.0

[a]25 and 75% percentile thresholds.
[b]20 and 80% percentile thresholds.

TABLE 5. Prediction for Sepsis Survival on Admission: Selected Oxidative Activity Classification Parameters of Table 4A

Parameter	DHR assay	Data column	Parameter change	
			Survivors	Non-survivors
1 = DHR LOW DHR GRANULO	ex vivo	30	0	—
2 = SSC LYMPHO	+ TNF-α	4	0	—
3 = FSC LYMPHO	+ FMLP	3	0	—
4 = % MONO OF VITAL CELLS	+ FMLP	20	0	+
5 = FSC ALL DHR MONO	+ FMLP	21	0	+
6 = DHR/FSC HIGH DHR GRANULO	+ FMLP	37	0	—
7 = DHR/FSC ALL DHR GRANULO	+ FMLP	43	0	—
8 = FSC LYMPHO	+ (T+ F)	3	0	—
9 = SSC LYMPHO	+ (T+ F)	4	0	—
10 = SSC/FSC LYMPHO	+ (T+ F)	5	0	—
11 = DHR/FSC LOW DHR MONO	+ (T+ F)	13	0	+

TABLE 6. Prediction for Sepsis Survival on Admission: Selected Serine Proteinase Parameters of Table 4B

Parameter	Data column	Parameter Change	
		Survivors	Nonsurvivors
1 = FSC LOW SER LYMPHO	9	0	—
2 = SSC LOW SER LYMPHO	10	0	—
3 = SER.PROT LOW SER LYMPHO	12	0	+
4 = SER/FSC LOW SER LYMPHO	13	0	+
5 = FSC HIGH SER LYMPHO	16	0	—
6 = SSC HIGH SER LYMPHO	17	0	—
7 = SER.PROT HIGH SER LYMPHO	19	0	+
8 = FSC LOW SER MONO	23	0	+
9 = SSC LOW SER MONO	24	0	+
10 = SSC/FSC LOW SER MONO	25	0	+
11 = FSC HIGH SER MONO	30	0	+
12 = SSC HIGH SER MONO	31	0	+
13 = SSC/FSC HIGH SER MONO	32	0	+
14 = % LOW SER GRAN/VITAL CELLS	35	0	—
15 = SER/FSC LOW SER GRANULO	41	0	+
16 = % HIGH SER GRAN/VIT GRANULO	43	0	+
17 = SER HIGH SER GRANULO	47	0	—

TABLE 7. DHR 123 Assay: Standardized Means, SEM, and Classification Thresholds of Data Columns Selected for Classification

	Patients	Ex Vivo	TNF-α	fMLP					TNF-α + FMLP			
Data Column[a]	(n)	30	4	3	20	21	37	43	3	4	5	13
Survivors	13	1.000	1.000	1.000	1.000	1.000	1.000	1.000	1.000	1.000	1.000	1.000
SEM[b]		.108	.044	.035	.144	.039	.059	.212	.031	.035	.020	.035
Non survivors	5	.887	.811	.9961	.9691	.403	.808	.707	.843	.807	.919	1.064
SEM		.196	.051	.151	.540	.336	.257	.389	.059	.057	.025	.124
Classification threshold[c]		<.660	<.928	<.839	>1.600	>1.117	<.731	<.313	<.947	<.929	<.938	>1.224

[a]seeTable.5.
[b]SEM = standard error mean.
[c]Values classified as: − or: + if: < or > as percentile threshold.

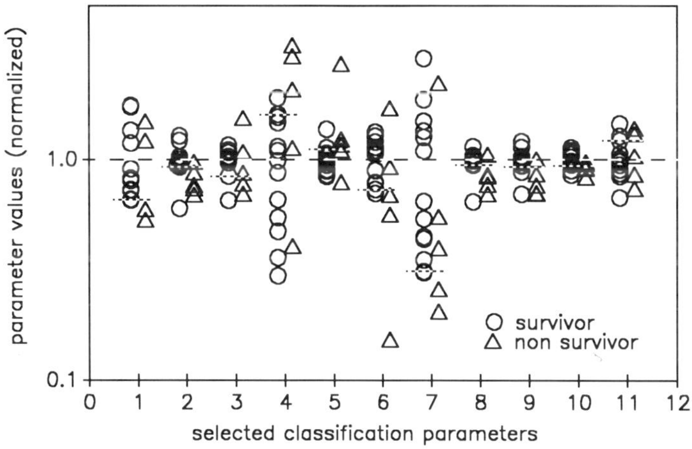

Fig. 4. Display of the selected classification parameters of Table 7 for survivors and nonsurvivors of sepsis. The percentile thresholds are indicated as dotted lines. Values below the threshold are classified as − for percentiles below the 1.0 normalization line while values above the threshold are classified as + for percentiles above the 1.0 normalization line. All other values are classified as 0. Although it seems visually impossible to reach unambiguous classifications according to the aspect of the data clouds, the triple matrix classification of the the same data (Figure 5) is clearly possible. This is due to the inherent error tolerance of the CLASSIF1 classifier for a certain degree of aberrations of the patient sample mask from the reference classification masks.

not look very informative in a traditional display (Fig. 4). The same data as a triple matrix display (Fig. 5), however, show immediately the classification potential of the selected parameters. The classification is error tolerant, as can be seen from the CLAS.COIN data column in which the patient classification mask coincides (1.00) in only 5 of 18 patients with the reference classification masks. Patient 17 is misclassified with nine positional hits for survivor and two hits for nonsurvivor. The plus sign in pattern position 8 counts as a hit for survivor because the nonsurvivor reference classification mask shows a minus sign at this location.

Following the recognition of the usefulness of DHR 123 and serine proteinase assays for individual patient risk assessment in sepsis, it seems important to optimize the number of performed assays. It was checked whether the DHR 123 assay in combination with the serine proteinase assay increased the correct prediction to 100% for survivors and nonsurvivors on the day of admission (Table 4). Furthermore the various DHR 123 assays were checked as single assays as well as in all possible combinations. The results (Table 8) indicate that the combination of the DHR 123 and the serine proteinase assay does not improve the overall result (Table 8, column 4). They also show that the combination of the data from the separate TNF-α and fMLP assays (column 12) classifies even better than all DHR 123 assays together (column 3) with regard to high recognition (100%, upper part) and low multiplicity (1.0, middle part). The optimum percentile threshold in most classifications is 20–80 or 25–75%.

The two unknown test-set patients were correctly classified by the DHR 123 classifier, showing robustness for this restricted number of patients.

NR.	REFERENCE CLASSIFIC.	ABBREVIATION	COIN	REF.CLASSIFIC.MASKS
1	SURVO	aN	1.00	00000000000
2	DEADO	aT	1.00	---++-----+

REC. NR.	DATAB: S8LEARN.BI4 DATAB.RECORD LABEL	CLASSIF1-CLASSIFIC.	CLAS COIN	SAMPLE CLASSIF.MASKS . = no value
1	KE1277...aN.........	aN	1.00	00000000000
2	KE1292...aN.........	aN	.64	0--0-0+--00
3	KE1296...aN.........	aN	.82	+000+++000+
4	KE1298...aN.........	aN	1.00	+0000+00000
5	KE1299...aN.........	aN	.82	000000-0-00
6	KE1343...aN.........	aN	.64	000+0-000-+
7	KE1344...aN.........	aN	.82	0-0000000-0
8	KE1349...aN.........	aN	1.00	0++-000++00
9	KE1367...aN.........	aN	.91	0000000000+
10	KE1378...aN.........	aN	1.00	00000000000
11	KE1379...aN.........	aN	.73	-00++00000-
12	KE1386...aN.........	aN	.91	0++000-+++0
13	KE1398...aN.........	aN	1.00	0000-0000+0
14	KE1301...aT.........	aT	.55	0--+-++--0+
15	KE1334...aT.........	aT	.73	0-+++----0+
16	KE1376...aT.........	aT	.73	---0+-0----
17	KE1380...aT.........	aN	.82	-000000+0-0
18	KE1382...aT.........	aT	.82	--0++-----0

Fig. 5. Reclassification of the learning set with the selected classification parameters of Table 7. The learning set classifies quite well with 17 of the 18 patient samples correctly classified (Table 4A). This is quite unexpected if the visual aspect of the same data (Figure 4) is considered. The apparent fact of unambiguous and mostly correct classification shows the high categorization effect of the triple matrix transformation of the original data-base parameter values.

DISCUSSION

The results indicate the significant potential of DHR 123 and protease activity determinations for the risk assessment of surviving and nonsurviving septic patients in intensive care units. Owing to the restricted number of patients in this pilot study, additional cases are required to establish definitive classifiers for practical use in hospitals. However, according to the present experience, asymptotic classification results are usually reached when between 20 and 40 representative patients (Valet et al., 1993) are available per classification state.

At this stage, the overall recognition results may decrease but the classifiers established so far have been quite robust upon the addition of new patients. A major reason for this is probably that all noninformative parameters, that is, the classification "noise," are entirely removed during the CLASSIF1 learning process. Additionally, the pattern from a certain number of parameters on, for example, >10 parameters, inherently provides a high stability even if some of the parameters have incorrect val-

TABLE 8. Optimization of Prediction for Sepsis Survival

Days	Serine Prot.	Oxid All Ass.	All Oxid + Serine	Oxid ex vivo	Oxid TNF-α	Oxid FMLP	Oxid (T+F)	Oxid ex+T	Oxid ex+F	Oxid ex+ (T+F)	Oxid T+F	Oxid F+ (T+F)	Oxid ex+T+F	Oxid T+ F+(T+F)
						A. Average Recognition Index (ARI, %)[a]								
0	90.0	90.0	90.0	90.0	80.0	80.3	90.0	86.2	86.2	90.0	100.0	100.0	100.0	100.0
1	100.0	95.8	95.8	85.8	100.0	100.0	100.0	100.0	100.0	90.0	100.0	95.8	100.0	100.0
2	83.3	100.0	100.0	100.0	100.0	100.0	100.0	100.0	100.0	100.0	100.0	100.0	100.0	100.0
3	100.0	100.0	100.0	100.0	100.0	100.0	100.0	100.0	100.0	100.0	100.0	100.0	100.0	100.0
						B. Average Multiplicity Index (AMI)[b]								
0	1.00	1.00	1.00	1.00	1.00	1.00	1.00	1.00	1.00	1.08	1.00	1.00	1.00	1.00
1	1.00	1.00	1.00	1.00	1.00	1.00	1.00	1.00	1.00	1.04	1.00	1.00	1.04	1.04
2	1.00	1.00	1.00	1.00	1.00	1.00	1.00	1.00	1.00	1.00	1.00	1.00	1.00	1.00
3	1.00	1.00	1.00	1.00	1.00	1.00	1.00	1.00	1.00	1.00	1.00	1.00	1.00	1.00
						C. Lower Percentile Threshold (%)[c]								
0	20	25	25	25	20	20	20	25	15	15	25	25	25	25
1	20	25	25	25	20	15	30	30	20	15	30	25	25	30
2	20	25	30	20	20	20	20	25	20	25	25	25	25	25
3	20	25	30	20	20	20	25	25	20	25	25	25	25	20

[a]ARI = sum of recognition values in diagonal of confusion matrix divided by number of classification states (%).

[b]AMI = sum of recognition values in all horizontal lines of the confusion matrix divided by number of classification states and by 100 for data normalization. Optimum 1.00, that is, no multiple classifications (acceptable range 1.00–1.20).

[c]Upper threshold = 100 -– lower percentile (%).

ues. The classifiers are not overdetermined in a statistical sense because the number of classification parameters (Tables 5, 6) is smaller than the total number of classified patients (Table 4). Furthermore, the parameter means (Table 7) of surviving and nonsurviving patients are significantly different in most instances, that is, no "noise" is classified.

The comparison between the traditional data display (Fig. 4, Table 7) and the triple matrix classification matrix (Fig. 5) indicates the importance of regarding flow cytometric results as a pattern rather than as a sequence of single values being compared, for example, by their position in relation to confidence limits. The same was true for the recent evaluation of clinical chemistry values in another intensive care condition, the capillary leak syndrome (CLS) following pediatric cardiac surgery (Tarnok et al., 1997). In this affection, it was impossible to distinguish between the causative role of either surgery or pre-existing humoral or cellular conditions of the organism. CLASSIF1 data pattern analysis clearly indicated a pre-existing humoral or cellular condition, whereas the modalities of surgery seem to be less important. The selected parameter pattern suggests a latent infection as an important CLS risk factor; that is, a preoperatively curable affection is identified by CLASSIF1 data analysis. Pattern analysis in this context provides for the first time individual patient risk assessment, therapy monitoring, and a scientific hypothesis for a difficult clinical condition.

Another important aspect concerns the type of selected classification parameters. Although the usual evaluation in, for example, immunophenotyping, DNA distribution, or thrombocyte activation analysis is preferentially oriented toward the extraction of percent positive/negative, or percent cells in particular evaluation quadrants, the entire CLASSIF1 experience shows so far that the percent cell frequency parameters carry usually only a low information content. This is equally true for the prognosis of septic patients. Only 3 out of 28 parameters (10.7%) listed in Tables 5 and 6 represent percent cell frequency values. The additionally surprising fact in this evaluation is that light scatter parameters contain quite meaningful information although they are in most evaluations utilized only for gating purposes. It is likely that the data pattern evaluation instead of single value analysis will lead to a significant reorientation in the evaluation of flow cytometric data in the future.

The pattern evaluation of the sepsis data provides furthermore efficient means for the optimization of cell assays. Protease and DHR 123 assays provide similar information in this context (Table 4), with the DHR 123 assay being slightly superior (Table 8). The calculation of single and combined assay results in all permutations shows that instead of initially five only two assays, that is, the TNF-α and the fMLP DHR 123 stimulation assay, are really required for the risk assessment of septic patients (Table 8). Both assays indicate the correct prognosis for all patients even on day 0, that is, on admission. The rigorous optimization of cellular assays provides a maximum of information with a minimum of effort.

ACKNOWLEDGMENT

This work was partially financed by a grant from the Sonderforschungsbereich (SFB) 207 of the Deutsche Forschungsgemeinschaft (DFG).

REFERENCES

American College of Chest Physicians/Society of Critical Care Medicine Consensus Conference (1992): Definitions for sepsis and organ failure and guidelines for the use of innovative therapies in sepsis. Crit Care Med 20:864–874.

Briviba K, Roussyn I, Sharov VS, Sies H (1996): Attenuation of oxidation and nitration reactions of peroxynitrite by selenomethionine, selenocysteine and ebselen. Biochem J 319:13–15

Ganesh S, Klingel S, Kahle H, Valet G (1995): Flow cytometric determination of aminopeptidase activities in viable cells using fluorogenic rhodamine110 substrates. Cytometry 20:334–340.

Klingel S, Rothe G, Kellermann W, Valet G (1994): Flow cytometric determination of cysteine and serine proteinase activities in living cells with rhodamine110 substrates. Meth Cell Biol 41:449–459.

Kooy NW, Lewis SJ (1996): The peroxynitrite product 3-nitro-L-tyrosine attenuates the hemodynamic responses to angiotensin II in-vivo. Eur J Pharmacol 315:165–170.

Löhrke B, Wegner J, Viergutz T, Dietl G, Ender K (1996): Flow cytometric analysis of oxidative and proteolytical activities in tissue associated phagocytes from normal and hypertrophic muscles. Anal Cell Pathol 9:281–293

Rothe G, Kellermann W, Valet G (1990): Flow cytometric parameters of neutrophils function as early indicators of sepsis- or trauma-related pulmonary or cardiovascular organ failure. J Lab Clin Med 115:52–61.

Rothe G, Klingel S, Assfalg-Machleidt I, et al. (1992): Flow cytometric analysis of protease activities in vital cells. Biol Chem Hoppe-Seyler 373:547–554.

Rothe G, Valet G (1988): Phagocytosis, intracellular pH, and cell volume in the multifunctional analysis of granulocytes by flow cytometry. Cytometry 9:316–324

Rothe G, Valet G (1993a): Measurement of mononuclear phagocyte cathepsin B/L activity with (*N*-benzylcarbonyl-ArgArg)$_2$-rhodamine110. In Robinson P (ed). *Handbook of Flow Cytometric Methods.* New York: Wiley-Liss, pp 202–203.

Rothe G, Valet G (1993b): Measurement of neutrophil elastase activity with (N-benzylcarbonyl-AlaAla)$_2$-rhodamine110. In Robinson P (ed). *Handbook of Flow Cytometric Methods.* New York: Wiley-Liss, pp 200–201.

Rothe G, Valet G (1993c): Measurement of phagosomal hydrogen peroxide production with dihydrorhodamine123. In Robinson P (ed). *Handbook of Flow Cytometric Methods.* New York: Wiley-Liss, pp 155–156.

Sobreira C, Davidson M, King MP, Miranda AF (1996): Dihydrorhodamine-123 identifies impaired mitochondrial respiratory chain function cultured cells harboring mitochondrial DNA mutations. J Histochem Cytochem 44:571–579.

Tarnok A, Hambsch J, Borte M, Valet G, Schneider P (1997): Immunological and serological discrimination of children with and without post-surgical capillary leak syndrome. In Faist E (ed). *4th Congress on Immune Consequences of Sepsis, Trauma and Shock.* Bologna: Monduzzi Editore, pp 845–849.

Wang DS, Youngson C, Wong V, et al. (1996): NADPH-Oxidase and a hydrogen peroxide sensitive K+ channel MAG function as an oxygen sensor complex in airway chemoreceptors and small cell lung carcinoma cells. Proc Natl Acad Sci USA 93:13182–13187.

Wenisch C, Narzt E, Sessler DI, et al. (1996a): Mild intraoperative hypothermia reduces production of reactive oxygen intermediates by polymorphonuclear leukocytes. Anesth Analg 82:810–816.

Wenisch C, Parschalk B, Zedwitzliebstein K, Graninger W, Rieger A (1996a): Dysregulation of the polymorphonuclear leukocyte—candida spp, interaction in HIV positive patients. AIDS 10:983–987.

Wersto RP, Rosenthal ER, Crystal RG, Spring KR (1996): Uptake of fluorescent dyes associated with the functional expression of the cystic fibrosis transmembrane conductance regulator in epithelial cells. Proc Natl Acad Sci USA 93:1167–1172.

Williams MS, Henkart PA (1996): role of reactive oxygen intermediates in TCR-induced death of T-cell blasts and hybridomas. J Immunol 157:2395–2402.

Woodman RC, Newburger PE, Anklesaria P, Erickson RW, Rae J, Cohen MS, Curnutte JT (1995): A new

X-linked variant of chronic granulomatous disease characterized by the existence of a normal clone of respiratory burst-competent phagocytic cells. Blood 85:231–241.

Valet G, Raffael A, Mororder L, Wünsch E, Ruhenstroth-Bauer G (1981): Fast intracellular pH determination in single cells by flow-cytometry. Naturwissenschaften 68:265–266.

Valet G, Valet M, Tschöpe D, et al. (1993): White cell and thrombocyte disorders: Standardized, self-learning flow cytometric list mode data classification with the CLASSIF1 program system. Ann NY Acad Sci 677:233–251.

Zingarelli B, Day BJ, Crapo JD, Salzman AL, Szabo C (1997): The potential role of peroxynitrite in the vascular contractile and cellular energetic failure in endotoxic shock. Br J Pharmacol 120:259–267.

13

Alterations in the Nonspecific Immune Response Following Severe Traumatic Injury

George F. Babcock
University of Cincinnati College of Medicine and Shriners Burns Institute, Cincinnati, Ohio

Following severe trauma, particularly after a blunt, penetrating, or thermal injury, a large number of pathophysiological changes occur. Included among these changes are a large number of alterations in the immune response, both hyperresponsive and hyporesponsive in nature. The sequelae of hyperimmune responsiveness include systemic inflammatory response syndrome (SIRS), which can lead to multiple organ dysfunction syndrome (MODS), which may in turn lead to multiple organ failure (MOF). Multiple organ failure is a major cause of death in these patients. Because of the detrimental nature of hyperimmune responsiveness, it could be inferred that hypoimmune responsiveness would be more desirable. However, this is not necessarily the case. The importance of the immune suppression (hypoimmune responsiveness) observed following trauma can best be appreciated by the large increase in infections observed in these patients. In fact, infection is now the leading cause of death in the burn trauma group (Baker et al., 1984). The infections encountered are often of the opportunistic type, with *Pseudomonas aeruginosa, Staphylococcus epidermidis,* and *S. aureus* being the most commonly isolated bacteria, and *Candida albicans* dominates the fungi isolated (Neely et al., 1991). The infections are reminiscent of recurrent bacte-

Phagocyte Function: A Guide for Research and Clinical Evaluation, Edited by J. Paul Robinson and George F. Babcock. ISBN 0-471-12364-1

rial infections described in patients with chronic granulomatous disease and inherited leukocyte adhesion deficiencies (Palliser and Hancock, 1995; Ross, 1986; Rosen and Law, 1989). For these reasons it could be theorized that cells of the phagocytic series, granulocytes and perhaps monocytes, would be candidates for the immunosuppression observed following severe traumatic injuries.

In this chapter, the effects of various types of trauma, burn, blunt, and penetrating, on phagocytic cells are explored. Severe injuries of this type produce very extensive physiological changes, many of which are not completely characterized. The reticuloendothelial system is not the only part of the immune system altered by catastrophic injury, but instead represents one important part of a complex and dysfunctional host defense system. The emphasis is placed here on alterations in the expression of important cell surface receptors and functional capabilities. Because of the ease of obtaining peripheral blood, which is a natural single cell suspension, studies of this type are ideal for flow or image cytometric analysis. Much of the data presented was generated using these methodologies.

STUDIES OF FUNCTIONAL ALTERATIONS OF NEUTROPHILS AND MONOCYTES FOLLOWING VARIOUS TYPES OF TRAUMA

Although infections and sepsis were known for some time to be sequelae following major traumatic injuries, their importance was not fully appreciated until the emergence of better methods of resuscitation, ventilation, and immediate debridement and grafting (in the case of severe burns), which have reduced the number of early postinjury deaths. Early studies by the laboratories of Alexander, Rittenbury, and Grogan (Alexander and Moncrief, 1966; Rittenbury and Hancock, 1967; Alexander, 1968; Alexander et al., 1968; Rittenbury, 1970; Grogan, 1976; Grogan and Miller, 1973) pointed to the fact that phagocytosis and killing of bacteria by human neutrophils were altered following major trauma, including burns and selected surgical procedures. These observations have also been confirmed in animal models (Niedworok and Greybowski, 1979). Animal studies examining phagocytosis and killing of bacteria by monocytes following thermal injury have indicated depressed killing in some cases (Loose and Turinsky, 1979), but enhanced killing in others (Harmon et al., 1976). Chemotaxis of both neutrophils and monocytes was also found to be depressed in burn and nonburn trauma patients (Warden et al., 1974; Grogan, 1976; Altman et al., 1977; Fikrig et al., 1977; Bjerknes et al., 1989; Tellado and Christou, 1991). Part of the reduced chemotaxis in neutrophils could be attributed to the drugs given to the burn patients (Warden et al., 1975). However, Nelson et al. (1987) found reduced expression in the receptor for C5a, an important receptor in the chemotactic response.

The decrease in microbial killing by neutrophils observed following injury can be explained, at least in part, by a reduction in the neutrophil's ability to mediate oxygen-dependent killing, that is, an oxidative burst (Curreri et al., 1973; Heck et al., 1975; Allen and Pruitt, 1982; Dobke et al., 1989). Heck et al. (1979) suggested that neutrophils isolated from burn patients possessed a defect in nicotinamide-adenine dinucleotide (NADH) and nicotinamide-adenine phosphodinucleotide (NADPH) regulation. Similar results were obtained by El-Maallem and Fletcher (1981), who found that following the trauma of major surgery, neutrophil microbicidal capacity was markedly impaired, and this impairment was associated with the loss of myleoperoxidase. Lanser et al. (1986) reported that serum from trauma patients reversibly suppressed superoxide production by neutrophils from normal individuals. A factor in the serum was found to inhibit membrane depolarization, which is necessary for the production of bactericidal oxygen species. Bjerknes et al. (1989) demonstrated that following major thermal injury, alkalization of the phagolysosomes did not occur, indicating that the oxygen-independent microbicidal mechanisms were altered.

It is obvious from the above studies that following major trauma, thermal injury, or even major surgery the normal physiological homeostasis is altered, including many components of the immune system. Neutrophils and in some cases monocytes appear to be defective in their ability to migrate to the inflammatory site and engulf microorganisms and destroy microorganisms. The defect(s) leading to this reduced microbicidal activity appears complex and may occur at several levels. What is currently not clear is the exact mechanism leading to these changes, the importance of each individual change, and the relationship of any particular change, if any, to infection. A few studies have shown a strong correlation between a particular immune parameter and infections or sepsis. Alexander et al. (1979) reported that abnormal neutrophil bactericidal function had the strongest correlation with morbidity and mortality from infections following severe multisystem trauma. Defective opsonization was associated with a high risk of bacteremia only when there was a coexisting abnormality of neutrophil function. Lanser et al. (1986) found that the chemiluminescence response of normal neutrophils was of value in detecting resistance to sepsis when tested with sera from trauma patients. Polk et al. (1988) reported similar findings. Other data suggested that the percentage of monocytes expressing CD14 correlated with the clinical outcome of severely injured patients (Heinzelmann et al., 1996). A summary of the alterations in neutrophil or monocyte functions following severe injury appears in Table 1.

CELLULAR AND MOLECULAR STUDIES OF NEUTROPHILS AND MONOCYTES FOLLOWING MAJOR TRAUMA

The advent of monoclonal antibodies, flow and image cytometry, and molecular techniques have allowed investigators to define trauma-associated problems at a new level. Although many questions remain unanswered, many of the observations made by early investigators can now be explained in more mechanistic terms.

Alterations in Cytokine Production

It would be expected that following severe injury, cytokine production may be altered. Many investigators have reported such alterations following severe injury and trauma, including both increases and decreases in cytokine production. In most instances

TABLE 1. A Summary of Studies by Various Investigators Reporting Functional Alterations of Neutrophils and/or Monocytes Following Severe Trauma

Study	Neutrophil or Monocyte Abnormality
Alexander and Moncrief (1966), Rittenbury and Hanback (1967), Alexander (1968), Alexander et al. (1968), Rittenbury (1970), Grogan (1976), Grogan and Miller (1973)	Reduced killing of bacteria
Warden et al. (1974), Grogan et al. (1976), Altman et al. (1977), Fikrig et al. (1977), Bjerknes et al. (1989), Tellado and Christou (1981)	Reduced chemotaxis
Curreri et al. (1973), Heck et al. (1975), Allen and Pruitt (1982), Dobke et al. (1989)	Reduced oxidative burst
Heck et al. (1979)	Defects in leukocyte NADH and NADPH
El-Mallem and Fletcher (1981)	Loss of myeloperoxidase
Lanser et al. (1986)	Suppressed superoxide production
Bjerknes et al. (1989)	Altered oxygen-independent microbial killing

the cause and effect relationship and the relationship of a particular cytokine to infection still remain obscure. Both the pro-inflammatory and anti-inflammatory responses have been reported to be increased or decreased following trauma depending upon the variables of the particular study. The pro-inflammatory cytokines mediate capillary permeability, capillary thrombosis, and hemorrhagic tissue necrosis (Wardle, 1993; Lowery, 1993). It has been hypothesized that overproduction of these cytokines is responsible for the pathophysiology associated with systemic inflammatory response syndrome (SIRS). Complications following SIRS can lead to multiple organ dysfunction syndrome (MODS). Various investigators have reported increased production of pro-inflammatory cytokines by monocytes/macrophages following severe injury.

The pro-inflammatory cytokine tumor necrosis factor-alpha (TNF-α) has been reported to play a pivotal role in sepsis syndrome and shock following major injury (Spooner et al., 1992). Padalino et al. (1996) reported an increase in TNF-α levels in patients following trauma, with the greatest increases observed in patients with multiple organ dysfunction. However, Fabian et al. (1995) reported that TNF-α was at or below the threshold of detection following trauma and could only be detected following activation of peripheral blood with lipopolysaccharide. Even after lipopolysaccharide stimulation, monocytes from control samples produced more TNF-α than did trauma samples. Most studies have reported a significant increase in interleukin 6 (IL-6) production following thermal injury (Nijsten et al., 1991). In critically ill patients with a variety of diseases, including polytrauma, IL-6 levels in the plasma were also found to be increased, especially in those individuals with sepsis (Delogu et al., 1996). Patients who succumbed to their infections had higher levels of

IL-6 than those who survived. In the study cited above, a similar pattern for interleukin 8 (IL-8) production was also observed. These results are in accordance with those observed following major burns where the serum level of IL-8 was found to be 60 times higher than in healthy controls (Vindenes et al., 1995). Donnelly et al. (1993) found that IL-8 was elevated in bronchoalveolar lavage fluid in patients at high risk of adult respiratory distress syndrome (ARDS), compared to the non-ARDS group. Plasma levels of IL-8 were elevated but not significantly different between the ARDS and non-ARDS group. It is probable that the large amount of IL-8 found following severe thermal and nonthermal injury may contribute to the sustained activation of neutrophils and the associated sequelae rather than the increased rates of infection in these patients. However, other investigators have found suppressed levels of IL-8 both at the cytokine synthesis and mRNA level following major burn and nonburn trauma (Schinkel et al., 1996). Although interleukin 1 (IL-1) is associated with the acute phase response, in one study it was reported to be induced in only a small percentage of trauma patients (O'Nuallain et al., 1995). In studies by Schinkel et al. (1996) and Rodrick et al. (1986), the production of IL-1 by monocytes was essentially unaltered or actually suppressed following severe nonthermal trauma.

In opposition to the pro-inflammatory cytokines are the anti-inflammatory, which are potentially immunosuppressive. The lack of a pro-inflammatory cytokine response following a severe injury must be viewed in light of the anti-inflammatory response. Faist et al. (1987) and Miller et al. (1988) demonstrated that prostaglandin E2 (PGE_2) production by monocytes increased following severe trauma and was responsible for much of the immunosuppression observed in these patients. Antonacci et al. (1984) investigated arachidonic acid metabolism in burn patients and found that their monocytes produced increased quantities of PGE_2. In addition these investigators reported that burn patients' peripheral blood mononuclear cells were more sensitive to inhibition by PGE_2 than those from normal controls. Grbic et al. (1991) also reported monocytes from burn patients suppressed lymphocyte proliferation, an effect that was at least in part mediated by PGE_2. Miller-Graziano et al. (1991) demonstrated that transforming growth factor beta (TGF-β) produced by monocytes was elevated in immunosuppressed trauma patients. This finding was important for two reasons: TGF-β is very immunosuppressive and it also enhances monocyte (PGE_2) production. Levels of interleukin 10 (IL-10), which under the appropriate conditions can also be a very immunosuppressive cytokine, varied in trauma patients depending on other immune parameters (Miller-Graziano, 1995). Patients with suppressed lymphocyte responses or elevated macrophage TNF-α had depressed IL-10 levels.

An important cytokine for neutrophil hematopoiesis and development, granulocyte/macrophage–colony-stimulating factor (GM-CSF), was found to be depressed after both major trauma and burns (Iyengar et al., 1986; Moore et al., 1990). The level of colony-stimulating factor was shown to be decreased with severity of injury. Correlating with the decrease in colony-stimulating factor was a blunted bone marrow response with fewer myeloid stem cells. It could be hypothesized that the reduced levels of GM-CSF could be regulated by increases in the Th2 type cytokine IL-4 or PGE_2, both of which suppress GM-CSF production. However, Ertel and co-workers (1996) have found that following severe nonthermal injury IL-4 levels essentially remained stable. Other investigators have reported enhanced or depressed levels of the

Th2 type cytokines. Several studies have been reported that indicated an enhanced level of IL-4 (or Th2 cells) following cardiac surgery (Markewitz et al., 1996), endotoxemia (Zimmer et al., 1997) and sepsis (Docke et al., 1997), shock (Ayala and Chaudry, 1997), and severe trauma (DePiro and Howdieshell, 1997; Kim et al., 1997). De et al. (1997) reported that Th1- and Th2-type cytokines, including IL-4, were suppressed following major trauma.

Several important points can be derived from the above studies: (1) following major traumatic injury numerous components of the neutrophil/monocyte/macrophage system are altered; (2) the timing at which the studies are performed in relation to the clinical course often reflect the data obtained; (3) following these severe injuries, the immune response appears biphasic. The first phase of the response appears as a strong upregulation of pro-inflammatory cytokines followed closely by an increase in anti-inflammatory cytokines. In the second phase of the response a suppression of the pro-inflammatory cytokines occurs whereas the anti-inflammatory may either stay increased or become decreased. It is during this second phase that infection often occurs. A summary of these studies appears in Table 2. Taken in total the data suggest, but do not prove, that increases in anti-inflammatory cytokines that are of major importance in curtailing SIRS may also enhance infections.

Studies from Our Laboratory

Our laboratory has been interested in the post-burn/trauma alterations in the nonspecific immune response that occur during the second or immunosuppressive phase of the immune response and predispose seriously injured patients to infections. We have

TABLE 2. A Summary of Studies by Various Investigators Reporting Alterations in Cytokine Production by Neutrophils and/or Monocytes Following Severe Trauma

Study	Cytokine or Factor
Padalino et al. (1996)	TNF-α increased
Fabian et al. (1995)	TNF-α no increase
Nijsten et al. (1991), Delogu et al. (1996)	IL-6 increased
Delogu et al. (1996), Vindenes et al. (1995), Donnelly et al. (1993)	IL-8 increased
Schinkel et al. (1996), Rodrick et al. (1986)	IL-1 unaltered or reduced
Faist et al. (1987), Miller-Graziano et al. (1988), Antonacci et al. (1984)	PGE-2 increased
Miller-Graziano et al. (1991)	TGF-β
Miller-Graziano et al. (1995)	IL-10 variable or increased
Iyengar et al (1986), Moore et al. (1990)	GM-CSF reduced
Ertel et al. (1996)	IL-4 unchanged
Markewitz et al. (1996), Zimmer et al. (1997), Docke et al. (1997), Ayala and Chaudry, (1997) DiPiro and Howdieshell (1997), Kim et al. (1997)	IL-4 increased

TABLE 3. Monoclonal Antibodies Tested and Their Specificity

Cluster Designation	Specificity	Expression Following Trauma
CD10	Neutral endopeptidase	Unchanged
CD11a	β2 integrin (lymphocyte function antigen, LFA-1, binds ICAM-1,2,3)	Unchanged or decreased
CD11b	β2 integrin (CR3 receptor, binds iC3b, fibronectin, ICAM-1)	Decreased
CD11c	β2 integrin (gp 150/95, binds fibronectin, ICAM-1)	Decreased
CD14	LPS receptor	Variable
CD15	Fucosyl *N*-acetyllactosamine	Increased
CD16	FcRIII for IgG	Decreased
CD18	Beta chain of CD11a,b,c	Decreased
CD32	FcRII for IgG	Decreased
CD33	gp67	Unchanged
CD35	CR1 receptor for C3b, C4	Decreased or increased
CD62L	L-selectin, binds sialyl-Lewis x, glyCAM-1	Decreased
CD64	FcRI for IgG	Increased

analyzed a number of functional properties and cell-surface receptors on neutrophils following both thermal and nonthermal trauma (Babcock et al., 1993). Table 3 lists the receptors examined, all of which are important for the proper function of the neutrophil. Of the cell-surface receptors examined, two groups of receptors stand out as having consistently altered expression (lower). One group consists of two of the three members of the β2 integrin family, CD11b and CD11c. The second group consists of the two of the Fc receptors expressed on neutrophils, CD16 and CD32.

The importance of the many changes in receptor expression appearing in Table 3 is not apparent unless correlated with important clinical parameters. When a retrospective analysis was performed comparing patient variables with changes in receptor expression, a close association was observed between a reduced expression of CD11b and CD16 and infections (Babcock et al., 1990). It appeared that the change in percentage and absolute number of $CD11b^{+}CD16^{+}$ neutrophils was more important than the actual value obtained. An increase in the percentage of neutrophils expressing no or very low CD11b and CD16 correlated with the onset of infections in severely burned patients (Fig. 1). Infection in these studies was defined as two consecutive positive blood cultures. In trauma patients, an increase in the percentage of neutrophils expressing no or very low CD11b and CD16 correlated with the severity of injury as judged by the relationship between the APACHE II score and the number of infectious complications per patient (White-Owen et al., 1992) (Fig. 2).

The CD11b/CD18 complex and CD16 are very important for regulating the response of neutrophils to a number of extracellular effector molecules including potential pathogens. The CD11b/CD18 complex on neutrophils has been demonstrated to be important for chemotaxis (Anderson et al., 1986), adhesion (Butcher, 1991;

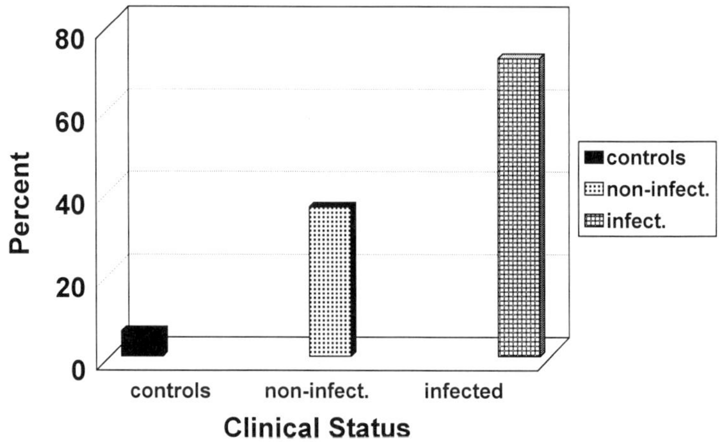

Fig. 1. The mean percentage of CD11b/CD16-negative (or very low expressing) neutrophils isolated from thermal trauma patients. Three groups are displayed: controls, which are healthy individuals; noninfected, which are patients who were not infected at the time of sampling; and infected, which are culture-positive patients. Noninfected patients were both culture-negative and free of the clinical symptoms of sepsis.

Vadas and Gamble, 1990), iC3b-mediated phagocytosis (Anderson et al., 1986; Rosen and Law, 1990), oxidative burst (Anderson et al., 1986; Nathan et al., 1989), and intracellular killing (Anderson et al., 1986). The CD16 molecule is the FcγRIII and is important for immunoglobulin-mediated phagocytosis (Babcock et al., 1997). Suppression of any of these functions would reduce the ability to combat possible infectious microorganisms. It could be postulated, just on the basis of the altered CD11b expression, that these changes could predispose the severely injured patients to infections.

Simultaneously with the measurements of neutrophil receptor expression described above, an investigation of the functional properties of neutrophils isolated from patients with severe traumatic injury was also instigated. Following severe thermal injury or other types of traumatic injury, the percentage of cells capable of phagocytosis was reduced by 49% compared with the controls (Babcock et al., 1997). However, when these were examined phenotypically there were large differences in the phagocytic capacity between $CD11b^+CD16^+$, $CD11b^+CD16^-$, and $CD11b^-CD16^-$ neutrophils (Fig. 3). Almost all the CD11b+CD16+ were phagocytic, very similar to what is observed on examining neutrophils isolated from healthy individuals. Neutrophils that were $CD11b^+CD16^-$ were slightly less phagocytic, 80% of control values, whereas $CD11b^-CD16^-$ neutrophils were 22% of control values.

The ability of neutrophils isolated from patients with thermal injuries to undergo an oxidative burst following stimulation with phorbol myristate acetate (PMA) was examined using 2′,7′-dichlorofluorescin diacetate and flow cytometry to monitor

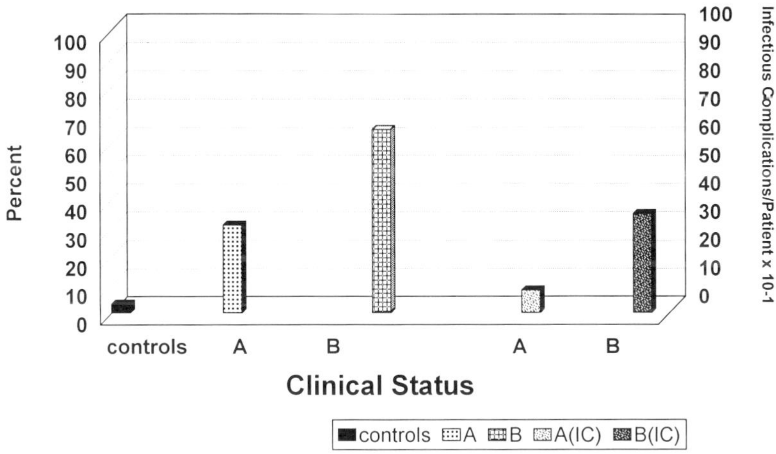

Fig. 2. The percentage of CD11b/CD16-negative neutrophils in nonburn trauma patients at time of admission. Controls consisted of neutrophils from healthy individuals. Group A patients had initial APACHE II scores with a mean of 14.2 (range 10–18). Group B patients had an initial APACHE II scores with mean of 21 (range 19–25). Groups A(IC) and B(IC) use the *y*-axis on the right and represent the mean number of infectious complications per patient per hospital stay $\times 10^{-1}$.

H_2O_2 production (Babcock et al., 1997). As with phagocytosis, the overall oxidative response of neutrophils was reduced following thermal injury. However, the response of $CD11b^+CD16^+$ neutrophils was actually slightly above that of healthy volunteers (115% of controls) whereas $CD11b^-CD16^-$ displayed a much lower response, 60% of controls (Fig. 4).

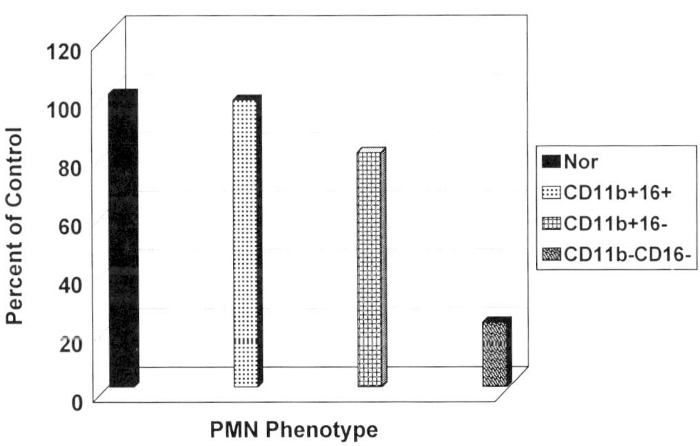

Fig. 3. The mean percentage of neutrophils of varying phenotypes capable of phagocytosis following thermal injury. All cells were tested in the presence of autologous serum. Nor represent neutrophils from healthy individuals. The $CD11b^-/CD16^+$ phenotype was not observed in sufficient numbers of cells to allow testing.

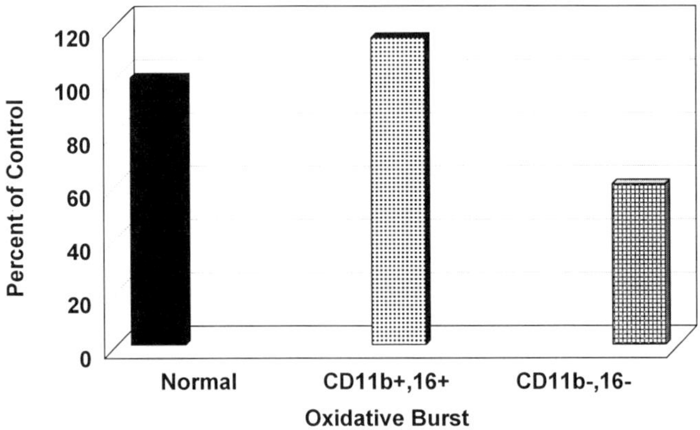

Fig. 4. The oxidative burst response of neutrophil subpopulations isolated from burn patients. Cells were activated with PMA and the oxidative burst monitored using dihydrorhodamine 1,2,3. Neutrophils from healthy clones were adjusted to 100% for comparative purposes. Neutrophil phenotypes other than those shown were not tested.

These results strongly suggest that the neutrophil dysfunction observed following severe trauma was correlated with the altered expression of CD11b. The reduced expression of CD16 may contribute to the overall neutrophil dysfunction but is probably not the major contributor. In general neutrophils isolated from patients expressing the $CD11b^{+}CD16^{+}$ phenotype responded in a nearly normal manner, those with the $CD11b^{+}CD16^{-}$ phenotype were slightly suppressed, and those with the $CD11b^{-}CD16^{-}$ phenotype were extremely dysfunctional. However, neutrophils with the $CD11b^{-}CD16^{+}$ phenotype were not observed, preventing complete analysis of all possible phenotype combinations.

Because many of the functional abnormalities observed in neutrophils isolated from trauma or thermally injured patients could be attributed to a defect in CD11b expression, we investigated the response of neutrophils from burn patients to inflammatory stimuli that normally upregulate this complex on the cell surface (Babcock and Rodeberg, 1997; Rodeberg et al., 1997). Neutrophils from burn patients were stimulated with various concentrations of lipopolysaccharide, fMLP, PMA, or A23187 and the cells analyzed for the expression of CD11b. Burn patients were unable to upregulate CD11b on their neutrophils to the same degree exhibited by neutrophils from healthy patients following stimulation with any of the agonists (Fig. 5). Additional studies indicated that this defect in CD11b regulation was not due to differences in the ability to bind lipopolysaccharide, the level of CD14 receptor expression, or soluble factors in the serum of burn patients. However, this defect can be partially explained by a decrease in the total quantity of CD11b present within intracellular granules (Rodeberg et al., 1997). Neutrophils from burn patients also displayed abnormalities in the ability to upregulate CD35 (CR1) following stimulation with agonist (Fig. 6). This decrease in the ability to increase CD11b and other receptors may provide an explanation for some of the functional abnormalities observed in

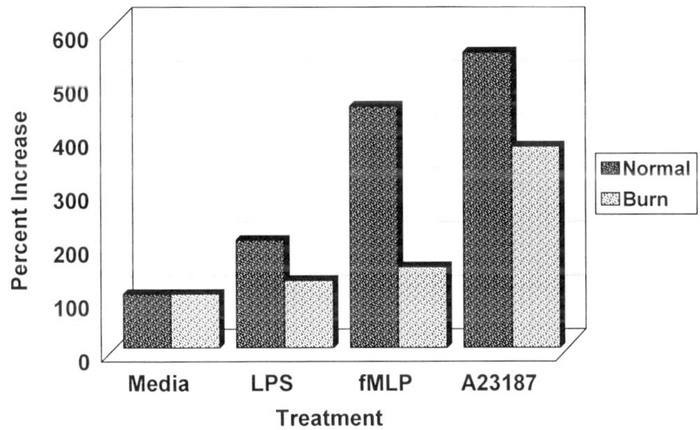

Fig. 5. A comparison between the expression of CD11b on neutrophils isolated from burn patients and on those from healthy individuals in response to several stimuli (LPS (100 ng/ml), fMLP (10^{-8} *M*), PMA (100 ng/ml), and A23187 (10^{-6} *M*)) incubated for 30 min at 37°C. Data are expressed as percentage increase from the control.

burn patient neutrophils. It also suggests that there may be a more global defect in the neutrophils following thermal injury, possibly a deficiency in granular proteins.

REVERSING IMMUNOSUPPRESSION FOLLOWING TRAUMA

If, as hypothesized, defects in neutrophils and or monocytes following thermal injury lead to immunosuppression and increased infections, then these patients would be candidates for various immunotherapy protocols. Several investigators have report-

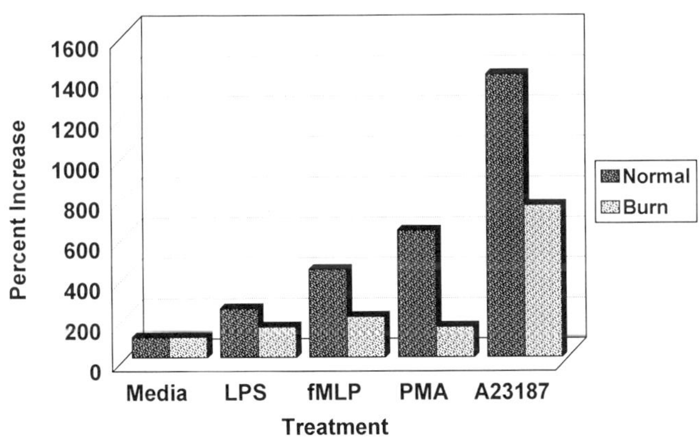

Fig. 6. A comparison between the expression of CD35 on neutrophils isolated from burn patients and on those from healthy individuals in response to several stimuli (LPS (100 ng/ml), fMLP (10^{-8} *M*), and A23187 (10^{-6} *M*)) incubated for 30 min at 37°C. Data are expressed as percentage increase from the control.

ed studies using immunomodulation both in vitro and in vivo to correct injury-based defects. Various approaches to intervention therapy in sepsis have been suggested, including cytokines, anti-cytokine monoclonal antibodies, anti-receptor monoclonal antibodies, receptor antagonists, anti-endotoxin monoclonal antibodies, inhibitors of inflammation, and many others (Bone, 1991).

One promising approach reported by several investigators is the utilization of colony-stimulating factors to modulate the mortality or cellular response following severe trauma or under septic conditions. Studies by Reimund et al. (1997) in a clinically relevant animal model of post-surgical peritonitis found that G-CSF combined with antibiotics reduced the mortality to 20% from 60% with antibiotics alone. These investigators also found a decrease in the level of TNF-α from macrophages and suggested that this is one the possible mechanisms evoked by the G-CSF. In two additional studies, G-CSF was shown to improve survival and reduce burn wound infection in a murine model (O'Reilly et al., 1994; Gamelli et al., 1995). GM-CSF has been employed in a number of studies using animal models of burn or trauma. Survival was significantly enhanced in a thermal injury model of posttraumatic sepsis (Austin et al., 1995; Molly et al., 1995). In studies using burn patients, Pellegrin et al. (1997) found that GM-CSF restored the respiratory burst in suppressed neutrophils following thermal injuries, and Ono et al. (1993) utilized G-CSF to enhance the suppressed chemiluminescence response following burn injury.

Our laboratory has performed experiments in vitro, using neutrophils isolated from burn or trauma patients, designed to determine if the expression of and/or ability to upregulate CD11b could be restored with colony-stimulating factors. PMNs isolated from burn patients were treated with G-CSF, GM-CSF, or IL-8 and the expression of CD11b was examined. As previously stated, the percentage of neutrophils expressing normal levels of CD11b was reduced following thermal injury. Neutrophils were isolated from burn patients, incubated with 30 IU of G-CSF or GM-CSF, and analyzed for the expression of CD11b, CD11c, and CD14 (White-Owen et al., 1993). Both G-CSF and GM-CSF significantly increased the percentage of CD11b and CD11c (Fig 7). Following treatment with G-CSF or GM-CSF the percentage of neutrophils expressing CD11b increased from 48 to 100% and the percentage expressing CD11c increased from 43 to 61% with G-CSF and to 80% with GM-CSF. G-CSF had little effect on β2 integrin receptor density whereas GM-CSF increased the density of CD11b and CD11c (Fig. 8).

The effects of G-CSF and GM-CSF on the expression of the lipopolysaccharide receptor, CD14, and the Fcγ receptors were also investigated. Neither CSF altered the percentage of CD14 on PMNs isolated from healthy individuals or burn or trauma patients. Treatment in vitro with 50 IU of either G-CSF or GM-CSF increased the density of CD14 on patients' neutrophils as assessed by an increase of 47 and 111% in the mean channel fluorescence (Fig. 9). Following both thermal and nonthermal traumatic injuries, the expression of the CD32 in addition to CD16 was significantly altered. The effects of the colony-stimulating factors on the expression of CD16 (FcRI-II), CD32 (FcRII), and CD64 (FcRI) on neutrophils following severe trauma were investigated. Thermal injury and to a lesser extent traumatic injury produced a significant reduction in the percentage of neutrophils that express CD16 or CD32 (Table

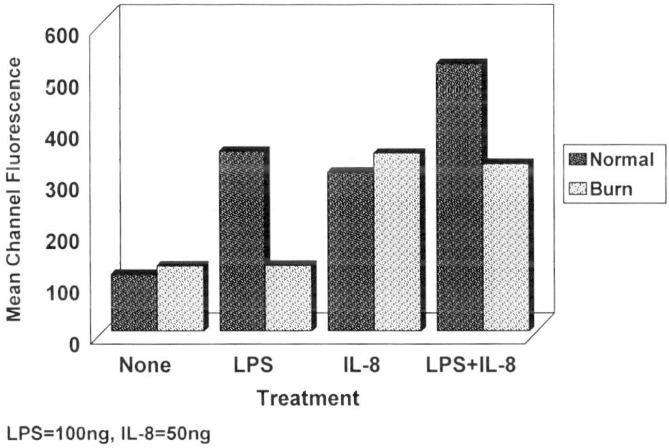

Fig. 7. A comparison between the expression of CD11b on neutrophils isolated from burn patients and on those from healthy individuals in response to treatment with LPS and/or IL-8 incubated for 30 min at 37°C. Data are expressed as mean channel fluorescence.

4). However, CD64 expression, which is normally very low in neutrophils, was significantly upregulated in burn patients (28%) and trauma patients (79%). The upregulation of this receptor is considered a marker for activation of neutrophils. Treatment of patient neutrophils in vitro with 30 IU of G-CSF or GM-CSF was ineffective, with the exception of CD32, which approached normal levels following incubation with GM-CSF but not G-CSF. Patients' neutrophils that did not demonstrate an initial upregulation for CD64 following injury did not respond to treatment in vitro with G-CSF but displayed an increase in expression of this receptor following incubation with GM-CSF.

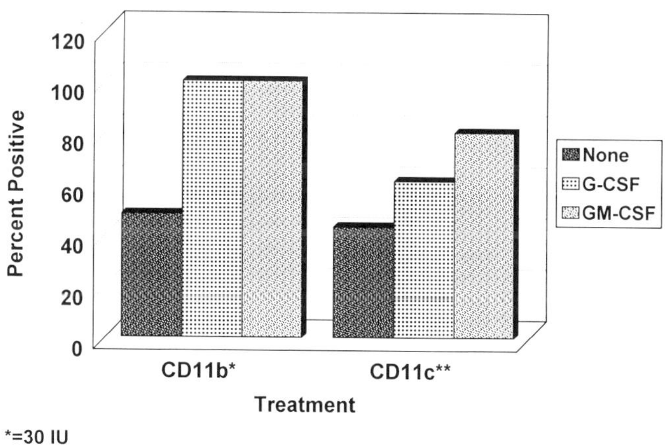

Fig. 8. The effects G CSF and GM-CSF on the expression of CD11b and CD11c on neutrophils from burn patients. Results are expressed as percentage of cells positive for the specific antigen.

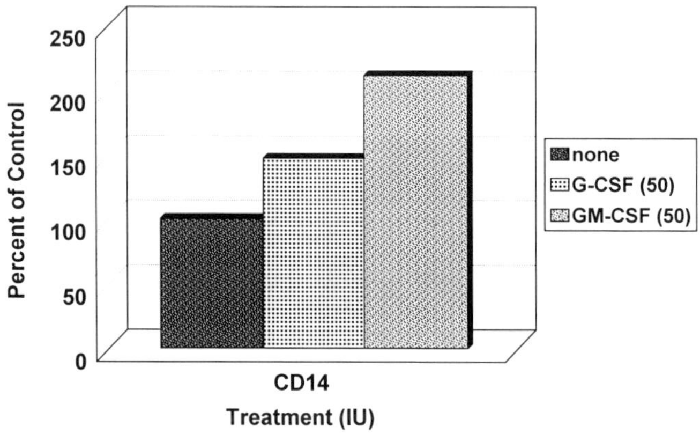

Fig. 9. The effects of G-CSF and GM-CSF on the expression of CD14 on neutrophils from burn patients. The results are expressed as a percentage of the control. Mean channel fluorescence values obtained from the control cell populations were set at 100% and compared to values obtained from the experimental cell populations.

In an additional series of experiments, the effect of IL-8 on CD11b receptor density and upregulation was examined in vitro using neutrophils from burn patients (Fig.10). As in previous experiments, lipopolysaccharide was unable to upregulate CD11b receptors on neutrophils isolated from burn patients, but significant upregulation was observed with normal neutrophils. Treatment of neutrophils with IL-8 increased the density of CD11b approximately threefold on neutrophils from healthy individuals or burn patients. Treatment with lipopolysaccharide further increased the density of CD11b on neutrophils from healthy individuals nearly fivefold. However, lipopolysaccharide failed to enhance further the density of CD11b over that of IL-8 on neutrophils from burn patients. These results suggests that the signaling pathway

TABLE 4. The Effects of Thermal and Traumatic Injury on the Expression of the Neutrophil Fcγ Receptors.

		Mean % Positive		
Subjects	Treatment[a]	CD16	CD32	CD64
Normals	None	91	87	3
Burn	None	28	54	28
Trauma	None	54	62	79
All patients[b]	G-CSF[c]	61	67	5
All patients	GM-CSF[d]	58	91	19

[a]Neutrophils were incubated with the appropriate CSF or with medium alone.
[b]For this study the data from the G-CSF and GM-CSF burn and trauma patients were combined.
[c]Neutrophils were incubated with 30 IU of G-CSF.
[d]Neutrophils were incubated with 30 IU of GM-CSF.

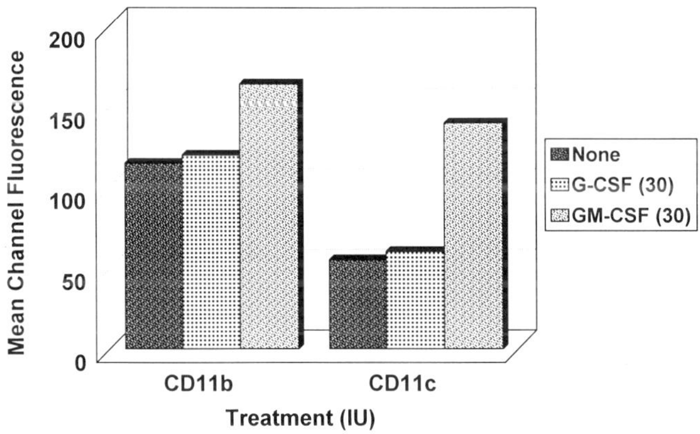

Fig. 10. The effects of G-CSF and GM-CSF on the cell surface receptor density expression of CD11b on neutrophils from burn patients. Results are expressed as mean channel fluorescence values.

for lipopolysaccharide upregulation of CD11b is perturbed and can not be corrected by IL-8. However, IL-8 was not able to induce CD11b expression on neutrophils that did not express this receptor.

SUMMARY

Following many types of severe injuries, alterations occur in the immune response. A hyperresponsiveness occurs immediately following injury, which is associated with many of the sequelae of SIRS, MODS, and MOF. The second phase, which occurs several days to several weeks after injury, often correlates with an increase in infections. It is our belief that a combination of factors, including the production of IL-4 and PGE_2, leads to a decrease in the production of GM-CSF. This decreased production of GM-CSF promotes an alteration in the functional capabilities of neutrophils, including reduced phagocytosis and oxidative burst response. Neutrophils leaving the bone marrow approximately 5 or more days post-injury do not activate/differentiate in a normal fashion. This type of dysfunction is especially seen at the receptor level and demonstrated by a greatly decreased expression of the CD11b molecule. Upregulation of the CD11b receptors is greatly suppressed following stimulation with lipopolysaccharide, fMLP, PMA, and A23187. Our current model for trauma-induced alterations in neutrophil function leading to increased infections appears in Figure 11. It is felt that increased PGE_2 from monocytes and IL-4 from T cells reduces the production of GM-CSF. Concurrently large number of neutrophils are migrating from the bone marrow in response to the inflammatory response induced by the trauma. The lack of sufficient GM-CSF produces neutrophils that are not fully activated and have reduced microbicidal properties. The events in turn lead to an increase in the rate of infections. The results from our laboratory as well as those from others suggest that

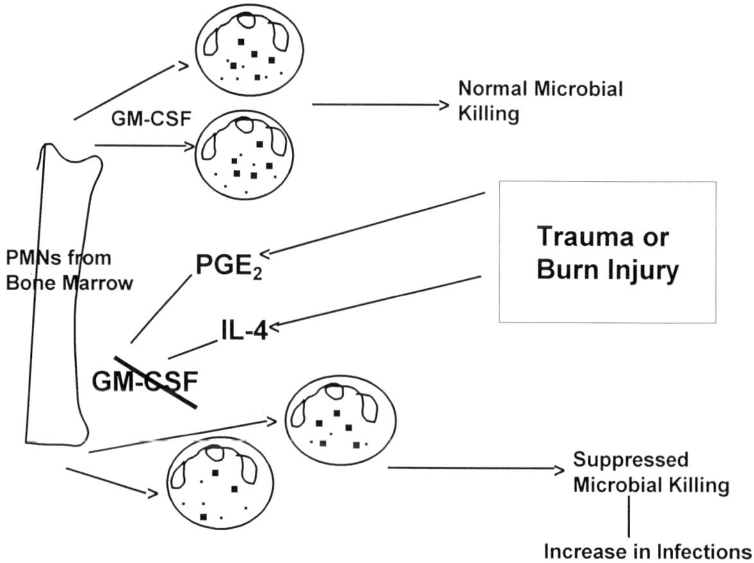

Fig. 11. Proposed model for thermal or nonthermal trauma-induced alterations in neutrophil function leading to increased infections.

GM-CSF and G-CSF may be useful in restoring the patient's host defense mechanisms, especially the neutrophils, to a normal level of function.

ACKNOWLEDGMENTS

The author would like to acknowledge the expertise of those who performed most of the experiments, including Dr. Cathy White-Owen, Dr. Dave Rodeberg, Ann Fiddler-Taylor, and Sharon Hartmann. The author would also like to thank Susan Dawes and James Cornelius for performing the flow cytometry. These studies were supported in part by grants from the Shriners of North America.

REFERENCES

Alexander JW (1968): Effect of thermal injury upon the early resistance to infection. J Surg Res 8:128–137.

Alexander JW, Hegg M, Altemeier WA (1968): Neutrophil function in selected surgical disorders. Ann Surg 168:447–458.

Alexander JW, Moncrief JA (1966): Alterations of the immune response following severe thermal injury. Arch Surg 93:75–83.

Alexander JW, Stinnett JD, Ogle CK, Ogle JD, Morris MJ (1979): A comparison of immunologic profiles and their influence on bacteriemia in surgical patients with a high risk of infection. Surgery 86:94–104.

Allen RC, Pruitt BA (1982): Humoral-phagocyte axis of immune defense in burn patients. Arch Surg 117:133–140.

Altman LC, Furukawa CT, Klebanoff SJ (1977): Depressed mononuclear leukocyte chemotaxis in thermally injured patients. J Immunol 119:199–205.

Anderson DC, Miller LJ, Schmalstieg FC, Rothlein R, Springer TA (1986): Contributions of the Mac-1 glycoprotein family to adherence dependent granulocyte functions: structure-function assessments employing subunit-specific monoclonal antibodies. J Immunol 137:15–27.

Antonacci AC, Calvano SE, Reaves LE, et al. (1984): Autologous and allogeneic mixed-lymphocyte responses following thermal injury in man: the immunomodulatory effects of interleukin 1, interleukin 2 and a prostaglandin inhibitor, WY-18251. Clin Immunol Immunopathol 30:304–320.

Austin OM, Redmond HP, Watson WG, Cunney RJ, Grace PA, Bouchier-Hayes D (1995): The beneficial effects of immunostimulation on posttraumatic sepsis. J Surg Res 59:446–449.

Ayala A, Chaudry I (1997): The anti-inflammatory mediator response: the murine Th2 response following shock and sepsis. Shock 7(suppl 1):85.

Babcock GF, Alexander JW, Warden GD (1990): Flow cytometric analysis of neutrophil subsets in thermally injured patients developing infection. Clin Immunol Immunopathol 54:117–125.

Babcock GF, Rodeberg DA (1996): The effect of thermal injury and exposure on β2 integrin and CD14 expression. In Faist E, Baue AE, Schildberg FW (eds). *The Immune Consequences of Trauma, Shock and Sepsis—Mechanisms and Therapeutic Approaches,* Vol 1. Lengerich, Germany: PABST Science Publishers, pp 202–210.

Babcock GF, Rodeberg DA, White-Owen CL (1996): Changes in neutrophil function following major trauma or thermal injury. In Faist E, Baue AE, Schildberg FW (eds). *The Immune Consequences of Trauma, Shock and Sepsis—Mechanisms and Therapeutic Approaches,* Vol 1. Lengerich, Germany: PABST Science Publishers, pp 194–201.

Babcock GF, White-Owen CL, Alexander JW, Warden GD (1993): Characteristics of neutrophil dysfunction. In Faist E, Meakins J, Schildberg FW (eds). *Host Defense Dysfunction in Trauma, Shock and Sepsis,* Berlin, Heidleberg: Springer-Verlag, pp 95–105.

Baker C, Miller C, Trunkey D (1984): Predicting fatal sepsis in burn patients. J Trauma 19:64–68.

Bjerknes R, Vindenes H (1989): Neutrophil dysfunction after thermal injury: alteration of phagolysosomal acidification in patients with large burns. Burns 15:77–81.

Bjerknes R, Vindenes H, Pitkanen J, Ninnemann J, Laerum O, Abyholm F (1989): Altered polymorphonuclear neutrophilic granulocyte functions in patients with large burns. J Trauma 29:847–855.

Bone RC (1991): A critical evaluation of new agents for treatment of sepsis. J Am Med Assoc 266:1686–1691.

Butcher EC (1991): Leukocyte-endothelial cell recognition: three (or more) steps to specificity and diversity. Cell 76:1033–1036.

Curreri PW, Heck EL, Browne L, Baxter C (1973): Stimulated nitroblue tetrazolium test to assess neutrophil antibacterial function: prediction of wound sepsis in burn patients. Surgery 74:6–13.

De A, Kodys D, Pellegrini J, Puyana J, Miller-Graziano C (1997): Characteristics of post trauma anergic T cells and their clinical relevance. Shock 7(suppl 1):85.

Delogu G, Marandol M, Paoletti F, Marano M, Trinceieri V, De Simone C (1996): Critically ill patients with gram-negative sepsis have increased levels of interleukin-6 and -8 but not interleukin-1. In Faist E, Baue AE, Schildberg FW (eds). *The Immune Consequences of Trauma, Shock and Sepsis—Mechanisms and Therapeutic Approaches,* Vol 1. Lengerich, Germany: Pabst Science Publishers, pp 378–388.

DiPiro J, Howdieshell T (1997): Eosinophilia with sepsis: more evidence of swich to Th2 cytokine profile. Shock 7(suppl 1):86.

Dobke MK, Deitch EA, Harner TJ, Baxter CR (1989): Oxidative activity of polymorphonuclear leukocytes after thermal injury. Arch Surg 124:856–859.

Docke W, Lenz A, Savat R, Randow F, Vold H (1997): Immunodepression in septic disease is associated with a switch to type 2 cytokine pattern: *In vitro* reversal of sepsis-/sepsis mediator-induced type 2 switch by IL-2. Shock 7(suppl 1):86.

Donnelly SC, Strieter RM, Kunkel SL, et al. (1993): Interleukin-8 and development of adult respiratory distress syndrome in at-risk patient groups. Lancet 341:643–647.

El-Maallem H, Fletcher J (1981): Effects of surgery on neutrophil granulocyte function. Infect Immun 32:380–341.

Ertel W, Schook FA, Trentz O (1996): The role of anti-inflammatory mediators for the control of systemic inflammation following severe injury. In Faist E, Baue AE, Schildberg FW (eds). *The Immune Consequences of Trauma, Shock and Sepsis—Mechanisms and Therapeutic Approaches,* Vol 1. Lengerich, Germany: Pabst Science Publishers, pp 453–470.

Fabian TC, Croce MA, Fabian MJ, et al. (1995): Reduced tumor necrosis factor production in endotoxin-spiked whole blood after trauma: experimental results and clinical correlation. Surgery 118:63–72.

Faist E, Mewew A, Baker CC, et al. (1987): Prostaglandin E2 (PGE2)-dependent suppression of interleukin-2 (IL-2) production in patients with majory trauma. J Trauma 27,28:827–947.

Fikrig SM, Karl SC, Suntharalingam K (1977): Neutrophil chemotaxis in patients with burns. Ann Surg 186:746–748.

Gamelli RL, He LK, Liu H (1995): Recombinant human granulocyte colony-stimulating factor treatment improves macrophage suppression of granulocyte and macrophage growth after burn and burn wound infection. J Trauma 39:1141–1146.

Grbic JT, Mannick JA, Gough DB, Rodrick ML (1991): The role of prostaglandin E2 in the immune suppression following injury. Ann Surg 214:253–263.

Grogan JB (1976): Altered neutrophil phagocytic function in burn patients. J Trauma 16:734–738.

Grogan JB (1976): Suppressed *in vitro* chemotaxis of burn neutrophils. J Trauma 16:985–988.

Grogan JB, Miller RC (1973): Impaired function of polymorphnuclear leukocytes in patients with burns and other trauma. Surg Gynecol Obstet 137:784–788.

Harmon JW, Skornik WA, McDonald, Dressler DP (1976): Pulmonary bacterial defense. Effect of the burn wound on transfer of alveolar macrophage activation in rats by parabiosis. Am J Surgery 131:447–451.

Heck EL, Browne L, Curreri PW, Baxter CR (1975): Evaluation of leukocyte function in burn individuals by in vitro oxygen consumption. J Trauma 11:486–489.

Heck EL, Edgar MA, Masters BS, Baxter CR (1979): The role of NADH–NADPH oxidase activity in the leukocyte function of burned patients. J Trauma 19:49–51.

Heinzelmann M, Mercer-Jones M, Cheadle WG, Polk HC (1996): CD124 expression in injured patients correlates with outcome. Ann Surg 224:91–96.

Iyengar VG, Peterson VM, Rundus C, Robinson WA (1986): Postburn serum inhibits *in vitro* production of colony-stimulating factor by mononuclear peripheral blood cells. Int J Cell Cloning 4:472–482.

Kim C, Dremer J, Van Donnersmarck G, de Wall Malefyt R, Faist E (1997): The profile of TH1/TH2 responses post major trauma. Shock 7(suppl 1):87.

Lanser ME, Brown GE, Mora R, Coleman W, Siegel JH (1986): Trauma suppresses superoxide production by normal neutrophils. Arch Surg 121:157–162.

Loose L, Turinsky J (1979): Macrophage dysfunction after burn injury. Infect Immun 26:157–162.

Lowery SF (1993): Cytokine mediators of immunity and inflammation. Arch Surg 128:1235–1241.

Markewitz A, Lang S, Hultner L, Wienhold C, Reichart B (1996): An imblance in T-helper cell subsets alters immune response after cardiac surgery. Eur J Cardio-Thoracic Surg 10:61–67.

Miller-Graziano CL, De AK, Kodys K (1995): Altered IL-10 levels in trauma patients' M phi and T lymphocytes. J Clin Immunol 15:93–104.

Miller-Graziano CL, Fink M, Wu JY, Szabo, Kodys D (1988): Mechanisms of altered monocyte prostaglandin E_2 production in severely injured patients. Arch Surg 123:293–299.

Miller-Graziano CL, Szabo G, Griffey D, Mehta B, Kodys D, Catalano D (1991): Role of elevated monocyte transforming growth factor beta (TGF beta) production in posttrauma immunosuppression. J Clin Immunol 1:95–102.

Molly RG, Holzheimer R, Nestor M, Collins K, Mannick JA, Rodrick ML (1995): Granulocyte-macrophage colony-stimulating factor modulates immune function and improves survival after experimental thermal injury. Br J Surg 82:770–776.

Moore FA, Peterson BM, Moore EE, Rundus C, Poggetti R (1990): Inadequate granulopoiesis after major torso trauma: a hematopoietic regulatory paradox. Surgery 108:667–674.

Nathan C, Srimal S, Farber C, et al. (1989): Cytokine-induced respiratory burst of human neutrophils: dependence on extracellular matrix proteins and CD11/CD18 integrins. J Cell Biol 109:1341–1349.

Neely AN, Childress CM, Maley MP, Holder IA (1991): Causes of colonization of autografted burn wounds. J Burn Care Rehabil 12:294–299.

Nelson RD, Hasslen SR, Ahrenholz DH, Lolem LD (1987): Mechanism of loss of human neutrophil chemotaxis following thermal injury. J Biochem Res:496–502.

Niedworok J, Grzybowski A (1979): Activity of the reticuloendothelial system in the course of experimental burn disease. Arch Immunol Ther Exp 27:547–551.

Nijsten MW, Hack CE, Helle M, ten Duis HG, Klasen HJ, Aarden LA (1991): Interleukin-6 and its relation to the humoral immune response and clinical parameters in burned patients. Surgery 109:761–767.

Ono Y, Kunii O, Kobayashi K, Kanegasaki S (1993): Evaluation of opsonophagocytic dysfunctions in severely burned patients by luminol-dependent chemiluminescence. Microbiol Immunol 37:563–571.

O'Nuallain EM, Puri P, Mealy D, Reen DJ (1995): Induction of interleukin-1 receptor antagonist (IL-1ra) following surgery is associated with major trauma. Clin Immunol Immunopathol 76:96–101.

O'Reilly M, Silver GM, Gamelli RL, Davis JH, Hebert JC (1994): Dose dependency of granulocyte-macrophage colony stimulating factor for improving survival following burn wound infection. J Trauma 36:486–490.

Padalino P, Gardinali M, Catania A, et al. (1996): Tumor necrosis factor and soluble tumor necrosis factor receptor I in multiple organ dysfunction syndrome. In Faist E, Baue AE, Schildberg FW (eds). *The Immune Consequences of Trauma, Shock and Sepsis—Mechanisms and Therapeutic Approaches,* Vol 1. Lengerich, Germany: Pabst Science Publishers, 368–377.

Pallister CJ, Hancock JT (1995): Phagocytic NADPH oxidase and its role in chronic granulomatous disease. Br J Biomed Sci 52:149–156.

Pellegrin K, Peterson V, Thurman G, Cusack N, Hartfor C, Ambruso D (1997): rhGM-CSF enhances production of PMN superoxide anion in burned patients: a randomized, double-blind, pilot study. Shock 7(suppl):165.

Polk HC, George CD, Hershman Mj, Wellhausen SR, Cheadle WG (1988): The capacity of serum to support neutrophil phagocytosis is a vital host defense mechanism in severely injured patients. Ann Surg 207:686–692.

Reimand KP, Weitzel F, Lorenz W, et al. (1997): Granulocyte-colony-stimulating factor prophylaxis and treatment in postoperative peritonitis: animal experiments and a successful case report. In Faist E, Baue AE, Schildberg FW (eds). *The Immune Consequences of Trauma, Shock and Sepsis—Mechanisms and Therapeutic Approaches,* Vol 2. Lengerich, Germany: PABST Science Publishers, pp 493–503.

Rittenbury MS (1970): The response of the reticuloendothelial system to thermal injury. Surg Clin N Am 50:1227–1123.

Rittenbury MS, Hanback LD (1967): Phagocytic depression in thermal injuries. J Trauma 7:5223–5240.

Rodeberg DA, Bass RC, Alexander JW, Warden GD, Babcock GF (1997): Neutrophils from burn patients are unable to increase the expression of CD11b/CD18 in response to inflammatory stimuli. J Leukoc Biol 61:575–582.

Rodrick ML, Wood JJ, O'Mahony JB, et al. (1986): Mechanisms of immunosuppression associated with severe nonthermal traumatic injuries in man: production of interleukin 1 and 2. J Clin Immunol 6:310–318.

Rosen H, Law S (1989): The leukocyte cell surface receptors(s) for the iC3bi product of complement. Curr Top Microbiol Immunol 153:99–122.

Rosen H, Law SKA (1990): The leukocyte cells surface receptor (s) for the iC3b product of complement. Curr Top Microbiol Immunol 153:99–122.

Ross GD (1986): Clinical and laboratory features of patients with an inherited deficiency of neutrophil membrane complement receptor type 3 (CR3) and the related membrane antigens LFA-1 and p150,95. J Clin Immunol 6:107–113.

Schinkel C, Zimmer S, Kremer JP, et al. (1996): Synthesis patterns of proinflammatory cytokines in states of trauma. In Faist E, Baue AE, Schildberg FW (eds). *The Immune Consequences of Trauma, Shock and Sepsis—Mechanisms and Therapeutic Approaches,* Vol 1. Lengerich, Germany: Pabst Science Publishers, pp 438–452.

Spooner CE, Markowitz NP, Sarvolatz LD (1992): The role of tumor necrosis factor in sepsis. Clin Immunol Immunopathol 62:11–17.

Tellado JM, Christou NV (1991): Critically ill anergic patients demonstrate polymorphonuclear neutrophil activation in the intravascular compartment with decreased cell delivery to inflammatory foci. J Leukoc Biol 50:547–553.

Vadas MA, Gamble JR (1990): Regulation of the adhesion of neutrophils to endothelium. Biochem Pharmacol 40:1683–1687.

Vindenes H, Ulvestad E, Bjerknes R (1995): Increased levels of circulating interleukin-8 in patients with large burns: relation to burn size and sepsis. J Trauma 39:636–640.

Warden G, Mason A, Pruitt B (1975): Suppression of leukocyte chemotaxis *in vitro* by chemotherapeutic agents used in management of thermal injuries. Ann Surg 181:863–869.

Warden GD, Mason AD, Pruitt BA (1974): Evaluation of leukocyte chemotaxis *in vitro* in thermally injured patients. J Clin Invest 54:1001–1004.

Wardle EN (1993): Cytokines in relation to ICU problems. Clin Intensive Care 4:183–189.

White-Owen C, Alexander JW, Babcock GF (1992): Reduced expression of neutrophils CD11b and CD16 after severe traumatic injury. J Surg Res 52:22–26.

White-Owen C, Hartmann S, Alexander JW, Babcock GF (1993): Reduced PMN β_2 integrins after trauma: a possible role for colony-stimulating factors. Clin Exp Immunol 92:477–481.

Zimmer S, Pollard V, Marshall G, Garofalo R, Prough D, Herndon D (1997): TH1/TH2 response in human experimental endotoxin. Shock 7(suppl 1):86.

14

Human Monocytes and Macrophages: Characterization in Health and Inflammatory Disease States

Prue H. Hart
Flinders University of South Australia, Adelaide, South Australia

Julie Rowe
Flinders University of South Australia, Adelaide, South Australia

John J. Finlay Jones
Flinders University of South Australia, Adelaide, South Australia

Mononuclear phagocytes constitute a heterogeneous cell population distributed among three main compartments: bone marrow, in which precursor cells divide and mature into monocytes; blood, where monocytes circulate transitionally; and peripheral tissues, where they migrate and differentiate into functionally active macrophages. Resident macrophages in tissue are constitutively distributed throughout the normal body and display regional heterogeneity, for example, alveolar macrophages, Kupffer cells, microglia. Elicited "inflammatory" monocytes/macrophages are also found at sites of tissue inflammation. These cells have left the circulation in response to "inflammatory" chemotactic signals and are less differentiated than resident tissue macrophages, for example, macrophages from synovial fluid of patients with inflammatory arthritides, and peritoneal monocytes/macrophages during peritonitis.

Phagocyte Function: A Guide for Research and Clinical Evaluation, Edited by J. Paul Robinson and George F. Babcock. ISBN 0-471-12364-1

With variations in activation, their differentiative state, and their physical location, monocytes and macrophages are functionally heterogeneous, which may be reflected in quantitative and qualitative differences in the macromolecular composition of their plasma membranes.

This chapter concentrates on the phenotypic and functional characterization in vitro of peripheral blood monocytes and their counterparts in tissue and at inflammatory sites. It highlights the functional differences detected between blood monocytes and their extravasated counterparts. The activities of the cells selected for study reflect the tissue location from which the cells were isolated and the clinical requirement to modulate the activity during inflammatory disease. This study shows that one can not assume that the responses of tissue and inflammatory monocytes/macrophages to potential therapeutic agents will be similar to those of circulating monocytes.

In our studies, monocytes and tissue/inflammatory macrophages were isolated from the same donor whenever possible to accommodate differences between individuals. Monocytes/macrophages were isolated from synovial fluid from patients with inflammatory arthritis (predominantly rheumatoid arthritis) and represented cells from a site of chronic inflammation. For cells from a site of acute inflammation, monocytes/macrophages were isolated from the peritoneal cavities of patients on continuous ambulatory peritoneal dialysis (CAPD) with peritonitis. As an example of resident tissue macrophages, the self-replicating human alveolar macrophage (Tarling et al., 1987) obtained by bronchoalveolar lavage was studied.

MONOCYTES: SUBPOPULATIONS AND PROINFLAMMATORY MEDIATORS

In normal healthy individuals, monocytes constitute 15–30% of circulating mononuclear cells. There have been considerable study and debate concerning the existence of monocyte subsets or whether variations in monocyte size and activity reflect a continuum of cells of different maturation status (Elias et al., 1985). There has been a report that approximately 20–30% of monocytes do not express MHC class II, and it is these cells that are attracted by chemoattractants to inflammatory sites (Cox et al., 1994). These cells have an increased capacity for an oxidative burst and for adherence to extracellular matrices (Cox et al., 1994). Other researchers have identified a monocyte-like group of blood cells that they hypothesize are dendritic cell precursors (Thomas et al., 1994; O'Doherty et al., 1994). Thomas and her colleagues identified a population of cells that were $CD33^{bright}$, but $CD14^{dim}$ and negative for CD16 and B7 expression. O'Doherty and his group identified two subpopulations of dendritic cell precursors among the $CD33^{bright}$, $CD14^{dim}$ cells, namely, CD11c-negative cells that they hypothesize are migrating from the bone marrow and CD11c-positive cells that they believe may be moving away from tissues. It has been hypothesized that IL-4 and granulocyte-macrophage colony-stimulating factor (GM-CSF) are responsible, at least in part, for the differentiation of monocyte precursors/monocytes to dendritic cell precursors (Peters et al., 1991; Sallusto and Lanzavecchia, 1994). A study of murine bone marrow-derived macrophages concluded that different activating agents can induce macrophage populations with different functional activities; for example, poly I:C and β1,3 glucan stimulate the development of cytocidal and inflammatory macrophages, respectively (Laszlo et al., 1993).

Monocytes and macrophages have the capacity to synthesize and secrete a large range of pro-inflammatory mediators if appropriately stimulated. Well-known stimulants include lipopolysaccharide and muramyl dipeptide. However, disease-specific antigens have been recognized. Several cytokines have also been identified to prime monocytes/macrophages for increased pro-inflammatory mediator production upon receipt of a second activating signal, for example, interferon-γ (Hart et al., 1989a). Monocytes/macrophages express receptors (constitutively or inducibly) for many of the pro-inflammatory cytokines that they produce, and by an autocrine mechanism these cells can express relatively quickly a significant inflammatory response. Studies in experimental mice injected with lipopolysaccharide suggest that tumor necrosis factor α (TNF-α) may be the first monokine produced (within 1–2 hr), followed rapidly by the appearance of interleukin-1 (IL-1) and IL-6. An autocrine/paracrine mechanism of action of TNF-α is suggested by the regulatory effects of an anti-TNF-α antibody on IL-1 and IL-6 production (for a review, see Vassalli, 1992). The monocyte/macrophage also contains regulatory mechanisms (constitutive and inducible) to control an escalating inflammatory reaction. For example, monocytes and macrophages produce an IL-1 receptor antagonist (IL-1ra) and cleave cytokine receptors from their membrane, thus providing molecules that compete with cell surface receptors for binding of cytokines in the pericellular environment and limit the functional activity of cytokines. In studies of human inflammation, soluble receptors for TNF-α and IL-1 have received the most attention (Leeuwenberg et al., 1994; Colotta et al., 1993). The unstable nature of mRNA for inflammatory cytokines has also helped to prevent overzealous macrophage inflammatory responses (Kruys et al., 1989).

MACROPHAGES AT A SITE OF CHRONIC INFLAMMATION: RHEUMATOID AND OTHER INFLAMMATORY ARTHRITIDES

Rheumatoid arthritis is associated with a marked inflammatory reaction involving cells from synovial fluid, synovial membranes, cartilage, and periarticular tissues. The chronic inflammatory reaction can lead to destruction of the functioning joint. Of the mononuclear cells infiltrating the synovium, activated macrophages are central to the destructive processes because of their ability to produce many molecules including proteases, prostanoids, IL-1, TNF-α, IL-6, IL-8, and granulocyte-macrophage colony-stimulating (GM-CSF) that can damage tissue or, in turn, attract into synovial tissue and then activate other cells (e.g., neutrophils and T lymphocytes) to produce proinflammatory molecules (Panayi, 1993).

Studies of cytokines that contribute to the pathogenesis of rheumatoid arthritis and to inflammatory arthritis in animals have focused on the production and function of IL-1 and TNF-α. Although IL-1 is also produced by synovial fibroblasts (Dalton et al., 1989) and TNF-α by lymphocytes (Vassalli, 1992), monocytes and macrophages are the principal producers of these cytokines in the inflamed joint. Intra-articular injection of IL-1 or TNF-α in rabbits induces an inflammatory response, with synergistic effects detected upon co-administration. Anti-TNF-α ameliorates joint disease in murine collagen-induced arthritis (Williams et al., 1992) and TNF-α transgenic mice develop arthritis at 4 weeks of age (Keffer et al., 1991). Furthermore, a recent study reported that treatment for 8 weeks with chimeric monoclonal antibodies to TNF-α resulted in significant clinical and laboratory test improvement of 20 patients with active rheumatoid arthritis (Elliott et al., 1993).

Many believe that the challenge for new therapies for rheumatoid and other forms of inflammatory arthritis is the regulation of the production and activity of IL-1, TNF-α, and other macrophage-derived pro-inflammatory molecules. One of the recognized activities of glucocorticoids is the regulation of pro-inflammatory mediators such as IL-1 and TNF-α. However, although they are among the most potent and widely used anti-inflammatory agents, glucocorticoids affect the properties of many cell types and tissues, often with undesirable side effects.

Interleukin-4, IL-10, and IL-13 have a suppressive effect on human monocyte functions that are associated with a pro-inflammatory response (Hart et al., 1989b, 1995a, b). Furthermore, IL-4 and glucocorticoids suppress pro-inflammatory monokine production by different mechanisms (Hart et al., 1990), suggesting that IL-4 may be used in therapy alone, or as a supplement to therapy with lower doses of steroid. Interleukin-4, IL-10, and IL-13 have been characterized as products of "T helper type-2" (Th2) lymphocytes; Th2-type cytokines are involved in modulating B lymphocyte responses, humoral immunity, and downregulation of monocyte/ macrophage responses. In contrast, upon stimulation, Th1 cells produce interferon-γ, IL-2, and TNF-β and are involved in the development of delayed-type hypersensitivity and inflammatory and cytotoxic macrophage responses (Mosmann and Coffman, 1989). These patterns of cytokine production were first observed for murine T-cell clones. As is the case for hyperimmunized mice, similar divisions of human T-cell clones (CD4 and CD8) into type 1 and type 2 cytokine profiles have now been detected for allergic individuals (Del Prete, 1992) or individuals responding to an infectious agent, for example, *Mycobacterium leprae* (Salgame et al., 1991). Other T-helper lymphocyte clones (Th0) may produce both Th1- and Th2-type cytokines (O'Garra and Murphy, 1994). Interleukin-4, IL-10, and IL-13 (type 2 cytokines) downregulate monocyte production of IL-1, TNF-α, IL-6, IL-8, and prostaglandin E2 (PGE_2) and stimulate the production of IL-1ra, a molecule that binds to, but does not stimulate downstream signaling from, the type 1 IL-1 receptors (Arend, 1991). The regulatory effects of IL-4, IL-10, and IL-13 on monocyte activity suggested that these molecules, or agonists, may be used in therapy as anti-inflammatory agents. However, if cytokines are to be used for immunotherapy, they must have appropriate regulatory properties on extravasated inflammatory and tissue macrophages rather than just on blood monocytes.

Phenotypic Characterization of Synovial Fluid Macrophages

Increased expression of MHC class II antigens by macrophages from synovial tissue and synovial fluid from patients with rheumatoid and other forms of inflammatory

TABLE 1. Phenotype of CD14⁺ Cells from Peripheral Blood and Synovial Fluid of Patients with Rheumatoid Arthritis

	Fluorescence Intensity[a]	
Surface Marker (Ab)	PBMC	SFMC
Nil (X63)	19 ± 9	16 ± 3
MHC class II (FMC 52)	580 ± 59	1829 ± 237[b]
IL-4 receptor (M57)	25 ± 5	23 ± 3
p55 TNF receptor (HTR-9)	66 ± 9	27 ± 7[b]
p75 TNF receptor (UTR-1)	306 ± 111	120 ± 20[b]

[a]Arbitrary units; mean ± SEM, $n=3$; PBMC = peripheral blood mononuclear cells; SFMC = synovial fluid mononuclear cells.
[b]Significantly different from PBMC, $p < 0.05$.

arthritis has been reported previously (Klareskog et al., 1981) and provides confirmation of the chronic inflamed nature of the joints involved. Both synovial fluid and synovial tissue macrophages express decreased levels of CD11a, the α_1 chain of LFA-1, but express normal levels of CD11b and c, and CD18 (Emmrich et al., 1990). Table 1 shows the results of phenotyping synovial fluid macrophages and peripheral blood monocytes isolated from three patients with rheumatoid arthritis. As expected, the expression of MHC class II was increased on the inflammatory joint cells. Receptors for IL-4 as detected by the antibody M57 were unchanged on the two cell populations (Hart et al., 1993a). However, receptors for TNF-α, both the p55 receptor recognized by HTR-9 and the p75 receptor recognized by UTR-1, were reduced on synovial fluid CD14⁺ cells (unpublished). We hypothesize that the reduced expression of these receptors reflects the protease content of inflammatory synovial fluid and may explain the discrepancy with a previous report that TNF-α receptor expression is up-regulated on rheumatoid synovium (Brennan et al., 1992).

The Effect of Th2 Cytokines on TNF-α, IL-1β, and IL-1ra Production by Synovial Fluid Mononuclear Cells

A series of studies examined the ability of IL-4, IL-10, and IL-13 to regulate cytokine production and expression of certain surface markers on mononuclear cells isolated from both the synovial fluid and peripheral blood of patients with rheumatoid arthritis (Hart et al., 1993a, 1995a, b). Cells isolated from a few patients with psoriatic arthritis or ankylosing spondylitis were also examined. However, no differences in response were detected for cells from patients with these different forms of inflammatory arthritis or undergoing different anti-inflammatory treatments.

Synovial fluid cells (mean macrophage concentration 41%, range 18–62%) and peripheral blood mononuclear cells (mean macrophage concentration 34%, range 22–43%) were isolated from 12 patients with inflammatory arthritis, including 10 patients with rheumatoid arthritis, and incubated with or without IL-4, IL-10, or

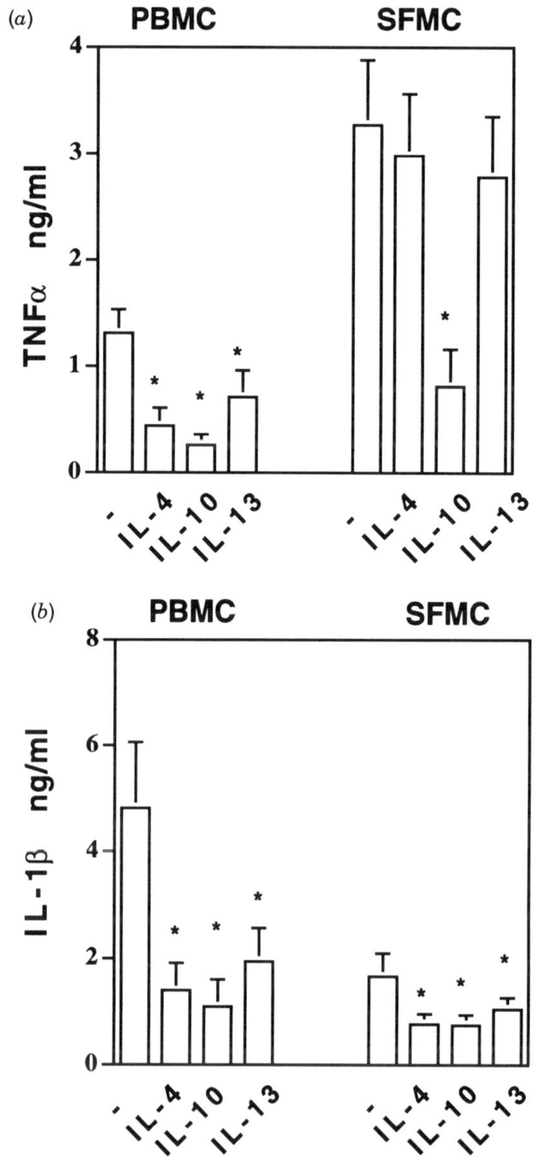

Fig. 1. The effect of IL-4, IL-10, and IL-13 on (A) TNF-α and (B) IL-1β production by LPS-stimulated peripheral blood mononuclear cells (PBMC) and synovial fluid-derived mononuclear cells (SFMC). Cells from 12 patients were incubated for 22 h with LPS (500 ng/ml) alone or together with IL-4 (10 ng/ml), IL-10 (10 ng/ml), or IL-13 (10 ng/ml). The mean + SEM is shown. An asterisk indicates a significant effect of the cytokines on TNF-α production. *Source:* Clinical and Experimental Immunology 99:333, 334.

IL-13, each at 10 ng/ml, for 22 h. In the absence of an exogenous stimulus, synovial fluid mononuclear cells produced 0.35 ± 0.17 ng/ml TNF-α. For peripheral blood mononuclear cells (PBMC), 0.37 ± 0.14 ng/ml TNF-α (n = 8) was detected. These levels were too low to detect significant regulatory effects of IL-4, IL-10, or IL-13. In response to lipopolysaccharide (500 ng/ml), synovial fluid cells produced greater levels of TNF-α (Fig. 1A). Interleukin-4 and IL-13 significantly inhibited lipopolysaccharide-induced TNF-α production by mononuclear cells harvested from peripheral blood but not from synovial fluid. Increased concentrations of IL-4 or IL-13 were unable to have a greater inhibitory effect. In contrast, IL-10 decreased TNF-α production by both lipopolysaccharide-stimulated peripheral blood and synovial fluid cells (Fig. 1A). In contrast to TNF-α production, the same synovial fluid mononuclear cells produced lower levels of lipopolysaccharide-induced IL-1β than detected for peripheral blood-derived mononuclear cells (Fig. 1B). In contrast to the results with TNF-α, IL-4, IL-10, and IL-13 significantly inhibited IL-1β production by lipopolysaccharide-stimulated synovial fluid mononuclear cells. This result suggests that the synovial fluid cells express functional receptors for IL-4 and IL-13 and that the inability of these cytokines to regulate TNF-α production efficiently reflects an inefficient post-receptor signaling event. Alternatively, there may be two distinct IL-4 receptors as suggested by Rigley and colleagues (Rigley et al., 1991) and by He and Malek (1995). It is notable that TNF-α production by synovial fluid cells could be downregulated by other mediators, namely, IL-10 (Fig. 1A) and dexamethasone (data not shown).

The effect of IL-4 and IL-10 on the production of the anti-inflammatory molecule IL-1ra was investigated. Interleukin-4 efficiently induced IL-1ra production by both peripheral blood monocytes and synovial fluid macrophages (Fig. 2). Thus IL-4 would not only reduce IL-1 production but also reduce the activity of the remaining IL-1. Interleukin-10 enhanced the production of IL-1ra by PBMC, but this increase was minimal in contrast to the effect of IL-4. Furthermore, an effect of IL-10 was not detectable on synovial fluid cells.

The Effect of Interleukin-10 on MHC Class II Expression by Synovial Fluid Macrophages

In the experiments described above, the ability of the Th2 cytokines to regulate mediator production by a population of mononuclear cells was investigated. Although lymphocytes were present, the responses were ascribed to monocytes/macrophages because TNF-α, IL-1β, and IL-1ra production by nonmyeloid cells is minimal in comparison to that by monocytes/macrophages. However, phenotypic responses to these cytokines by the same cells were also investigated by flow cytometry, a procedure that allowed the specific measurement of responses by macrophages selected according to expression of the surface marker CD14.

Increased expression of MHC class II is characteristic of inflammatory macrophages (Klareskog et al., 1981). Interleukin-10 has been shown to suppress monocyte expression of MHC class II antigens, a function that may be responsible for the IL-10-induced reduction of antigen-specific T-cell proliferation (De Waal Malefyt et al.,

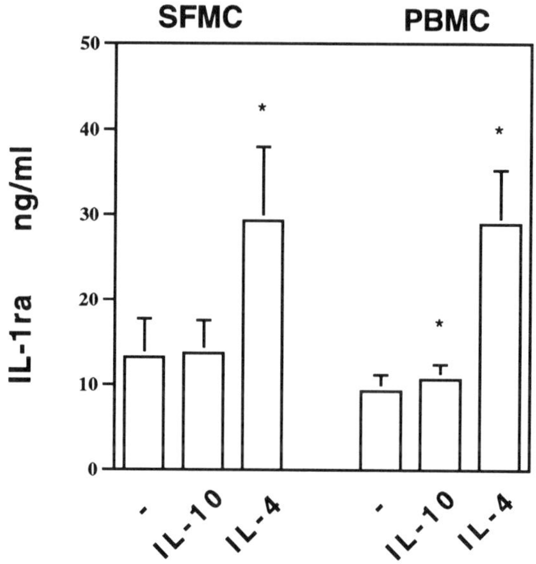

Fig. 2. The effect of IL-10 and IL-4 on the production of IL-1ra by LPS-stimulated synovial fluid (SFMC) and peripheral blood mononuclear cells (PBMC). Cells from seven patients with RA were incubated for 22 h with LPS (500 ng/ml) alone or together with IL-10 (10 ng/ml) or IL-4 (10 ng/ml). The mean + SEM is shown. An asterisk indicates a significant effect of IL-10 or IL-4 (by paired *t* test) on IL-1ra production. *Source:* Immunology 84:540.

1991). We investigated IL-10 regulation of MHC class II expression by synovial fluid macrophages, distinguished from lymphocytes by both larger size (i.e., forward and side scatter on flow cytometry) and CD14 expression (Hart et al., 1995a). The cells had also been exposed to lipopolysaccharide because its addition was necessary to measure sensitively Th2-cytokine regulation of monocyte/macrophage production of IL-1β and TNF-α (see preceding section); however, lipopolysaccharide was without significant effect on the binding of FMC52, an antibody that binds to an epitope common to HLA-DR, -DP and -DQ (Hart et al., 1995a). Interleukin-10 at 10 ng/ml significantly suppressed MHC class II expression by both blood monocytes and synovial fluid macrophages. However, at 1 and 0.1 ng/ml, IL-10 significantly inhibited antigen expression on peripheral blood-derived cells but not on cells derived from synovial fluids (Fig. 3).

As a positive response to IL-4 and IL-13, the induction of CD23, the FcεRII receptor, was investigated on synovial fluid and peripheral blood monocytes/ macrophages from the same donor. Although significant, the extent of induction by these cytokines was greatly reduced on the inflammatory macrophages (Hart et al., 1993a, 1995b). This result shows that the poor ability of IL-4 and IL-13 to regulate TNF-α production by synovial fluid cells is not unique to TNF-α production and suggests that regulation of TNF-α and CD23 may be linked. Like the regulation of IL-1 production, the control of CD14 expression by IL-4 was not influenced by monocyte differentiation and activation, as measured on synovial fluid macrophages (Hart et al., 1993a).

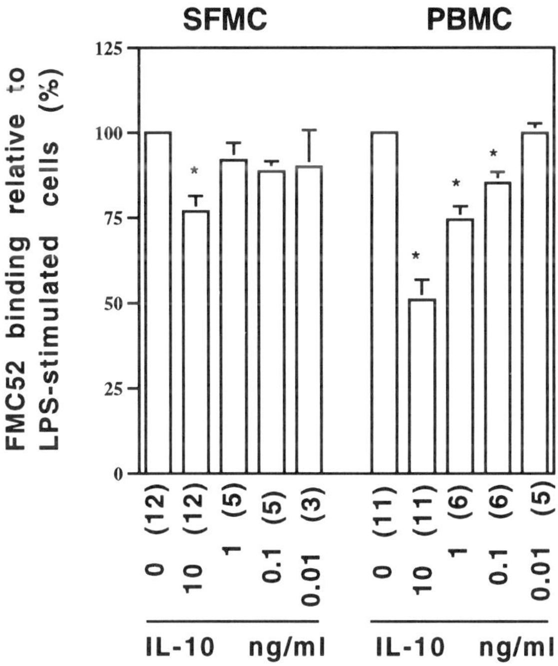

Fig. 3. The effect of decreasing concentrations of IL-10 on MHC class II expression by LPS-stimulated synovial fluid (SFMC) and peripheral blood mononuclear cells (PBMC). After incubation of 22 h with IL-10, the fluorescence intensity of binding of FMC52 to cells from *n* patients was calculated as a percentage of the intensity of antibody binding by LPS-stimulated cells. The mean + SEM is shown. An asterisk indicates a significant effect (by paired *t* test) of IL-10 on MHC class II expression. *Source:* Immunology 84:540.

The Effect of Th2 Cytokines on Expression of Tumor Necrosis Factor Receptors by Synovial Fluid Cells

In the experiments described above, IL-4 not only suppressed IL-1 production by synovial fluid macrophages but also enhanced the production of IL-1ra. It has recently been reported that IL-4 also induces the production and shedding of the type II IL-1 receptor, a non-signaling or "decoy" receptor (Colotta et al., 1993). In consideration of the TNF system and the potential of Th2 cytokines to regulate inflammatory mediator production and activity simultaneously, the effect of IL-4 and IL-10 on monocyte and synovial macrophage TNF receptor expression was examined. As stated above there are two types of TNF receptor, with approximate molecular weights of 55 and 75 kDa. They share approximately 25% overall amino acid sequence identity in the extracellular domain, but an absence of homology between intracellular domains of the two receptors suggests that they signal different response pathways. The p55 receptor is expressed primarily on epithelial cells, whereas myeloid cells express both p55 and p75 receptors (for a review, see Tartaglia and Goeddel, 1992). The p75 form is the predominant TNF receptor on monocytes and macrophages (Gehr et al.,

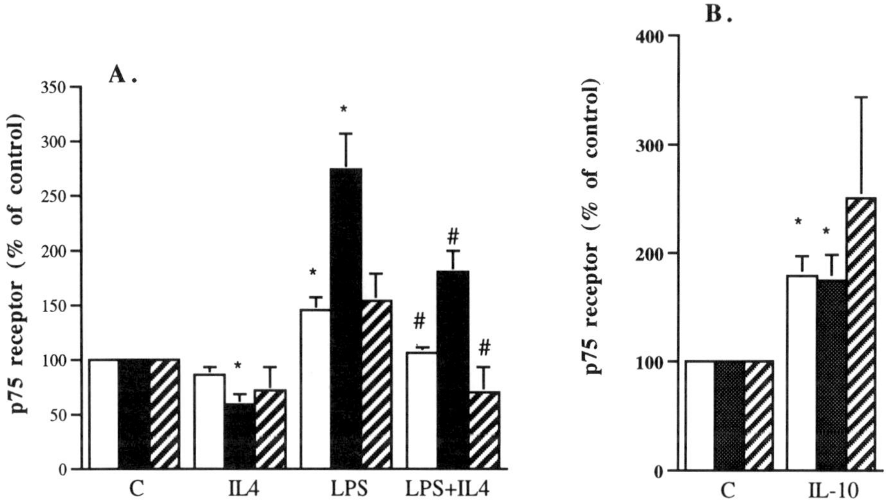

Fig. 4. The effect of IL-4 and IL-10 on p75 TNF receptor expression. Monocytes were incubated for 17 h with (A) IL-4 (10 ng/ml) in the presence or absence of LPS (500 ng/ml) and (B) IL-10 (10 ng/ml). Expression of the p75 TNF receptor on monocytes was examined by flow cytometry (open bars, n = 4–11). Alternatively, cells were examined for receptor mRNA by RT-PCR (hatched bars, n = 4). Supernatants were assayed by ELISA for soluble receptor expression (solid bars, n = 7–12). The results for each cell population were normalized to that of the control cells. The mean + SEM is shown. An asterisk denotes significantly different from control, a hash denotes significantly different from LPS, $p < 0.05$.

1992) and is recognized by the antibody UTR-1 (Dr M Brockhaus, Hoffmann-La Roche, Basel, Switzerland). Both types of receptors are shed from the cell surface, where the soluble form of each receptor inhibits TNF-α action in vitro and in vivo.

Interleukin-4 suppressed the expression by monocytes of the p75 (Fig. 4A) but not p55 TNF receptor (data not shown). Lipopolysaccharide induced the expression of the p75 receptor and enabled a more accurate assessment of IL-4′s suppressive effects (Fig. 4A). The inhibitory effects of IL-4 were determined, in part, at the mRNA level with reduced levels of p75 mRNA detected by RT-PCR in IL-4-treated monocyte populations (Fig. 4A). By ELISA, reduced levels of soluble p75 TNF receptors were detected in the supernatants of IL-4-treated cells (Fig. 4A). In contrast, IL-10 stimulated the expression of the p75 TNF receptor, as well as the release of increased levels of the soluble form of this receptor (Fig. 4B). Figure 5 shows the extent of binding of an anti-p75 antibody to synovial fluid macrophages from five patients with rheumatoid arthritis. As for monocytes, IL-4 was inhibitory and IL-10 stimulatory. Levels of soluble p75 receptor were similarly regulated (data not shown). Thus although IL-4 and IL-10 were unable efficiently to regulate TNF-α production and MHC class II expression, respectively, by synovial fluid cells, the regulation of the expression of the TNF receptors was similar on monocytes and synovial fluid cells.

It was hoped that a regulatory point for Th2 cytokines on TNF receptor expression may have been at the level of shedding, such that decreased receptor expression was

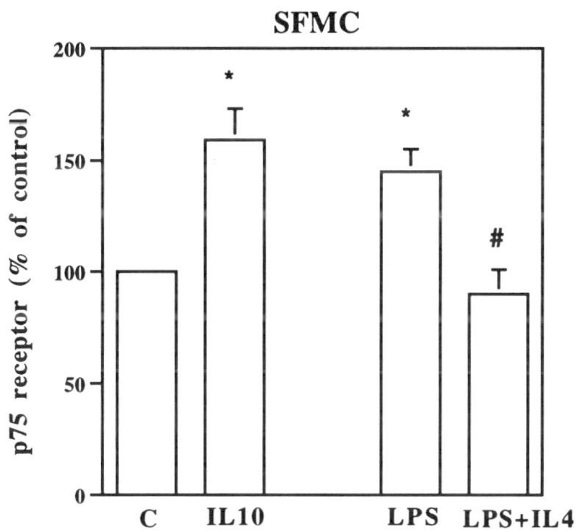

Fig. 5. The effect of IL-4 and IL-10 on p75 TNF receptor expression by synovial fluid monocytes/macrophages. Mononuclear cells isolated from the synovial fluid of 5 patients with rheumatoid arthritis were cultured with IL-10 (10 ng/ml) or LPS (500 ng/ml) ± IL-4 (10 ng/ml). The expression of the p75 TNF receptor was examined by binding of the receptor-specific antibody UTR-1 and flow cytometry. The fluorescence intensity for each cell preparation was normalized to that of the control cells. The mean + SEM is shown. An asterisk denotes significantly different from control, a hash denotes significantly different from LPS, $p < 0.05$.

seen in the same cultures as increased levels of soluble receptors. These results suggest, however, that the expression on monocytes and macrophages of TNF receptors and the shedding of soluble forms of the receptor are linked.

The Future of Th2 Cytokines in the Treatment of Inflammatory Forms of Arthritis

Table 2 summarizes the regulatory effects of IL-4, IL-10, and IL-13 on inflammatory activities of peripheral blood monocytes and synovial fluid macrophages. Initially, the potential of IL-4 as an anti-inflammatory agent was dismissed because of its inability to regulate TNF-α production. However, IL-4, and we assume IL-13, but not IL-10, can stimulate synovial fluid macrophage production of IL-1ra. IL-4, IL-10, and IL-13 can reduce synovial macrophage IL-1β production. Thus, to complement the effects of the other, these studies suggest that combined, rather than selective, use of the Th2 cytokines, particularly IL-4 and IL-10, may be beneficial in the treatment of chronic inflammatory disease such as rheumatoid arthritis.

The question remains as to why chronic inflammatory macrophages, unlike monocytes, are not responsive to IL-4 for suppression of TNF-α, and are less responsive to IL-10 for decreased expression of MHC class II. The challenge remains to understand the mechanisms involved and to question the potential use of agonists of the Th2 cytokines. To understand better the effects of Th2 cytokines on monocytes that have differentiated or been activated to different extents, we have studied peritoneal

TABLE 2. Comparison of the Regulatory Effects of IL-4, IL-10, and IL-13 on Mononuclear Cells from Peripheral Blood (PBMC) and Synovial Fluid (SFMC) from Patients with Inflammatory Arthritis

	IL-4		IL-10		IL-13	
	PBMC	SFMC	PBMC	SFMC	PBMC	SFMC
TNF-α	↓↓	-	↓↓	↓↓	↓↓	-
IL-1β	↓	↓	↓	↓	↓	↓
IL-1ra	↑↑↑	↑↑↑	↑	-	Not done	Not done
MHC class II	Not done	Not done	↓↓	↓	Not done	Not done
p75 TNF receptor	↓	↓	↑	↑	Not done	Not done

monocytes/macrophages from patients with peritonitis (see following section). We have also established an in vitro culture system in which monocytes are differentially matured and activated as a response to the cytokine environment (see below).

MACROPHAGES AT A SITE OF ACUTE INFLAMMATION: CHARACTERIZATION OF INFLAMMATORY PERITONEAL MACROPHAGES

There are few studies of the phenotype and function of human monocytes/macrophages at a site of infection. The harvest of cells from "spent" dialyzate of patients on continuous ambulatory peritoneal dialysis (CAPD) with peritonitis provides a source of cells from a site of acute inflammation. We hypothesized that peritoneal macrophages isolated from CAPD patients during peritonitis were appropriate for functional studies of the control of extravasated inflammatory cells by cytokines, drugs, and other potential anti-inflammatory agents. To support the hypothesis that such peritoneal cells were "inflammatory" and assuming that a phenotypic profile reflects functional status, expression of a large range of surface markers was investigated on cells isolated from both peripheral blood and from inflammatory dialyzates obtained from the overnight peritoneal effluent drained from CAPD patients on the morning immediately after the detection of peritonitis (manifest as "cloudy" dialyzate) and hospitalization.

Phenotypic Characterization of Inflammatory Peritoneal Macrophages

Peritoneal cells were studied from nine hospitalized CAPD patients (Hart et al., 1992). In each experiment, peripheral blood monocytes from a healthy donor were used as a direct control for macrophage expression of activation markers, as previous studies had suggested little phenotypic difference in monocytes between controls and CAPD patients (Lewis and Norris, 1990).

Peritoneal macrophages were isolated (to >90% purity) from effluents by centrifugation on two consecutive Lymphoprep density gradients. Upon selection of $CD14^+$ cells, there was increased expression by peritoneal macrophages of CD14, ICAM-1 (CD54), FcγRI (CD64), FcγRII (CDw32), FcγRIII (CD16), transferrin receptors (CD71), and tissue factor (Hart et al., 1992). Increased expression of MHC class II was marginally significant. There was no detectable expression of either the p55 (CD25) or p70 chains of the IL-2 receptor on peritoneal macrophages. The expression of the complement receptors CR1 (CD35) and CR3 (CD11b, CD18) was reduced. We were unable to reproduce the phenotypic changes on macrophages by incubation of monocytes for 16 h with "spent" peritoneal dialyzate from one of the CAPD patients with peritonitis (20% v/v), or with uremic molecules.

Many of the phenotypic changes seen on the peritoneal macrophages were suggestive of a response to IFN-γ [increased MHC class II expression (Blanchard and Djeu, 1991), increased ICAM-1 (Hutchinson et al., 1991), and increased FcγRI (Fanger et al., 1989)]. Others were suggestive of a response to GM-CSF [increased FcγRIII and FcγRII (Blanchard and Djeu, 1991)]. Other increases in marker expression may have resulted from the action of less well characterized cytokines, for example, expression of tissue factor, which is induced by macrophage procoagulant inducing factor (Ryan and Geczy, 1986). There have been inconsistent reports of the effects of inflammatory cytokines on CD14 expression. However, inflammation per se is stimulatory (Lewis and Norris, 1990).

From this study, we concluded that macrophages isolated from the peritoneal effluents of CAPD patients with peritonitis were not abnormal, but reflected a population phenotypically very similar to monocytes exposed in vitro to a selection of inflammatory cytokines and other mediators involved in immunological responses (Hart et al., 1992).

Inflammatory Peritoneal Macrophages and Peripheral Blood Monocytes: Functional Differences in Cytokine Production and Responsiveness to Th2 Cytokines

As for the phenotypic studies described above, mononuclear cells (90 ± 7% macrophages, mean ± SD, $n = 7$) were isolated from overnight (approximately 8 h) peritoneal effluents from patients on CAPD with peritonitis. Very low (<0.5 ng/10^6 cells) to undetectable levels of IL-1β or TNF-α were secreted by human peritoneal macrophages when cultured in vitro under lipopolysaccharide-free conditions. Unlike macrophages from synovial fluid, which produced relatively higher levels of TNF-α compared with blood monocytes when stimulated with lipopolysaccharide for 18 h, the peritoneal cells produced levels of TNF-α and PGE_2 that were comparable to those produced by monocytes (Hart et al., 1989b, 1991, 1995a, b). In contrast to blood monocytes, there was reduced secretion, rather than production, of IL-1β for peritoneal cells from patients on CAPD (Hart et al., 1993b). Unlike synovial fluid cells, and like monocytes, the peritoneal cells responded to IL-4 for decreased TNF-α production (Hart et al., 1991). Interleukin-4 also significantly suppressed IL-β and PGE_2 production by human peritoneal monocytes/macrophages (Hart et al., 1991).

Thus acute inflammatory cells, unlike their counterparts from a site of chronic inflammation (see above), retain the ability to respond to IL-4 for decreased TNF-α, suggesting that differentiation, rather than activation, may influence the extent of IL-4 responsiveness. To gain a greater understanding of the controlling influences on IL-4 responsiveness, a model of monocyte differentiation/activation in different in vitro environments was established.

TISSUE MACROPHAGES: CHARACTERIZATION OF ALVEOLAR MACROPHAGES

One of the principal functions of alveolar macrophages is in host defense against foreign organisms and particles that enter the body via the airways. Alveolar macrophages play a role in early inflammation by production of chemotactic signals that signal the influx of varying immune and inflammatory cells into alveoli. By secretion of bioactive lipids, alveolar macrophages can alter vascular permeability of the airway tissue. If appropriately stimulated, alveolar macrophages can secrete proteases that signal the remodeling of lung tissue (Shapiro et al., 1993). Several studies have now concluded that alveolar macrophages play an important role in the pathogenesis of asthma (for example, Catena et al., 1993).

Phenotypic Characterization of Alveolar Macrophages

The major limitation to analysis of surface markers on alveolar macrophages has been the autofluorescence expressed by these cells (Rossman and Douglas, 1988). Recent advances to overcome this problem have included the use of new fluorescent dyes, for example, a tandem dye of R-phycocrythrin linked to a cyanine fluorochrome, Cy5, that emit beyond the autofluorescence of alveolar macrophages (>600 nm) (Viksman et al., 1994). In comparison with blood monocytes from the same healthy donors, alveolar macrophages express significantly lower levels of CD11b, CD14, CD18, CD33, CD35, CD44, and HLA class I. In contrast, levels of expression of CD71, HLA-DR, and HLA-DQ are greater on alveolar macrophages (Viksman et al., 1994). In interstitial lung disease, the phenotype of alveolar macrophages varies, with increased CD14 expression by cells from patients with extrinsic allergic alveolitis, decreased CD13 expression by cells from patients with idiopathic pulmonary fibrosis, and a change to more monocyte-like cells for pulmonary cells from patients with sarcoidosis (Hoogsteden et al., 1989).

Alveolar Macrophages and Peripheral Blood Monocytes: Functional Differences in Cytokine Production and Responsiveness to Histamine

Histamine has been implicated in the inflammation, bronchoconstriction, vasodilation, and excessive mucus secretion of asthma. There is increasing evidence that inflammatory cytokines are also involved in the pathophysiology of asthma. In biopsied bronchial mucosa from asthmatics, increased TNF-α levels have been detected consistently in mast cells and macrophages (Bradding et al., 1994; Ackerman et al., 1994). Alveolar macrophages from asthmatics are activated and in contrast to blood monocytes produce in vitro higher levels of TNF-α, IL-1, IL-6, and IL-8 (Barnes, 1991).

Histamine dose-dependently suppresses TNF-α production by activated human monocytes and reflects the histamine-induced increase of intracellular cAMP levels in these cells (Vannier et al., 1991; Hart et al., 1993c). Thus there appeared an inconsistency in the alveolar compartment, especially of asthmatics, with increased levels of both histamine and TNF-α detected. In this study, the effects of histamine on TNF-α production by alveolar macrophages were examined and compared with the response by PBMC isolated from the same donors. Alveolar macrophages constituted 78% ± 11% (±SD, n = 56) of cells obtained from bronchoalveolar lavage fluid obtained from healthy male volunteers. Peripheral blood mononuclear cells and alveolar macrophages were cultured for 22 h with or without lipopolysaccharide (500 ng/ml) and histamine (10^{-5} M). Histamine alone had no effect on TNF-α production by PBMC or alveolar macrophages. However, histamine suppressed TNF-α production by lipopolysaccharide-stimulated PBMC from 1.7 ± 1.8 ng/10^6 monocytes (±SD, n = 9) to 0.2 ± 1.1 ng/10^6 cells but was without effect on lipopolysaccharide-stimulated alveolar macrophages (38.2 ± 12.3 to 38.7 ± 4.2 ng/10^6 cells, n = 12). Similar results were obtained when bioactive TNF-α levels were measured in the culture supernatants by an L929 cytotoxicity assay. Histamine was unable to regulate the production of other inflammatory cytokines (IL-1α, IL-1β, IL-6) by monocytes and macrophages. The inability of histamine to suppress TNF-α production by alveolar macrophages was unique to these cells. Histamine suppressed TNF-α production by lipopolysaccharide-stimulated synovial fluid mononuclear cells and cells isolated from the dialyzates of CAPD patients with peritonitis, that is, cells from chronically and acutely inflamed sites, respectively. Histamine also suppressed TNF-α production by decidual macrophages and macrophages isolated from breast milk, that is, macrophages from noninflamed sites. In contrast to histamine, both IL-10 and PGE_2 suppressed TNF-α production by lipopolysaccharide-stimulated alveolar macrophages.

Unlike its effect on monocytes (enriched to >90% from PBMC by elutriation) and PBMC, histamine was unable to increase cAMP levels in lipopolysaccharide-stimulated alveolar macrophages. However, upon the addition of the phosphodiesterase inhibitor 3-isobutyl-1-methyl-xanthine (IBMX, 400 μM), histamine-induced increases in cAMP were detected and suggested that alveolar macrophages may have greater levels of phosphodiesterase than do blood monocytes. The possibility of increased histamine receptor expression in the presence of IBMX was examined, but competitive radioligand binding studies with ^{3}H-histamine suggested that histamine receptors were present on alveolar macrophages and that their numbers were not increased with exposure to IBMX. This work is ongoing as we attempt to identify the reasons why alveolar macrophages are unable to accumulate cAMP in response to histamine that would signal reduced TNF-α production.

This study provides a further example of the inability to assign the same responses by the more easily accessible blood monocytes to their tissue counterparts. The lack of regulation by histamine of TNF-α production by activated alveolar macrophages may be responsible, at least in part, for the inflammatory component in the airways of asthmatics.

MONOCYTE ACTIVATION AND DIFFERENTIATION IN VITRO: A MODEL FOR STUDYING FUNCTIONAL RESPONSES BY MONOCYTES/MACROPHAGES FROM SITES OF ACUTE AND CHRONIC INFLAMMATION AND PERIPHERAL TISSUES

Studies of the regulation of cell activity by Th2 cytokines or histamine are limited by the number and heterogeneity in the synovial fluids, peritoneal dialyzates, or bronchoalveolar lavage fluids available. Furthermore, resident cells from noninflamed joints or peritoneal and pulmonary cavities are not available in sufficient numbers. We hypothesized that by culture of monocytes for 7 days in the presence of GM-CSF, M-CSF, or IL-4 compared with no cytokine, cells would activate and differentiate to a degree that was characteristic of the environment in which they were cultured. It was hypothesized that exposure to GM-CSF would assist a transition to inflammatory cells because GM-CSF is characteristic of an inflammatory pathological condition, for example, rheumatoid arthritis (Rutherford et al., 1993; Falk and Vogel, 1990). In contrast, M-CSF would provide a signal for monocyte differentiation in a simulated homeostatic tissue environment (Rutherford et al., 1993). Monocytes were cultured in IL-4 because it has been reported that IL-4 assists in the development of monocytes into antigen-presenting accessory cells (Peters et al., 1991; Sallusto and Lanzavecchia, 1994). In this study, the phenotype of the cultured cells, the level of expression of MHC class II, and the lipopolysaccharide-induced levels of TNF-α provided a measure of the biological and inflammatory status of the culture environment.

Phenotype of Cultured Cells

Monocytes from five donors (purified to >90% by elutriation) were cultured for 7 days under non-adherent conditions, and after harvest were stained with marker specific antibodies and analyzed by flow cytometry. Analysis of forward and side scatter plots showed that cells resulting from culture under all conditions were much larger than the monocytes from which they were derived. The phenotypic profile of the harvested cells is presented in Table 3 and confirms that the cells were activated and differentiated under distinct influences. Relative to cells incubated in the absence of cytokines, GM-CSF-treated cells expressed increased CD11b (CR3) and decreased CD71, the transferrin receptor. M-CSF-treated cells expressed increased levels of CD14 and decreased levels of MHC class II and CD71. In contrast, IL-4-treated cells expressed increased levels of CD11b, MHC class II, and CD71, but decreased levels of CD14. There were no differences between 7-day-cultured cells in the expression of CD16 (FcγRIII), MHC class I, or the p75 or p55 TNF receptors (Hart et al., 1995c).

TABLE 3. Phenotype of Peripheral Blood Monocytes Cultured for 7 Days in 10% Fetal Calf Serum and the Presence of GM-CSF, M-CSF, IL-4, or No Cytokine

	Fluorescence Ratio[a]			
Surface Marker (Ab)	No Cytokine	GM-CSF	M-CSF	IL-4
CD14 (FMC 32)	1.81	1.92	2.27[b]	1.30[b]
CD16 (Leu 11b)	1.15	1.07	1.09	1.02
CD11b (CR3) (OKM-1)	8.85	20.42[b]	8.75	17.52[b]
MHC class I (FMC 16)	26.72	20.32	16.01	27.15
MHC class II (FMC 52)	14.83	16.85	6.03[b]	34.68[b]
CD71 (transferrin R) (5E9C11)	4.14	2.62[b]	2.71[b]	6.26[b]
p75 TNF receptor (UTR-1)	1.68	1.66	1.76	1.65
p55 TNF receptor (HTR-9)	1.21	1.02	1.04	0.99

[a]Binding of specific Ab/binding of X63; $n = 5$.
[b]Significant ($p < 0.05$) versus control by Fisher's least significant difference test.

Functional Responses to Interleukin-4, Interleukin-10, and Histamine by Cultured Cells

Monocytes/macrophages cultured for 7 days in the absence or presence of GM-CSF, M-CSF, or IL-4 responded to IL-10, but not IL-4, for diminished lipopolysaccharide-induced TNF-α production. This finding was consistent regardless of whether the cells were cultured for 7 days in serum-free medium or medium supplemented with 10% fetal calf serum. The very same cells responded to both IL-10 and IL-4 for decreased IL-1β production, which suggested that the expression of the relevant cytokine receptors were not altered with differentiation. The responses by cultured monocytes to IL-10 and IL-4 were very similar to those previously reported for synovial fluid cells (see above), notably an inability to respond to IL-4 for decreased TNF-α production. Compared with freshly isolated monocytes, cultured monocytes (regardless of conditions) responded to IL-4 with a lesser increase in CD23 expression. However, the IL-4-induced decreases in CD14 expression were not significantly altered by differentiation/activation for 7 days in vitro.

In a further study of IL-10 responsiveness by cultured cells, monocytes cultured for 7 days in GM-CSF-containing medium expressed increased levels of MHC class II. However, these same cells responded inefficiently to IL-10 for decreased MHC class II expression (Hart et al., 1995c). Thus the cells cultured for 7 days with GM-CSF and showing increased MHC class II expression and lipopolysaccharide-induced TNF-α production can be likened to synovial fluid macrophages from patients with chronic inflammatory arthritis. Both responded poorly to IL-4 and IL-10 for decreased TNF-α and MHC class II expression, respectively. These studies suggest that the biological function in vivo of IL-4, a cytokine that is generally not considered to circulate at significant levels (Needleman et al., 1992), may not be as a regulator of

TNF-α production. In contrast, responses to IL-10 by monocytes and cultured cells were dissimilar only under more extreme inflammatory conditions and thus many of the functions of IL-10 reported for blood monocytes may be relevant in vivo in the control of tissue macrophage activity. Monocytes cultured in GM-CSF or M-CSF for 7 days retained their ability to respond to histamine for decreased TNF-α production, suggesting that conditions for differentiation of monocytes to alveolar macrophages were not optimal. Studies are ongoing using surfactant and surfactant proteins as possible environmental agents to signal the appropriate maturation of blood monocytes.

CONCLUSIONS

This review demonstrates that the regulation of production of pro-inflammatory mediators, in particular TNF-α, varies between peripheral blood monocytes and inflammatory and tissue macrophages. However, in the experiments described, subpopulations of cells were not selected for analysis. For changes with inflammation in surface marker expression, the development of subsets of cells was not seen. However, with cytokine expression, it remains unknown whether a majority or minority of the cells was responsible for the changes detected. It would be advantageous to examine the production of the inflammatory cytokines by individual cells by measurement of intracytoplasmic levels by flow cytometry (de Caestecker et al., 1992; Chikanza et al., 1992). In our studies, monocytes/macrophages from sites of acute and chronic inflammation were distinguished functionally by their response to IL-4. It is interesting that surface antigens on monocytes are being recognized that distinguish cells from early acute and chronic inflammatory conditions; for example, under inflammatory conditions the interaction of extravasating monocytes with extracellular matrix induces the expression of an antigen comprising a heterodimer of MRP8 and MRP14, two Ca^{2+}-binding proteins related to the S-100 protein family (Mahnke et al., 1995).

Synovial fluid macrophages differed from blood monocytes in their responses to Th2-type cytokines, namely IL-4, IL-10, and IL-13 (see above). In further studies, responses to IL-4 by intestinal lamina propria mononuclear cells from patients with inflammatory bowel disease (Schreiber et al., 1995) and by alveolar macrophages from patients with asthma (Chanez et al., 1994) differed from those detected for blood monocytes. Thus it remains clear that for development of immunotherapy for control of inflammation at tissue sites, the responses by blood monocytes and tissue and inflammatory macrophages must not be assumed to be similar. With differentiation and activation, the number and type of receptors for regulatory agents on monocytes may alter. Alternatively, there may be significant metabolic changes and alteration to intracellular regulators of transcription and translation. This chapter also covers the development of an in vitro model to examine such changes with monocyte differentiation under various environmental conditions; this model may help us to predict the effects of potential immunotherapeutic agents on inflammatory tissue macrophages.

REFERENCES

Ackerman V, Marini M, Vittori E, Bellini A, Vassalli G, Mattoli S (1994): Detection of cytokines and their cell sources in bronchial biopsy specimens from asthmatic patients. Chest 105:687–696.

Arend WP (1991): Interleukin 1 receptor antagonist. A new member of the interleukin 1 family. J Clin Invest 88:1445–1451.

Barnes PJ (1991): Biochemistry of asthma. Trends Biochem Sci 16:365–369.

Blanchard DK, Djeu JY (1991): Differential modulation of surface antigens on human macrophages by IFN-gamma and GM-CSF: effect on susceptibility to LAK lysis. J Leukoc Biol 50:28–34.

Bradding P, Roberts JA, Britten KM, et al. (1994): Interleukin-4, -5, and -6 and tumor necrosis factor-α in normal and asthmatic airways: Evidence for the human mast cell as a source of these cytokines. Am J Respir Cell Mol Biol 10:471–480.

Brennan FM, Gibbons DL, Mitchell T, Cope AP, Maini RN, Feldmann M (1992): Enhanced expression of tumor necrosis factor receptor mRNA and protein in mononuclear cells isolated from rheumatoid arthritis synovial joints. Eur J Immunol 22:1907–1912.

Catena E, Mazzarella G, Peluso GF, Micheli P, Cammarata A, Marsico SA (1993): Phenotypic features and secretory pattern of alveolar macrophages in atopic asthmatic patients. Monaldi Arch Chest Dis 48:6–15.

Chanez P, Vignola AM, Paul-Eugene N, et al. (1994): Modulation by interleukin-4 of cytokine release from mononuclear phagocytes in asthma. J Allergy Clin Immunol 94:997–1005.

Chikanza IC, Corrigal V, Kingsley G, Panayi GS (1992): Enumeration of interleukin-1 alpha and beta producing cells by flow cytometry. J Immunol Meth 154:173–178.

Colotta F, Re F, Muzio M, Bertini R, et al. (1993): Interleukin-1 type II receptor: a decoy target for IL-1 that is regulated by IL-4. Science 261:472–475.

Cox CA, Campbell MA, Borkedes SS, Stockley RA, Campbell EJ (1994): A discrete subpopulation of human monocytes expresses a neutrophil-like proinflammatory (P) phenotype. Am J Physiol 267:L775–L785.

Dalton BJ, Connor JR, Johnson WJ (1989): Interleukin-1 induces interleukin-1α and interleukin-1β gene expression in synovial fibroblasts and peripheral blood monocytes. Arth Rheum 32:279–287.

de Caestecker MP, Telfer BA, Hutchinson IV, Ballardie FW (1992): The detection of intracytoplasmic interleukin-1α, interleukin-1β and tumour necrosis factor α expression in human monocytes using two colour immunofluorescence flow cytometry. J Immunol Meth 154:11–20.

Del Prete G (1992): Human Th1 and Th2 lymphocytes: their role in the pathophysiology of atopy. Allergy 47:450–455.

De Waal Malefyt R, Haanen J, Spits H, et al. (1991): Interleukin 10 (IL-10) and viral IL-10 strongly reduce antigen-specific human T cell proliferation by diminishing the antigen-presenting capacity of monocytes via downregulation of class II major histocompatibility complex expression. J Exp Med 174:915–924.

Elias JA, Chien P, Gustilo KM, Schreiber AD (1985): Differential interleukin-1 elaboration by density-defined human monocyte subpopulations. Blood 66:298–301.

Elliott MJ, Maini RN, Feldmann M, et al. (1993) Treatment of rheumatoid arthritis with chimeric antibodies to tumor necrosis factor α. Arth Rheum 36:1681–1690.

Emmrich D, Gebauer F, Weseloh G, Gluckert K, Kinne R (1990): Phenotype analysis of "inflammatory" macrophages in rheumatoid arthritis by two-colour cytofluorometry and immunohistology. Agents Actions 29:95–97.

Falk LA, Vogel SN (1990): Differential production of IFN-α/β by CSF-1-and GM-CSF-derived macrophages. J Leukocyte Biol 48:43–49.

Fanger MW, Shen L, Graziano RF, Guyre PM (1989): Cytotoxicity mediated by human Fc receptors for IgG. Immunol Today 10:92–99.

Gehr G, Gentz R, Brockhaus M, Loetscher H, Lesslauer W (1992): Both tumor necrosis factor receptor types mediate proliferative signals in human mononuclear cell activation. J Immunol 149:911–917.

Hart PH, Ahern MJ, Jones CA, Jones KL, Smith MD, Finlay-Jones JJ (1993a): Synovial fluid macrophages and blood monocytes differ in their responses to IL-4. J Immunol 151:3370–3380.

Hart PH, Ahern MJ, Smith MD, Finlay-Jones JJ (1995a): Comparison of the suppressive effects of interleukin-10 and interleukin-4 on synovial fluid macrophages and blood monocytes from patients with inflammatory arthritis. Immunology 84:536–542.

Hart PH, Ahern MJ, Smith MD, Finlay-Jones JJ (1995b): Regulatory effects of IL-13 on synovial fluid macrophages and blood monocytes from patients with inflammatory arthritis. Clin Exp Immunol 99:331–337.

Hart PH, Cooper RL, Finlay-Jones JJ (1991): IL-4 suppresses IL-1β, TNF-α and PGE_2 production by human peritoneal macrophages. Immunology 72:344–349.

Hart PH, Jones CA, Finlay-Jones JJ (1992): Peritoneal macrophages during peritonitis. Phenotypic studies. Clin Exp Immunol 88:484–491.

Hart PH, Jones CA, Finlay-Jones JJ (1995c): Monocytes cultured in cytokine-defined environments differ from freshly isolated monocytes in their responses to IL-4 and IL-10. J Leukoc Biol 57:909–918.

Hart PH, Jones CA, Jones KL, Finlay-Jones JJ (1993b): Reduced secretion of IL-1β by peritoneal cells from patients on continuous ambulatory peritoneal dialysis. Immunol Cell Biol 71:99–107.

Hart PH, Jones CA, Jones KL, et al. (1993c): Cis-urocanic acid stimulates human peripheral blood monocyte prostaglandin E_2 production and suppresses indirectly tumor necrosis factor-α levels. J Immunol 150:4514–4523.

Hart PH, Vitti GF, Burgess DR, Whitty GA, Piccoli DS, Hamilton JA (1989b): Potential antiinflammatory effects of interleukin 4: Suppression of human monocyte tumor necrosis factor α, interleukin 1 and prostaglandin E_2. Proc Natl Acad Sci USA 86:3803–3807.

Hart PH, Whitty GA, Burgess DR, Croatto M, Hamilton JA (1990): Augmentation to glucocorticoid action on human monocytes by interleukin-4. Lymphokine Res 9:147–153.

Hart PH, Whitty GA, Piccoli DS, Hamilton JA (1989a): Control by IFN-γ and PGE_2 of TNF-α and IL-1 production by human monocytes. Immunology 66:376–383.

He Y-W, Malek TR (1995): The IL-2 receptor γ_c chain does not function as a subunit shared by the IL-4 and IL-13 receptors. Implication for the structure of the IL-4 receptor. J Immunol 155:9–12.

Hoogsteden HC, Van Dongen JJ, Van Hal PT, Delahaye M, Hop W, Hilvering C (1989): Phenotype of blood monocytes and alveolar macrophages in interstitial lung diseases. Chest 95:574–577.

Hutchinson P, Kraft N, Atkins RC (1991): Increased ICAM-1 expression on interferon gamma (IFNγ) stimulated mononuclear phagocytes. Cytometry 5(suppl): 108 (abstr).

Keffer J, Probert L, Cazlaris H, et al. (1991): Transgenic mice expressing human tumour necrosis factor: a predictive genetic model of arthritis. EMBO J 10:4023–4031.

Klareskog L, Farsum, U, Malmnaes Tjernlund U, Labelitz D, Wigren A (1981): Appearance of anti-HLA-DR-reactive cells in normal and rheumatoid synovial tissue. Scand J Immunol 14:183–192.

Kruys V, Marinx O, Shaw G, Deschamps J, Huez G (1989): Translational blockade imposed by cytokine-derived UA-rich sequences. Science 245:852–855.

Laszlo DJ, Henson PM, Remigio LK, et al. (1993): Development of functional diversity in mouse macrophages. Mutual exclusion of two phenotypic states. Am J Pathol 143:587–597.

Leeuwenberg JFM, Jeunhomme TMAA, Buurman WA (1994): Slow release of soluble TNF receptors by monocytes in vitro. J Immunol 152:4036–4043.

Lewis SL, Norris PJ (1990): Monocyte/macrophage function in CAPD. Contrib Nephrol 85:1–9.

Mahnke K, Bhardwaj R, Sorg C (1995): Heterodimers of the calcium-binding proteins MRP8 and MRP14 are expressed on the surface of human monocytes upon adherence to fibronectin and collagen. Relation to TNF-α, IL-6, and superoxide production. J Leukoc Biol 57:63–71.

Mosmann TR, Coffman RL (1989): T_{h1} and T_{h2} cells: different patterns of lymphokine secretion lead to different functional properties. Annu Rev Immunol 7:145–173.

Needleman BW, Wigley FM, Stair RW (1992): Interleukin-1, interleukin-2, interleukin-4, interleukin-6, tumor necrosis factor α, and interferon-γ levels in sera from patients with scleroderma. Arth Rheum 35:67–72.

O'Doherty U, Peng M, Gezelter S, et al. (1994): Human blood contains two subsets of dendritic cells, one immunologically mature and the other immature. Immunology 82:487–493.

O'Garra A, Murphy K (1994): Role of cytokines in determining T-lymphocyte function. Curr Opin Immunol 6:458–466.

Panayi GS (1993): The immunopathogenesis of rheumatoid arthritis. Br J Rheumatol 32(suppl 1):4–14.

Peters JH, Ruppert J, Gieseler RKH, Najar HM, Xu H (1991): Differentiation of human monocytes into CD14 negative accessory cells: Do dendritic cells derive from the monocytic lineage? Pathobiology 59:122–126.

Rigley KP, Thurstan SM, Callard RE (1991): Independent regulation of interleukin 4 (IL-4)-induced expression of human B cell surface CD23 and IgM: functional evidence for two IL-4 receptors. Int Immunol 3:197–203.

Rossman MD, Douglas SD (1988): The alveolar macrophage: receptors and effector cell function. In Daniele RP (ed). *Immunology and Immunologic Diseases of the Lung.* Chicago: Blackwell Scientific Publishers, pp 167–187.

Rutherford MS, Witsell A, Schook LB (1993): Mechanisms generating functionally heterogeneous macrophages: chaos revisited. J Leukoc Biol 53:602–618.

Ryan J, Geczy CL (1986): Characterization and purification of mouse procoagulant-inducing factor. J Immunol 137:2864–2870.

Salgame P, Abrams JS, Clayberger C, et al. (1991): Differing lymphokine profiles of functional subsets of human CD4 and CD8 T cell clones. Science 254:279–282.

Sallusto F, Lanzavecchia A (1994): Efficient presentation of soluble antigen by cultured human dendritic cells is maintained by granulocyte/macrophage colony-stimulating factor plus interleukin 4 and downregulated by tumor necrosis factor α. J Exp Med 179:1109–1118.

Schreiber S, Heinig T, Panzer U, et al. (1995): Impaired response of activated mononuclear phagocytes to interleukin 4 in inflammatory bowel disease. Gastroenterology 108:21–33.

Shapiro SD, Kobayashi DK, Ley TJ (1993): Cloning and characterization of a unique elastolytic metalloproteinase produced by human alveolar macrophages. J Biol Chem 268:23824–23829.

Tarling JD, Lin H-S, Hsu S (1987): Self-renewal of pulmonary alveolar macrophages: evidence from radiation chimera studies. J Leukoc Biol 42:443–446.

Tartaglia LA, Goeddel DV (1992): Two TNF receptors. Immunol Today 13:151–153.

Thomas R, Lipsky PE (1994): Human peripheral blood dendritic cell subsets. Isolation and characterization of precursor and mature antigen-presenting cells. J Immunol 153:4016–4028.

Vannier E, Miller LC, Dinarello CA (1991): Histamine suppresses gene expression and synthesis of tumor necrosis factor α via histamine H2 receptors. J Exp Med 174:281–284.

Vassalli P (1992): The pathophysiology of tumor necrosis factors. Annu Rev Immunol 10:411–452.

Viksman MY, Liu MC, Schleimer RP, Bochner BS (1994): Application of a flow cytometric method using autofluorescence and a tandem fluorescent dye to analyze human alveolar macrophage surface markers. J Immunol Methods 172:17–24.

Williams RO, Feldmann M, Maini RN (1992) Anti tumor necrosis factor ameliorates joint disease in murine collagen-induced arthritis. Proc Natl Acad Sci USA 89:9784–9788.

15

The Role of Monocytes/Macrophages in Wound Healing

David G. Greenhalgh
Shriners Hospitals for Children, Northern California, Sacramento, California

One may ask, why do we heal? Just because one suffers a wound, why does the body proceed with tissue repair? Clearly, the pain felt with injury triggers a response, but insensate wounds heal just as well. Why do some wounds heal rapidly while other wounds remain chronically open? Why do some people heal with minimal scarring whereas others with similar wounds develop keloids? These questions have been asked for centuries. The answers are still not completely known. With the ability to examine wounds microscopically, evidence started to suggest that the cells themselves must play a role in the regulation in tissue repair. This chapter examines how one group of cells, the monocytes and macrophages, contributes to the healing of a wound. For the sake of simplicity, the distinction between monocytes and macrophages is ignored here. During tissue repair, resident macrophages are activated. Similarly, monocytes are recruited into the wound. Distinguishing the various cell types is very difficult and will not be attempted. Once macrophages are present in the wound, the difference in origin is probably not of major significance.

Wound healing involves set patterns that are consistent for all types of tissue repair. Three types of tissue repair occur in all wounds: (1) creation of a scar, (2) re-epithelialization, and (3) contraction. Although all three types of tissue repair occur simultaneously, they are usually considered separately. Scar formation also occurs in

Phagocyte Function: A Guide for Research and Clinical Evaluation, Edited by J. Paul Robinson and George F. Babcock. ISBN 0-471-12364-1

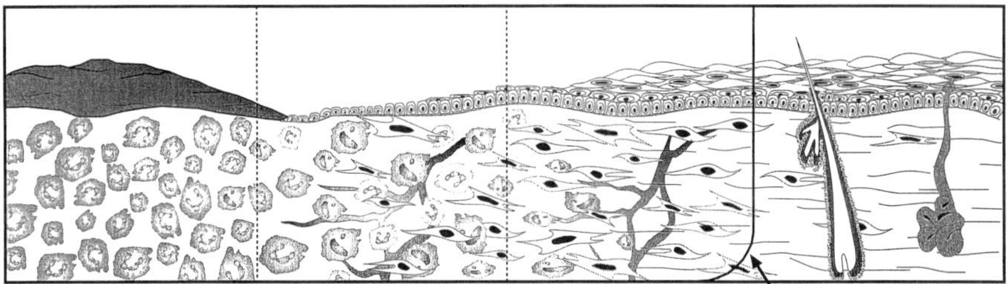

Fig. 1. Stages in scar formation during wound healing.

consistent phases that were originally descriptive for an incision: (1) inflammatory (lag) phase, (2) proliferative (collagen) phase, and (3) maturation phase. The duration of each phase depends on the type and condition of wound. In an incision, for example, the inflammatory phase lasts only a few days. In a large, ungrafted burn wound, the inflammatory phase lasts as long as the wound remains uncovered by epithelium, which can be a matter of months if no treatment is instituted. For a large wound, there is a continuum of each of these phases. In the center and open part of the wound, inflammation is active. Away from the center of the wound, collagen deposition is actively proceeding, and scar maturation is occurring at the wound edge (Fig. 1). Certain events must occur for one phase to end and the next proceed. Macrophages have been found to partake in the initiation and termination of each of the phases of tissue repair.

There is also a clear pattern for which cell type predominates in a wound. Repeated studies have documented that neutrophils are the dominant cell in the first few days after injury (Ross and Benditt, 1961; Schafer et al., 1994). Macrophages are the most common cell type between 3 and 10 days in small wounds. Lymphocyte populations peak a little after macrophages. Fibroblasts tend to dominate as neovascularization occurs, a necessary event because synthesis of the extracellular matrix requires oxygen and nutrients. Eventually, with maturation, cellularity and vascularity diminish to lead to the ultimate scar.

In the 1970s, investigators tried to determine which cell types were necessary for regulating the healing process. Simpson and Ross (1972) eliminated neutrophils from wounds and demonstrated that wounds healed normally unless they developed frank infections. These findings suggested that neutrophils do not play a major role in regulating tissue repair; instead, they probably are more essential for defense against invading organisms. When macrophages were eliminated, however, Leibovich and Ross (1975) clearly demonstrated that healing was significantly impaired. Their landmark paper has since been supported by many investigations that suggest that macrophages are one of the major regulators of the healing response (Diegelmann et al., 1981; DiPietro, 1995). When macrophages were found to play a role in wound healing, several studies were performed to determine whether macrophage stimulation would augment tissue repair. Some investigators reported that systemic macrophage stimulation would lead to augmented wound healing (Hunt et al., 1984; Danon et al., 1989) but not all investigators found that nonspecific macrophage stimulated healing (Greenhalgh and Gamelli, 1987). Macrophage activation was found to impair healing in at least one study (Greenhalgh et al., 1986). Others have demonstrated that lymphocytes have profound influences on tissue repair.

The current philosophy is that there is an active communication between all cell types in the wound that lead to the healing response. Earlier investigations focused on how growth factors and cytokines turned on the healing response. As time progressed, investigators realized that healing involved not only proliferation of cells and synthesis of proteins, but also elimination of cells and degradation of extracellular matrix proteins. The determination of how extracellular messages are transferred to intracellular responses has been a focus of recent work. Macrophages play an integral part in all of the stimulatory and inhibitory phases of tissue repair. The role of macrophages in the initiation and termination of each phase of tissue repair is covered here. It is important to remember that macrophages have a multiplicity of tasks that frequently appear to have opposing actions. One should not think of macrophages as one cell type but instead as a class of cells that have different functions that vary depending on the environmental stimuli. Although a great deal of information is known about different types of lymphocytes, little is known about different classes of macrophages. The different activities probably are as much a response to changes in conditions of the wound as to the presence of different cell subclasses. What causes the different responses or subclasses probably is the result of interactions with different types of cytokines, wound conditions, and extracellular matrix.

MACROPHAGES IN THE INFLAMMATORY PHASE OF WOUND HEALING

Macrophages are recruited and activated almost as soon as injury has occurred. Products of tissue damage and local mediators are released to attract inflammatory cells. The vascular response, vasodilatation and increased permeability, facilitates migration into the injured site. These vessel wall functions are the result of many different mediators such as norepinephrine, histamines, bradykinins, serotonin, and many other cellular products. Although in the past it was felt that the vasculature has a passive role, evidence now clearly indicates that the endothelial cells play a predominant role in directing inflammatory cell traffic. Although adhesion molecules are activated on inflammatory cells, they are also expressed on endothelial cells to help target leukocytes to the site of injury (Jutila, 1994).

Monocytes are attracted to the wound, along with other inflammatory cells, by the release of various mediators. Not only are these mediators released by resident cells but many growth factors are also released from the alpha granules of platelets as part of platelet aggregation in the wound. As each cell type enters the wound, it in turn releases various cytokines to attract more cells. Chemoattractants for monocytes and macrophages include plasma protein products (C5, fibrinopeptides, IgG fragments, α-thrombin), many cellular products [transforming growth factor-β (TGF-β), platelet-derived growth factor (PDGF), the inflammatory cytokine *gro* (Iida and Grotendorst, 1990), monocyte-chemoattractant proteins, macrophage inflammatory proteins, platelet factor 4, leukotriene B_4], extracellular matrix protein fragments (fragments of collagen, elastin, fibronectin), and bacterial products (endotoxin, fMLP) (Riches, 1996). Although many of macrophages are attracted from distant sites by chemoattractants, resident macrophages also are stimulated to proliferate and in some tissues are the predominant wound macrophages.

After arriving at the wound site, monocytes and macrophages are activated by various mediators. Activation is necessary for the majority of macrophage functions. As for recruitment, many bacteriologic by-products, such as endotoxin and fMLP, activate macrophages. Lipopolysaccharide interacts with lipopolysaccharide-binding protein (LBP) and together they bind to the cell surface CD14 receptor (Sweet and Hume, 1996). Binding to the CD14 receptor leads to activation of various cell signaling pathways. Conditions in the wound may also activate macrophages. Extracellular matrix protein fragments and other debris, including nonviable cells, are involved in activation. Low tissue oxygen, low pH, and elevated lactate levels stimulate macrophage cytokine release that leads to increase angiogenesis (Knighton et al., 1983; Jensen et al., 1986). Many different cytokines and growth factors act as switches to stimulate different macrophage activities.

An initial and essential function of macrophages/monocytes is for nonspecific immune activity. Macrophages are one class of many inflammatory cells that directly attack invading organisms. As for other leukocytes, macrophages release cytotoxic agents (such as oxygen free radicals and proteases) that help with the killing of microorganisms. Unfortunately, these cytotoxic agents are not selective in what they destroy and may lead to local tissue destruction that extends the depth of injury. With excessive injury, as for an extensive burn, the release of cytotoxic agents and pro-inflammatory mediators (cytokines) contributes to the systemic inflammatory response syndrome (SIRS) and multiple organ failure (MOF) (Mateo et al., 1994; Fong and Lowry, 1996). Macrophages play an important role in phagocytosis of microorganisms and other nonessential debris. They also are essential for specific immune function. Macrophages participate as antigen presenting cells, an essential role for both humoral and cell-mediated immunity.

During the inflammatory phase of healing, macrophages release many cytokines and growth factors that attract fibroblasts and stimulate angiogenesis. Fibroblasts and invading endothelial cells are required for the second phase of healing, the proliferative or collagen phase. To assist with migration, Brown and co-workers (1993) have demonstrated that macrophages are also capable of producing embryonic fibronectins that may produce a better environment for cellular migration. For effective progression to the proliferative phase of healing, the inflammatory response must be down-regulated. In a closed incision, there is a marked decrease in inflammatory cells after 4–5 days. In an open wound, such as a burn, inflammation continues as long as the wound remains exposed to the environment. If one examines the open wound histologically, the different phases of healing are seen to co-exist. In the center of the wound, active inflammation predominates. Toward the wound edge collagen deposition appears to increase. In areas covered with an epithelium, wound maturation dominates. These findings, along with the observation that wounds covered with a graft have much less inflammation, suggest that the migrating keratinocytes play a role in downregulating inflammatory cells. One may ask what happens to inflammatory cells when inflammation resolves? It does not appear that inflammatory cells return to the circulation. Instead, it appears that keratinocytes may induce apoptosis (programmed cell death) in wound inflammatory cells. Studies in our laboratory suggest that inflammatory cell apoptosis follows the leading edge of keratinocyte as it migrates across the wound surface (Brown et al., 1997). The signal produced by keratinocytes to induce apoptosis in inflammatory cells has still to be elucidated.

Macrophages are also involved in the breakdown of the provisional matrix of fibrin and fibronectin. Fibrin and other early matrix proteins must be removed to allow for the development of a more permanent collagenous scar. In a carefully performed study, Schafer et al. (1994) determined which cells in a healing wound were involved in early fibrinolytic activity. Little activity was found in neutrophils; however, macrophages/monocytes were found to express urokinase-type plasminogen activator (uPA) and the uPA receptor. Fibroblasts also expressed uPA but lacked the uPA receptor. The activity peaked at 6–10 days, the time one would expect healing to shift from the inflammatory phase to collagen synthesis (proliferative phase). Eliminating the initial fibrinous matrix, an essential step for the normal progression of healing, appears to be regulated by macrophages.

MACROPHAGES IN THE PROLIFERATIVE PHASE OF WOUND HEALING

There are two major functions during the proliferative phase: deposition of collagen and angiogenesis. Macrophages produce growth factors that stimulate fibroblast chemotaxis and proliferation and extracellular matrix deposition. Prior to the identification of specific growth factors, several investigators attempted to stimulate healing by the nonspecific activation of macrophages. In 1980, Leibovich and Danon demonstrated that glucan, a product of yeast cell walls, augmented healing by increasing early macrophage infiltration into a wound. These findings were confirmed in 1988 by Browder et al., who improved incisional strength with systemic treatment with glucan. Not all stimulators of phagocytic activity have improved healing. Systemic stimulation with muramyl dipeptide, the active component of mycobacterium cell wall, did not affect incisional strength (Greenhalgh and Gamelli, 1987), and the immunomodulator *Corynebacterium parvum* impaired incisional strength (Greenhalgh et al., 1986). These studies suggest that macrophages can be activated to stimulate or inhibit tissue repair.

To support the concept that macrophages augment tissue repair, several investigators have isolated macrophages from wounds to examine their in vitro activities. The "wound macrophages" were found to have activities that differed from monocytes isolated from serum or macrophages isolated from other sites (Fukasawa et al., 1988,

1990). As early as 1978, Leibovich demonstrated that macrophages released a product that increased fibroblast proliferation. Several investigators have since found confirming data that reveal that macrophages have the potential to attract fibroblasts into the wound, stimulate their proliferation, and increase collagen synthesis (Greenhalgh et al., 1990).

The in vitro studies stimulated the great interest in isolating and purifying macrophage by-products for clinical use. At the same time, recombinant techniques were developed that allowed the mass production of growth factors. Growth factors are cytokines that specifically function, at least in part, as mitogenic stimulators. Macrophages have been found to release numerous growth factors that augment fibroblast activities, including platelet-derived growth factors (PDGF), fibroblast growth factors (FGF), insulin-like growth factors (IGF), epidermal growth factor/transforming growth factor-α (EGF/TGF-α), transforming growth factor-β (TGF-β), and many others. The list of growth factors and cytokines that are released from macrophages increases at all times. Almost all of the growth factors have been applied to wounds, especially wounds with impaired healing, and have resulted in improvement in all forms of tissue repair (Greenhalgh et al., 1990, 1993; Brown et al., 1994).

The second component of the proliferative phase of tissue repair is angiogenesis. Early after injury, the wound is a profoundly hypoxic and acidic environment (Knighton et al., 1983; Jensen et al., 1986). Extracellular matrix synthesis requires not only oxygen but also nutrients to fuel the high metabolic needs for tissue repair (Niinikoski, 1977; Jonsson, 1991). Most cell types can not tolerate such an environment. Macrophages tolerate hypoxia and acidosis. The addition of wound macrophages to corneal tissue leads to angiogenesis (Knighton et al., 1983). Hypoxia and elevated lactate levels have also been found to stimulate the release of angiogenesis factors. The release of these angiogenesis factors leads to the in-growth of new capillaries, which in turn allows for extracellular matrix synthesis.

Macrophages do not always augment tissue repair. In 1980, Korn et al. discovered that supernatants from both unstimulated and activated cultured mononuclear cells resulted in suppressed fibroblast proliferation. In 1983, Hart et al. found that activated peripheral blood mononuclear cells inhibited fibroblast collagen synthesis. Macrophages also release collagenases and elastases that break down the extracellular matrix. It is apparent that macrophages play a role in augmenting and downregulating the healing process.

Just as macrophages are involved in the shift from the inflammatory to the proliferative phases of tissue repair, they appear to participate in the transition from the proliferative to the maturation phase of tissue repair. Desmouliere et al. (1995) reported that fibroblasts and vascular cells undergo apoptosis when the wound closes. The source of the message that "tells" the fibroblasts to initiate apoptosis is not known but may involve macrophages. Even if they do not initiate the signal for apoptosis, macrophages do play a major role in programmed cell death. Macrophages phagocytize the cellular fragments of apoptotic cells in a process that does not incite inflammation.

MACROPHAGES IN THE MATURATION PHASE OF WOUND HEALING

Wound maturation is marked by a balance between collagen synthesis and degradation. This balance is essential because any imbalance leads to pathological healing. Too little collagen deposition or excessive breakdown leads to wound breakdown and infectious complications. Scurvy is the classic example, where collagen synthesis is impaired whereas collagenase activity persists, leading to healing failure and dehiscence of previously healed wounds. If the balance is shifted to greater collagen deposition, then manifestations of excessive scarring occur. In the skin wound, hypertrophic scars or keloids result. In other tissues, fibrosis, cirrhosis, arthritis, or other pathological changes lead to significant morbidity.

As previously mentioned, macrophages produce the messages to upregulate fibroblast proliferation and extracellular matrix production. Macrophages also participate in the other side of the collagen synthesis/breakdown regulation scheme. Macrophages have been found to produce various collagenases after activation with various agents, including endotoxin and other inflammatory mediators (Werb and Gordon, 1975b; Wahl et al., 1974). Wahl and co-workers demonstrated that collagenase synthesis was regulated by prostaglandins and cyclic adenosine monophosphate (cAMP) (McCarthy et al., 1980; Wahl et al., 1985). Werb and Gordon (1975a) also found that activated macrophages released elastase. These findings clearly support the supposition that macrophages play a key role in the regulation of wound maturation.

An obvious question arises. Why do some macrophages stimulate healing and collagen deposition, whereas others appear to downregulate the process? No one has answered this provocative question. Studies clearly show how complicated the healing process is. In a healing wound, cells interact with hundreds of different cytokines and growth factors. The interactions of cells with the various types of extracellular matrix also lead to different cellular responses. The different activities of the cells leads one to speculate that macrophages have different populations with opposing functions. This hypothesis parallels the lymphocyte experience with many subsets of helper, suppressor, null, natural killer, and other populations. Initially lymphocytes were divided into B and T cells. With more sophisticated detection techniques, many subclasses were identified. In studies with macrophages, investigators have not attempted to separate these phagocytes with the same degree of sophistication. It may very well be that many subclasses of macrophages exist if one were to look for them. The other possibility is that the macrophage is a very adaptable cell that changes function in different environments or after different stimuli. Future studies will clarify these concerns.

CONCLUSION

Macrophages are attracted to a wound from the very onset of injury. These cells are involved in all phases of tissue repair and appear to assist with the transition from one phase to another. The potential of manipulating macrophages to alter the healing process (either enhance wound closure or prevent scarring) has been investigated for 30–40 years. Although a greater understanding of the role of macrophages has been gained, our ability to control healing through nonspecific immune modulation, application of growth factors, or downregulating scarring through the use of steroids has been only mildly successful. A great number of unanswered questions regarding the role of macrophages remain.

ACKNOWLEDGMENT

These studies were supported by grants from the National Institutes of Health (GM50959) and The Shriners of North America.

REFERENCES

Browder W, Williams D, Lucore P, Pretus H, Jones E, McNamee R (1980); Effect of enhanced macrophage function on early wound healing. Surgery 104:224–230.

Brown DL, Kao WWY, Greenhalgh DG (1997): Apoptosis down-regulates inflammation under the advancing epithelial wound edge: Delayed patterns in diabetes and improvement with topical growth factors. Surgery 121:372–380.

Brown LF, Dubin D, Lavigne L, Logan B, Dvorak HF, van de Water L (1993): Macrophages and fibroblasts express embryonic fibronectins during cutaneous wound healing. Am J Pathol 142:793–80.

Brown RL, Breeden MP, Greenhalgh DG (1994): PDGF and TGF-α act synergistically to improve healing in the genetically diabetic mouse. J Surg Res 56:562–570.

Danon D, Kowatch MA, Roth GS (1989): Promotion of wound repair in old mice by local injection of macrophages. Proc Natl Acad Sci USA 86:2018–2020.

Desmouliere A, Redard M, Darby I, Gabbiani G (1995): Apoptosis mediates the decrease in cellularity during the transition between granulation tissue and scar. Am J Pathol 146:56–66.

Diegelmann RF, Cohen IK, Kaplan AM (1981): The role of macrophages in wound repair: A review. Plast Reconstr Surg 63:107–113.

DiPietro LA (1995): Wound healing: The role of the macrophage and other immune cells. Shock 4:233–240.

Fong Y, Lowry SF (1996): Cytokines and the cellular response to injury and infection. In Wilmore DW, Cheung LY, Harken AH, Holcroft JW, Meakins JL (eds), *American College of Surgeons—Care of the Surgical Patient.* New York: Scientific American, Inc., pp 1–21.

Fukasawa M, Bryant SM, DiZerega GS (1988): Incorporation of thymidine by fibroblasts: Evidence for complex regulation by postsurgical macrophages. J Surg Res 45:460–466.

Fukasawa M, Campeau JD, Yanagihara DL, Rodgers KE, DiZerega GS (1990): Regulation of proliferation of peritoneal tissue repair cells by peritoneal macrophages. J Surg Res 49:81–87.

Greenhalgh DG, Gamelli RL (1987): Immunomodulators and wound healing. J Trauma 27:510–514.

Greenhalgh DG, Gamelli RL, Foster RS, Jr, Chester A (1986): Inhibition of wound healing by *Corynebacterium parvum.* J Surg Res 41:209–214.

Greenhalgh DG, Hummel RP, III, Albertson S, Breeden MP (1993): The synergistic actions of platelet-derived growth factor and the insulin-like growth factors *in vivo.* The enhancement of tissue repair in genetically diabetic mice. Wound Rep Reg 1:69–81.

Greenhalgh DG, Sprugel KH, Murray MJ, Ross R (1990): PDGF and FGF stimulate healing in the genetically diabetic mouse. Am J Pathol 136:1235–1246.

Hart PH, Powell LW, Cooksley WGE, Halliday JW (1983): Mononuclear cell factors that inhibit fibroblast collagen synthesis. Scand J Immunol 18:51–58.

Hunt TK, Knighton DR, Thakral KK, Goodson WH, III, Andrews WS (1984): Studies on inflammation and wound healing: Angiogenesis and collagen synthesis stimulated in vivo by resident and activated wound macrophages. Surgery 96:48–54.

Iida N, Grotendorst GR (1990): Cloning and sequencing of a new gro transcript from activated human monocytes: Expression in leukocytes and wound tissue. Mol Cell Biol 10:5569–5599.

Jensen JA, Hunt TK, Scheunstuhl H, Banda MJ (1986): Effect of lactate, pyruvate, and pH on secretion of angiogenesis and mitogenesis factors by macrophages. Lab Invest 54:574–578.

Jonsson K, Jenson JA, Goodson WH III, et al. (1991): Tissue oxygenation, anemia, and perfusion in relation to wound healing in surgical patients. Ann Surg 214:605–613.

Jutila MA (1994): Function and regulation of leukocyte homing receptors. J Leukoc Biol 55:133–140.

Knighton DR, Hunt TK, Scheunstuhl H, Halliday BJ, Werb Z, Banda MJ (1983): Oxygen tension regulates the expression of angiogenesis factor by macrophages. Science 221:1283–1285.

Korn JH, Halushka PV, LeRoy EC (1980): Mononuclear cell modulation of connective tissue function. J Clin Invest 65:543–554.

Leibovich SJ (1978): Production of macrophage-dependent fibroblast-stimulating activity (M-FSA) by murine macrophages. Exp Cell Res 113:47–56.

Leibovich SJ, Danon D (1980): Promotion of wound repair in mice by application of glucan. J Reticuloendothel Soc 27:1–11.

Leibovich SJ, Ross R (1975): The role of the macrophage in wound repair. A study with hydrocortisone and antimacrophage serum. Am J Pathol 78:71–100.

Mateo RB, Reichner JS, Albina JE (1994): Interleukin-6 activity in wounds. Am J Physiol 266:R1840–R1844.

McCarthy JB, Wahl SM, Rees JC, Olsen CE, Sandberg AL, Wahl LM (1980): Mediation of macrophage collagenase production by 3′-5′ cyclic adenosine monophosphate. J Immunol 124:2405–2409.

Niinikoski J (1977): Oxygen and wound healing. Clin Plast Surg 4:361–374.

Riches DWH (1996): Macrophage involvement in wound repair, remodelling, and fibrosis. In Clark RAF (ed), *The Molecular and Cellular Biology of Wound Repair.* 2nd ed. New York: Plenum Press, pp 95–141.

Ross R, Benditt EP (1961): Wound healing and collagen formation. J Biophys Biochem Cytol 11:677–700.

Schafer BM, Maier K, Eickhoff U, Todd RF, Kramer MD (1994): Plasminogen activation in healing human wounds. Am J Pathol 144:1269–1280.

Simpson DM, Ross R (1972): The neutrophilic leukocyte in wound repair: A study with antineutrophil serum. J Clin Invest 51:2009–2023.

Sweet MJ, Hume DA (1996): Endotoxin signal transduction in macrophages. J Leukocyte Biol 60:8–26.

Wahl LM, Wahl SM, Mergenhagen SE, Martin GR (1974): Collagenase production by endotoxin-activated macrophages. Proc Natl Acad Sci USA 71:3598–3601.

Wahl SM, Wahl LM (1985): Regulation of macrophage collagenase, prostaglandin, and fibroblast-activating-factor production by anti-inflammatory agents: Different regulatory mechanisms for tissue injury and repair. Cellular Immunol 92:302–312.

Werb Z, Gordon S (1975a): Elastase secretion by stimulated macrophages. J Exp Med 142:361–377.

Werb Z, Gordon S (1975b): Secretion of specific collagenase by stimulated macrophages. J Exp Med 142:346–360.

16

Monocytes and Macrophages in Solid Organ Transplantation: Important Implications for the Future of Transplant Immunology

John F. Valente
University of Cincinnati College of Medicine, Cincinnati, Ohio

Cora K. Ogle
University of Cincinnati College of Medicine and Shriners Burns Institute, Cincinnati, Ohio

J. Wesley Alexander
University of Cincinnati, College of Medicine, Cincinnati, Ohio

The central role of the monocyte/macrophage as a phagocyte, antigen-presenting cell, and regulator of immune responses is well described, and is perhaps such an accepted premise that the dynamic impact of cells from the monocyte/macrophage line in transplantation immunology has been overlooked. The macrophage, however, appears to exert a surprising degree of influence over several important phenomena of transplant immunology, including tolerance, chronic allograft rejection, and the transfusion effect. In this chapter, we briefly review the accepted, traditional role of the macrophage in transplantation immunology. We then discuss some of the evidence

Phagocyte Function: A Guide for Research and Clinical Evaluation, Edited by J. Paul Robinson and George F. Babcock. ISBN 0-471-12364-1

supporting a greater role for the macrophage in chronic rejection (a cause of graft loss not altered by current therapy) and in the induction of tolerance (perhaps the most important goal facing transplant immunologists and physicians in the next decade). We conclude with a discussion of some of the clinical research protocols in which an application of flow cytometric analysis to the study of macrophage phenotype and function may become relevant.

MACROPHAGE DEVELOPMENT AND DIVERSITY

Antigen-presenting cells (APCs), especially macrophages and dendritic cells, are of central importance in transplantation. Recent evidence suggests that these APCs, along with granulocytes, are derived from a common myeloid precursor in the bone marrow, and that all three cell lines develop under the direction of granulocyte/macrophage colony-stimulating factor (GM-CSF) (Inaba et al., 1993, Szabolcs et al., 1996). The production of macrophages from bone marrow precursor cells is a dynamic and highly regulated process. The differentiation of myeloid lineage-restricted progenitor cells is initially controlled by interleukin (IL)-3, which drives the progenitors along a maturation pathway that promotes responsiveness to specific lineage-restricted colony-stimulating factors (CSFs) (Rutherford and Schook, 1992). IL-6 activates stem cells from the G0 cell cycle phase to the G1 phase, and IL-3 then further drives these cells to proliferation (Ikebuchi et al., 1987). The CSFs determine the specific mature cell type produced. Production of monocytes/macrophages from bone marrow progenitor cells is normally controlled by CSF-1 or macrophage (M)-CSF. Different functional phenotypes (based on patterns of mRNA transcription) have been observed for M-CSF- and GM-CSF-derived macrophage colonies (Witsell and Schook, 1991), as well as different surface markers (DeNichilo and Burns, 1993). Granulocyte/macrophage CSF upregulates the M-CSF receptor, and stimulation through this receptor drives cells from the G1 cell cycle phase to the S-phase. This, in part, explains the synergistic effects on bone marrow progenitor cell growth shown to occur between different CSFs. Such synergy is also noted between CSFs and tumor necrosis factor (TNF), IL-1, and IL-6. Numerous other factors contribute to the development of monocyte/macrophage cell regulatory functions, including IL-3, GM-CSF, G-CSF, IL-13, transforming growth factor beta (TGF-β), and the prostaglandins. The combination of IL-3 and GM-CSF in particular appears to elicit a myeloid progenitor cell with natural suppressor activity. Inflammatory mediators such as TNF-α, IL-6, and even lipopolysaccharide (which acts partially through CD14 stimulation) also have a profound influence on the regulatory functions of the derived myeloid cells. Thus a functional link is formed between the regulation of myelopoiesis and immune responses through the production of these cytokines. Therefore, it becomes important to distinguish between normal or "steady-state" myelopoiesis and responsive or "induced" myelopoiesis during an inflammatory response.

Once released from the bone marrow, monocytes circulate with a half-life of 1–3 days, prior to migration into peripheral sites where they become fixed tissue macrophages. Differences in peripheral blood monocytes following administration of

M-CSF or GM-CSF have also been shown (Schmid et al., 1995). Both CSFs produced monocytes with the usual markers and an increased cell size, but a subpopulation of $CD14^{bright}CD16^{-}$ cells was unregulated for CD14 by M-CSF, but downregulated for CD14 by GM-CSF. Furthermore, although GM-CSF was superior in stimulating in vitro antibody-dependent cytotoxicity, monocytes derived under the influence of M-CSF were far more cytotoxic than those derived by GM-CSF. During inflammation, margination and migratory events are mediated by a combination of well described chemoattractant and adhesion molecule effects, but the specific control and degree of predetermination (if any) of monocyte destiny under normal circumstances is poorly defined. Once macrophages become residents of a particular tissue, they develop locally specific responses depending on their microenvironment and degree of activation. For instance, alveolar macrophages are normally suppressive, requiring a stimulus (e.g., GM-CSF) to induce inflammation, whereas peritoneal macrophages are essentially exudative cells. Kupffer cells, microglia cells, and Langerhans cells exemplify the functional and phenotypic specialization undergone by monocytes as they become fixed tissue macrophages (Rutherford et al., 1993).

The development of dendritic cells in the bone marrow is less well defined, but may involve the influence of a number of cytokines with hematopoietic potential, especially GM-CSF in combination with TNF-α (Santiago-Schwarz et al., 1992) or IL-4 (Kiertscher and Roth, 1996). Often recognized as potent activators of naive lymphocytes, dendritic cells (like macrophages) are a diverse family of cells whose ultimate functions depend on tissue site and cell surface markers (Hill et al., 1994). In this regard, dendritic cells express a degree of functional diversity similar to that of macrophages. Certain dendritic cells may engender Th2 responses (or actually be suppressive), possibly due to the type and degree of co-stimulatory molecule expression, a feature also shared by their myeloid lineage cousin, the macrophage.

MACROPHAGE ANTIGEN-PRESENTING CELL FUNCTIONS AND RECRUITMENT DURING GRAFT INFLAMMATION

Two classical pathways of T cell activation exist. The route common to most responses is the indirect pathway, where alloantigen is expressed, following phagocytosis, as a short peptide in conjunction with the major histocompatibility antigens (MHC) on the surface of a macrophage/APC and presented to a T lymphocyte. In transplant immune responses a direct pathway involving donor-derived APCs expressing foreign MHC to recipient T cells also exists.

Because macrophages and dendritic cells bear both MHC-I and MHC-II, they utilize both proteosome (ubiquitin-labeled internal proteins) and lysosome (digestion of ingested external antigens) enzymatic machinery. Antigens are expressed in the cleft of the MHC molecules and engage the T cell via its T-cell receptor (TCR) along with important co-stimulatory signals (see next section). Once selected and activated, the allospecific T-cell clones proliferate and assault the allograft. During this rejection process, other cells are recruited as part of the inflammatory infiltrate and are themselves activated by endogenous and exogenous signals. Certain factors, such as

RANTES, are expressed by activated T cells and can recruit macrophages to sites of inflammation, drawing an important link between allospecific T cell responses and nonallospecific macrophage responses. Once recruited to a site (e.g., allograft) of inflammation, macrophages are quickly activated. One mediator commonly associated with allograft rejection is IFN-γ, secreted by T cells, and perhaps one of the most potent activators of macrophages. Thus both the recognition and the rejection of foreign tissue are the result of a progressive and cooperative interaction between macrophages and lymphocytes.

ROLE OF THE MACROPHAGE IN TRANSPLANTATION IMMUNOLOGY

Apart from the well recognized role as an APC, the macrophage is able to regulate cellular immune responses through both cell–cell contact and secreted factors. Contact-associated events chiefly mediate the antigen presentation process, utilizing a number of co-stimulatory molecules (Table 1) that are important in lymphocyte proliferation. The presentation of antigen without co-stimulation can lead to apoptosis of the T-cell clone. Macrophage-derived mediators can be stimulatory (IL-1, IL-6, IL-12, IL-15) or inhibitory (TGF-β, IL-10, PGE_2) to lymphocyte function, and T-cell products can in turn either activate (IFN-γ) or inhibit (IL-4, IL-10, IL-13) macrophage functions (Doherty et al., 1993). Key factors in the overall outcome of these cytokine regulatory networks are the Th1/Th2 paradigm and the functional commitment of the macrophage. The family of cytokine mRNAs transcribed by an activated T helper cell can be classified as Th1 (IL-2, IL-12, IFN-γ), promoting a cellular immune response, or Th2 (IL-4, IL-6, IL-10), promoting a humoral immune response. In transplantation, Th1 cells are derived when antigen presentation is received with appropriate co-stimulation and IFN-γ, leading to the development and proliferation of alloreactive clones of cytotoxic T lymphocytes (thus T cytotoxic cells can also be of the Tc1 or Tc2 variety). A Th2 response is induced by altered co-stimulatory events and/or IL-10, and thought to induce a tolerogenic stimulus. Whether a shift from a Th1/Tc1 to a Th2/Tc2 repertoire will be protective of the allograft or lead to alloreactive antibody-mediated graft damage (one mechanism of chronic rejection) depends on unknown factors. Furthermore, macrophages can be stimulatory or sup-

TABLE 1. Macrophage Co-stimulatory Molecules

Macrophage Cell Surface Molecule	Lymphocyte Counterpart
LFA-3	CD-2
B7-1 (CD80)	CD28
B7-2 (CD86)	CD28
Mac-1 (CD11b)	ICAM-1 (CD54)
Integrinβ2 (CD18)[a]	ICAM-1 (CD54)
CD1	TCR-CD3

[a]Binds CD54 as a heterodimer with Mac-1.

pressive to lymphocyte function. Suppression of lymphoproliferation is mediated partly by nitric oxide (NO) production, which is stimulated by IFN-γ (Angulo et al., 1995). Thus an activating stimulus promotes secretion of downregulatory mediators from suppressor macrophages. These responses can in turn be reversed by downregulatory influences on the suppressor macrophages themselves. One example is the ability of IL-4 to inhibit the NO production and NO-mediated suppression by macrophages, a "suppression of suppression" or double negative of sorts in the regulatory influence of macrophages (Al-Ramadi, 1992). As one would predict, both IL-10 and TGF-β also suppress production of NO from activated macrophages, and IL-4, IL-10, and TGF-β are synergistic in this effect (Oswald et al., 1992).

The Macrophage in Chronic Rejection

Chronic rejection remains an essentially unaltered and incompletely understood phenomenon in transplantation, resulting in the loss of a significant number of allografts each year. Alloreactive antibody-mediated events and nonspecific cellular responses (macrophage mediated) are thought to be the pathways through which graft injury occurs. Use of differential display techniques for mRNA detection have shown the presence of macrophage-related transcripts in chronically rejecting allografts, including a macrophage-specific lectin (galactose/*N*-acetylgalactosamine) and a ubiquitin-like gene (Utans et al., 1994). With an experimental model of chronic rejection, examination of mRNA transcripts expressed in the allograft over time shows a pattern of macrophage-specific events, such as the expression of RANTES at 2 weeks, followed by the upregulation of transcripts for inducible nitric oxide synthase, IL-6, and TNF-α at 16 weeks, followed later by infiltration of macrophages, which correlates with the histological changes of chronic rejection (Nadeau et al., 1995). Interestingly, this study also showed some evidence of macrophage regulatory events, with an elevation of TGF-β in the isograft controls along with a subsequent reduction in the nitric oxide synthase *i*NOS. The timing of macrophage recruitment is seen very early following engraftment, with MCP-1 detected at 7 days (Russell et al., 1996) and RANTES peaking at 14 days. In one experimental study of liver transplantation, recipient monocytes migrated into the allograft and transformed into macrophages within the first week of engraftment (Yamamoto et al., 1994). Thus chronic rejection is a process that most likely starts early following transplantation with the recruitment of macrophages and subsequently becomes a progressive and destructive process initiated by these cells. One report indicates that the macrophages associated with chronic rejection express and complex surface markers differently from those present in the nonspecific infiltrating cells associated with acute rejection (Goebeler et al., 1994). In one study, the presence of activated macrophages in dysfunctional transplanted organs (as detected by the presence of thromboxane synthase using immunohistochemistry) correlated negatively with allograft survival, regardless of the histological diagnosis (Ramos et al., 1995). As anticipated, these authors also showed that the use of a macrophage/monocyte specific inhibitor, γ-lactone, prevented both the macrophage infiltration and the subsequent functional and histological changes of chronic rejection in this model. An experimental immunosuppressive agent, deoxyspergualin, inhibits macrophage infiltration and local proliferation within the allograft (Kerr et al., 1994). In addition, various commonly used immunosuppressive agents differentially affect the expression of the macrophage differentiation antigen RM3/1 (expressed by macrophages infiltrating rejecting allografts) (Roth et al., 1994). Such agents might be employed to clinical advantage in transplantation, but this would require development of monitoring techniques for the presence and phenotype of infiltrating macrophages within the allograft.

The Macrophage in Transplantation Tolerance

Transplant tolerance can be defined as the lasting acceptance of foreign (transplanted) antigens with the retention of normal immune reactivity to other antigens, and thus would provide all the benefits sought in transplantation with few of the risks. The significant clinical improvements made in transplantation in the last decade have further underscored the fact that certain complications such as chronic rejection, the development of malignancy, and other adverse effects of immunosuppressive drugs remain essentially unaltered problems. Accordingly, there has been a resurgence of interest in the development of strategies to induce uniform and stable tolerance to transplanted organs. Few patients have developed documented tolerance at present, and these appear to have done so spontaneously. Cells of donor origin can be found in the blood, skin, or other tissues in transplant recipients (Hisanaga et al., 1996; Starzl et al., 1993), with tolerance in kidney recipients most often noted in donor-recipient pairs who share some degree of HLA match (living related-donor transplants). An immunoregulatory impact of such cells is more difficult to document, but suppression of donor-specific cytotoxic T-lymphocyte (CTL) activity has been convincingly shown (Burlingham et al., 1995). This case reported by Sollinger's group is an important "experiment of nature," documenting the presence of an immunologically functional cell of donor origin in the circulation of a tolerant recipient, and supporting the functional veto concept without actual clonal deletion in peripheral transplant tolerance in humans. The exact mechanism of this veto effect is poorly understood, but the induction of alloantigen-specific anergy is one process that shares many similarities with the findings in this case.

How such anergy is maintained, the necessity (if any) of a regulatory donor-derived cell in this process, and the mechanism by which lymphocyte cytokine profiles and responses remain altered to maintain tolerance are not fully understood. Whether suppression of mixed lymphocyte responses (as noted in Starzl's patients) or only CTL responses represent different mechanism or a continuum of the same spectrum of immunologic events is also not known. Some of these effects are associated with donor-specific cell infusions in selected patients (Grailer et al., 1991), prompting a recent revival of support for donor cell-based therapies to induce tolerance. Clinical protocols using blood or bone marrow cell infusions have shown variable reductions in acute rejection, rejection severity, or donor-specific immune responsiveness (Alexander et al., 1992, Bayle et al., 1995, Barber et al., 1991), along with an increase in allograft survival, but have fallen far short of the goal of uniform and reliable tolerance.

Numerous experimental models suggest that donor-specific cell infusions could be used to induce tolerance reliably. In a variety of models, Monaco and others have demonstrated the tolerogenic effects of whole or fractionated bone marrow or splenocytes (Blakely et al., 1995; Hartner et al., 1986). In these studies the dose and timing of cells and the immunosuppressive regimen have profound effects on outcome. A large-animal model using peripheral blood mononuclear cells has also been described (Granger et al., 1994). Thomas and others have shown in outbred animal models that tolerance induction using a marrow-derived cell with veto functions can be accomplished (Thomas et al., 1994). Bone marrow cells derived from the myeloid lineages such as macrophages, suppressor cells, or dendritic cells can "tolerize" or functionally counteract or delete cytotoxic T cells specific for allogenic antigens. Although lymphocytes can be suppressive or even act to delete CTLs, clearly more than one population of cells with tolerizing function can exist, even concurrently (Wood et al., 1991). Because donor-specific cells used in bone marrow infusions are clearly of a myeloid lineage, approaches using the induction of myelopoietic chimerism in transplantation are reasonable (and successful), as shown by Ildstad and others (Li et al., 1994). Though laboratory success in tolerance induction has been striking, human application was often not forthcoming owing to the real limitations on timing, cell sources, and drug toxicity in clinical transplantation. The development of strategies to enhance or enrich bone marrow-derived cells with tolerogenic potential for use in clinical transplantation is therefore a relevant and important goal. In order to accomplish this, the development and functional phenotype of the desired cell must be considered.

Marrow-derived populations contain stem cells and lineage-specific progenitor cells. Because myeloid lineage, antigen presenting capability, and plasticity/engraftment of the donor cell all have an influence on the tolerogenic effect, the bone marrow is an important source for the harvest of potentially tolerogenic cells. The mere presence of chimerism does not assure the induction or maintenance of tolerance, however. Therefore, the function and exact phenotype of the chimeric cell are central to its role in tolerogenesis. In this light, the prolonged effect on the recipient's immune system may or may not require a prolonged survival of the cell involved. Although dendritic cells are known to develop in response to GM-CSF and have the potential to induce a tolerogenic response, they express CD8 and their myeloid lineage has been confirmed. The myeloid lineage of the macrophage is clear, but the cell surface marker pattern of certain macrophages can include CD4 (Wallgren et al., 1995). The determination of phenotype (especially when cultured, activated, or precursor cells are examined) requires a thorough assessment using a panel of markers because the distinction between myeloid and lymphoid origin is not always easy.

One strategy to enhance or enrich potentially tolerogenic cells is by in vitro culture of bone marrow. The incidence of preprogrammed myeloid lineage cells with tolerogenic functions from fresh bone marrow is an infrequent but naturally occurring event. The derivation of such cells should be possible in culture, and enhancing the conditions supporting this development would result in cells capable of inducing prolonged allograft survival. GM-CSF favors the development of suppressor cells (Moore et al., 1992), a function possibly limited to actively dividing cells (Sugiura et al., 1988). Interactions between marrow cells and lymphocytes in culture show the development of suppressor activity and the impact of Th2 cytokines on macrophage function (Oswald et al., 1992). The potential benefits of the co-culture of myeloid and lymphoid cells for the development of tolerogenic functions have not been well described.

FLOW CYTOMETRIC ANALYSIS OF MACROPHAGES IN CLINICAL TRANSPLANTATION

Macrophages present several predictable and occasionally difficult problems for analysis of cell-mediated events in transplant immunology by flow cytometric techniques, including small population sizes, autofluorescence, cell–cell adherence, nonspecific antibody binding, technique-induced alterations in surface marker expression, and size/granularity alterations with activation. Macrophage activation in particular causes a number of these problems to worsen, especially nonspecific binding as Fc receptor expression is enhanced. Some strategies for addressing these issues are known, such as the use of an overabundance of blocker to limit nonspecific antibody binding (Hendrzak et al., 1994), or established techniques to reduce autofluorescence, but other problems remain. As examination of needle aspiration cytology specimens is more commonly applied to the evaluation of allograft infiltrates in clinical transplantation, the development of standardized techniques for the flow cytometric examination of macrophages (as well as an evaluation of their state of activation and intracellular cytokine stores) will become increasingly relevant. Some potential roles for flow cytometric analysis of macrophages in both clinical and research applications in transplantation are outlined in Table 2.

There are at least two current protocols in clinical transplantation where the application of flow cytometric analysis of macrophages is relevant. The use of needle aspiration cytology to assess allograft infiltrates in acute rejection is a potentially useful tool under current study. The assessment of the phenotype, activation, and cytokine contents of infiltrating lymphocytes must be correlated with histopathology and clinical outcome before reliance on architectural features obtained using a core biopsy can be bypassed. Once standardized, aspirates could be used to identify macrophage subsets as well as their cytokine contents and thus serve as a means to diagnose or predict chronic rejection. Another ongoing field of investigation involves the use of donor-specific bone marrow infusions to augment allograft survival. Inasmuch as the cells that confer a survival advantage in animal studies demonstrate both functional and phenotypic differences from model to model, defining the lineage, phenotypic markers, and functional characteristics of the relevant cells in a human model remains an important task.

As the role of the macrophage in transplantation immunology is further explored, macrophage monitoring and functional assessment may find a role in the clinical management of transplant recipients. As we have briefly reviewed, in two key areas of ongoing research in transplantation, chronic rejection and tolerance induction, macrophages or other myeloid lineage cells may play prominent roles. Other aspects

TABLE 2. Flow Cytometrics in Macrophage Analysis in Transplantation

Specimen Sources
Graft core biopsies or needle aspiration cytology
Peripheral blood mononuclear cells
Alveolar lavage macrophages (lung transplantation)
Bone marrow aspirates
Cadaveric donor tissues (research or clinical applications)
Peritoneal dialysate (exudative macrophages)
Studies
Enzymatic activation / NO synthesis
Intracellular cytokine levels
Immunophenotyping/ differentiation markers/receptor levels
HLA and functionally specific markers
Detection of apoptosis
Cell cycle analysis
Assessment of drug effects
Applications
Detection of cells associated with chronic rejection
Prognostic information (degree of macrophage infiltration/activation)
Detection of chimerism
Assessment of myelopoietic development under immune influences
Sorting/enriching cells for therapeutic use in tolerogenic protocols
Phenotypic characterization of myeloid progenitors
Study of macrophage:Th1/Th2 interactions
Role of macrophages in peripheral lymphocyte clonal selection
Therapeutic monitoring of macrophage-specific drugs
Infectious disease risk prognosis

of transplant immunology, including acute rejection, the transfusion effect, peripheral lymphocyte anergy induction or deletion, and responses to infection in the immunocompromised host are also influenced by macrophage development and function.

REFERENCES

Alexander JW, Babcock GF, First MR, et al. (1992): The induction of immunologic hyporesponsiveness by preoperative donor-specific transfusions and cyclosporine in human cadaveric transplantation. Transplantation 53:423–427.

Al-Ramadi BK, Meissler JJ, Huang D, Eisenstein TK (1992): Immunosuppression induced by nitric oxide and its inhibition by interleukin-4. Eur J Immunol 22:2249–2254.

Angulo I, Rodriguez R, Garcia B, Medina M, Navarro J, Subiza JL (1995): Involvement of nitric oxide in bone marrow-derived natural suppressor activity. J Immun 155:15–26.

Barber WH, Mankin JA, Laskow DA, et al. (1991): Long-term results of a controlled prospective study with transfusion of donor-specific bone marrow in 57 cadaveric renal allograft recipients. Transplantation 51:70–75.

Bayle F, Masson D, Zaoui P, et al. (1995): Beneficial effect of one HLA haplo- or semi-identical transfusion versus three untyped blood units on alloimmunization and acute rejection episodes in first renal allograft recipients. Transplantation 59:719–723.

Blakely ML, Shaffer D, Ohzato H, Gottschalk R, VanderWerf WJ, Monaco AP (1995): Indefinite survival following small intestinal transplantation after intrathymic injection of the donor with recipient-type splenocytes in a rat model. Transplantation 59:309–311.

Burlingham WJ, Grailer AP, Fechner JH, et al. (1995): Microchimerism linked to cytotoxic T lymphocyte functional unresponsiveness (clonal anergy) in a tolerant renal transplant recipient. Transplantation 59:1147–1155.

DeNichilo MO, Burns GF. (1993): Granulocyte-macrophage and macrophage colony-stimulating factors differentially regulate αv integrin expression on cultured human macrophages. Proc Natl Acad Sci USA 90: 2517–2521.

Doherty TM, Kastelein R, Menon S, Andrade S, Coffman RL. (1993): Modulation of murine macrophage function by IL-13. J Immun 151:7151–7160.

Goebeler M, Roth J, Burwinkel F, Vollmer E, Bocker W, Sorg C (1994): Expression and complex formation of S100-like proteins MRP8 and MRP14 by macrophages during renal allograft rejection. Transplantation 58:355–361.

Grailer AP, Sollinger HW, Kawamura T, Burlingham WJ (1991): Donor-specific cytotoxic T lymphocyte hyporesponsiveness following renal transplantation in patients pretreated with donor-specific transfusions. Transplantation 51:320–324.

Granger DK, Matas AJ, Jenkins MK, Moss AA, Chen SC, Almond PS (1994): Prolonged survival without posttransplant immunosuppression in a large animal model. Surgery 116:236–241.

Hartner WC, DeFazio SR, Maki T, Markees TG, Monaco AP, Gozzo JJ (1986): Prolongation of renal allograft survival in antilymphocyte-serum-treated dogs by postoperative injection of density-gradient-fractionated donor bone marrow. Transplantation 42:593–596.

Hendrzak JA, Wallace PK, Morahan PS (1994): Optimizing the detection of cell surface antigens on elicited or activated mouse peritoneal macrophages. Cytometry 17:349–356.

Hill S, Coates JP, Kimber I, Knight SC (1994): Differential function of dendritic cells isolated from blood and lymph nodes. Immun 83:295–301.

Hisanaga M, Hundrieser J, Boker K, et al. (1996): Development, stability, and clinical correlations of allogeneic microchimerism after solid organ transplantation. Transplantation 61:40–45.

Ikebuchi K, Wong GG, Clark SC, Ihle JN, Hirai Y, Ogawa M (1987): Interleukin-6 enhancement of interleukin-3-dependent proliferation of multipotent hemopoietic progenitors. Proc Natl Acad Sci USA 84:9035.

Inaba K, Inaba M, Deguchi M, et al. (1993): Granulocytes, macrophages, and dendritic cells arise from a common major histocompatibility complex class II-negative progenitor in mouse bone marrow. Proc Natl Acad Sci USA 90:3038–3042.

Kerr PG, Nikolic-Paterson DJ, Lan HY, Tesch G, Rainone S, Atkins RC (1994): Deoxyspergualin suppresses local macrophage proliferation in rat renal allograft rejection. Transplantation 58:596–601.

Kiertscher SM, Roth MD (1996): Human CD14+ leukocytes acquire the phenotype and function of antigen-presenting dendritic cells when cultured in GM-CSF and IL-4. J Leukoc Biol 59:208–218.

Li H, Kaufman CL, Ildstad ST (1994): Allogeneic chimerism induces donor-specific tolerance to simultaneous islet allografts in nonobese diabetic mice. Surgery 118:192–198.

Moore SC, Theus SA, Barnett JB, Soderberg LSF (1992): Cytokine regulation of bone marrow natural suppressor cell activity in the suppression of lymphocyte function. Cell Immunol 141:398–408.

Nadeau KC, Azuma H, Tilney NL (1995): Sequential cytokine dynamics in chronic rejection of rat renal allografts: Roles for cytokines RANTES and MCP-1. Proc Natl Acad Sci USA 92:8729–8733.

Oswald IP, Gazzinelli RT, Sher A, James SL (1992): IL-10 synergizes with IL-4 and transforming growth factor-β to inhibit macrophage cytotoxic activity. J Immunol 148:3578–3582.

Ramos EL, Barri YM, Croker BP, Clapp WL, Peterson JC, Wilcox CS (1995): Thromboxane synthase expression in renal transplant patients with rejection. Transplantation 59:490–494.

Roth J, Goebeler M, Erpenstein U, Sorg C (1994): Differential regulation of the macrophage-specific surface antigen Rm3/1 by cyclosporine, azathioprine, and dexamethasone. Transplantation 57:127–133.

Russell ME, Wallace AF, Hancock WW, et al. (1996): Upregulation of cytokines associated with macrophage activation in the Lewis-to-F344 rat transplantation model of chronic cardiac rejection. Transplantation 59:572–578.

Rutherford MS, Schook LB (1992): Differential immunocompetence of macrophages derived using macrophage or granulocyte-macrophage colony-stimulating factor. J Leukoc Biol 51:69–76.

Rutherford MS, Witsell A, Schook LB (1993): Mechanisms generating functionally heterogeneous macrophages: chaos revisited. J Leukoc Biol 53:602–618.

Santiago-Schwarz F, Belilos E, Diamond B, Carsons SE (1992): TNF in combination with GM-CSF enhances the differentiation of neonatal cord blood stem cells into dendritic cells and macrophages. J Leukoc Biol 52:274–281.

Schmid I, Baldwin GC, Jacobs EL, et al. (1995): Alterations in phenotype and cell-surface antigen expression levels of human monocytes: differential response in vivo administration of rhM-CSF or rhGM-CSF. Cytometry 22:103–110.

Starzl TE, Demetris AJ, Trucco M, et al. (1993): Chimerism and donor-specific nonreactivity 27 to 29 years after kidney allotransplantation. Transplantation 55:1272–1277.

Sugiura K, Inaba M, Ogata H, et al. (1988): Wheat germ agglutinin-positive cells in a stem cell-enriched fraction of mouse bone marrow have potent natural suppressor activity. Proc Natl Acad Sci USA 85:4824–4826.

Szabolcs P, Avigan D, Gezelter S, et al. (1996): Dendritic cells and macrophages can mature independently from a human bone marrow-derived, post-colony-forming unit intermediate. Blood 87:4520–4530.

Thomas, JM, Vobanac ICM, Carver FM, Kasten-Jolly J, Haisch CE, Gross U, Smith JP (1994): Veto cells in transplant tolerance. Clin Transplant 8:195–203.

Utans U, Liang P, Wyner LR, Karnovsky MJ, Russell ME (1994): Chronic cardiac rejection: Identification of five upregulated genes in transplanted hearts by differential mRNA display. Proc Natl Acad Sci USA 91:6463–6467.

Wallgren AC, Karlsson-Parra A, Korsgren O (1995): The main infiltrating cell in xenograft rejection is a CD4+ macrophage and not a T lymphocyte. Transplantation 60:594–601.

Witsell AL, Schook LB (1991): Macrophage heterogeneity occurs through a developmental mechanism. Proc Natl Acad Sci USA 88:1963.

Wood ML, Monoaco AP, Gottschalk R (1991): Characterization of spleen cells capable of inducing unresponsivness in ALS-treated mice. Transplantation 51:208–213.

Yamamoto H, Kaneda K, Kamada N, Wake K (1994): Morphometric analysis of infiltrating cells that adhere to the sinusoidal wall and migrate into the space of disse in the orthotypically transplanted rat liver. Transplantation 58:8–13.

INDEX